HOW TO COUNT
AN INTRODUCTION TO COMBINATORICS
Second Edition

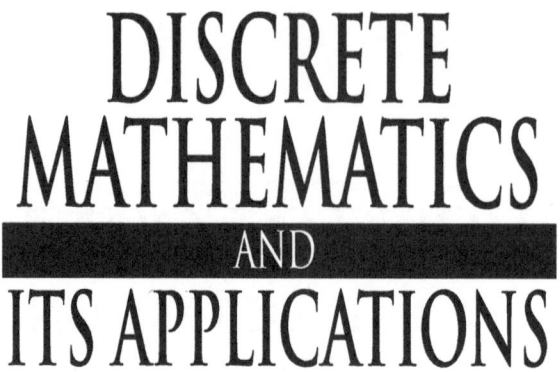

DISCRETE MATHEMATICS AND ITS APPLICATIONS

Series Editor
Kenneth H. Rosen, Ph.D.

R. B. J. T. Allenby and Alan Slomson, and How to Count: An Introduction to Combinatorics, Third Edition

Juergen Bierbrauer, Introduction to Coding Theory

Francine Blanchet-Sadri, Algorithmic Combinatorics on Partial Words

Richard A. Brualdi and Dragoš Cvetković, A Combinatorial Approach to Matrix Theory and Its Applications

Kun-Mao Chao and Bang Ye Wu, Spanning Trees and Optimization Problems

Charalambos A. Charalambides, Enumerative Combinatorics

Gary Chartrand and Ping Zhang, Chromatic Graph Theory

Henri Cohen, Gerhard Frey, et al., Handbook of Elliptic and Hyperelliptic Curve Cryptography

Charles J. Colbourn and Jeffrey H. Dinitz, Handbook of Combinatorial Designs, Second Edition

Martin Erickson, Pearls of Discrete Mathematics

Martin Erickson and Anthony Vazzana, Introduction to Number Theory

Steven Furino, Ying Miao, and Jianxing Yin, Frames and Resolvable Designs: Uses, Constructions, and Existence

Mark S. Gockenbach, Finite-Dimensional Linear Algebra

Randy Goldberg and Lance Riek, A Practical Handbook of Speech Coders

Jacob E. Goodman and Joseph O'Rourke, Handbook of Discrete and Computational Geometry, Second Edition

Jonathan L. Gross, Combinatorial Methods with Computer Applications

Jonathan L. Gross and Jay Yellen, Graph Theory and Its Applications, Second Edition

Jonathan L. Gross and Jay Yellen, Handbook of Graph Theory

David S. Gunderson, Handbook of Mathematical Induction: Theory and Applications

Darrel R. Hankerson, Greg A. Harris, and Peter D. Johnson, Introduction to Information Theory and Data Compression, Second Edition

Darel W. Hardy, Fred Richman, and Carol L. Walker, Applied Algebra: Codes, Ciphers, and Discrete Algorithms, Second Edition

Daryl D. Harms, Miroslav Kraetzl, Charles J. Colbourn, and John S. Devitt, Network Reliability: Experiments with a Symbolic Algebra Environment

Silvia Heubach and Toufik Mansour, Combinatorics of Compositions and Words

Titles (continued)

Leslie Hogben, Handbook of Linear Algebra

Derek F. Holt with Bettina Eick and Eamonn A. O'Brien, Handbook of Computational Group Theory

David M. Jackson and Terry I. Visentin, An Atlas of Smaller Maps in Orientable and Nonorientable Surfaces

Richard E. Klima, Neil P. Sigmon, and Ernest L. Stitzinger, Applications of Abstract Algebra with Maple™ and MATLAB®, Second Edition

Patrick Knupp and Kambiz Salari, Verification of Computer Codes in Computational Science and Engineering

William Kocay and Donald L. Kreher, Graphs, Algorithms, and Optimization

Donald L. Kreher and Douglas R. Stinson, Combinatorial Algorithms: Generation Enumeration and Search

C. C. Lindner and C. A. Rodger, Design Theory, Second Edition

Hang T. Lau, A Java Library of Graph Algorithms and Optimization

Elliott Mendelson, Introduction to Mathematical Logic, Fifth Edition

Alfred J. Menezes, Paul C. van Oorschot, and Scott A. Vanstone, Handbook of Applied Cryptography

Richard A. Mollin, Advanced Number Theory *with Applications*

Richard A. Mollin, Algebraic Number Theory

Richard A. Mollin, Codes: The Guide to Secrecy from Ancient to Modern Times

Richard A. Mollin, Fundamental Number Theory with Applications, Second Edition

Richard A. Mollin, An Introduction to Cryptography, Second Edition

Richard A. Mollin, Quadratics

Richard A. Mollin, RSA and Public-Key Cryptography

Carlos J. Moreno and Samuel S. Wagstaff, Jr., Sums of Squares of Integers

Dingyi Pei, Authentication Codes and Combinatorial Designs

Kenneth H. Rosen, Handbook of Discrete and Combinatorial Mathematics

Douglas R. Shier and K.T. Wallenius, Applied Mathematical Modeling: A Multidisciplinary Approach

Alexander Stanoyevitch, Introduction to Cryptography with Mathematical Foundations and Computer Implementations

Jörn Steuding, Diophantine Analysis

Douglas R. Stinson, Cryptography: Theory and Practice, Third Edition

Roberto Togneri and Christopher J. deSilva, Fundamentals of Information Theory and Coding Design

W. D. Wallis, Introduction to Combinatorial Designs, Second Edition

W. D. Wallis and John George, Introduction to Combinatorics

Lawrence C. Washington, Elliptic Curves: Number Theory and Cryptography, Second Edition

DISCRETE MATHEMATICS AND ITS APPLICATIONS

Series Editor KENNETH H. ROSEN

HOW TO COUNT
AN INTRODUCTION TO COMBINATORICS
Second Edition

R.B.J.T. Allenby

University of Leeds

UK

Alan Slomson

University of Leeds

UK

CRC Press
Taylor & Francis Group
Boca Raton London New York

CRC Press is an imprint of the
Taylor & Francis Group, an **informa** business

A CHAPMAN & HALL BOOK

First published in paperback 2024

First published 2011
by Chapman & Hall/CRC
2385 NW Executive Center Drive, Suite 320, Boca Raton FL 33431

and by Chapman & Hall/CRC
4 Park Square, Milton Park, Abingdon, Oxon, OX14 4RN

CRC Press is an imprint of Taylor & Francis Group, LLC

© 2011, 2024 Taylor & Francis Group, LLC

Reasonable efforts have been made to publish reliable data and information, but the author and publisher cannot assume responsibility for the validity of all materials or the consequences of their use. The authors and publishers have attempted to trace the copyright holders of all material reproduced in this publication and apologize to copyright holders if permission to publish in this form has not been obtained. If any copyright material has not been acknowledged please write and let us know so we may rectify in any future reprint.

Except as permitted under U.S. Copyright Law, no part of this book may be reprinted, reproduced, transmitted, or utilized in any form by any electronic, mechanical, or other means, now known or hereafter invented, including photocopying, microfilming, and recording, or in any information storage or retrieval system, without written permission from the publishers.

For permission to photocopy or use material electronically from this work, access www.copyright.com or contact the Copyright Clearance Center, Inc. (CCC), 222 Rosewood Drive, Danvers, MA 01923, 978-750-8400. For works that are not available on CCC please contact mpkbookspermissions@tandf.co.uk

Trademark notice: Product or corporate names may be trademarks or registered trademarks and are used only for identification and explanation without intent to infringe.

Publisher's Note
The publisher has gone to great lengths to ensure the quality of this reprint but points out that some imperfections in the original copies may be apparent.

Library of Congress Cataloging-in-Publication Data

Allenby, R. B. J. T.
  How to count : an introduction to combinatorics / R.B.J.T. Allenby, Alan Slomson. -- 2nd ed.
    p. cm. -- (Discrete mathematics and its applications)
  "A CRC title."
  First published as: an introduction to combinatorics, 1991.
  Includes bibliographical references and index.
  ISBN 978-1-4200-8260-9 (hardcover : alk. paper)
  1. Combinatorial analysis. I. Slomson, A. B. II. Slomson, A. B. Introduction to combinatorics. III. Title. IV. Series.

QA164.S57 2011
511'.6--dc22
                                                                                2010024420

ISBN: 978-1-4200-8260-9 (hbk)
ISBN: 978-1-03-291977-5 (pbk)
ISBN: 978-0-429-11312-3 (ebk)

DOI: 10.1201/9781439895153

Visit the Taylor & Francis Web site at
http://www.taylorandfrancis.com

and the CRC Press Web site at
http://www.crcpress.com

Table of Contents

Preface to the Second Edition, xi

Acknowledgments, xiii

Authors, xv

Chapter 1 ▪ What's It All About?	1
1.1 WHAT IS COMBINATORICS?	1
1.2 CLASSIC PROBLEMS	2
1.3 WHAT YOU NEED TO KNOW	14
1.4 ARE YOU SITTING COMFORTABLY?	15

Chapter 2 ▪ Permutations and Combinations	17
2.1 THE COMBINATORIAL APPROACH	17
2.2 PERMUTATIONS	17
2.3 COMBINATIONS	21
2.4 APPLICATIONS TO PROBABILITY PROBLEMS	28
2.5 THE MULTINOMIAL THEOREM	34
2.6 PERMUTATIONS AND CYCLES	36

Chapter 3 ▪ Occupancy Problems	39
3.1 COUNTING THE SOLUTIONS OF EQUATIONS	39
3.2 NEW PROBLEMS FROM OLD	43
3.3 A "REDUCTION" THEOREM FOR THE STIRLING NUMBERS	47

Chapter 4 ▪ The Inclusion–Exclusion Principle	51
4.1 DOUBLE COUNTING	51
4.2 DERANGEMENTS	58
4.3 A FORMULA FOR THE STIRLING NUMBERS	60

Chapter 5 ▪ Stirling and Catalan Numbers   63
- 5.1 STIRLING NUMBERS   63
- 5.2 PERMUTATIONS AND STIRLING NUMBERS   68
- 5.3 CATALAN NUMBERS   71

Chapter 6 ▪ Partitions and Dot Diagrams   81
- 6.1 PARTITIONS   81
- 6.2 DOT DIAGRAMS   83
- 6.3 A BIT OF SPECULATION   89
- 6.4 MORE PROOFS USING DOT DIAGRAMS   92

Chapter 7 ▪ Generating Functions and Recurrence Relations   95
- 7.1 FUNCTIONS AND POWER SERIES   95
- 7.2 GENERATING FUNCTIONS   98
- 7.3 WHAT IS A RECURRENCE RELATION?   101
- 7.4 FIBONACCI NUMBERS   103
- 7.5 SOLVING HOMOGENEOUS LINEAR RECURRENCE RELATIONS   109
- 7.6 NONHOMOGENEOUS LINEAR RECURRENCE RELATIONS   114
- 7.7 THE THEORY OF LINEAR RECURRENCE RELATIONS   120
- 7.8 SOME NONLINEAR RECURRENCE RELATIONS   124

Chapter 8 ▪ Partitions and Generating Functions   127
- 8.1 THE GENERATING FUNCTION FOR THE PARTITION NUMBERS   127
- 8.2 A QUICK(ISH) WAY OF FINDING p(n)   132
- 8.3 AN UPPER BOUND FOR THE PARTITION NUMBERS   142
- 8.4 THE HARDY–RAMANUJAN FORMULA   145
- 8.5 THE STORY OF HARDY AND RAMANUJAN   147

Chapter 9 ▪ Introduction to Graphs   151
- 9.1 GRAPHS AND PICTURES   151
- 9.2 GRAPHS: A PICTURE-FREE DEFINITION   152
- 9.3 ISOMORPHISM OF GRAPHS   154
- 9.4 PATHS AND CONNECTED GRAPHS   163
- 9.5 PLANAR GRAPHS   168
- 9.6 EULERIAN GRAPHS   178
- 9.7 HAMILTONIAN GRAPHS   182
- 9.8 THE FOUR-COLOR THEOREM   188

Chapter 10 ▪ Trees	199
10.1 WHAT IS A TREE?	199
10.2 LABELED TREES	204
10.3 SPANNING TREES AND MINIMAL CONNECTORS	210
10.4 THE SHORTEST-PATH PROBLEM	217

Chapter 11 ▪ Groups of Permutations	223
11.1 PERMUTATIONS AS GROUPS	223
11.2 SYMMETRY GROUPS	229
11.3 SUBGROUPS AND LAGRANGE'S THEOREM	235
11.4 ORDERS OF GROUP ELEMENTS	240
11.5 THE ORDERS OF PERMUTATIONS	242

Chapter 12 ▪ Group Actions	245
12.1 COLORINGS	245
12.2 THE AXIOMS FOR GROUP ACTIONS	247
12.3 ORBITS	249
12.4 STABILIZERS	250

Chapter 13 ▪ Counting Patterns	257
13.1 FROBENIUS'S COUNTING THEOREM	257
13.2 APPLICATIONS OF FROBENIUS'S COUNTING THEOREM	259

Chapter 14 ▪ Pólya Counting	267
14.1 COLORINGS AND GROUP ACTIONS	267
14.2 PATTERN INVENTORIES	270
14.3 THE CYCLE INDEX OF A GROUP	274
14.4 PÓLYA'S COUNTING THEOREM: STATEMENT AND EXAMPLES	277
14.5 PÓLYA'S COUNTING THEOREM: THE PROOF	281
14.6 COUNTING SIMPLE GRAPHS	285

Chapter 15 ▪ Dirichlet's Pigeonhole Principle	293
15.1 THE ORIGIN OF THE PRINCIPLE	293
15.2 THE PIGEONHOLE PRINCIPLE	294
15.3 MORE APPLICATIONS OF THE PIGEONHOLE PRINCIPLE	297

CHAPTER 16 • Ramsey Theory	303
16.1 WHAT IS RAMSEY'S THEOREM?	303
16.2 THREE LOVELY THEOREMS	310
16.3 GRAPHS OF MANY COLORS	314
16.4 EUCLIDEAN RAMSEY THEORY	315

CHAPTER 17 • Rook Polynomials and Matchings	319
17.1 HOW ROOK POLYNOMIALS ARE DEFINED	319
17.2 MATCHINGS AND MARRIAGES	332

SOLUTIONS TO THE A EXERCISES, 339

BOOKS FOR FURTHER READING, 419

INDEX FOR NOTATION 421

INDEX 423

Preface to the Second Edition

We explain the aims and range of this book in Chapter 1, which we strongly urge you to read before any of the subsequent chapters.

This second edition is a considerably expanded version of the first edition, originally published in 1991. It has 100% more authors, and correspondingly more material. Nonetheless, it does not cover the whole of the subject, and the coverage of the topics that are included is not comprehensive. Our choice of topics has mainly been determined by personal taste, but the emphasis is on *counting problems*, that is, questions about how many different arrangements there are of a particular type. We will not have fully succeeded in our aim of providing an attractive introduction to these topics if the book does not leave you wanting more. To this end we have added a list of suggestions for further reading.

The first edition was based on a course of 22 lectures given to students at the University of Leeds. We think the material included in this expanded edition could be covered in around 40 lectures. We emphasize, however, that it is our hope that the book can be read independently by anyone wishing for an accessible introduction to the topics that it covers. To this end, and unlike most other texts, we have, as far as possible, set exercises *in pairs* and provided an extensive answer section in which a *complete solution* is given to one exercise in each pair. Of course, diligent readers will tackle the exercises before looking to the answer section (this being an essential part of the learning process), but, if they are *really* stuck, the full solution should prove to be of value.

R.B.J.T Allenby
Alan Slomson

Acknowledgments

We have not originated any of the mathematics in this book. We have tried to give references to the papers where the main theorems were first proved, but a lot of the book covers standard material that has been known for some time, and where locating original sources is not easy. In writing biographical notes about many of the mathematicians whose work we discuss, we have made much use of the excellent Web site: *The MacTutor History of Mathematics Archive,* http://www-history.mcs.st-andrews.ac.uk, based at the University of St Andrews.

We have also benefited from the expository texts listed in our suggestions for further reading. Each of us has taught much of the material to students from the University of Leeds. We are grateful to our colleagues in Leeds who first allowed us to teach in an area of mathematics away from our research interests and to the students whose comments helped to shape our teaching.

Modern technology has made the writing of this book much easier than it was for the first edition twenty years ago. However, technology does have some disadvantages. It turns out that the template used to typeset the book has a mind of its own. This has meant that in some places the layout of the pages, and, in particular, the position of the diagrams and tables is not as we would have wished. We apologize for this, but it has been outside our control.

Previous experience leads us to believe that not everything in the book is correct. We hope that the errors are mostly minor misprints that have escaped our attention and that there are no major blunders. We encourage readers who detect errors or have questions about anything in this book to contact us.

Authors

R.B.J.T. Allenby received his PhD from the University of Wales and MSc Tech from the University of Manchester. He taught mathematics at the University of Leeds from 1965 to 2007. He is the author of *Rings, Fields and Groups: An Introduction to Abstract Algebra* (Edward Arnold, 1983); *Linear Algebra* (Edward Arnold, 1995), and *Number and Proofs* (Edward Arnold, 1997) and, with E. J. Redfern, of *Introduction to Number Theory with Computing* (Edward Arnold, 1989).

Alan Slomson received his MA and DPhil from the University of Oxford. He taught mathematics at the University of Leeds from 1967 to 2008. He is the author of the first edition of this book (Chapman and Hall, 1991) and, with John Bell, of *Models and Ultraproducts* (North Holland Publishing Company, 1969). He is currently the secretary of the United Kingdom Mathematics Trust.

CHAPTER 1

What's It All About?

1.1 WHAT IS COMBINATORICS?

Mathematics is a problem-solving activity, and the ultimate source of most mathematics is the external, nonmathematical world. Mathematical concepts are developed to help us tackle problems arising in this way. The abstract mathematical ideas that we use soon assume a life of their own and generate further problems, but these are more technical problems whose connection with the external world is more remote.

Since the time of Isaac Newton and until quite recently, almost the entire emphasis of applied mathematics has been on continuously varying processes, modeled by the mathematical continuum and using methods derived from the differential and integral calculus. In contrast, combinatorics concerns itself mainly with *finite* collections of *discrete* objects. With the growth of digital devices, especially computers, discrete mathematics has become more and more important.

The way mathematics has developed creates a difficulty when it comes to teaching and learning the subject. It is generally thought best to begin with the most basic ideas and then gradually work your way up to more and more complicated mathematics. This seems more sensible than being thrown in at the deep end and hoping you will learn to swim before you drown. However, this logical approach often obscures the historical reasons why a particular mathematical idea was developed. This means that it can be difficult for the student to understand the real point of the subject.

Fortunately, combinatorics is different. The starting point usually consists of problems that are easy to understand even if finding their solutions is not straightforward. They tend to be concrete problems that can be understood by those who do not know any technical mathematics. In this chapter we list a number of these problems that gave rise to much of the mathematics explained in the remainder of this book.

So what sorts of problems does combinatorics address? As combinatorics finds its origins in statistical, gambling, and recreational problems, a rough answer is: anywhere where knowing "How many?" (the answer to which may be "zero") is of interest. So, for example, the techniques we describe (or generalized versions of them) have been used in design of experiments, for example, the testing of crops (statistics), design of traffic routes (graph theory), construction of codes, arrangements of meetings (permutations and combinations), placing of people in jobs (rook polynomials), determination of certain chemical compounds (graph theory), production and school teaching schedules, as well as increasingly in mathematical biology in relation to the DNA code. There are also applications within mathematics, especially in the theory of numbers, and within computing, where questions of speed and complexity of working are important. Among recreational puzzles with a combinatorial flavor are the well-known Rubik's Cube, "magic" squares, Lucas's Tower of Hanoi, and the famous problem that was the origin of graph theory, that of crossing the bridges of Königsberg.

In many problems, not only a solution but an "optimal" solution is sought. For example, it would seem desirable to seek to maximize factory output or traffic flow or to minimize factory or computing costs or travel cost (by choosing the shortest route when visiting a succession of towns; this is the well-known traveling salesman problem). These aspects indicate the three basic problems of combinatorics: counting the number of solutions, checking existence (is there even *one* solution?), and searching for an optimal solution.

In this general introductory book we are not able to go into any of these topics in any great depth. Our aim is to give the reader the flavor of a broad range of interesting combinatorial ideas. At the end of the book we make suggestions for further reading where many of these ideas can be followed up.

1.2 CLASSIC PROBLEMS

In this section we list a number of classic and other interesting combinatorial problems that, later in the book, we show you how to solve. Many of these problems have been around for a long time. Accordingly, in most cases we have not tried to attribute the problems to their authors. Nevertheless, the present authors would be grateful for any enlightenment readers wish to provide in this regard.

We have selected at least one problem for all but one of the subsequent chapters, and our numbering of them matches the chapter numbers. So our first problem is called "Problem 2A" to indicate that it relates to Chapter 2 and is the first problem listed here from that chapter.

Many combinatorial problems arise from questions about probabilities. For, if a certain event can produce a finite set of equally likely different outcomes, of which some are deemed *favorable*, then we say that the probability of a favorable outcome is the fraction

$$\frac{\text{the number of favorable outcomes}}{\text{the total number of all outcomes}}.$$

So we can work out probabilities by counting the number of outcomes in the two sets occurring in this fraction. This sort of counting is mostly what this book is about. For example, consider the probability problem: "When throwing two standard dice, what is the probability that the total shown on the dice is 6?" The total number of outcomes when two dice are thrown is 36 as each face 1,2,3,4,5,6 on one die[*] can appear partnered by each one of the faces

---

[*] Being mathematicians, naturally we are pedantic. Although *dice* is often used as the singular term, strictly speaking, it is *one die* and *two or more dice*. Remember Julius Caesar's remark as he crossed the Rubicon, "The die is cast" ("Iacta alea est").

1,2,3,4,5,6 on the other die. The outcome 6 can be achieved in five different ways, namely, $1 + 5, 2 + 4, 3 + 3, 4 + 2$, and $5 + 1$. So the answer to the probability problem is 5/36.

The first problem we list is rather more complicated but one of direct interest to anyone who "invests" their money in a lottery.

PROBLEM 2A
What Is the Chance of a Jackpot?

The operators of "Lotto," the British National Lottery, advertize that each ticket has a 1 in 13,983,816 chance of winning a share of the jackpot. How is this probability calculated?

Next we have an old classic, whose answer usually comes as a surprise.

PROBLEM 2B
The Birthdays Problem

How many people do you need to have in a room before there is a better than 50% chance that at least two people share a birthday?

While solving Problems 2A and 2B is fairly straightforward, some counting problems are a bit more perplexing. Let us have a look at two involving food and money, the second being of rather more general interest than the first!

PROBLEM 3A
Hot Chocolates

A manufacturer of high-quality (and therefore high-priced) chocolates makes just six different flavors of chocolate and sells them in boxes of 10. He claims he can offer over 3000 different "selection boxes." If he is wrong, he will fall foul of the advertizing laws. Should he fear prosecution?

PROBLEM 3B
A Common Opinion

There is a widely held view that, in a truly random selection of six distinct numbers from among the numbers 1 to 49 (as in the British lottery), the chance that two consecutive numbers will be chosen is extremely small. Has this opinion any credibility?

The solution to both of these problems relies on the same general principles used to solve the following purely arithmetic problem. Notice how much less "cluttered" and more transparent the arithmetic problem is without the "real-life" trimmings!

PROBLEM 3C
Counting Solutions

How many solutions does the equation $x + y + z + t = 60$ have where x, y, z, and t are positive integers?

If, in this problem, there were only, say, three unknowns x, y, and z and the 60 were replaced by a 6, there would be little difficulty, as we could easily list all the solutions systematically, that is, (1,1,4), (1,2,3), (1,3,2), (1,4,1), (2,1,3), (2,2,2), (2,3,1), (3,1,2), (3,2,1), and (4,1,1). So, there are exactly 10 of them.

That was easy. But what if we return to four unknowns and reintroduce the 60 or change it to 600? How many solutions then? It is clear that the same technique of listing all possible answers could be tried; the difficulty would seem to be in making sure you count all the solutions once and once only. To be sure, in the case where there are only three unknowns, it is not too difficult, even by the above method, to determine the *number* of solutions of each equation $x + y + z + = n$ without actually *listing any of them*. But with more variables we surely need some new ideas. And after these new ideas are introduced in Chapter 3 you will be able to write down the answer to *every* problem of this sort (that is, with any number of variables and any n on the right-hand side) *immediately*.

In Chapter 3 we shall also see how Problem 3A can be reinterpreted as a problem of placing identical marbles in distinguishable boxes. This new point of view then generates a host of fascinating problems known as *occupancy problems*, one slight variant of which is the intriguing but very difficult problem involving the *partitions of an integer* (see Problem 6 and Chapters 6 and 8).

Here is a probability problem whose answer would confound most people's intuition.

PROBLEM 4
Snap!
Two fully shuffled standard packs of 52 cards are placed face down and side by side. One after another, pairs of cards, one from each pack, are turned over. What is the probability that, as all 52 pairs are turned over, at least one pair of cards will be identical?

If you have not seen this problem before, you will surely be intrigued by the answer. And for those who are happy to engage in a little gambling (neither of the authors does) there is a chance for readers with no conscience to use the counterintuitive result to make a little money on the side from their more susceptible friends! In Chapter 4 we introduce the *inclusion-exclusion principle* and show how it may be used to solve this problem.

Certain numbers arise in combinatorics so frequently that they often bear the names of their originators. The numbers that arise on putting distinct balls into identical cells, with no cell left empty, are named after James Stirling, who defined the *Stirling numbers* in a completely different context. Here is a problem from calculus (don't worry if you haven't yet met the ideas involved) to which the Stirling numbers provide an answer. We let θ be the operator $x(d/dx)$. Thus $\theta y = x(dy/dx)$, and

$$\theta^2 y = x\frac{d}{dx}\left(x\frac{dy}{dx}\right) = x\left(x\frac{d^2y}{dx^2} + \frac{dy}{dx}\right) = x^2\frac{d^2y}{dx^2} + x\frac{dy}{dx},$$

and you can check that, similarly,

$$\theta^3 y = x^3\frac{d^3y}{dx^3} + 3x^2\frac{d^2y}{dx^2} + x\frac{dy}{dx}.$$

PROBLEM 5A
Differential Operators

Find a formula for the coefficient of the term $x^k(d^ky/dx^k)$ in $\theta^n y$, for $1 \leq k \leq n$.

In Chapter 5 we also introduce the *Catalan numbers*, which arise in very many seemingly unrelated problems.

PROBLEM 5B
Walking East and North!

Suppose we have an $n \times n$ grid. How many paths are there, following edges of the grid, from the bottom left corner to the top right corner that may touch, but not go above, the diagonal shown in Figure 1.1?

Here is a problem that engaged the Swiss mathematician Leonhard Euler.* By a *triangulation* of a polygon, we mean a way of dividing the polygon in triangles by nonintersecting diagonals, that is, lines joining two vertices. For example, Figure 1.2 is an example of a triangulation of a regular hexagon.

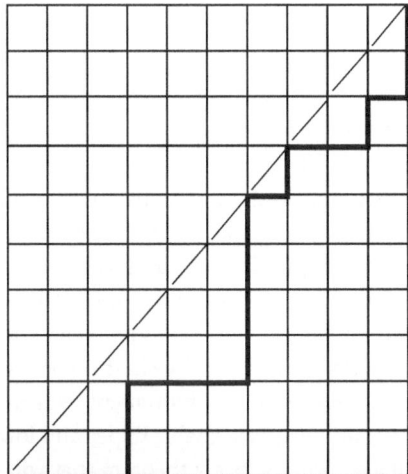

FIGURE 1.1

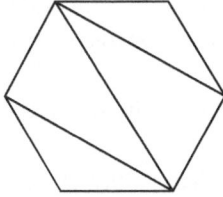

FIGURE 1.2

---

* See Chapter 8, Section 8.2, for a brief biography of Euler.

PROBLEM 5C
Chopping Up a Hexagon
How many different triangulations are there of a regular hexagon?

A classic combinatorial problem is that of counting *partitions*. Here is a problem of this type.

PROBLEM 6
Partitions
In how many different ways can the number 100 be expressed as the sum of positive integers?

To make this question precise we explain that we are here interested only in which numbers make up the sum and not the order in which they are written. So, for example, we regard the sum $50 + 24 + 13 + 13$ as representing the same way of expressing 100 as a sum of positive integers as does $13 + 24 + 50 + 13$. We call $50 + 24 + 13 + 13$ a *partition* of 100, and we let $p(n)$ be the number of different partitions of n. When n is small, the value of $p(n)$ can be calculated by listing all the possibilities. For example, we can see that $p(6) = 11$, from the list

$6, 5 + 1, 4 + 2, 4 + 1 + 1, 3 + 3, 3 + 2 + 1, 3 + 1 + 1 + 1, 2 + 2 + 2, 2 + 2 + 1 + 1,$

$2 + 1 + 1 + 1 + 1, 1 + 1 + 1 + 1 + 1 + 1$

of all the different ways of writing 6 as the sum of positive integers. Of course, to evaluate $p(6)$ in this way, we need to be satisfied that we have included all the possibilities. And while it is feasible to evaluate $p(6)$ in this way, it is hardly practicable for $p(100)$.

All this raises the question of finding a formula for $p(n)$. This problem was solved by two giants of mathematics, the Indian Srinivasa Ramanujan and the Englishman G. H. Hardy. Their proof is far too involved to reproduce in this book, but, in Chapters 6 and 8, we can experience a more modest sense of achievement by using some elementary calculus, believe it or not, to obtain fairly reasonable upper and lower bounds for $p(n)$ and related functions. We shall also prove a lovely theorem that will enable you to calculate quite a large number of the smaller $p(n)$ fairly quickly *by hand* and avoiding brute force!

Almost every mathematically inclined student will have heard of Fibonacci and his rabbits. At the heart of the story is the *Fibonacci sequence* 1, 1, 2, 3, 5, 8, 13, 21, 34, ... in which each integer after the first two is the sum of the previous two. If we let f_n be the nth term of this sequence, we can describe the sequence by saying that

$$f_1 = f_2 = 1 \text{ and, generally, for } n \geq 3, f_n = f_{n-2} + f_{n-1}.$$

The Fibonacci sequence has so many wonderful properties that a quarterly journal is produced to publicize them. One famous application is the test devised by Lucas[*] to

---
[*] Francois-Edouard-Anatole Lucas was born in Amiens on April 4, 1842, and died in Paris on October 3, 1891. He introduced the recreational game The Tower of Hanoi, which we discuss in Chapter 7. He died of an infection after being hit by a flying shard of a broken dinner plate.

check whether or not a number of the form $2^n - 1$ is a prime. In 1876 he proved, with only pencil and paper but lots of patience, that the 39-digit number

$$2^{127}-1 = 170,141,183,460,469,231,731,687,303,715,884,105,727$$

is prime. This number stood as the largest prime known to anyone until the 1950s when electronic calculators were applied to the task!

PROBLEM 7
The Fibonacci Numbers
Is there a formula explicitly giving the values of the Fibonacci numbers?

The Fibonacci numbers are defined by the *recurrence relation* $f_n = f_{n-2} + f_{n-1}$, which relates later numbers in the sequence to earlier numbers in the sequence. We study recurrence relations systematically in Chapter 7. One technique that is useful here is that of *generating functions*, which we describe in Chapter 8. This technique is particularly useful with some intriguing problems that arise if we place restrictions on the numbers that are allowed in a partition. Here is one suggested by experiments with small numbers.

PROBLEM 8
Special Partitions
Is it true that for each integer n the number of ways of writing n as the sum of odd positive integers is the same as the number of ways of writing n as the sum of positive integers that are all different?

In the following table we have listed the partitions of 12 into odd numbers and into different numbers.

11+1	12
9+3	11+1
9+1+1+1	10+2
7+5	9+3
7+3+1+1	9+2+1
7+1+1+1+1+1	8+4
5+5+1+1	8+3+1
5+3+3+1	7+5
5+3+1+1+1+1	7+4+1
5+1+1+1+1+1+1+1	7+3+2
3+3+3+3	6+5+1
3+3+3+1+1+1	6+4+2
3+3+1+1+1+1+1+1	6+3+2+1
3+1+1+1+1+1+1+1+1	5+4+3
1+1+1+1+1+1+1+1+1+1+1+1	5+4+2+1

Not many partitions occur in both lists, but both lists contain the same number of partitions. Is this a coincidence? It comes as a surprise that the answer is "no" and that the proof uses only rather simple manipulations of power series.

We now come to one of the best known of all combinatorial problems. The river Pregel flows through Königsberg* and divides the town, flowing around an island called the Kneiphof ("Beer Garden"). The city was connected by seven bridges as shown in the sketch map in Figure 1.3. It is said that the residents considered the following problem during their Sunday stroll.

PROBLEM 9A
The Bridges of Königsberg
Is there a route that would have taken the burghers of Königsberg over each of the bridges exactly once?

This problem was solved by Leonhard Euler in 1735. Euler's solution of this problem is regarded as the origin of the branch of mathematics known as *graph theory*. The "graphs" of graph theory are rather different from the "graphs" of functions you will be familiar with, but in both cases, *graph* abbreviates *graphical representation*. In the case of the Königsberg bridges problem, all that really matters is how the different parts of the city are connected by the bridges. So we can replace Figure 1.3 with the representation of the same situation shown in Figure 1.4, where the dots represent the land areas and the lines represent the bridges.

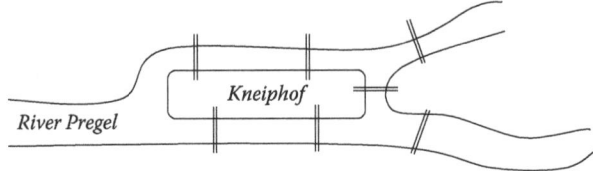

FIGURE 1.3

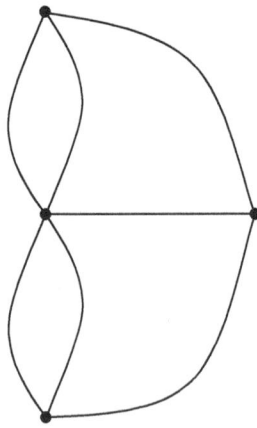

FIGURE 1.4

---

* Originally a medieval city, Königsberg became the capital of East Prussia in the fifteenth century. In 1945 it was ceded by Germany to the Soviet Union and was renamed Kaliningrad. It remains part of modern-day Russia.

Another classic problem of graph theory is the *utilities problem.* We state it in a slightly modernized form.

PROBLEM 9B
The Utilities Problem
Is it possible to connect up a house, a cottage, and a bungalow to supplies of electricity, gas, and cable television so that the pipes, wires, and cables do not cross each other?

In Figure 1.5 we show a failed attempt to solve this problem. But one failure does not prove that the problem does not have a solution. The issue is whether the graph consisting of two sets of three dots, where each dot in the first set is joined to each dot in the second graph, can be drawn in the plane so that the lines meet only at dots.

An even more famous problem in this area is whether when drawing maps we always need only four colors to ensure that countries with a common boundary can be colored differently. The question was first asked in 1852, but it took 124 years for a solution to emerge—although a "proof" proposed in 1879 stood for 11 years before a flaw in it was found.

PROBLEM 9C
The Four-Color Problem
Is it possible to color every map drawn in the plane with at most four colors so that adjacent countries are colored differently?

We show in Chapter 9 how this problem may be reworded as a problem about coloring the vertices of a graph. The *four-color theorem* says that the answer to this question is "yes." It was proved to most people's satisfaction by Kenneth Appel and Wolfgang Haken in 1977. Their proof involved a considerable use of computer time to check a large number of cases and so has not been accepted as a "mathematical proof" by everyone. We are not able to go into the technicalities in this book, but we are able to give a proof of the less ambitious claim that every map may be colored using at most five colors.

Graphs can also be used to represent chemical molecules, with the vertices representing atoms and the edges representing valency bonds. Two examples are shown in Figure 1.6.

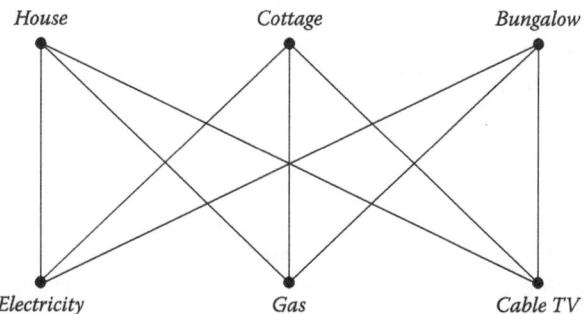

FIGURE 1.5

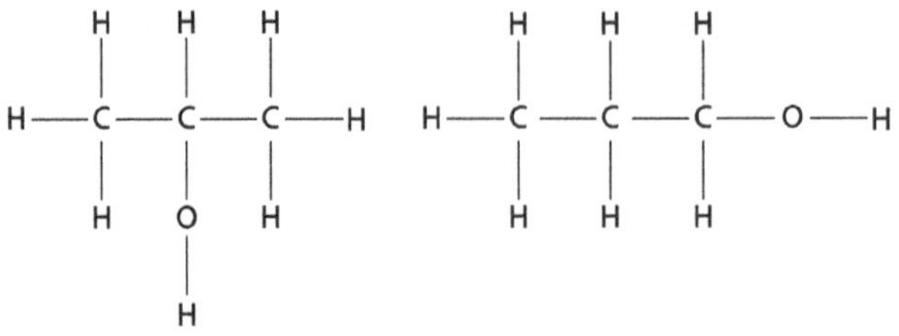

```
H   H   H              H   H   H
|   |   |              |   |   |
H—C—C—C—H          H—C—C—C—O—H
|   |   |              |   |   |
H   O   H              H   H   H
    |
    H
```

FIGURE 1.6

The graphs used to represent these molecules are of a special kind called *trees*. These graphs have lots of applications, and we study these in Chapter 10. The classic problem in this area is the following.

PROBLEM 10
Labeled Trees
How many different trees are there with n vertices labeled with the numbers $1,2,...,n$?

In Chapter 10 we give the answer originally found by the English mathematician Arthur Cayley.

Much of mathematics is about finding patterns. In order to prepare for our study of patterns, we look at symmetries of geometric figures. Symmetries give rise to mathematical structures called *groups*. You may already have met this concept, but in case not, we introduce groups from scratch in Chapter 11. Ideas from group theory can be used to solve a large range of problems. Here is one example.

PROBLEM 11
Shuffling Cards
What is the most effective way to shuffle a pack of cards?

We next look at the relationship between the coloring of figures and the symmetries of these figures. We prove an important theorem, the *orbit-stabilizer theorem*, in Chapter 12, but we find we have to develop the theory further in Chapter 13 in order to answer questions such as the next three.

PROBLEM 13A
Coloring a Chessboard
How many different ways are there to color the squares of a chessboard using two colors?

On a standard 8×8 chessboard, the squares are colored alternately black and white as shown in Figure 1.7i, but clearly they could be colored in many other ways. Alternative colorings are shown in Figure 1.7ii and iii. The problem is to decide exactly how many different colorings are possible.

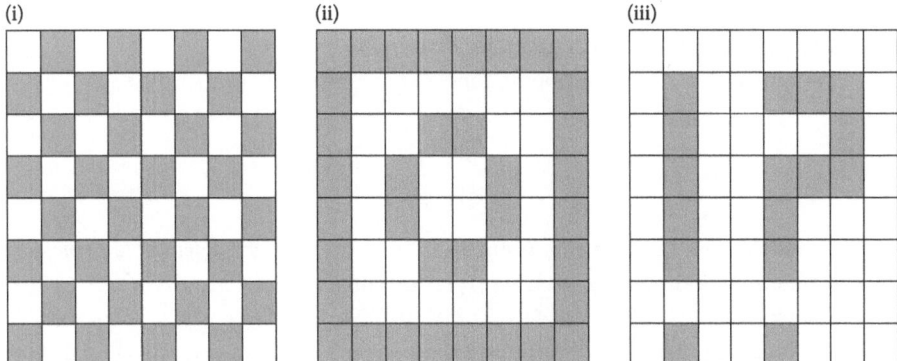

FIGURE 1.7

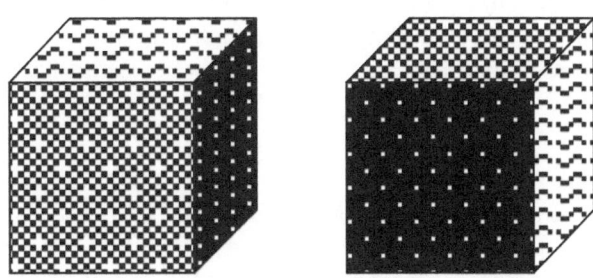

FIGURE 1.8

PROBLEM 13B
Varieties of Colored Cubes
In how many different ways can you color a cube using three colors?

Before we can answer this question we need to make clear what "different" means. The cube on the right in Figure 1.8 looks different from the one on the left. But perhaps they could be made to look the same by rotating the cube on the right about an appropriate axis. This is where the *symmetries* of a geometric figure come in.

This eventually leads us to a very powerful technique due to George Pólya for answering more complicated questions, such as the following.

PROBLEM 14
Counting Patterns Again
In how many different ways can you color a cube using one red, two white, and three blue faces?

We also show in Chapter 14 that Pólya's Counting Theorem enables us to count fairly readily the number of different simple graphs with a given number of vertices. Our next combinatorial principle arises in the context of number theory where it was used by Dirichlet to solve the following problem.

PROBLEM 15A
Rational Approximations to Irrational Numbers

Show that for each irrational number a, there exists a rational number p/q such that

$$\left| a - \frac{p}{q} \right| < \frac{1}{q^2}.$$

The principle used to solve this problem is called *Dirichlet's pigeonhole principle*, which is of deceptive simplicity but has surprising consequences. Here is one of a more recreational kind.

It is obvious that, using 18 dominoes, each of size 1 by 2, we can "cover" a 6 by 6 (square) board completely and without any two dominoes overlapping. An example is given in Figure 1.9.

Note that, in the figure, the two leftmost *columns* of dominoes are separated from the other four columns by a "fault line." That is, a knife could be dragged down the board cutting it in two *without* having to cut through any domino. At the same time, the board cannot be similarly split along any other row or column. So, the problem is:

PROBLEM 15B
Placing Dominoes "Faultlessly"

Can 18 dominoes be placed on a 6 by 6 board "faultlessly"?

We next come to another existence problem that is the starting point for a large area of study called *Ramsey theory* and that can also be reinterpreted in terms of graphs:

PROBLEM 16A
Friends at a Party

There are six people at a party. Each pair are either friends or strangers. We claim that, among the six, there are (at least) three people who are either all friends or all strangers. Are we correct?

A nice way to picture this problem is by using graphs, this time with *colored* edges. If we ask each pair of friends to hold opposite ends of a red string and each pair of

FIGURE 1.9

strangers (if they would be so kind) to hold opposite ends of a blue string, the problem becomes: Is it true that if the 15 edges of a graph with six vertices are colored either red or blue, then there is always either a red "triangle," that is, three edges forming a triangle that are all red, or a blue "triangle"? In brief, must there always exist a *monochromatic* triangle?

There are many other fascinating problems that are concerned with coloring (even infinitely many) *points*. For example:

PROBLEM 16B
Plane Colors

Let us be given a red/blue coloring of the points of the plane. That is, imagine that with each point of the plane there is associated one of those two colors. Suppose that someone now draws a particular triangle T in the plane. Is it always possible to move T to a position in the plane so that its vertices lie over three points *all of the same color*? In other words, can we find, in the plane, a *monochromatic* triangle congruent to T? If not, can we find one *similar* to T? We attempt no picture here, for obvious reasons!

In the final chapter of this book we deal with questions of the following kind.

PROBLEM 17A
Nonattacking rooks

Given the 5×5 board in Figure 1.10, in how many ways can 0, 1, 2, 3, 4, 5, or more nonattacking rooks (that is, no two rooks in the same row or column) be placed on the board so that *none of them lies on a black square*?

At first sight this seems rather a frivolous problem, but actually it is one with many practical applications. As you can see from Figure 1.11, it is equivalent to asking in how

FIGURE 1.10

	Job 1	Job 2	Job 3	Job 4	Job 5
Employee A					
Employee B					
Employee C					
Employee D					
Employee E					

FIGURE 1.11

many ways the five people A, B, C, D, and E can each be allocated one of the jobs 1 to 5, where a black square indicates that the person named in its row is unable to do the job named in its column.

Finding the answer leads to the introduction of *rook polynomials*. In most cases the key question is whether there is any solution at all rather than the number of solutions. One particularly intriguing version of this problem is the following, whose answer is certainly not obvious one way or the other.

PROBLEM 17B
Selecting Cards
Let 40 cards—10 red, 10 blue, 10 green, and 10 yellow, each set being numbered 1 to 10—be well shuffled and then dealt out into 10 groups of four. Is it always possible to pick one card from each group so that the 10 cards chosen will include exactly one of each number 1,2,…,10? (Obviously we do not insist that all the chosen cards are also of the same color.)

This problem is solved using Hall's marriage theorem with which we end Chapter 17.

1.3 WHAT YOU NEED TO KNOW

One of the advantages of following a course in combinatorics is that only a modest mathematical background is necessary in order to get started. This modest amount includes some basic set theory, which we remind you of below, a bit of elementary algebra, and, perhaps surprisingly (to help with two or three estimation problems), some calculus. In Chapter 11 we use *groups*, but we assume no previous knowledge of group theory, which we introduce as we need it.

Suppose A and B are sets. The *union* of A and B, written as $A \cup B$, is the set of elements that are in A or B or both. The *intersection* of A and B, written as $A \cap B$, is the set of elements that are both in A and in B. The *difference*, written $A \setminus B$, is the set of elements in A but not in B. That is, $A \setminus B = \{x : x \in A \text{ and } x \notin B\}$. This may be represented pictorially as in Figure 1.12, where the shading represents the set $A \setminus B$.

There is, unfortunately, no standard notation for the number of elements in a set, X. The notations $\overline{X}$, $|X|$, and $card(X)$ are all used. However, our preferred notation is $\#(X)$.

We use the symbols N, N$^+$, Z, Q, R, and C for the following sets of numbers:

N is the set of *natural numbers*, that is, N = $\{0,1,2,…\}$. N$^+$ is the set of *positive integers*, that is, N = $\{1,2,3,…\}$. Z is the set of *integers*, that is, Z = $\{…–2,–1,0,1,2,…\}$. Q is the set of *rational numbers*, R is the set of *real numbers*, and C is the set of *complex numbers*. We use R^2 for the points of the plane.

We use the "arrow" notation for functions. For example, the function that squares each number will be written as $x \mapsto x^2$. If the function f maps elements of the set D to the set C, we write $f : D \to C$ and call the set D the *domain* of f, and the set C the *codomain* of f.

We say that a function $f : D \to C$ is *injective* if f does not repeat values, that is, if for all x, $y \in D$, $x \neq y$ implies that $f(x) \neq f(y)$. We say that f is *surjective* if each element of C is a value of f, that is, if for each $y \in C$, there is some $x \in D$ such that $f(x) = y$. We say that f is *bijective*

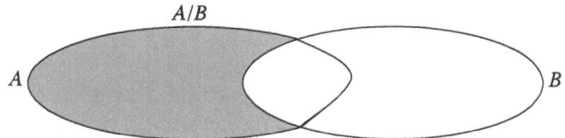

FIGURE 1.12

if it is both injective and surjective. (The alternative nomenclature that you may have met is *one-one* for injective, *onto* for surjective, and *one-one correspondence* for bijective.)

We assume that you are already familiar with the "sigma" notation for sums. For example, $\sum_{i=1}^{n} a_i$ represents the sum $a_1 + a_2 + \ldots + a_n$. On occasion we also, in a similar way, use the "pi" notation for products. So $\prod_{i=1}^{n} a_i$ represents the product $a_1 \times a_2 \times \ldots \times a_n$, or, omitting the multiplication signs, $a_1 a_2 \ldots a_n$.

Although we approach most of the topics discussed in this book through concrete problems, a strong aim of this book is to emphasize the importance of proof in mathematics. While this is a book that focuses on combinatorial methods, these are worthless if we are unable to prove that employing them will give the correct answer. Mathematics is in the very privileged position of being the only area of human knowledge where assertions made have the chance of being verified by unassailable proof – or shot down by counterexample! A course in combinatorics provides an ideal opportunity for paying special attention to methods of proof since, often, the reader will not have to make a huge mental effort to understand the meaning of the statements themselves. Accordingly, we offer no apology for paying careful attention to the majority of proofs themselves, the odd exceptions being proofs that are beyond the scope of this book or where even the most fastidious mathematician might say, "Clearly, this very same proof goes over to the general case."

We therefore largely assume that the reader is already familiar with the standard methods of proof. In particular, as we are frequently concerned with results that hold for all natural numbers, or for all positive integers, *proof by mathematical induction* is used a good deal. If you need an introduction to this topic, or a reminder about it, please consult one of the many books that cover this topic.[*]

1.4 ARE YOU SITTING COMFORTABLY?

Once upon a time there was a program on the radio called *Listen with Mother*. (In those days it was assumed that it would be the mother who would be at home with young children.) In the first program in 1950 the storyteller, Julia Lang, introduced the story she was about to tell by saying, "Are you sitting comfortably? Then we'll begin." Apparently this introduction was not planned, but it caught on and was used regularly until the program came to an end in 1982.[†]

---

[*] We mention just two, R. B. J. T. Allenby, *Numbers and Proofs*, Arnold, London, 1997, and Kevin Houston, *How to Think like a Mathematician*, Cambridge University Press, Cambridge, 2009.
[†] See Nigel Rees, *Sayings of the Century*, Allen & Unwin, London, 1984.

When it comes to reading mathematics, however, this is not an appropriate beginning. A mathematics book cannot be read like a novel, sitting in a comfortable chair, with a glass at your side. Reading mathematics requires you to be active. You need to be sitting at a table or a desk, with pencil and paper, both to work through the theory and to tackle the problems. A good guide is the amount of time it takes you to read the book. A novel can be read at a rate of about 60 pages an hour, whereas with most mathematics books you are doing well if you can read 5 pages an hour. (It follows that, even at 12 times the price, a mathematics book is good value for the money!)

Since the approach the book takes is to begin with problems and usually to use them to lead into the theory, we have posed a good number of *problems* in the text. The bth problem in Chapter a is labeled **Problem a.b**. These problems are immediately followed by their solutions, but you are strongly encouraged to try the problems for yourself before reading our solution.

At the end of most of the chapter sections, there are *exercises*. Some of these are routine problems to help consolidate your understanding, and some take the theory a bit further or are designed to challenge you. In most cases these problems occur in pairs, labeled A and B. Usually the B question is quite similar to the A question. The difference is that we have included solutions for the A questions at the back of the book but not for the B questions. The solutions to the A questions are usually written out in detail. This is intended to be helpful, but it will not achieve their purpose of helping you to learn the subject, if you give in to the temptation to read the solutions before making your own attempt at the exercises. The B questions, with no solutions, are there for those who cannot resist this temptation!

CHAPTER 2

Permutations and Combinations

2.1 THE COMBINATORIAL APPROACH

In Chapter 1 we gave examples of counting problems that we hope convinced you of their interest and importance. In this chapter we introduce two of the most basic ideas, counting *permutations* and counting *combinations*. These occur over and over again throughout this book. You may have already met these ideas in algebra in connection with the binomial theorem, but the combinatorial approach may be new to you. It can be hard to relearn a topic you are already familiar with but using a different approach. However, we encourage you to adopt the combinatorial approach, which gives more importance to counting methods than to algebraic manipulation, as this is the key to much of the rest of this book.

2.2 PERMUTATIONS

We begin with some problems that are very simple, but the ideas behind their solutions are of fundamental importance in many counting problems.

PROBLEM 2.1
Cayley's Café has the following menu:

Cayley's Café
Starters
Tomato Soup
Fruit Juice
Mains
Lamb Chops
Battered Cod
Nut Bake
Desserts
Apple Pie
Strawberry Ice

How many different three-course meals could you have?

Solution

You have two choices for your starter, and, whichever choice you make, you have three choices for your main course. This makes 2 × 3 = 6 choices for the first two courses.

```
Soup   Soup   Soup   Juice   Juice   Juice
 |      |      |      |       |       |
Chops  Cod   Bake   Chops   Cod    Bake
```

In each of these six cases you have two choices for your dessert, making 6 × 2 = 12 possibilities altogether. We can set them out in Figure 2.1, which makes it clear why the number of cases multiplies at each stage and why the final answer is the product of the number of choices at each stage.

So we obtain 2 × 3 × 2 = 12 as the total number of possible meals.

PROBLEM 2.2

In a race with 20 horses, in how many ways can the first three places be filled? (For simplicity, assume that there cannot be a dead heat.)

Solution

There are 20 horses, each of which could come first. Whichever horse comes first, there are 19 other horses that can come second. So there are 20 × 19 = 380 ways in which the first two places can be filled. In each of these 380 cases, there are 18 remaining horses that can come third. So there are 380 × 18 = 20 × 19 × 18 = 6840 ways in which the first three places can be filled.

We now consider the way in which these two problems are different and the way in which they are similar. In Problem 2.1 your choice of a starter did not affect the choice of the main course. Whether you chose the tomato soup or the fruit juice, you still have the choice of lamb chops, battered cod, or nut bake for your main course. And whatever your choices of starter and main course, you still have the same choices, apple pie or strawberry ice, for your dessert.

In Problem 2.2, the horse that wins the race cannot also come in second. So the possibilities for which horse comes in second vary according to which horse wins the race. However, the *number* of possibilities remains the same. Whichever horse wins the race, there are 19 horses that can come second, though which 19 horses these are varies according to which the winner is. Likewise, the possibilities for the third horse vary according to which two horses come in first and second, but, whichever these horses are, there always remain 18 horses each of which can come in third. It is

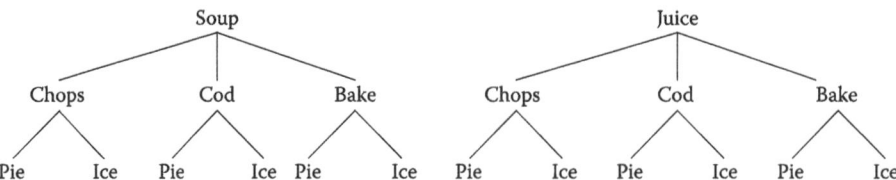

FIGURE 2.1

because the *number of choices* at each stage does not depend on the particular choices made earlier that we could use the same multiplication method to solve Problem 2.2 as we used to solve Problem 2.1, and thus 20 × 19 × 18 does indeed give the number of ways in which the first three positions in the race can be filled.

The multiplication principle we have used in these two problems is sufficiently important to be worth stating explicitly.

> **THE PRINCIPLE OF MULTIPLICATION OF CHOICES**
>
> If there are r successive choices to be made, and for $1 \leq i \leq r$, the ith choice can be made in n_i ways, then the total number of ways of making these choices is $n_1 \times n_2 \times \ldots \times n_r$.

Note that we can use the "pi" notation to write the product in the box as $\prod_{i=1}^{r} n_i$.

Although the principle of multiplication of choices applies equally to Problems 2.1 and 2.2, Problem 2.2 has an additional feature that frequently occurs in problems of this type. The successive choices were all being made from the set of 20 horses taking part in the race. So the number of horses left to choose from goes down by one at each successive stage. That is, in the notation we are using,

$$n_{i+1} = n_i - 1, \quad \text{for } 1 \leq i < r.$$

In such a case, if $n_1 = n$, then for $2 \leq i \leq r$, $n_i = n - i + 1$, so that the product $\prod_{i=1}^{r} n_i$ is $n(n-1)(n-2)\ldots(n-r+1)$. We can express this product more succinctly by making use of factorial notation. We have that

$$n(n-1)(n-2)\ldots(n-r+1) = \frac{n(n-1)(n-2)\ldots(n-r+1)(n-r)(n-r-1)\ldots \times 2 \times 1}{(n-r)(n-r-1)\ldots \times 2 \times 1} = \frac{n!}{(n-r)!}.$$

Since this situation occurs very frequently, we introduce some special terminology and notation to describe it. We call a choice of r objects from a set of n objects in which the *order* of choice is to be taken into account, a *permutation* of r objects from n. We let $P(n,r)$ be the number of different permutations of r objects from n. Of course, this makes sense only in the case where r and n are nonnegative integers with $r \leq n$. The above remarks yield the general formula for $P(n,r)$.

THEOREM 2.1

For all nonnegative integers r, n with $r \leq n$, $P(n,r) = n!/(n-r)!$

It is important to remember that $P(n,r)$ counts the number of ways of choosing r objects *in order* from a set of n objects. If the order does not matter, the number of choices is smaller, as we shall see in the next section.

In Problem 2.2 we considered only the number of different ways in which the first three positions could be filled. Suppose now we are interested in the number of different ways all 20 horses can finish in order (again, assuming no dead heats). We can see that this number is 20 × 19 × 18 × ... × 2 × 1, that is, 20! Note that this is the same as the number of ways of choosing 20 horses in order from a set of 20 horses, and so we could have obtained this by using Theorem 2.1, which tells us that this number is $P(20,20) = 20!/0! = 20!$ (Recall the standard convention that $0! = 1$.) Thus we have:

THEOREM 2.2

The number of different permutations of n objects is $n!$

The values of $n!$ grow very rapidly. Even for quite small values of n, the factorial $n!$ is very large. For example, $10! = 3,628,800$ and $100!$ is larger than 10^{157}.

PROBLEM 2.3

In how many ways can eight counters be placed on a square 8×8 chessboard in such a way that no two counters lie either in the same row or in the same column? Note that we can reword this problem as: In how many ways can eight rooks be placed on a chessboard so that no two rooks are "attacking" each other? This latter problem is generalized in Chapter 17.

Solution

Let us place one counter in each row in turn. For the first row there are eight columns in which the counter may be put. Having placed this counter, when it comes to the second row, there are just seven columns where we may place a counter, as it must not be in the same column as the counter in the first row. As we place counters in successive rows, the number of possible columns where the next counter may be placed goes down by one at each stage. So the total number of permissible arrangements of the counters is $8 \times 7 \times \ldots \times 2 \times 1$, that is, $8!(= 40,320)$. One of these arrangements is shown in Figure 2.2.

Clearly there are, more generally, $n!$ different ways to place n counters on the squares of an $n \times n$ chessboard so that there are neither two counters in the same row nor in the same column. In Exercise 2.2.4A you are asked to generalize this to the case of placing any number of counters on rectangular boards of any size.

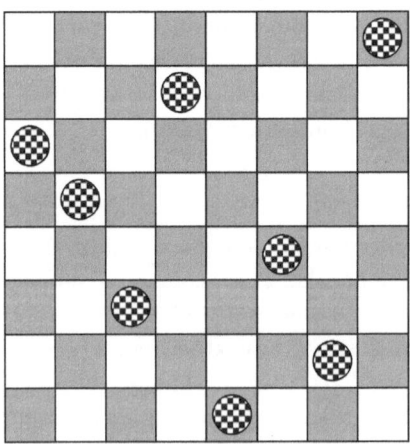

FIGURE 2.2

Exercises

2.2.1A Currently a €10 note has a "serial number" of the form X19298164502, that is, a letter followed by 11 digits. How many different serial numbers of this form are there? A Bank of England £10 note has a serial number of the form CD49000372, that is, two letters followed by eight digits. Are there more of these serial numbers than there are for a €10 note?

2.2.1B A personal identification number (PIN) consists of a sequence of four digits, each drawn from the set {0,1,2,3,4,5,6,7,8,9}, except that the first digit of a PIN cannot be 0. How many different PINs are there? How many different PINs are there in which no digit is repeated?

2.2.2A How many different sequences of length 10 are there in which each of the digits 0,1,2,3,4,5,6,7,8,9 is used once? How long would it take you to list them all if each sequence took one second to write down?

2.2.2B i. How many sequences are there of n digits in which all the digits are different?

ii. How many sequences are there of n digits in which no two consecutive digits are the same?

2.2.3A In three races there are 10, 8, and 6 horses running, respectively. You win a jackpot prize if you correctly predict the first 3 horses, in the right order (assuming no dead heats), in each race. How many different predictions can be made?

2.2.3B A password is a sequence of six characters, the first three being either an upper or a lowercase letter, the next being a digit, and the final two coming from the set {!,£,$,%,^,&,*,(,),_, + , = ,{,},[,], @,#,?} of 19 other symbols occurring on a standard keyboard. How many different passwords are there? How many are there if consecutive characters must be different? How many are there if all the characters must be different?

2.2.4A Let k, m, and n be positive integers with $k \leq m$, $k \leq n$, and $m \leq n$. In how many different ways may k counters be placed on the squares of an $m \times n$ grid so that no two counters are in the same row or in the same column?

2.2.4B In how many different ways may eight red and eight green counters be placed on the squares of an 8×8 chessboard so that there are not two counters on any one square and there is one red counter and one green counter in each row and column?

2.3 COMBINATIONS

Let us now count the number of ways of choosing a specified number of objects from a set when the order of selection does not matter. We tackle this problem by relating it to the problem of counting permutations, which we have already solved. (Reducing a new problem to a case that has already been solved is a common mathematical technique. It is said that many a mathematician who has learned how to make a cup of tea starting with an empty kettle will, when given a full kettle and asked to make tea, first empty the kettle to reduce the problem to one that they already know how to solve.) A couple of examples will make the line of approach clear.

PROBLEM 2.4

A team of three bowls players is to be selected from a squad of six players. How many different teams can be selected?

Solution

We have seen that we can choose three players, in order, from a squad of six players in $P(6,3) = 6 \times 5 \times 4 = 120$ ways. But, *and this is the key point*, there are not 120 different teams of three players. This is because the order in which we pick the members of the team does not matter. For example, choosing first Pat, then Chris, and then Sam leads to the same team as first choosing Chris, then Sam, and then Pat. Thus, each team can be chosen in more than one way. The number of ways in which three given players can be chosen in order is 3!, that is, 6. Since we get 120 when we count each team six times, the number of different teams is 120/6 = 20. Put another way, the number of different ways to pick three bowls players from six is $P(6,3)/3!$

The technique that we have used in this problem is used again, not only in the next problem, but in many other counting problems. We count the number of arrangements of a particular kind by counting them in such a way that each arrangement is counted more than once. We then adjust our answer to allow for the duplicate counting.

PROBLEM 2.5

How many different hands of 5 cards can be chosen from a pack of 52 cards?

Solution

We can choose 5 cards, in order, from a pack of 52 cards, in $P(52,5)$ different ways. But the order in which the cards are chosen does not affect the hand we end up with. The same hand of 5 cards can be arranged in order in 5! ways and so can be chosen, in order, in 5! ways. Thus $P(52,5)$ gives the number of 5-card hands when each hand is counted 5! times. Hence the number of different 5-card hands is

$$\frac{P(52,5)}{5!} = \frac{(52!/47!)}{5!} = \frac{52!}{5!47!} = \frac{52 \times 51 \times 50 \times 49 \times 48}{5 \times 4 \times 3 \times 2 \times 1} = 2,598,960.$$

We can now generalize the method used in these last two problems. We call a selection of r objects chosen from n objects, *when the order in which they are chosen does not matter*, a *combination* of r objects from n. We use the notation $C(n,r)$ for the number of different combinations of r objects from n. (Notice that the mathematical usage of *permutation*, where the order matters, and *combination*, where it does not, does not correspond to all the uses of these words in everyday life. In football pools permutations or "perms" are selections of football (otherwise known as "soccer") matches where the order does not matter. In a combination lock, the order of the numbers is important.)

The method that we used to solve Problems 2.4 and 2.5 leads us to the general formula for $C(n,r)$.

THEOREM 2.3

For all nonnegative integers r, n with $r \leq n$,

$$C(n,r) = \frac{n!}{r!(n-r)!}.$$

Proof

By Theorem 2.2, we know that a set of r objects can be ordered in $r!$ ways. Thus $P(n,r)$, the number of ways in which r objects can be chosen in order from a set of n objects, counts each set of r objects chosen from the given set of n objects $r!$ times. Hence, $C(n,r) = P(n,r)/r!$, and therefore, by Theorem 2.1, $C(n,r) = n!/[r!(n-r)!]$.

The numbers $C(n,r)$ are the well-known *binomial coefficients* that occur in the binomial theorem that we give as Theorem 2.6. There are several common alternative notations for these binomial coefficients. The number for which we have used the notation $C(n,r)$ is often written as $\binom{n}{r}$, or ${}_nC_r$ or C_r^n. We have chosen $C(n,r)$ because it is less cumbersome to print than $\binom{n}{r}$ and, unlike ${}_nC_r$ and C_r^n, it makes the numbers n and r easier to read. It also fits in with the standard mathematical notation, $f(x,y)$, for a function of two variables and also with the notation for two-dimensional arrays in many programming languages. Its disadvantage is that it ties the letter C to a particular meaning. To avoid this, the alternative notation $(n!r)$ was once suggested.*

It is worth noting that since $C(n,r)$ is the number of ways of choosing r objects from n, we must have $0 \leq r \leq n$. We allow the case $r = n$. In this case the formula gives $C(n,n) = n!/(0!n!) = 1$. This corresponds to the fact that an n-element set A has just one n-element subset, namely, the set A itself. We also have $C(n,0) = n!/(0!n!) = 1$, corresponding to the fact that there is just one subset of A that has zero elements, namely, the empty set $\emptyset$.

The formula for $C(n,r)$ given by Theorem 2.3 can be used to give algebraic proofs of many properties of the binomial coefficients. We prefer, however, to emphasize the combinatorial meaning of these numbers and to give combinatorial proofs whenever this is convenient. In line with this approach, we have given a combinatorial definition of the number $C(n,r)$. The alternative would have been to define $C(n,r)$ by the formula of Theorem 2.3. It would then have been necessary to prove that the number of r-element subsets of an n-element set is indeed $C(n,r)$. Our combinatorial approach is illustrated by our proofs of the next four theorems.

THEOREM 2.4

For all positive integers r, n with $r \leq n$, $rC(n,r) = nC(n-1,r-1)$.

Proof

Let X be an n-element set. We evaluate the sum of the numbers of elements in all the r-element subsets of X in two different ways.

---

* In *The Printing of Mathematics*, by T. W. Chaundy, P. R. Barrett, and Charles Batey, Oxford University Press, London, 1954. This book is out of date technologically as it was written in the days of hot-metal typesetting, but its advice to mathematical authors is still valuable.

First, as there are $C(n,r)$ subsets of X, each containing r elements, this sum is $C(n,r) \times r$, that is, $rC(n,r)$. Second, consider one particular object, say a, from the n-element set X. To obtain an r-element subset of X containing a we need to choose a further $r-1$ elements from the remaining $n-1$ elements of X. This can be done in $C(n-1,r-1)$ ways. Therefore, each of the n elements of X occurs in $C(n-1,r-1)$ different r-element subsets of X. Consequently, the sum of the numbers of elements in these sets is $n \times C(n-1,r-1)$. As these two different ways of obtaining this sum must lead to the same answer, it follows that $rC(n,r) = nC(n-1,r-1)$.

Note that we can deduce from Theorem 2.4 that $C(n,r) = (n/r)[C(n-1,r-1)]$. This enables us to give a direct, combinatorial proof that $C(n,r) = n!/[r!(n-r)!]$ without the need to consider permutations.

THEOREM 2.5

For all nonnegative integers r, n with $r \leq n$, $C(n,r) = C(n,n-r)$.

Proof

Deciding which r objects to select from a set of n objects amounts to exactly the same thing as deciding which $n-r$ objects *not* to select. Hence the number of ways of choosing r objects from n is the same as the number of ways of choosing $n-r$ objects from n.

The next theorem explains how the binomial coefficients get their name.

THEOREM 2.6
The Binomial Theorem

For all variables a, b, and each positive integer n,

$$(a+b)^n = a^n + C(n,1)a^{n-1}b + C(n,2)a^{n-2}b^2 + \ldots + b^n,$$

that is,

$$(a+b)^n = \sum_{r=0}^{n} C(n,r)a^{n-r}b^r, \text{ as } C(n,0) = C(n,n) = 1.$$

Proof
Consider the product

$$(a+b)(a+b)\ldots(a+b)$$

with n pairs of brackets. When we multiply out this product, each separate term that arises comes from choosing either a or b from each pair of brackets and then multiplying these a's and b's together. We obtain the term $a^{n-r}b^r$ each time we choose b from r of these pairs of brackets and a from the remaining $n-r$ pairs. Thus the number of terms of the form $a^{n-r}b^r$ that we obtain equals the number of ways of choosing r pairs of brackets from which to pick b, and this number is $C(n,r)$. Hence when we gather similar terms together, the coefficient of $a^{n-r}b^r$ is $C(n,r)$.

Of course, selecting r b's forces us to select $n-r$ a's. Repeating the argument with a and b interchanged shows that the coefficient of $a^{n-r}b^r$ is also $C(n,n-r)$, as Theorem 2.5 tells us it should be.

The idea that we have used in this combinatorial proof of the binomial theorem will play an important role later in this book (in Chapter 7). The algebraic expression $(a + b)^n$ is called a *binomial* (from the Latin *binomius* meaning "having two names"), and this is why the binomial coefficients were given their name. The binomial theorem is sometimes attributed to Isaac Newton, though the binomial coefficients were known and tabulated long before Newton's time. His main contribution in this area was to prove the form of this theorem that applies when the exponent n is not a positive integer.

Our next theorem about binomial coefficients leads to a very well-known method for calculating their values. We again emphasize that we give a combinatorial proof of this theorem. An algebraic proof, using the formula for $C(n,r)$, is very straightforward but hides the combinatorial meaning of the result.

THEOREM 2.7

For all positive integers r, n with $r \leq n$,

$$C(n+1,r) = C(n,r-1) + C(n,r).$$

Proof

Let X be a set containing $n + 1$ objects, and let a be one of the objects in the set X. We count the number of subsets of X containing r elements by separating them into the set, say Y, of those r-element subsets that include a and the set, say Z, of those r-element subsets that do *not* include a.

A subset of X in Y contains r elements one of which is a and a further $r-1$ elements chosen from the n-element set $X \setminus \{a\}$. Thus, there are $C(n,r-1)$ subsets in Y.

A subset of X in Z contains r elements none of which is a, and hence consists of r elements chosen from $X \setminus \{a\}$ and hence there are $C(n,r)$ of these.

Each r-element subset of X is either in Y or in Z, and none of them is in both. Hence the number of r-element subsets of X is the sum of the number of subsets in Y and the number in Z, that is, $C(n+1,r) = C(n,r-1) + C(n,r)$.

The numbers $C(n,r)$, for $0 \leq r \leq n$, are often displayed in a triangle formation, as in Figure 2.3. It then follows from Theorem 2.7 that each number in the $(n + 1)$th row (apart from those at the ends) is the sum of the two adjacent numbers in the row above. For example, the number 21 in the eighth row is the sum of 6 and 15 from the row above. This triangle is usually called *Pascal's triangle*, after the seventeenth-century French mathematician Blaise Pascal, although it was not originated by him.[*] The first 11 rows of Pascal's triangle are shown in Figure 2.3.

---

[*] We quote the following account of the matter from *The Backbone of Pascal's Triangle* by Martin Griffiths, United Kingdom Mathematics Trust (UKMT), Leeds, 2008 p. 10: "Pascal himself called it 'the arithmetical triangle', but after the mathematicians Pierre Rémond de Montmort and Abraham de Moivre referred to it in writing as 'the combinatorial triangle of Mr. Pascal' (in 1708) and 'Pascal's arithmetical triangle' (in 1730) respectively, the name stuck. However the Italian mathematician Nicolo Tartaglia actually published these numbers in 1556, and there is evidence that the Chinese mathematician Yang Hui was working with these numbers in the thirteenth-century (the Chinese do indeed use the term 'Yang Hui's triangle')."

```
                    1
                  1   1
                1   2   1
              1   3   3   1
            1   4   6   4   1
          1   5  10  10   5   1
        1   6  15  20  15   6   1
      1   7  21  35  35  21   7   1
    1   8  28  56  70  56  28   8   1
  1   9  36  84 126 126  84  36   9   1
1  10  45 120 210 252 210 120  45  10   1
```

FIGURE 2.3

There are innumerable relationships between the binomial coefficients that correspond to patterns that can be found within Pascal's triangle. For example, it follows from Theorem 2.5 that Pascal's triangle is symmetrical about its central vertical axis. Some other relationships are given in the next theorem and in the exercises at the end of this section.

THEOREM 2.8

For all positive integers k, n with $k \leq n$,

$$C(n+1, k+1) = C(n,k) + C(n-1,k) + \ldots + C(k,k).$$

Proof

Let $X = \{x_1, x_2, \ldots, x_n, x_{n+1}\}$. $C(n+1, k+1)$ is the number of subsets of X that contain $k+1$ of the elements of X. We can also count the number of these subsets in the following way. For $k+1 \leq r \leq n+1$ we let X_r be the set of all those $(k+1)$-element subsets, Y, of X such that r is the largest integer for which $x_r \in Y$. Thus, if $Y \in X_r$, then $x_r \in Y$ but $x_{r+1}, \ldots, x_{n+1} \notin Y$.

Clearly, the sets $X_{k+1}, \ldots, X_n, X_{n+1}$ are pairwise disjoint. Also, between them they include all the $(k+1)$-element subsets of X, as each subset of X containing $k+1$ elements must include at least one of the elements $\{x_{k+1}, \ldots, x_{n+1}\}$. Hence

$$\#(X) = \sum_{r=k+1}^{n+1} \#(X_r). \tag{2.1}$$

The sets in X_r contain x_r and k elements chosen from the set $\{x_1, \ldots, x_{r-1}\}$. Therefore, $\#(X_r) = C(r-1, k)$. We can therefore deduce from Equation 2.1 that $C(n+1, k+1) = \sum_{r=k+1}^{n+1} C(r-1, k)$, which, when we rewrite the terms on the right-hand side in reverse order, gives $C(n+1, k+1) = C(n,k) + C(n-1,k) + \ldots + C(k,k)$.

We conclude this section with a simple but intriguing application of Theorem 2.3 to number theory.

THEOREM 2.9

For each positive integer r, the product of any r consecutive positive integers is divisible by $r!$.

Proof

We need to prove that for all positive integers k, r the product of the r consecutive integers $k, k+1, k+2, \ldots, k+r-1$ is divisible by r. Now,

$$\frac{k(k+1)(k+2)\ldots(k+r-1)}{r!} = C(k+r-1, r),$$

by Theorem 2.3. Since this binomial coefficient gives the number of r-element subsets of a set of $k + r - 1$ elements, it must be an integer. So $k(k+1)(k+2)\ldots(k+r-1)$ is divisible by $r!$

Exercises

2.3.1A A mathematics course offers students the choice of three options from 12 courses in pure mathematics, two options from 10 courses in applied mathematics, two options from 6 courses in statistics, and one option from 4 courses in computing. In how many different ways can the students choose their eight options?

2.3.1B A cricket squad consists of six batsmen, eight bowlers, three wicketkeepers, and four all-rounders. The selectors wish to pick a team made up of four batsmen, four bowlers, one wicketkeeper, and two all-rounders. How many different teams can they pick?

The next three pairs of questions can all be answered by using the binomial theorem. However, you are encouraged to give combinatorial proofs in the style of those we have given for the theorems in this section.

2.3.2A Prove that a set of n elements has 2^n different subsets, and deduce that for each positive integer n, $\sum_{r=0}^{n} C(n,r) = 2^n$.

2.3.2B Prove that, for each positive integer n, $\sum_{r=0}^{n} C(n,r)^2 = C(2n,n)$.

2.3.3A Prove that, for all positive integers n, k, s with $s \leq k \leq n$, $C(n,k) C(k,s) = C(n,s) C(n-s, k-s)$.

2.3.3B Let X be a finite set. Prove that the number of subsets of X that contain an even number of elements is equal to the number of subsets of X that contain an odd number of elements. Deduce that for each positive integer n, $\sum_{r=0}^{n} (-1)^r C(n,r) = 0$.

2.3.4A Prove that, for each positive integer n, $\sum_{r=0}^{n} rC(n,r) = n2^{n-1}$.
[*Hint:* Let X be an n-element set. Note that as X has $C(n,r)$ subsets containing r elements, $\sum_{r=0}^{n} rC(n,r) = \sum_{A \subseteq X} \#(A)$. Also, we can calculate $\sum_{A \subseteq X} \#(A)$ by pairing off each subset A of X with its complement $X \backslash A$. How many pairs are there, and what is $\#(A) + \#(X \backslash A)$?]

2.3.4B Prove that for each positive integer n, $\sum_{r=0}^{n} r^2 C(n,r) = n(n+1)2^{n-2}$.
(*Hint:* First find $\sum_{r=0}^{n} r(r-1)C(n,r)$ by counting the ordered pairs (a, b), where a and b are chosen from an n-element set, in two ways. Then use the result of Exercise 2.3.4A.)

2.4 APPLICATIONS TO PROBABILITY PROBLEMS

We begin with a very simple problem.

PROBLEM 2.6

A fair coin is tossed 10 times. What is the probability of getting four heads?

Solution

Each toss can have one of two results, either a head or a tail. Thus there are altogether $2^{10} = 1024$ equally likely outcomes for a sequence of ten tosses. The number of these sequences that consist of four heads and six tails is the number of ways in which four of the ten tosses can be heads, that is, $C(10,4) = 210$. Hence the probability of getting four heads is 210/1024, which is 0.205 to three decimal places.

This calculation is straightforward enough, but what does it mean? And how do we know that we have solved the problem correctly? Fortunately, from the point of view of combinatorics we do not have to answer the difficult philosophical question as to what is meant by *probability*. Our work on this and similar probability problems begins after any philosophical work has been done, and the precise mathematical problem has been formulated. These problems will involve a set of events E, taken to be equally likely, together with a subset E_1. The probability of an event falling into the subset E_1 is defined to be ratio of the number of events in E_1 to the number of events in E, that is,

$$\frac{\#(E_1)}{\#(E)}.$$

Thus our task will be to calculate the numbers in this ratio, and we will not need to concern ourselves with its philosophical significance.

However, there remains the question of how we knew, from the formulation of the problem, which the sets E and E_1 of events were in this particular case. You may have noticed that the problem refers to a *fair coin*. This is intended to indicate that on any one throw of the coin the probability (whatever this means) of getting a head is exactly the same as the probability of getting a tail and that the outcome of any one toss is independent of what happens in earlier or later tosses. This means that any one sequence of outcomes is as likely as any other sequence. This is reflected in our solution, where we took the set of events, E, to be the set of all 1024 sequences of 10 results of a toss, each toss resulting in either a head or a tail.

For us, doing combinatorics, that is the end of the matter. If you wish to apply the answer to Problem 2.6 and similar calculations to practical situations, you need to know how realistic the assumption of a *fair coin* is and how statements of probability are to be interpreted. These are not easy questions, and so it is fortunate that, in this book, we can largely avoid answering them.

In general, the statement of a probability problem should indicate which set is to be taken as the set, E, of events. Thus the events in E are events that are to be regarded as equally likely. This indication is often done in a coded way. For example, some of these problems concern packs of cards *dealt at random*. This is intended to mean that any of the 52! ways in which the pack of 52 cards can be arranged is as likely as any other.

Hence the set E will consist of these 52! arrangements or will be derived from it in some straightforward way.

The codes used in this way to set up probability problems are analogous to the coded way of describing problems in mechanics, where such phrases as *a light string* or *a frictionless pulley* are intended to indicate what assumptions can be made in devising the mathematical model.

We are now in a position to answer Problem 2A from Chapter 1.

PROBLEM 2.7

In the British national lottery, 6 balls are drawn at random from a set of 49 balls, numbered 1, 2, ..., 49. To play, you need to buy one or more tickets. Each player selects six numbers from 1 to 49 on each of his or her tickets. You win the *jackpot* if on one of your tickets you have chosen the six numbers on the balls that are drawn. What is the probability that a particular ticket wins the jackpot in the national lottery?

Solution

Here, the set E consists of all possible ways of drawing 6 balls from 49. Thus $\#(E) = C(49, 6) = 13{,}983{,}816$, since the order in which the balls are drawn is irrelevant. The set E_1 consists of just the one case where the six numbers selected by the player correspond to the numbers on the six balls that are drawn. So $\#(E_1) = 1$. Thus the probability of a ticket winning the jackpot is $1/13{,}983{,}816$.

PROBLEM 2.8

You win a prize of £10 in the national lottery if precisely three of the six numbers that you select are included among the six numbers on the balls that are drawn. What is the probability that you win a prize of £10?

Solution

Here, E is again the set of all possible ways that 6 balls can be drawn from 49, so, as before, $\#(E) = C(49,6)$. E_1 is the number of ways that 6 balls can be drawn so that the numbers on 3 of them are included among your 6 numbers, and the numbers on the other 3 are included among the 43 numbers that you did not select. So to get a set of 6 numbers that is in E_1 we first need to choose 3 numbers from the 6 numbers on the chosen balls and then 3 numbers from the other 43 numbers. Thus $\#(E_1) = C(6,3) \times C(43,3)$, and hence the required probability is $[C(6,3) \times C(43,3)]/C(49,6) = 8{,}815/499{,}422$. This is approximately a 1 in 56.7 chance, or 0.01765 to five decimal places.

Card games are a rich source of probability problems of this type. To understand the examples that follow, you do not need to know the rules of the card games that are mentioned. All you need to know is that a standard pack contains 52 cards, divided up into 4 *suits*: spades (for which the symbol ♠ is used), hearts (♥), diamonds (♦), and clubs (♣). There are 13 cards in each suit. The *ranks* of the cards in each suit are 2,3,4,5,6,7,8,9,10, jack (often denoted by J), queen (Q), king (K), and ace (A).

In games such as bridge and whist there are four players, often called North, South, East, and West. In the initial deal, each player is dealt a hand of 13 cards. Thus the number of different hands a player may receive is the number of ways of choosing 13 cards from 52, that is, $C(52,13) = 635{,}013{,}559{,}600$. In bridge and whist the distribution of the cards between the different suits is important.

PROBLEM 2.9

How many different bridge hands are there with

i. Four spades, four hearts, four diamonds, and one club?
ii. Four spades, four hearts, three diamonds, and two clubs?

Solution

i. We can easily solve this problem by using the methods of this chapter. To choose a hand of the kind described, we first choose 4 spades from the 13 spades in the pack, which we can do in $C(13,4)$ ways, then 4 hearts, which can also be done in $C(13,4)$ ways, then 4 diamonds, also in $C(13,4)$ ways, and finally 1 club, which can be chosen in $C(13,1)$ ways. Hence using the principle of multiplication of choices, the total number of hands with 4 spades, 4 hearts, 4 diamonds, and 1 club is $C(13,4) \times C(13,4) \times C(13,4) \times C(13,1) = 4{,}751{,}836{,}375$.
ii. Similarly, the answer here is $C(13,4) \times C(13,4) \times C(13,3) \times C(13,2) = 11{,}404{,}407{,}300$.

A hand with four cards in each of three suits and one card in the fourth suit is said to have a 4–4–4–1 *suit distribution*. In general, a hand with an *a–b–c–d* suit distribution, with $a \geq b \geq c \geq d$, is one with a cards in one suit, b cards in a second suit, c cards in a third suit, and d cards in the fourth suit, irrespective of which suits these are. Of course, for a bridge or whist hand of 13 cards, we require that $a + b + c + d = 13$.

PROBLEM 2.10

What is the probability that a bridge hand dealt at random has the following suit distributions?

i. 4–4–4–1
ii. 4–4–3–2

Solution

i. We have calculated in Problem 2.9(i) the number of bridge hands with four cards in each of three specified suits and one card in a fourth suit. To get the number of all hands with a 4–4–4–1 suit distribution we need to multiply the answer to that problem by the number of ways in which we can specify the three suits with four cards each and the one suit with just one card. We can choose the three 4-card suits in $C(4,3) = 4$ ways, and having chosen these, the suit with one card is automatically determined. (Equivalently, the one suit containing one card may be chosen in $C(4,1) = 4$ ways, and the three suits each containing three cards are then automatically determined.) So there are $4 \times 4{,}751{,}836{,}375 = 19{,}007{,}345{,}500$ hands with a 4–4–4–1 suit distribution. To get the probability that a hand dealt at random has this suit distribution, we need to divide this number by the total number of bridge hands. Thus the required probability is 19,007,345,500/635,013,559,600. This probability is 0.030 to three decimal places.

ii. We need to multiply the answer to Problem 2.9(ii) by the number of ways of choosing the two suits with four cards each and the one suit with three cards, after which the suit with two cards is automatically determined. This multiplier is $C(4,2) \times C(2,1) = 12$. So the total number of such hands is $12 \times 11,404,407,300 = 136,852,887,600$, and hence the probability of dealing such a hand at random is 136,852,887,600/635,013,559,600, which is 0.216 to three decimal places.

PROBLEM 2.11

How many poker hands of five cards are there in which there is at least one suit with no cards in it?

Solution

We list the suit distributions with at least one suit with zero cards in it, and then use the method of Problem 2.10 to work out the total number of hands with these suit distributions, as follows:

Suit Distribution	Number of Hands
5–0–0–0	5,148
4–1–0–0	111,540
3–2–0–0	267,696
3–1–1–0	580,008
2–2–1–0	949,104
Total	1,913,496

You might think that there is a quicker way to solve Problem 2.11. It is easy to count the number of 5-card poker hands with, for example, no spades. Such a hand is obtained by choosing 5 cards from the 39 cards in the pack that are not spades. So there are $C(39,5) = 575,757$ poker hands with no spades. There is a similar number of hands with no hearts, with no diamonds, and with no clubs, respectively. So it seems that the total number of 5-card hands with a missing suit is $4 \times 575,757 = 2,303,028$. Unfortunately this quick solution gives an answer that is different from that we obtained in our solution to Problem 2.11. Where have we gone wrong?

It is not difficult to see where our mistake lies. Let V be the set of poker hands with at least one missing suit, and let V_S, V_H, V_D, V_C be those hands with no spades, hearts, diamonds, and clubs, respectively. Clearly, $V = V_S \cup V_H \cup V_D \cup V_C$. Our second calculation assumed that $\#(V) = \#(V_S) + \#(V_H) + \#(V_D) + \#(V_C)$, but this overlooks the fact that some of the hands in V are in more than one of the sets V_S, V_H, V_D, V_C. For example, a hand made up of three diamonds and two clubs but no spades and no hearts is in both V_S and V_H. Thus the sum $\#(V_S) + \#(V_H) + \#V_D + \#(V_C)$ counts some of the hands with a missing suit more than once. Thus it is no wonder that our "quick answer" is higher than the correct answer that we obtained in the solution to Problem 2.11.

To sum up this point, it is correct that if we have a collection of sets $X_1, \ldots, X_k$ that are *pairwise disjoint*, that is, $X_i \cap X_j = \emptyset$ for $1 \leq i < j \leq k$, then

$$\#\left(\bigcup_{i=1}^{k} X_i\right) = \sum_{i=1}^{k} \#(X_i), \tag{2.2}$$

but this is not true if the sets, X_i, are not pairwise disjoint. We discuss in Chapter 4 the modification that we need to make to Equation 2.2 in the cases where these sets are not pairwise disjoint.

Exercises

2.4.1A A ticket wins a fourth prize in the national lottery if precisely four of the six numbers selected on it are included among the six numbers on the balls that are drawn. What is the probability that a ticket wins a fourth prize?

2.4.1B In his autobiography, *What I Remember*, Adolphus Trollope[*] describes the Italian lottery as follows:

> Ninety numbers, 1–90, are always put into the wheel. Five only of these are drawn out. The player bets that a number named by him shall be one of these (*semplice estratto*); or that it shall be the first drawn (*estratto determinato*); or that two numbers named by him shall be two of the five drawn (*ambo*); or that three so named shall be drawn (*terno*). It will be seen, therefore, that the winner of an *estratto determinato*, ought, if the play were quite even, to receive ninety times his stake. But, in fact, such a player would receive only 75 times his stake, the profit of the Government consisting of this pull of 15 per 90 against the player. Of course, what he ought to receive in any of the other cases is easily (not by me, but by experts) calculable.

What would be fair odds for the *semplice estratto*, *ambo*, and *terno* bets?

2.4.2A A bag contains 50 red balls and 50 blue balls. Ten balls are drawn at random from the bag and not replaced. What is the probability that this sample will contain five red balls and five blue balls? (This question is connected with the reliability of opinion polls. See the solution for more about this.)

2.4.2B A bag contains $2n$ red balls and $2n$ blue balls. What is the probability that if $2n$ balls are drawn at random, the sample will consist of n red balls and n blue balls?

2.4.3A (This is Problem 2B from Chapter 1.) If there are n people in a room, what is the probability that at least two of them share a birthday? How large does n have to be before this probability becomes more than a half? (By "sharing a birthday" we mean that two people were born in the same month and on the same day in that month but not necessarily in the same year. For the purpose of this problem you should ignore leap years, so that there are 365 possible birthdays for each person. You should also assume that all 365 birthdays are equally likely. In fact, as the answer to Exercise 2.4.3B

---

[*] Thomas Adolphus Trollope, *What I Remember*, abridged by Herbert van Thal, William Kimber, London, 1973, pp. 189–190 (originally published in three volumes by R. Bentley & Son, London, 1887–1889).

indicates, if, as is actually the case, some birthdays are more likely than others, the probability of a coincidence increases.)

2.4.3B Suppose there are $2n$ balls in a bag of which a are red and b are blue, where $a + b = 2n$. One ball is removed at random from the bag and then replaced. Then a second ball is drawn at random from the bag and then replaced. Calculate the probability that either a red ball is drawn twice or a blue ball is drawn twice. Show that this probability is a minimum when $a = b = n$.

2.4.4A What is the probability that a bridge hand dealt at random has the following suit distributions?

i. 5–4–3–1   ii. 5–4–4–0   iii. 4–3–3–3

2.4.4B Suppose that in a given bridge deal North and South between them have nine spades. What is the probability that the remaining four spades in the other two hands are divided two-two?

2.4.5A *Poker Hands.* Poker hands, which in standard poker games consist of 5 cards drawn from the full pack of 52 cards, are classified as follows:
  i. *Flush:* five cards all of the same suit but not forming a sequence of consecutive ranks.
     For example, 5♣, 7♣, J♣, Q♣, K♣ is a flush.
  ii. *Four of a kind:* four cards of one rank and one other card, for example, 3♠, 3♥, 3♦, 3♣, J♠.
  iii. *Full house:* three cards of one rank and two cards of another rank, for example, 7♠, 7♥, 7♣, 10♦, 10♣.
  iv. *One pair:* two cards of one rank and three cards of three different ranks, for example, J♦, J♣, 4♠, 7♥, K♠.
  v. *Straight:* five cards of consecutive ranks but not all in the same suit (note that for this purpose an ace may count either low or high, so that both A, 2, 3, 4, 5 and 10, J, Q, K, A count as consecutive ranks), for example, 7♦, 8♦, 9♥, 10♠, J♥.
  vi. *Straight flush:* five cards of consecutive ranks (again, an ace may count either low or high) and in the same suit, for example, 4♥, 5♥, 6♥, 7♥, 8♥.
  vii. *Three of a kind:* three cards of one rank and two cards of two different ranks, for example, 9♥, 9♣, 9♦, 4♥, Q♣.
  viii. *Two pairs:* two cards of one rank, two cards of another rank, and a fifth card of a third rank, for example, 5♦, 5♣, 8♥, 8♦, J♦.
  ix. *Other hands:* all other hands that do not fall into any of the preceding categories.
     Calculate how many poker hands there are in each of the categories (a) to (e) above. Hence work out the probability that a poker hand dealt at random falls into each of the these categories.

2.4.5B Complete the calculation of Exercise 2.4.5A by working out how many poker hands there are that fall into the categories (f) to (i) above. Also work out the probability that a poker hand dealt at random falls into each of these categories.

2.4.6A We see, from the solution to Problem 2.6, that if a fair coin is tossed 10 times, then the probability of getting four heads is $C(10,4)/2^{10}$. More generally, if a fair coin is tossed n times, the probability of getting r heads, for $0 \leq r \leq n$ is $C(n,r)/2^n$. We note that the different probabilities, as k ranges from 0 to n, add up to 1, as we expect. This follows from the result of Exercise 2.3.2A, as we can deduce from this result that

$$\sum_{r=0}^{n} \frac{C(n,r)}{2^n} = \frac{1}{2^n} \sum_{r=0}^{n} C(n,r) = \frac{1}{2^n} \times 2^n = 1.$$

It can be shown that if the coin is not necessarily fair, so that the probability of getting a head is p and hence the probability of getting a tail is $1-p$, where p need not be equal to $\frac{1}{2}$, then the probability of getting r heads is $C(n,r)p^r(1-p)^{n-r}$. Show that these probabilities also add to 1, that is, show that for all values of p, $\sum_{r=0}^{n} C(n,r)p^r(1-p)^{n-r} = 1$.

2.4.6B A coin is biased so that the probability of getting a head is 0.6. If the coin is tossed five times, what is the probability of getting three heads?

2.5 THE MULTINOMIAL THEOREM

The multinomial theorem is a generalization of the binomial theorem, and, as with the binomial theorem, it can be approached either algebraically or combinatorially. We will look at it from a combinatorial viewpoint. We begin with the following, which generalizes Problem 2.6.

PROBLEM 2.12

A football team plays 38 games in a season. It has equal probabilities of winning, drawing, or losing each game. What is the probability that the team wins 20 games, draws 11, and loses only seven games?

Solution

The results obtained by the team can be regarded as a sequence of 38 symbols, each of which is either W, D, or L, indicating a win, a draw, and a loss, respectively. As the team has equal probabilities of achieving each of the three possible results, each sequence of 38 Ws, Ds, and Ls is equally likely. Hence, the required probability is the number of these sequences made up of 20 Ws, 11 Ds, and 7 Ls divided by the total number of sequences of 38 symbols. The second number is the easier to calculate. Since there are 3 choices for each symbol, there are altogether 3^{38} sequences of 38 symbols each of which is W, D, or L.

Next we count the number of these sequences in which there are 20 Ws, 11 Ds, and 7 Ls. We can construct such a sequence by first choosing the 20 positions in which the symbol W occurs. This involves choosing 20 positions from 38 and so may be done in $C(38,20)$ ways. This leaves 18 positions for the 11 Ds, which may therefore be chosen in $C(18,11)$ ways, and the remaining seven positions are then automatically filled by the Ls, that is, in just one way. Hence the total number of sequences is

$$C(38,20) \times C(18,11) = \frac{38!}{20!18!} \times \frac{18!}{11!7!} = \frac{38!}{20!11!7!},$$

and the required probability is $[38!/(20!11!7!)]/3^{38}$, which is 0.0008 to four decimal places.

We see from the form, 38!/(20!11!7), of our final expression for the number of sequences made up of 20, 11, and 7, respectively, of the three symbols that we would also obtain this answer if we had, for example, first placed the 11 Ds, then the 7 Ls, and finally the 20 Ws. It also suggests the general result, given as Theorem 2.10.

First, we generalize Problem 2.12. As we know, if we have k different objects, they can be arranged in order in $k!$ ways. However, if all the k objects are indistinguishable, all these arrangements look the same, and, from this point of view there is just one way to arrange the objects in order. In general, whenever we have a set comprising both distinguishable and indistinguishable objects, by "different" arrangements, we mean arrangements of those objects that can be distinguished from one another. For example, the letters $A,B,C,D,$ and E may be arranged in order in $5!$ different ways. However, if all we are interested in is whether a letter is a vowel (V) or a consonant (C), then the sequences A,B,C,D,E and E,B,C,D,A would both be recorded as V,C,C,C,V. Regarding the two vowels as indistinguishable and the three consonants as indistinguishable, there are just 10, that is, $C(5,2)$, different ways of arranging these five letters, corresponding to the number of ways of choosing the positions of the two Vs.

THEOREM 2.10

Suppose that for $1 \leq r \leq k$, we have n_r objects of type T_r, where the objects of any given type are indistinguishable but may be distinguished from the objects of any other type. Suppose also that we have n objects in total, so that $\sum_{r=1}^{k} n_r = n$. Then the number of different ways of arranging the n objects in order is $n!/(n_1!n_2!\ldots n_k!)$.

Proof

We have n positions to fill. We can choose n_1 of these for the objects of type T_1 in $C(n,n_1)$ ways. When these positions have been chosen, there remain $n-n_1$ positions to be filled, and hence positions for the n_2 objects of type T_2 may be chosen in $C(n-n_1,n_2)$ ways. Then the positions for the n_3 objects of type T_3 may be chosen in $C(n-n_1-n_2,n_3)$ ways, and so on. Therefore, the total number of different ways of choosing positions for the n objects is

$$C(n,n_1) \times C(n-n_1,n_2) \times C(n-n_1-n_2,n_3) \times \ldots \times C(n-n_1-n_2-\ldots-n_{k-1},n_k)$$

$$= \frac{n!}{n_1!(n-n_1)!} \times \frac{(n-n_1)!}{n_2!(n-n_1-n_2)!} \times \frac{(n-n_1-n_2)!}{n_3!(n-n_1-n_2-n_3)!} \times \ldots \times \frac{(n-n_1-n_2-\ldots-n_{k-1})!}{n_k!0!}$$

$$= \frac{n!}{n_1!n_2!\ldots n_k!},$$

after a lot of canceling. Having obtained the simplified formula by an algebraic argument, we should look for a combinatorial argument that shows why it is correct and that gives us a better understanding of why it is true.

Ignoring first the fact that the objects of each type are indistinguishable, we see that the n objects may be arranged in order in $n!$ different ways. However, since the n_1 objects of type T_1 are indistinguishable, we need to divide by $n_1!$ to allow for the fact that, once the positions of these objects have been chosen, the order in which each object of type T_1 is chosen does not matter. Likewise, we need also to divide by $n_2!$, $n_3!,\ldots,n_k!$ and hence the number of indistinguishable arrangements is $n!/(n_1!n_2!\ldots n_k!)$.

THEOREM 2.11
The Multinomial Theorem

The coefficient of the term $a_1^{n_1} a_2^{n_2} \ldots a_k^{n_k}$ in the expansion of $(a_1 + a_2 + \ldots + a_k)^n$, where $n_1 + n_2 + \ldots + n_k = n$, is $n!/(n_1!n_2!\ldots n_k!)$.

Proof

Each term in the expansion of $(a_1 + a_2 + \ldots + a_k)^n$ has the form $t_1 t_2 \ldots t_n$, where each t_r, for $1 \le r \le n$, is one of the symbols $a_1, a_2, \ldots, a_k$. The coefficient of $a_1^{n_1} a_2^{n_2} \ldots a_k^{n_k}$ is the number of such sequences in which, for $1 \le r \le n$, there are n_r occurrences of the symbol a_r. By Theorem 2.10 there are $n!/(n_1!n_2!\ldots n_k!)$ such sequences. This completes the proof.

Exercises

2.5.1A If 12 dice are thrown simultaneously, what is the probability that each of the faces from one to six comes up twice?

2.5.1B If 21 dice are thrown simultaneously, what is the probability that 1 comes up once, 2 comes up twice, 3 comes up three times, 4 comes up four times, 5 comes up five times, and 6 comes up six times?

2.5.2A In how many different ways can one arrange the sequence of letters in the word ABRACADABRA?

2.5.2B In how many different ways can one arrange the sequence of letters in the word PROPERISPOMENON ?

2.6 PERMUTATIONS AND CYCLES

In the final section of the chapter, we look at permutations in a new way that will turn out to be very fruitful later on. We have defined a *permutation* of n objects to be a way of choosing these objects in order. For example, one permutation of the set $\{1,2,3,4,5,6\}$ is the choice of these numbers in the order 6,1,3,5,4,2. We can think of this permutation as a reordering of the set.

In this way, we regard this permutation as a bijection (a one-one onto function) mapping the set $\{1,2,3,4,5,6\}$ to itself. In general, for any set X, by a *permutation* of X, we mean a bijection $f: X \to X$. Usually, however, we will be considering permutations of the set $\{1,2,3,\ldots,n\}$ consisting of the first n positive integers.

There are two commonly used notations for permutations. The first is a minor variant of the diagram of Figure 2.4. We drop the arrows and put the numbers inside brackets. So the permutation above would, in this notation, be written as

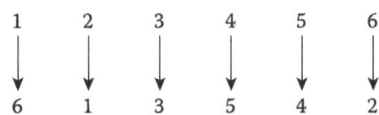

FIGURE 2.4

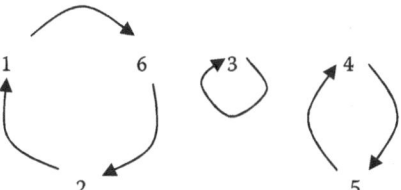

FIGURE 2.5

$$\begin{pmatrix} 1 & 2 & 3 & 4 & 5 & 6 \\ 6 & 1 & 3 & 5 & 4 & 2 \end{pmatrix}.$$

We call this the *bracket notation* for permutations. It is a rather cumbersome notation, and it is not very helpful when it comes to answering questions about permutations. For example, the bracket notation does not make it very clear how many times we need to repeat a permutation to get each number back to its starting point. So we introduce a second, more useful notation.

Since the domain and codomain of a permutation are the same set, we can represent them in a single diagram showing how the elements of the set are mapped by the permutation. For example, the above permutation may be represented by the diagram in Figure 2.5.

We can see from this diagram that the effect of the permutation is to cycle the numbers 1, 6, 2 in this order, to leave 3 fixed, and to cycle (or interchange) 4 and 5. Thus, the permutation is made up of three parts, namely, cycles of lengths 2 and 3 and one fixed point. We can represent this using a notation in which we write the numbers in three separate pairs of brackets, showing which cycle they are in, with the order of the numbers in each bracket showing how they are mapped. Thus in *cycle notation* we can write the permutation as

$$(1\ 6\ 2)(3)(4\ 5).$$

This is sometimes called the *disjoint cycle form* of the permutation because the numbers making up the different cycles form disjoint sets.

From the disjoint cycle form we see that the numbers 1, 6, and 2 are in a cycle of length 3 and so return to their initial position after we have carried out the permutation 3, 6, 9 ... times and, in general, any multiple of three times. Likewise, the numbers 4 and 5 form a cycle of length 2 and so return to their original position after we have carried out the permutation a multiple of two times. The number 3 is not moved, but we can think of it as forming a cycle of length 1. Thus, all the numbers are returned to their original positions for the first time after we have carried out the permutation six times.

Although the cycle notation for permutations makes it easy to answer such questions, we need to be careful with it. We are using a linear notation to represent cycles. For example, the bracket (1 6 2) represents the first cycle in Figure 2.5. We need to remember that it means that 2 is mapped back to 1, as this is not immediately obvious from the notation. Also, there is some arbitrariness in this notation. We could equally well have written it as (6 2 1) or (2 1 6), as these both represent the same cycle. Likewise, the cycle (4 5) of length 2 could also be written as (5 4). Although it is conventional to have first in each bracket the lowest number in the cycle, this is not essential. Furthermore, we could also have written the disjoint cycles in a different order, for example, as (4 5)(1 6 2)(3). We will need to remember this when counting permutations.

There is another notational point. The cycle (3), of length 1, that occurs in our permutation tells us that it maps 3 to itself. It is usual to omit cycles of length 1 when writing permutations in cycle notation. So normally we would write the above permutation as (1 6 2) (4 5). Although it is convenient to omit cycles of length 1, this does introduce another ambiguity. Thus, (1 6 2)(4 5) could be a permutation of the set {1,2,3,4,5,6} that fixes 3, but it could also be, for example, a permutation of {1,2,3,4,5,6,7,8} that fixes 3, 7, and 8. So it is really safe to omit cycles of length 1 only if it is clear from the context which set of numbers we are permuting.

We will make good use of the cycle notation for permutations later in the book, especially in Chapters 11, 12, 13, and 14.

Exercises

2.6.1A Write the following permutation in cycle notation:

$$\begin{pmatrix} 1 & 2 & 3 & 4 & 5 & 6 & 7 & 8 & 9 & 10 \\ 3 & 1 & 8 & 9 & 7 & 4 & 10 & 2 & 6 & 5 \end{pmatrix}.$$

2.6.1B Write the following permutation in cycle notation:

$$\begin{pmatrix} 1 & 2 & 3 & 4 & 5 & 6 & 7 & 8 & 9 & 10 \\ 5 & 2 & 8 & 10 & 7 & 6 & 4 & 9 & 3 & 1 \end{pmatrix}.$$

2.6.2A How many different permutations are there of the numbers {1,2,3,4,5,6,7,8,9,10} made up of
  i. Three disjoint cycles of lengths 2, 3, and 5?
  ii. Three disjoint cycles of which two are of length 3 and one is of length 4?

2.6.2B How many different permutations are there of the numbers {1,2,3,4,5,6,7,8,9,10} made up of
  i. Four disjoint cycles of lengths 1, 2, 3, and 4?
  ii. Four disjoint cycles of which three are of length 2 and one is of length 4?

2.6.3A If a permutation of the set $\{1,2,3,\ldots,n\}$ is chosen at random, what is the probability that it consists of a single cycle of length n?

2.6.3B If a permutation of the set $\{1,2,3,\ldots,n\}$ is chosen at random, what is the probability that it includes exactly one cycle of length 1?

CHAPTER 3

Occupancy Problems

3.1 COUNTING THE SOLUTIONS OF EQUATIONS

When we discussed Problem 3C in Chapter 1 we noted that attempting to determine even the *number* of solutions in nonnegative integers of an equation of the form $x + y + z = n$ by listing them would become impractical if the number of variables were substantially increased. Here, we first show how a simple *reinterpretation* of the problem allows us to solve *all* such problems instantly: Indeed we can simply write down the answer! To help you follow the method we begin by solving the problem in a particular case, "large" enough to make the listing method at best tiresome but "small" enough to fit the reinterpretation easily on the page. It should be fairly clear that the method employed is perfectly general,* that is, applicable in all circumstances.

PROBLEM 3.1

How many solutions are there in nonnegative integers of the equation $x + y + z + w + t = 14$?

Solution

It is easy to write down many solutions (x, y, z, w, t) of this equation, for example, (1,2,2,7,2) or (2,0,6,5,1) or (2,6,5,0,1) or (0,3,0,3,8). With each of these solutions we can associate a diagram of dots (•) and lines (|), as follows:

With (1,2,2,7,2) associate the diagram • | • • | • • | • • • • • • • | • • .

With (2,0,6,5,1) associate the diagram • • | | • • • • • • | • • • • • | • .

With (2,6,5,0,1) associate the diagram • • | • • • • • • | • • • • • | | • .

With (0,3,0,3,8) associate the diagram | • • • | | • • • | • • • • • • • • .

In the first diagram the four lines split the 14 dots into five groups of sizes 1, 2, 2, 7, and 2. In the second diagram the four lines split the 14 dots into five groups of sizes 2, 0,

---

* We commented on this attitude in Chapter 1, Section 1.3.

6, 5, and 1. In the third diagram the four lines split the 14 dots into five groups of sizes 2, 6, 5, 0, and 1. In the fourth diagram the four lines split the 14 dots into five groups of sizes 0, 3, 0, 3, and 8.

We readily see that each diagram faithfully represents the solution associated with it. It is equally clear that, conversely, any such diagram comprising four lines and 14 dots gives rise to a unique solution of the given equation; for example, the diagram

$$|\,|\bullet\bullet\bullet\bullet\bullet\bullet\bullet\bullet\bullet\bullet|\,|\bullet\bullet\bullet\bullet$$

corresponds to the solution (0,0,10,0,4). In this way we clearly obtain a matching of the set of solutions of the given equation with the set of all diagrams comprising four lines and 14 dots. So to solve Problem 3.1 we only need to work out how many diagrams of this type there are.

Now each diagram has 18 "places" each to be filled with a line or a dot. So the total number of diagrams is just the number of ways of choosing (out of 18 places) the 4 places in which to put a line, the remaining 14 places being filled automatically with dots. It follows that the number of different solutions, in nonnegative integers, of the given equation is just the binomial coefficient $C(18,4)$. Perhaps you are thinking that we could, instead, have put 14 dots in the 18 places, filling the remaining 4 places with lines. Then the number of solutions of our equation would, by the same argument, be $C(18,14)$. Fortunately, by Theorem 2.5, $C(18,4) = C(18,14)$.

Despite our general wish to impress on you that checking one example doesn't provide a general proof, we claim here that the method applied to our particular example is, transparently, just as valid when applied generally. We can therefore give a brief proof of the general result.

THEOREM 3.1

For each positive integer k and each nonnegative integer n, the number of nonnegative integer solutions, $(x_1, x_2, \ldots, x_k)$, of the equation $x_1 + x_2 + \ldots + x_k = n$ is $C(n + (k-1), k-1)$ or, equivalently, $C(n + (k-1), n)$.

Proof

As we have seen from the example above, the nonnegative integer solutions of the equation $x_1 + x_2 + \ldots + x_k = n$ are in one–one correspondence with sequences made up of n dots and $k-1$ lines, and there are $C(n + (k-1), k-1)$ such sequences.

We are now able to solve the following two problems from Chapter 1.

PROBLEM 3.2

(This is Problem 3A of Chapter 1.) A manufacturer of high-quality (and therefore high-priced!) chocolates makes just six different flavors of chocolate and sells them in boxes of 10. He claims he can offer more than 3000 different "selection boxes." If he is wrong, he will fall foul of the advertizing laws. Should he fear prosecution?

Solution

To make up a box of 10 chocolates the manufacturer has to decide how many chocolates of each of the six flavors to include. So the number of different selections is the number of nonnegative integer solutions of the equation $x_1 + x_2 + x_3 + x_4 + x_5 + x_6 = 10$. By Theorem 3.1, this is $C(15,5)$, which is equal to 3003. The manufacturer can therefore sleep soundly.

To solve the next problem we need a slight modification of Theorem 3.1 to cover the case where we consider only *positive* integer solutions.

THEOREM 3.2

Let k, n be positive integers with $n \geq k$. Then the number of *positive* integer solutions, $(x_1, x_2, \ldots, x_k)$, of the equation $x_1 + x_2 + \ldots + x_k = n$ is $C(n-1, k-1)$, or, equivalently, $C(n-1, n-k)$.

Proof

Let $X_i = x_i - 1$ for $1 \leq i \leq k$. Then, $X_1 + X_2 + \ldots + X_k = x_1 + x_2 + \ldots + x_k - k$ and, for $1 \leq i \leq k$, $x_i \geq 1 \Leftrightarrow X_i \geq 0$. Therefore, $(x_1, x_2, \ldots, x_k)$ is a solution in positive integers of the equation

$$x_1 + x_2 + \ldots + x_k = n \tag{3.1}$$

if and only if $(X_1, X_2, \ldots, X_k)$ is a solution in nonnegative integers of the equation

$$X_1 + X_2 + \ldots + X_k = n - k. \tag{3.2}$$

So the number of positive integer solutions of Equation 3.1 is the same as the number of nonnegative integer solutions of Equation 3.2. Hence, by Theorem 3.1, the number of positive integer solutions of Equation 3.1 is $C((n-k) + (k-1), k-1)$, that is, $C(n-1, k-1)$.

PROBLEM 3.3

(This is Problem 3B of Chapter 1.) There is a widely held view that, in a truly random selection of six distinct numbers from among the numbers 1 to 49 (as in the British national lottery), the chance that two consecutive numbers will be chosen is extremely small. Has this opinion any validity?

Solution

The total number of ways of choosing six distinct numbers from the numbers 1 to 49 is, of course, $C(49,6)$. Rather than count the number of selections in which consecutive numbers occur, we count those in which consecutive numbers do *not* occur. Thus, we count the number of sextuples of positive integers, $(t_1, t_2, t_3, t_4, t_5, t_6)$, satisfying

$$0 < t_1 < t_2 < t_3 < t_4 < t_5 < t_6 < 50$$

and

$$t_2 - t_1,\ t_3 - t_2,\ t_4 - t_3,\ t_5 - t_4,\ \text{and } t_6 - t_5 \text{ are all greater than 1.} \tag{3.3}$$

We see that the number of such sextuples is the same as the number of sextuples satisfying $-1 < t_1 < t_2 < t_3 < t_4 < t_5 < t_6 < 51$, where each of the numbers $t_1 - (-1)$, $t_2 - t_1$,

t_3-t_2, t_4-t_3, t_5-t_4, t_6-t_5, and $51-t_6$ is at least 2. (*All* the differences now being on equal footing, namely, "greater than or equal to 2," makes life a little easier!) Now, putting $T_1 = t_1-(-1)$, $T_2 = t_2-t_1$, $T_3 = t_3-t_2$, $T_4 = t_4-t_3$, $T_5 = t_5-t_4$, $T_6 = t_6-t_5$, and $T_7 = 51-t_6$, we see that $T_1 + T_2 + \ldots + T_7 = 52$. Hence the number of sextuples $(t_1,t_2,t_3,t_4,t_5,t_6)$ satisfying Equation 3.3 is the same as the number of positive integer solutions, $(T_1,T_2,T_3,T_4,T_5,T_6,T_7)$, of the equation $T_1 + T_2 + \ldots + T_7 = 52$ with $T_i \geq 2$ for $1 \leq i \leq 7$. Copying the method used in the proof of Theorem 3.2, by putting $S_i = T_i-2$, for $1 \leq i \leq 7$, this is the same as the number of nonnegative integer solutions of $S_1 + S_2 + \ldots + S_7 = 52-(7 \times 2) = 38$. By Theorem 3.1, this number is $C(38 + (7-1), 7-1)$, that is, $C(44,6)$.

Hence the probability that a lottery draw (of 6 balls from 49) will contain *no* pair of successively numbered balls is

$$\frac{C(44,6)}{C(49,6)} = \frac{44!}{6!38!} \Big/ \frac{49!}{6!43!} = \frac{44 \times 43 \times 42 \times 41 \times 40 \times 39}{49 \times 48 \times 47 \times 46 \times 45 \times 44} = \frac{22,919}{45,402} = 0.505$$

to three decimal places. So the probability of a pair of consecutive numbers appearing when six (different) numbers are drawn from numbers 1 to 49 is, approximately, $1-0.505 = 0.495$. In other words, it is only slightly *less* likely than not!

Another variation on Problem 3.1 is the following.

PROBLEM 3.4

How many nonnegative integer solutions are there of the inequality $x + y + z + w + t < 14$?

Solution

An unthinking attempt at this problem might say: "Let us find the number of solutions to each of the equations $x + y + z + w + t = n$ for $n = 0,1,2,\ldots,13$ and then add up the answers." There is nothing actually wrong with this approach, and, by Theorem 3.1, this gives the answer as $C(4,4) + C(5,4) + \ldots + C(17,4)$.

However, the problem can be answered more briefly as follows. To each solution, say (a,b,c,d,e), of the equation $x + y + z + w + t = n$, with $0 \leq n \leq 13$, there is a solution, (a,b,c,d,e,f), of the equation $x + y + z + w + t + u = 13$ with $f = 13-n$, and vice versa, with the newly introduced term, f, taking up the slack. Hence the number of nonnegative integer solutions of $x + y + z + w + t < 14$ is the same as the number of nonnegative solutions of $x + y + z + w + t + u = 13$. It follows immediately from Theorem 3.1 that this number is $C(18,5)$.

The reader who wonders if (or, perhaps better, *why*) these answers are the same should reread Theorem 2.8.

The second method used in the solution of Problem 3.4 clearly leads to the following further generalization of Theorem 3.1.

THEOREM 3.3

Let k, n be positive integers. Then the number of nonnegative integer solutions of the inequality $x_1 + x_2 + \ldots + x_k < n$ is $C(n + k-1, k)$ or, equivalently, $C(n + k-1, n-1)$.

Exercises

3.1.1A How many solutions are there in nonnegative integers of the inequality $x + y + z + t \leq 20$?

3.1.1B How many solutions are there in nonnegative integers of the inequality $x + y + z + t < 20$?

3.1.2A How many integer solutions are there of the equation $x + y + z + t + w = 14$ with $x \geq 5$, $y \geq -3$, $z \geq 2$, $t \geq -7$, and $w \geq 4$?

3.1.2B How many integer solutions are there to the equation $x + y + z = 20$ with $x > -4$, $y > 1$, and $z > 4$?

3.1.3A A manufacturer makes marbles that are identical except for their color, which can be red, blue, green, or yellow. In how many different ways can the manufacturer make up a pack of 50 marbles?

3.1.3B i. Each soccer team in the English Premier League plays 38 matches during the season. At the end of the season, the league table shows how many of these matches each team has won, drawn, and lost. How many combinations of these results are possible for any one team?

ii. What is the answer to the above question if the league table shows the results of the 19 matches that the team played at home separately from the 19 matches that it played away?

3.1.4A In a certain cricket match the 11 batsmen of one team between them scored 200 runs. In how many different ways could these runs be distributed between the batsmen?

3.1.4B How many different boxes can our chocolate manufacturer of Problem 3.2 supply if he guarantees that there is at least one chocolate of each flavor in each box?

3.1.5A In how many ways can 20 identical balls be placed in four distinct cups such that each cup has an even number of balls? (Count 0 as even.)

3.1.5B In how many ways can 20 identical balls be placed in four distinct cups such that each cup has an odd number of balls?

3.1.6A In the *Everwin* lottery, 10 numbers are drawn at random from the set $\{1,2,\ldots,100\}$. What is the probability that at least two consecutive numbers are drawn?

3.1.6B What is the smallest positive integer k such that if k numbers are drawn at random from the set $\{1,2,\ldots,100\}$, it is more likely than not that at least two of the numbers drawn are consecutive?

3.2 NEW PROBLEMS FROM OLD

The diagrams we drew in connection with Problem 3.1 were very helpful. But the solution to Problem 3.2 suggests a reinterpretation that leads to host of related and intriguing questions, several of which might have remained hidden, and some of which are a little more tricky to answer!

To reinterpret Problem 3.1 in the light of the solution to Problem 3.2, imagine five (distinct) boxes labeled x, y, z, w, and t. The solution $(4,0,1,3,6)$ of the equation $x + y + z + w + t = 14$ corresponds to placing 14 (identical) balls in the five boxes so that there are 4 balls in box x, 0 balls in box y, 1 ball in box z, 3 balls in box w, and 6 balls in box t. See Figure 3.1.

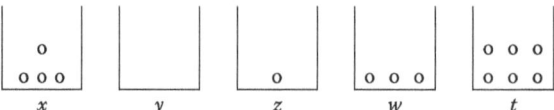

FIGURE 3.1

Consequently, Problem 3.1 can be recast as: In how many ways can 14 identical balls be placed in five distinct boxes?

In calling the balls "identical" and the boxes "distinct" we mean that we are interested only in how many balls there are in each box, but we want to distinguish, for example, having four balls in box *x* and zero in box *y*, from having zero balls in box *x* and four balls in box *y*. This problem has a number of variations according to whether we regard the balls as being identical or distinct and whether we regard the boxes as identical or distinct. We can also vary the problem by deciding whether we wish to include cases where some of the boxes may be empty. In this way we have eight different problems about placing balls in boxes: In how many ways can *n* (identical or distinct) balls be placed in *k* (identical or distinct) boxes with at least one ball in each box or with some boxes allowed to be empty?

We now give examples of all the eight possible cases of occupancy problems. It would be nice if, in the following examples, we could use the same values of *n* and *k* throughout. But, in order that our examples be neither too large to list easily nor too small to appear general enough, we shall choose *n* and *k* suitably in each case.

1a. *Placing* three *distinct balls in* two *distinct boxes*

We use a, b, c to represent the three distinct balls, and { }, [ ] to represent the two distinct boxes. Since for each ball there are two choices of boxes, there are $2 \times 2 \times 2 = 2^3 = 8$ different ways to allocate the balls to the boxes:

{a,b,c}, [ ]; {a,b}, [c]; {a,c}, [b]; {b,c}, [a]; {a}, [b,c]; {b}, [a,c]; {c}, [a,b]; { }, [a,b,c].

1b. *Placing* three *distinct balls in* two *distinct boxes, with no box empty*

We see from the above list that of the eight ways of assigning three distinct balls to two distinct boxes, there are two where a box is empty, and hence there are six cases with no empty boxes.

2a. *Placing* four *distinct balls in* two *identical boxes*

Here we use a, b, c, d for the distinct balls and { } for each of the identical boxes. We see that there are eight ways the balls can be placed in the two boxes:

{a,b,c,d},{ }; {a,b,c},{d}; {a,b,d},{c}; {a,c,d},{b}; {b,c,d},{a}; {a,b},{c,d}; {a,c},{b,d}; {a,d},{b,c}.

2b. *Placing* four *distinct balls in* two *identical boxes, with no box empty*

We see from (2a) that we need to discard the case {a,b,c,d},{ }, leaving seven ways of placing the balls.

3a. *Placing* six *identical balls in* three *distinct boxes*

As the balls are identical, all that matters is the number of balls in each box. As we have seen, the number of ways to do this is the same as the number of nonnegative integer solutions of the equation $x + y + z = 6$. So Theorem 3.1 tells us that there are $C(8,2) = 28$ ways to place the balls in the boxes.

3b. *Placing* six *identical balls in* three *distinct boxes, with no box empty*

The number of ways to do this is the same as the number of positive integer solutions of the equation $x + y + z = 6$. So by Theorem 3.2 there are $C(5,2) = 10$ ways to place the balls in the boxes.

4a. *Placing* six *identical balls in* three *identical boxes*

Because the balls are identical, we need only indicate the *number* of balls in each box—and this we can do most conveniently via Figure 3.2. There are therefore seven solutions to this problem.

4b. *Placing* six *identical balls in* three *identical boxes, with no box empty*

We see from Figure 3.2 that there are just three solutions in this case.

It is convenient, at this point, to introduce some notation. As is standard, we use $S(n,k)$ for the number of (different) ways of placing n distinct balls in k identical boxes so that no box is empty. We use $p_k(n)$ for the number of ways of placing n identical balls in k identical boxes, where some boxes may be empty.* It would be more consistent to use, say, $P(n,k)$, but there are historical reasons why these notations follow different patterns.

The letter S used in the notation $S(n,k)$ commemorates the Scottish mathematician James Stirling who introduced these numbers in a different context, which is explained in Chapter 5 where we give some biographical information about Stirling. So we call the numbers $S(n,k)$ *Stirling numbers*. We shall have to wait until Chapter 4 to determine an explicit formula giving their values for all values of n and k.

We now present a summary of the eight cases.

[ ]	[ ]	[ ]
6	0	0
5	1	0
4	2	0
4	1	1
3	3	0
3	2	1
2	2	2

FIGURE 3.2

---

* Beware: Some authors use the same notation for the case where no box can be empty.

TABLE 3.1

Case	Balls	Boxes	Empty Boxes?	Number of Arrangements
1a	Distinct	Distinct	Yes	k^n
1b	Distinct	Distinct	No	$k!S(n,k)$
2a	Distinct	Identical	Yes	$S(n,1) + S(n,2) + \ldots + S(n,k)$
2b	Distinct	Identical	No	$S(n,k)$
3a	Identical	Distinct	Yes	$C(n+k-1,k-1)$
3b	Identical	Distinct	No	$C(n-1,k-1)$
4a	Identical	Identical	Yes	$p_k(n)$
4b	Identical	Identical	No	$p_k(n) - p_{k-1}(n)$

We now explain where the entries in the last column of Table 3.1 come from.

Case 1a. With n distinct balls to be placed in k distinct boxes, there are k boxes in which we could place each ball. Hence the total number of ways to place the n balls in the k boxes is, using the principle of multiplication of choices, $k \times k \times \ldots \times k = k^n$.

Case 1b. We can view the problem of finding the number of ways of placing n distinct balls in k distinct boxes, with no box empty, in the following way. First, place n distinct balls in k *identical* boxes, with no box empty. We have used $S(n,k)$ for the number of ways of doing this. Then, distinguish the boxes by placing k different labels on the boxes. There are k choices of box for the first label, $k-1$ for the second label, and so on. So the k labels may be placed on the boxes in $k \times (k-1) \times \ldots = k!$ ways. Combining the $S(n,k)$ choices with $k!$ choices, we see that the total number of choices is $S(n,k) \times k!$. Of course, it still remains to find a formula for $S(n,k)$.

Case 2a. If we have k identical boxes, some of which are allowed to be empty, then we may place all balls in one box, or share them between two boxes, or three boxes, and so on. (Since the boxes are identical, exactly which boxes we use or leave empty is irrelevant.) Clearly, then, the total number of ways to do this $S(n,1) + S(n,2) + \ldots + S(n,k)$.

Case 2b. This is just the notation we introduced above, with the formula for $S(n,k)$ yet to be determined.

Cases 3a and 3b. As we have already noted, these cases are covered by Theorems 3.1 and 3.2.

Case 4a. At this stage this is just a matter of notation, as introduced above. We shall discuss the problem of counting the number of ways to place identical balls in identical boxes in Chapter 6.

Case 4b. The number of ways of placing n identical balls in k identical boxes so that no box is empty is clearly obtained by taking the total number of ways of placing the balls in the k boxes and subtracting the number of cases where at most $k-1$ boxes are used. Thus the number of such arrangements is therefore $p_k(n) - p_{k-1}(n)$.

We have already observed that, if the balls are identical and the boxes are identical, all that matters is how many balls there are in each box. So deciding how to place n balls in

k boxes, with some boxes possibly empty, is the same as deciding how to write n as the sum of k nonnegative integers. As the boxes are identical, the order in which we put these k nonnegative integers is irrelevant. Thus we can view the seven rows in Figure 3.2 as corresponding to the following seven ways of writing 6 as the sum of three nonnegative integers, namely, $6 + 0 + 0, 5 + 1 + 0, 4 + 2 + 0, 4 + 1 + 1, 3 + 3 + 0, 3 + 2 + 1$, and $2 + 2 + 2$. Since the zeros are somewhat redundant, we usually omit them. We view the problem in this way in Chapter 6.

In Table 3.1 we have carefully distinguished the cases according to whether the balls are regarded as distinct or identical and likewise for the boxes, and whether or not the boxes can be empty. In counting problems arising from real situations, you may need to think carefully about which category the situation falls into, and hence which line of Table 3.1 is relevant.

Exercises

3.2.1A A teacher has 30 identical chocolate bars and 30 nonidentical pupils. She gives the pupils an algebra test each day for six school weeks, that is, for 30 days in all, and gives a chocolate bar to the pupil with the highest score in each test. In how many different ways can the chocolate bars be distributed between the pupils?

3.2.1B There were 20 different birds and animals in the Caucus race. When the race had finished the Dodo said, "*Everybody* has won and all must have prizes." Alice had a box of 40 identical comfits in her pocket. In how many ways could she distribute these to the birds and animals so that each of them received at least one?*

3.2.2A In how many ways can we place four identical black marbles and six distinct nonblack marbles in five distinct boxes, some of which might be empty?

3.2.2B In how many ways can eight identical black marbles and ten distinct nonblack marbles be placed in five distinct boxes if there is to be at least one black and one nonblack marble in each box?

3.2.3A A manufacturer makes identical transparent marbles and also marbles in 20 different colors. In how many ways can he make up a bag of 20 marbles, given that the bag may contain up to 20 transparent marbles but not two nontransparent marbles that have the same color?

3.2.3B A manufacturer makes white chalk and also chalk in 12 other colors. In how many ways can he make up a box containing 12 sticks of chalk of which at least six must be white and in which there must not be two nonwhite sticks of chalk with the same color?

3.3 A "REDUCTION" THEOREM FOR THE STIRLING NUMBERS

Before we obtain a general formula for the Stirling numbers, $S(n,k)$, we can, at least, readily find the values for small values of n and k, not by laboriously listing all the ways of placing distinct balls in identical boxes, but by establishing a relationship between $S(n,k)$

---

* Based on *Alice's Adventures in Wonderland* by the mathematician Lewis Carroll (1832–1898). Comfit is an old-fashioned term for what in England is now called a *sweet* and in the United States a *piece of candy*. Also, Carroll used *animals* to mean "mammals."

and $S(n-1,k-1)$ and $S(n-1,k)$, which is analogous to the formula $C(n,k) = C(n-1,k-1) + C(n-1,k)$ for the binomial coefficients.

Next, we have this theorem:

THEOREM 3.4

For all positive integers n, k with $2 \le k < n$, $S(n,k) = S(n-1,k-1) + kS(n-1,k)$.

Proof

We divide the different ways of placing the n distinct balls $b_1, b_2, \ldots, b_n$ into k identical boxes, with no box empty, into two disjoint classes: (i) those where ball b_n is in a box on its own and (ii) those where it isn't.

The number of placings of type (i) is the same as the number of ways of placing the remaining $n-1$ balls into $k-1$ other boxes, with no box empty. (There is no need to take into account which box ball b_n is in since all the boxes are identical.) There are $S(n-1,k-1)$ such placings.

We now consider the placings of type (ii). Since the ball b_n is not alone in its cell, even when this ball is removed, all the boxes are nonempty. So a placing of this type is obtained by first placing the $n-1$ balls $b_1, \ldots, b_{n-1}$ in k boxes so that no box is empty and then deciding into which of the k boxes to place b_n. Since the balls are distinct, we obtain a different placing of the n balls corresponding to each of the $S(n-1,k)$ placings of the balls $b_1, \ldots, b_{n-1}$ and each of the k choices of box for ball b_n. Hence, there are $S(n-1,k) \times k$, that is, $kS(n-1,k)$, placings of this type.

Since each placing of the n balls into k boxes, with no box empty, is of one of these two types, and no placing is of both types, it follows that $S(n,k) = S(n-1,k-1) + kS(n-1,k)$.*

Using the above formula for $S(n,k)$ we can determine (although for large values of n and k it could be a very long job!) the value of any particular $S(n,k)$ if only we know the values of $S(n',k')$ for the relevant values of n', k' with $n' < n$ or $k' < k$. In particular, we need some values to enable us to make a start. These are given by part (i) of the next theorem.

THEOREM 3.5

   i. For all positive integers n, $S(n,1) = S(n,n) = 1$.
   ii. For all positive integers n, with $n \ge 2, S(n,2) = 2^{n-1} - 1$.

Proof

   i. There is clearly only one way of placing n balls in one box. There is also only one way of placing n distinct balls in n identical boxes, with no box empty, since the only possible arrangement is to have each ball in a separate box.
   ii. Suppose that we have a set $X = \{b_1, b_2, \ldots, b_n\}$ of n distinct balls to place in two boxes so that neither box is empty. X has 2^n subsets. So there are 2^n possibilities for the set of balls that could be placed in one box. If the balls making up the

---

* The reader with a good memory will recall not only that the formula of Theorem 3.4 is analogous to that of Theorem 2.7 but also that the proofs follow the same strategy of partitioning the set to be counted into two disjoint subsets, each of which is then counted separately.

subset Y of X are placed in this box, then the set $Z = X \backslash Y$ consists of the balls that have to go in the other box. Thus we get a different arrangement for each pair $\{Y,Z\}$ of subsets of X, where $Z = X \backslash Y$. There are $\frac{1}{2}(2^n) = 2^{n-1}$ such pairs. But this includes the arrangement where one of the boxes is empty. Hence there are $2^{n-1} - 1$ ways of placing the n balls in two boxes with neither box empty.

Notes:
1. In the spirit of this book we have given a combinatorial proof of (ii). The reader may be interested to check that this result may also be proved from Theorem 3.4 using mathematical induction.
2. Since, for each positive integer n, $S(n,1) = 1$, Theorem 3.4 would be true for $k = 1$ and $n \geq 2$, if $S(n-1,0) = 0$, giving $1 = S(n,1) = S(n-1,0) + S(n-1,1) = 0 + 1$. So we stipulate that, for each positive integer t, $S(t,0) = 0$

We now have sufficient information to continue, as far as we have the stamina and patience, Table 3.2, showing the value of $S(n,k)$ for small values of n and k. For example, the value 966 for $S(8,3)$ is obtained from the formula $S(n,k) = S(n-1,k-1) + kS(n-1,k)$ of Theorem 3.4. This gives $S(8,3) = S(7,2) + 3 \times S(7,3) = 63 + 3 \times 301 = 966$. The shaded "triangle" should help you to remember the method of evaluation.

Exercises

3.3.1A Use Theorem 3.4 to complete Table 3.2 for $n = 9, 10$ and $1 \leq k \leq n$.

3.3.1B Prove that, for all integers $n > 1$, $n! < S(2n,n) < (2n)!$.

3.3.2A Show that, for all integers $n > 2$, $S(n,1) + S(n,2) + \ldots + S(n,n) < n!$. (The left-hand side is the number of ways of placing n distinct balls in any number of identical boxes.)
(*Hint:* If, say, $n = 10$, one placement of 10 distinct objects, say the integers 1, 2,…,10, into four boxes would be [1 3 9 6 4], [2 10], [5], [8 7]. What has this to do with 10!?)

3.3.2B Show, by means of a combinatorial argument, that

$$S(n+1,k+1) = C(n,k)S(k,k) + C(n,k+1)S(k+1,k) + \ldots + C(n,n)S(n,k).$$

3.3.3A Show that, for $1 \leq k \leq n$, $k^n = C(k,1)1!S(n,1) + C(k,2)2!S(n,2) + \ldots + C(k,k)k!S(n,k)$.

TABLE 3.2

					k			
$S(n,k)$	1	2	3	4	5	6	7	8
1	1							
2	1	1						
3	1	3	1					
4	1	7	6	1				
n 5	1	15	25	10	1			
6	1	31	90	65	15	1		
7	1	63	301	350	140	21	1	
8	1	127	966	1701	1050	266	28	1

3.3.3B Prove that, for all positive integers n,
$$0!S(n,1)-1!S(n,2) + \ldots + (-1)^{n-1}(n-1)! S(n,n) = 0.$$
(*Hint*: Multiply through the equality in Theorem 3.4 by $(k-1)!$ and do a sum)

3.3.4A Let $X = \{1,2,3,4,5,6,7,8,9,10\}$ and let $Y = \{1,2,3,4,5\}$. How many surjective (onto) functions, $f: X \to Y$ are there with domain X and codomain Y?

3.3.4B Which is the least positive integer k such that, with $X = \{1,2,3,\ldots,k\}$ and $Y = \{1,2,3,4,5\}$, at least one-third of the functions $f: X \to Y$ are surjective?

3.3.5A Show that, for all positive integers $n \geq 1$, $S(n+1,n) = C(n+1,2) = [n(n+1)]/2$.
(*Hint for this and the next exercise*: Either give a direct combinatorial proof that $S(n+1,n) = C(n+1,2)$ or use the reduction formula of Theorem 3.4 to give a proof using mathematical induction that $S(n+1,n) = [n(n+1)]/2$.)

3.3.5B i. Show that, for all positive integers $n \geq 1$,
$$S(n+2,n) = \frac{n(n+1)(n+2)(3n+1)}{4!}.$$

ii. Find and prove the analogous formula for $S(n+3,n)$.

3.3.6A Prove that, for all integers $n \geq 3$,
$$S(n,3) = \frac{1}{2}(3^{n-1} - 2^n + 1).$$

3.3.6B Prove that, for all integers $n \geq 4$,
$$S(n,4) = \frac{1}{24}(4^n) - \frac{1}{6}(3^n) + \frac{1}{4}(2^n) - \frac{1}{6}.$$

3.3.7A Show that, for $m \geq n-1$, m 0s and n 1s may be arranged in a line so that no two 1s are adjacent in $C(m+1,n)$ ways.

3.3.7B Show that the number of sets of r elements that can be chosen from the set $\{1,2,\ldots,n\}$ and that contain no two consecutive numbers is $C(n-r+1,r)$. (*Hint*: Reduce this to the previous problem.)

3.3.8A Let S be a bag containing balls numbered 1 to n. You are to select r of them by choosing one at a time and returning it to the bag before making another selection. If the order of your choice is not taken into consideration, in how many ways can the r balls can be chosen? (For example, if $n = 2$ and $r = 4$, there are five ways of making the choice—and these are 1,1,1,1; 1,1,1,2; 1,1,2,2; 1,2,2,2; and 2,2,2,2.)

3.3.8B By a *nondecreasing integer* we mean one such as 1,123,555,699 where the digits involved do not decrease as you read from left to right. Determine the number of nondecreasing integers less than 1,000,000.

CHAPTER 4

The Inclusion–Exclusion Principle

4.1 DOUBLE COUNTING

In this chapter we return to the difficulty that we met in Chapter 2, Section 2.4. We saw there that the formula

$$\#(A \cup B \cup C \ldots) = \#(A) + \#(B) + \#(C) + \ldots$$

holds provided that the sets $A, B, C, \ldots$ are disjoint but not if there is an overlap among them. In this section we develop the formula that holds in the case where the sets are not disjoint. The general formula is given in Theorem 4.2. Our main application of it will be to obtain a formula for the Stirling numbers, $S(n,k)$. We begin with the easiest situation to understand, namely, where we have just two sets, say A and B.

In the sum $\#(A) + \#(B)$ we count each member of A once and each member of B once. So the members of the set $A \cap B$ get counted twice, once through their membership of A and once through their membership of B. To obtain the number of elements in $A \cup B$ we therefore need to subtract the number of elements that are counted twice, namely, those that are both in A and in B. This is shown in Figure 4.1. Thus we obtain:

THEOREM 4.1

For all sets A and B,

$$\#(A \cup B) = \#(A) + \#(B) - \#(A \cap B).$$

The next problem illustrates the use of this theorem.

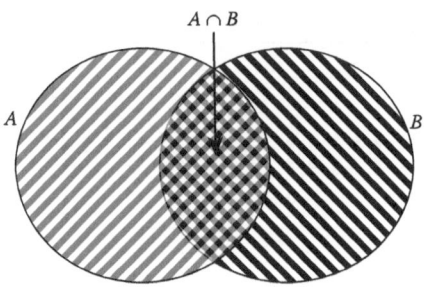

FIGURE 4.1

PROBLEM 4.1

How many integers are there in the range from 1 to 1,000,000 that are divisible by 2 or 3 or both?

Solution

We let D_2 and D_3 be the sets of those integers in the range from 1 to 1,000,000 that are divisible by 2 and by 3, respectively. By Theorem 4.1, $\#(D_2 \cup D_3) = \#(D_2) + \#(D_3) - \#(D_2 \cap D_3)$. Clearly, $\#(D_2) = 500{,}000$ and $\#(D_3) = 333{,}333$. Now, an integer is divisible by both 2 and 3 if and only if it is divisible by 6. Hence $\#(D_2 \cap D_3) = 166{,}666$. Therefore $\#(D_2 \cup D_3) = 500{,}000 + 333{,}333 - 166{,}666 = 666{,}667$.

Theorem 4.1 can easily be extended to deal with the case of three sets. All we need do is to write $A \cup B \cup C$ as $(A \cup B) \cup C$ and then apply Theorem 4.1, making use of some elementary set algebra. This gives

$$\#(A \cup B \cup C) = \#((A \cup B) \cup C)$$
$$= \#(A \cup B) + \#(C) - \#((A \cup B) \cap C)$$
$$= \big(\#(A) + \#(B) - \#(A \cap B)\big) + \#(C) - \#((A \cup B) \cap C),$$

using Theorem 4.1 twice. Now, by the distributive law for unions and intersections of sets, $(A \cup B) \cap C = (A \cap C) \cup (B \cap C)$. Hence, using Theorem 4.1 again,

$$\#((A \cup B) \cap C) = \#(A \cap C) + \#(B \cap C) - \#((A \cap C) \cap (B \cap C)),$$

and therefore, as $(A \cap B) \cap (B \cap C) = A \cap B \cap C$, we deduce that

$$\#(A \cup B \cup C) = (\#(A) + \#(B) + \#(C)) - (\#(A \cap B) + \#(A \cap C) + \#(B \cap C)) + \#(A \cap B \cap C).$$

Now let us see what this formula means. The three terms in the first pair of brackets count the members of A, B, and C separately. Thus, when we add them up, we have counted elements that are in two of the sets twice, and those in all three are counted three times. We take account of this double counting by subtracting the terms in the second pair of brackets. However, when we do this, an element that is in all three of the

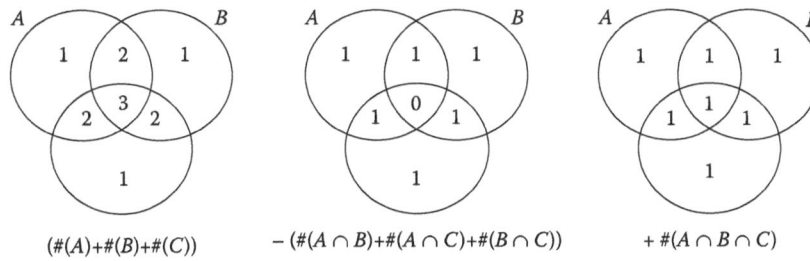

FIGURE 4.2

sets is discounted three times. To compensate for this we have to add on the final term $\#(A \cap B \cap C)$. We can illustrate the stages of this process by the diagrams in Figure 4.2.

We see from Figure 4.2 that our formula does indeed count each element of $A \cup B \cup C$ exactly once. Notice that this is accomplished by alternately including elements in the count and excluding them.

It is now not difficult to see how to extend this formula to deal with the general case of the number of elements in a union of n sets. The only complication is with the notation needed to deal with this general situation. The theorem gets its name from the inclusion and exclusion processes corresponding to the alternate + and − signs.

THEOREM 4.2
The Inclusion–Exclusion Theorem

For all sets $A_1, A_2, \ldots, A_n$,

$$\#(A_1 \cup A_2 \cup \ldots \cup A_n) = \big(\#(A_1) + \#(A_2) + \ldots + \#(A_n)\big)$$
$$- \big(\#(A_1 \cap A_2) + \#(A_1 \cap A_3) + \ldots + \#(A_{n-1} \cap A_n)\big)$$
$$+ \big(\#(A_1 \cap A_2 \cap A_3) + \#(A_1 \cap A_2 \cap A_4) + \ldots + \#(A_{n-2} \cap A_{n-1} \cap A_n)\big)$$
$$\vdots$$
$$(-1)^{n+1} \#(A_1 \cap A_2 \cap \ldots \cap A_n).$$

Comment

A word about the notation is necessary before we give a proof of this theorem. The notation above is intended to indicate that the second pair of large brackets on the right-hand side contains terms corresponding to each pair of sets, the third pair of large brackets contains terms corresponding to each triple of sets, and so on. If we want to be more explicit about this, at the price of complicating our formula, we can write the equation in Theorem 4.2 as follows:

$$\#\left(\bigcup_{i=1}^{n} A_i\right) = \sum_{k=1}^{n} (-1)^{k+1} \left(\sum_{1 \leq i_1 < \ldots < i_k \leq n} \#(A_{i_1} \cap A_{i_2} \cap \ldots \cap A_{i_k}) \right). \quad (4.1)$$

Here the inequalities $1 \leq i_1 < \ldots < i_k \leq n$ under the summation symbol are intended to indicate that the sum is taken over all choices of integers $i_1, i_2, \ldots, i_k$ satisfying these inequalities. In other words, we sum over all the k-element subsets of the set $\{1,2,\ldots,n\}$.

We are now ready for the proof.

Proof

Take an element $x \in A_1 \cup \ldots \cup A_n$. We calculate how many times x is included by the formula on the right-hand side of Equation 4.1.

Suppose that x is an element of exactly m of the given sets, say $A_{j_1},\ldots,A_{j_m}$. Then $x \in A_{i_1} \cap \ldots \cap A_{i_k}$ if and only if $\{i_1,\ldots,i_k\} \subseteq \{j_1,\ldots,j_m\}$. This can happen only if $1 \le k \le m$, in which case the set $\{j_1,\ldots j_m\}$ has $C(m,k)$ subsets of size k. Thus x is counted $C(m,k)$ times in the sum $\sum_{1 \le i_1 < \ldots < i_k \le n} \#(A_{i_1} \cap \ldots \cap A_{i_k})$ that occurs on the right-hand side of our equation. Thus, taking account of the alternating signs, x is altogether counted $\sum_{k=1}^{m}(-1)^{k+1}C(m,k)$ times. Now $\sum_{k=0}^{m}(-1)^{k}C(m,k)=0$ (see Exercise 2.3.3B).

It follows that $\sum_{k=1}^{m}(-1)^{k+1}C(m,k)=C(m,0)=1$, and thus each element of $A_1 \cup \ldots \cup A_n$ is counted just once by the formula, which therefore gives the correct value for $\#(A_1 \cup \ldots \cup A_n)$.

We can now use the inclusion–exclusion theorem to modify our calculation in Chapter 2 of the number of five-card hands with at least one suit with no cards in it (Problem 2.11).

PROBLEM 4.2

How many hands of five cards are there in which there is at least one suit with no cards in it?

Solution

We let V_S, V_H, V_D, and V_C be those five-card hands with no spades, hearts, diamonds, or clubs, respectively. We wish to calculate $\#(V_S \cup V_H \cup V_D \cup V_C)$. By the inclusion–exclusion theorem this is equal to

$$\big(\#(V_S)+\#(V_H)+\#(V_D)+\#(V_C)\big)-\big(\#(V_S \cap V_H)+\ldots\big)+\big(\#(V_S \cap V_H \cap V_D)+\ldots\big)$$
$$-\#(V_S \cap V_H \cap V_D \cap V_C).$$

A hand in V_S contains 5 cards drawn from the 39 cards that are not spades. Thus $\#(V_S) = C(39,5)$, and similarly for each of the terms in the first pair of large brackets of the above expression. A hand in $V_S \cap V_H$ consists of 5 cards drawn from the 26 cards that are neither spades nor hearts. Thus $\#(V_S \cap V_H) = C(26,5)$, and similarly for the other terms in the second pair of brackets. These terms correspond to all the ways of choosing two suits from the four suits in the pack, so there are $C(4,2)$ terms in the second pair of brackets. In the same way we see that there are $C(4,1)$ terms in the third pair of brackets, each of them equal to $C(13,5)$. Finally, of course, there are no five-card hands that contain no card of any suit and so $\#(V_S \cap V_H \cap V_D \cap V_C) = 0$. Therefore

$$\#(V_S \cup V_H \cup V_D \cup V_C) = C(4,1)C(39,5) - C(4,2)C(26,5) + C(4,3)C(13,5)$$
$$= 4 \times 575{,}757 - 6 \times 65{,}780 + 4 \times 1{,}287 = 1{,}913{,}496,$$

which agrees with our earlier solution to Problem 2.11 in Chapter 2.

If you compare the solutions to Problems 2.11 and 4.2, you will see that, in this case, the use of the inclusion–exclusion theorem does not save a lot of work, as in the solution to Problem 2.11 we needed to consider only five different suit distributions. However, with larger hands, where there are many more possible suit distributions, the use of inclusion–exclusion theorem saves a lot of work. This is illustrated by Exercise 4.1.3A.

The next problem involves an application of the inclusion–exclusion theorem of a different kind.

PROBLEM 4.3

A set of n objects is sampled at random with replacement. What is the probability that, after s samples have been drawn, each object has been sampled at least once?

Solution

"Sampling with replacement" means that the set from which the samples are drawn remains the same throughout. It occurs in many different contexts. For example, both bird watchers and train spotters can be thought of as sampling with replacement. Also, when you throw dice, you are sampling the numbers on the dice with replacement, since the set of numbers that can come up on each throw stays the same. Likewise, tossing a coin can be thought of as sampling with replacement the two sides of the coin. These cases are different from problems involving, for example, hands of cards, where, when a card has been drawn from the pack, it cannot be selected again.

Sampling *at random* is intended to mean that any sequence of s samples is as likely to be drawn as any other. So the required probability in Problem 4.3 is just the ratio

$$\frac{\#(\text{Sequences of } s \text{ samples in which each object occurs})}{\#(\text{All sequences of } s \text{ samples})}. \tag{4.2}$$

We will use $\theta(n,s)$ for this ratio. The number in the denominator of Equation 4.2 is very easy to calculate. As we are sampling with replacement, each time we draw a sample, we have n objects to choose from. Hence a sequence of s samples can be selected in n^s ways.

To calculate the number in the numerator of Equation 4.2, it is easier first to calculate the number of sequences of s samples in which at least one of the objects does *not* occur. We let A_i be the set of those sequences of s samples in which the ith object is missing. The number in the numerator of Equation 4.2 is therefore $n^s - \#(\bigcup_{i=1}^{n} A_i)$. Now, by the inclusion–exclusion theorem,

$$\#\left(\bigcup_{i=1}^{n} A_i\right) = \sum_{k=1}^{n} (-1)^{k+1} \left(\sum_{1 \leq i_1 < \ldots < i_k \leq n} \#(A_{i_1} \cap \ldots \cap A_{i_k}) \right).$$

Here $A_{i_1} \cap \ldots \cap A_{i_k}$ is the set of those sequences of s samples in which k of the objects do not occur, and hence in which each sample may be chosen in $(n-k)$ ways. It follows that $\#(A_{i_1} \cap \ldots \cap A_{i_k}) = (n-k)^s$. Now, there are $C(n,k)$ ways of choosing the k numbers $i_1, \ldots, i_k$ from the set $\{1, 2, \ldots, n\}$. Consequently, the numerator of Equation 4.2 is equal to $n^s - \sum_{k=1}^{n}(-1)^{k+1}C(n,k)(n-k)^s$. We can rewrite this as $\sum_{k=0}^{n}(-1)^{k}C(n,k)(n-k)^s$. So the required probability is given by

$$\theta(n,s) = \frac{1}{n^s}\sum_{k=0}^{n}(-1)^{k}C(n,k)(n-k)^s.$$

PROBLEM 4.4
How many times do you need to toss a fair coin to have a probability greater than 0.99 of throwing at least one head and at least one tail?

Solution
As we noted above, tossing a coin can be viewed as sampling with replacement from the two-element set {H,T}, where H = head and T = tail. So, it is the case of Problem 4.3 with $n = 2$. Consequently, the probability that both a head and a tail are thrown in s tosses is

$$\theta(2,s) = \frac{1}{2^s}\sum_{k=0}^{2}(-1)^{k}C(2,k)(2-k)^s = \frac{1}{2^s}(2^s - 2) = 1 - \frac{1}{2^{s-1}}.$$

(Note that in this particularly simple case, we didn't really need to use the general formula. With s tosses, there are 2^s sequences of possible outcomes, of which just two, HH...H and TT...T, do not include at least one head and at least one tail.)

We therefore seek the least s such that $1 - (1/2^{s-1}) > 0.99$. This inequality is equivalent to $2^s > 200$. The smallest integer s satisfying this last inequality is $s = 8$.

There is no straightforward way to simplify the formula for $\theta(n,s)$ in general. However, in two special cases we can do this.

THEOREM 4.3
a. For all positive integers n, s, if $s < n$, then

$$\theta(n,s) = \frac{1}{n^s}\sum_{k=0}^{n}(-1)^{k}C(n,k)(n-k)^s = 0. \quad (4.3)$$

b. For all positive integers n,

$$\theta(n,n) = \frac{1}{n^n}\sum_{k=0}^{n}(-1)^{k}C(n,k)(n-k)^n = \frac{n!}{n^n}. \quad (4.4)$$

Proof
a. When $s < n$ there cannot be a sequence of s samples that includes all the n objects. So in this case $\theta(n,s) = 0$.
b. When $s = n$, a sequence of s samples containing each of the n objects must contain each of them just once, and hence is just a permutation of those objects. Now there are $n!$ permutations of n objects, and so $\theta(n,n) = n!/n^n$.

Exercises

4.1.1A How many integers are there in the range from 1 to 1,000,000 that are either perfect squares or perfect cubes or both?

4.1.1B How many integers are there in the range from 1 to 1,000,000 that are divisible by 7 or 11 or both?

4.1.2A How many integers are there in the range from 1 to 1,000,000 that are divisible by none of 2, 3, 5, and 7?

4.1.2B How many integers are there in the range from 1 to 1,000,000 that are not *powers*, that is, not of the form n^k where k, n are positive integers with $k \geq 2$?

4.1.3A How many bridge hands (hands of 13 cards drawn from the standard pack of 52 cards) are there that include at least one void suit?

4.1.3B If a pack consists of cards from p suits with n cards in each suit, find a formula for the probability that a hand of n cards from this pack includes at least one card from each suit.

4.1.4A Euler's ϕ-function plays an important role in number theory. For each positive integer n, $\phi(n)$ is the number of integers in the range from 1 to n that have no prime factors in common with n. For example, $\phi(12) = 4$ since there are four positive integers in the range from 1 to 12 that have no prime factors in common with 12, namely, 1, 5, 7, and 11.

Use the inclusion–exclusion theorem to show that, if the distinct prime factors of n are $p_1, p_2, \ldots, p_k$, then

$$\phi(n) = n\left(1 - \frac{1}{p_1}\right)\left(1 - \frac{1}{p_2}\right)\cdots\left(1 - \frac{1}{p_k}\right).$$

4.1.4B Calculate the values of (a) $\phi(1,000,000)$ and (b) $\phi(7!)$, that is, $\phi(5040)$.

4.1.5A How many times do you need to throw a standard die so that there is a probability greater than 0.5 that each of the numbers 1, 2, 3, 4, 5, and 6 is thrown at least once?

4.1.5B If two coins are tossed, there are four equally likely outcomes, HH, HT, TH, and TT. How many times do you need to toss two coins so that there is a probability greater than 0.9 that each of these outcomes occurs at least once?

4.1.6A Suppose that the sets X_i for $1 \leq i \leq n$ are pairwise disjoint, and each contains p elements. Let $X = \bigcup_{i=1}^{n} X_i$. Show that if X is sampled with replacement at random, the probability that a sequence of s samples contains at least one element from each of the sets X_i is $\theta(n,s)$; that is, it is independent of p.

4.1.6B A bag contains equal numbers of red, green, blue, and yellow balls. One ball is drawn at random from the bag and then replaced. If this is done eight times, what is the probability that at least one ball of each color is drawn from the bag?

4.1.7A Cards are drawn at random from a standard pack and then replaced. How many times do you need to do this to have a probability of more than 0.99 that you have drawn at least one card of each rank?

4.1.7B Cards are drawn at random from a standard pack and then replaced. How many times do you need to do this to have a probability of more than 0.99 that you have drawn at least one card of each suit?

4.2 DERANGEMENTS

We now come to the "Snap" problem from Chapter 1, there called Problem 4. As we said, the answer may come as a surprise.

PROBLEM 4.5

Two fully shuffled standard packs of 52 cards are placed face down and side by side. One after another, pairs of cards, one from each pack, are turned over. What is the probability that, as all 52 pairs are turned over, at least one pair of cards will be the same?

Comment

Before reading the solution below, try to estimate this probability or, better still, solve the problem for yourself.

Solution

We can calculate the required probability, say p_{52}, by assuming that the order of the cards in one of the packs is fixed and then counting the number of arrangements of the second pack where there is at least one coincidence with the first pack. Then p_{52} is this number divided by the total number of arrangements of the cards in the second pack, which is, of course, 52!.

For $1 \leq i \leq 52$ we let A_i be the set of those arrangements of the second pack in which the ith card coincides with the ith card of the first pack. Then

$$p_{52} = \frac{\#\left(\bigcup_{i=1}^{52} A_i\right)}{52!}.$$

Given $1 \leq i_1 < \ldots < i_k \leq 52$, $A_{i_1} \cap \ldots \cap A_{i_k}$ is the set of those arrangements of the second pack in which the card in each position $i_1, \ldots, i_k$ is identical to the card in the same position in the first pack. The other $(52-k)$ cards can be arranged in any order in the remaining $(52-k)$ positions. Therefore $\#(A_{i_1} \cap \ldots \cap A_{i_k}) = (52-k)!$. Hence, by the inclusion–exclusion theorem,

$$\#\left(\bigcup_{i=1}^{52} A_i\right) = \sum_{k=1}^{52}(-1)^{k+1} \sum_{1 \leq i_1 < \ldots < i_k \leq 52} \#(A_{i_1} \cap \ldots \cap A_{i_k})$$

$$= \sum_{k=1}^{52}(-1)^{k+1} C(52,k)(52-k)!$$

The good thing is that this last expression simplifies because

$$C(52,k)(52-k)! = \frac{52!}{k!(52-k)!} \times (52-k)! = \frac{52!}{k!},$$

and therefore

$$p_{52} = \frac{\#\left(\bigcup_{i=1}^{52} A_i\right)}{52!} = \frac{1}{52!}\sum_{k=1}^{52}(-1)^{k+1}\frac{52!}{k!} = \sum_{k=1}^{52}(-1)^{k+1}\frac{1}{k!}. \qquad (4.5)$$

Of course, for a pack of n cards, all we would need to do is to substitute n for 52 throughout the solution. Hence the probability, p_n, for packs of n cards is given by Equation 4.5 with 52 replaced by n.

We can easily relate the number given by Equation 4.5 to the number e. Using the series for e^x in the case where $x = -1$, we have $e^{-1} = \sum_{k=0}^{\infty}[(-1)^k/k!]$, and hence

$$1 - e^{-1} = \sum_{k=1}^{\infty}\frac{(-1)^{k+1}}{k!}. \qquad (4.6)$$

Comparing Equations 4.5 and 4.6, we see that p_{52} is given by the first 52 terms of the infinite series in Equation 4.6. This series converges so rapidly that $1 - e^{-1}$ provides a very good approximation to the value of p_{52}. More generally, as we have already noted, if p_n is the probability of a coincidence with two packs of n cards, p_n is given by the first n terms of the series for $1 - e^{-1}$. The rapid convergence of this series means that once we have six cards in the pack, the probability of at least one coincidence does not change appreciably as the number of cards in the pack increases. This is shown by Table 4.1, which gives the values of p_n for $1 \leq n \leq 13$ and also the value of $1 - e^{-1}$, all to nine decimal places.

So, the remarkable, almost unbelievable, truth is that a coincidence of at least one pair of identical cards occurs in almost two cases in every three and that once we have at least six cards, the number of cards in the pack hardly affects this probability.

An arrangement of n objects in which no object occupies its original position is called a *derangement* of those objects. Problem 4.5 involves counting the number of

TABLE 4.1

n	p_n
1	1.000000000
2	0.500000000
3	0.666666667
4	0.625000000
5	0.633333333
6	0.631944444
7	0.632142857
8	0.632118056
9	0.632120811
10	0.632120536
11	0.632120561
12	0.632120559
13	0.632120559
$1-e^{-1}$	0.632120559

arrangements of the second pack that are *not* derangements of the first pack. Using the method of that problem we can easily arrive at the following result.

THEOREM 4.4

The number of derangements of a set of n objects is

$$n! \sum_{k=0}^{n} \frac{(-1)^k}{k!}.$$

The series in Theorem 4.4 consists of the first $n + 1$ terms of the infinite series for e^{-1}, hence the number of derangements of a set of n objects is approximately $n!e^{-1}$, and this approximation is very accurate for $n \geq 6$.

Exercises

4.2.1A If letters to 10 different people are placed, at random, in 10 envelopes addressed to the same people, what is the probability that every letter is put into the wrong envelope?

4.2.1B In our village we have a baker, a grocer, a publican, and a policeman. They are called Mr. Bun, Mr. Sugar, Mr. Pale-Ale, and Mr. Copper, but Mr. Bun is not the baker, Mr. Sugar is not the grocer, Mr. Pale-Ale is not the publican, and Mr. Copper is not the policeman. How many possibilities remain for the last names of the baker, grocer, publican, and policeman?

4.2.2A For the purpose of this and the next question, by an *anagram* of a given word (or, more generally, a string of letters), we mean a rearrangement of the letters so that in each position there is a change of letter. Thus when all the letters of a word are different, an anagram is just the same as a derangement of its letters, but when some letters are repeated, not all the derangements are anagrams. For example, the word *NOON* has just the one anagram *ONNO*, and the word *BALL* has the two anagrams *LLAB* and *LLBA*, whereas *TEE* has no anagrams at all. (Of course, we do not insist that an anagram of a word must be another word of our language.)

How many anagrams do each of the following words have?
   i. *ROBOT*   ii. *ANAGRAM*   iii. *TENNESSEE*

4.2.2B i. How many anagrams does *AABBCC* have?
   ii. How many anagrams does a string of letters made up of r As, s Bs, and t Cs have?

4.3 A FORMULA FOR THE STIRLING NUMBERS

The Stirling numbers, $S(n,k)$, were introduced in Chapter 3. They give the number of ways of placing n distinct balls in k identical boxes, so that no box is empty. Theorem 3.4 gives a recurrence relation* satisfied by these numbers. We can now use the argument of Problem 4.3 to obtain an explicit formula for them.

---

* See Chapter 7, Section 7.3, for a discussion of what is meant by a "recurrence relation."

THEOREM 4.5

For all integers n, k with $1 \leq k \leq n$,

$$S(n,k) = \frac{1}{k!} \sum_{s=0}^{k-1} (-1)^s C(k,s)(k-s)^n. \tag{4.7}$$

Proof

We have noted in Chapter 3 that $k!S(n,k)$ is the number of different ways of placing n distinct balls in k distinct boxes so that no box is empty and that this is the same as the number of surjective functions from a set with n elements to a set with k elements. So we can obtain a formula for $S(n,k)$ by calculating this number and then dividing by $k!$

We can take $A = \{1,\ldots, n\}$ to be our set of n elements, and we let $B = \{1,\ldots, k\}$ be our set of k elements. There are altogether k^n different mappings from A to B. We count the number of these mappings that are not surjective. For $1 \leq i \leq k$, we let X_i be the set of those mappings $f: A \to B$ that do not take the value i. So we are aiming to calculate the value of $\#(\bigcup_{i=1}^{k} X_i)$. Naturally, we do this by using the inclusion–exclusion theorem, which tells us that

$$\#\left(\bigcup_{i=1}^{k} X_i\right) = \sum_{s=1}^{k} (-1)^{s+1} \sum_{1 \leq i_1 < \ldots < i_s \leq k} \#(X_{i_1} \cap \ldots \cap X_{i_s}). \tag{4.8}$$

The argument we now use is one that we have seen before. There are $C(k,s)$ ways to choose the numbers $i_1,\ldots,i_s$ in the range from 1 to k. For each of these choices, $X_{i_1} \cap \ldots \cap X_{i_s}$ is the set of those functions $f: A \to B$ that do not take any of the values $i_1,\ldots, i_s$. So for each function f in this set, and for each integer j with $1 \leq j \leq n$, there are $(k-s)$ choices for $f(j)$. Hence $\#(X_{i_1} \cap \ldots \cap X_{i_s}) = (k-s)^n$. Therefore, by Equation 4.8,

$$\#\left(\bigcup_{i=1}^{k} X_i\right) = \sum_{s=1}^{k} (-1)^{s+1} C(k,s)(k-s)^n,$$

and hence the number of surjective functions $f: A \to B$ is

$$k^n - \sum_{s=1}^{k} (-1)^{s+1} C(k,s)(k-s)^n = \sum_{s=0}^{k} (-1)^s C(k,s)(k-s)^n.$$

It therefore follows that

$$S(n,k) = \frac{1}{k!} \sum_{s=0}^{k} (-1)^s C(k,s)(k-s)^n. \tag{4.9}$$

When $s = k$, $(k-s)^n = 0$, the last term of the sum in Equation 4.9 is 0, and so Equation 4.9 is equivalent to Equation 4.7 that we were aiming to prove. So this completes the proof.

Exercises

4.3.1A Use the formula in Equation 4.7 to calculate the values of $S(4,2)$ and $S(8,4)$, and check that the values you get agree with those given in Table 3.2.

4.3.1B Use the formula in Equation 4.7 to calculate the values of $S(7,k)$ for $1 \leq k \leq 7$, and check that the values you get agree with those given in Table 3.2.

4.3.2A Use the recurrence relation $S(n,k) = S(n-1,k-1) + kS(n-1,k)$ of Theorem 3.4, and the fact that $S(n,1) = 1$ for each positive integer, to give a proof of Theorem 4.5 by mathematical induction.

(Use induction on n to prove that for every positive integer $n \geq 1$, for integers k with $1 \leq k \leq n$, the formula in Equation 4.7 of Theorem 4.5 holds. This is a good example of a case where *once you know which formula it is you want to prove*, proof by mathematical induction is a useful tool. As you will see, it enables us to prove Equation 4.7 just by algebraic manipulations. However, it wouldn't be much use if all we had was the recurrence relation, and we didn't have any idea what the correct formula was.)

4.3.2B Verify that Theorem 4.5 gives the same answers as does Theorem 3.5 of Chapter 3; namely, that for all positive integers n, $S(n,1) = S(n,n) = 1$, and for $n \geq 2$, $S(n,2) = 2^{n-1} - 1$.

CHAPTER 5

Stirling and Catalan Numbers

5.1 STIRLING NUMBERS

One of the pleasures of studying mathematics lies in discovering how the same ideas, numbers, and so on keep cropping up in different settings. In the case of (collections of) numbers, in particular, many are named after famous mathematicians, often their originators or principal investigators. Apart from the well-known Fibonacci numbers, there are numbers named after Fermat, Bernoulli, Euler, Mersenne, Lucas, Carmichael, Ramsey (whom we meet in Chapter 16), and many more.*

In this chapter we consider just two classes of numbers, one named after the Scottish mathematician Stirling, the other after the French/Belgian mathematician Catalan. The former class was, indeed, introduced by Stirling, while the latter class was essentially introduced by Euler, with other mathematicians studying them before Catalan's time. So, what problems gave rise to these numbers?

In his book *Methodus Differentialis* (1730; following a paper of the same name in 1719) James Stirling[†] introduced methods by which he could quickly determine, accurate to many decimal places, the approximate value of various slowly converging series. For example, he was able to calculate $\sum_{n=1}^{\infty}(1/n^2)$ to 17 places of decimals. (Proceeding naively would require, approximately, adding the first 1,000,000,000 terms!) To do this Stirling needed to determine the coefficients $a_{n,k}$ arising when the so-called *falling factorial polynomial* $x(x-1)(x-2)\ldots(x-[n-1])$ is written in the form $a_{n,n}x^n - a_{n,n-1}x^{n-1} + \ldots + (-1)^{n-1}a_{n,1}x$ (the minus signs being chosen so that all the terms $a_{n,k}$ are positive) as well as the converse problem

---

* Not forgetting Erdös! Paul Erdös, who was born in Budapest on March 26, 1913, was a prolific and towering figure in combinatorics. He wrote 1475 research papers, mostly in collaboration. Altogether he had some 485 coauthors. He died in Warsaw on September 20, 1996. *Your* Erdös number, if it exists, is the least integer n such that there is a sequence of people $E_0, E_1, \ldots, E_n$ where E_0 is Erdos, E_n is you, and for each k, $0 \leq k < n$, E_k has coauthored a published research paper with E_{k+1}. There is an excellent biography of Erdös, *The Man Who Loved Only Numbers*, by Paul Hoffman, Fourth Estate, London, 1998.
† James Stirling was born in Scotland in 1692. He was a student in Glasgow and Oxford, and then went to Venice. He settled in London in 1724. In his later life Stirling turned his attention to engineering. He was employed by the Scottish Mining Company and moved back to Scotland in 1735 to Leadhills, Lanarkshire, a village near the company's lead mines. He died in Edinburgh on December 5, 1770.

of finding the coefficients $b_{n,k}$ arising when x^n is expressed in terms of the falling factorial polynomials, so that

$$x^n = b_{n,n}x(x-1)..(x-[n-1]) + b_{n,n-1}x(x-1)...(x-[n-2]) + \ldots + b_{n,2}x(x-1) + b_{n,1}x.$$

The definition of falling factorial polynomials given above is fairly easy to grasp, but clearly a succinct notation would be a help. So for each integer $k > 0$ we use $[x]_k$ for the falling factorial polynomial $x(x-1)\ldots(x-[k-1])$. Note that the suffix k indicates the degree of the polynomial.

PROBLEM 5.1
Calculate the values of the numbers $a_{n,k}$ for $1 \leq n \leq 5$ and $1 \leq k \leq n$.

Solution
$[x]_1 = x$
$[x]_2 = x(x-1) = x^2 - x$
$[x]_3 = x(x-1)(x-2) = x^3 - 3x^2 + 2x$
$[x]_4 = x(x-1)(x-2)(x-3) = x^4 - 6x^3 + 11x^2 - 6x$
$[x]_5 = x(x-1)(x-2)(x-3)(x-4) = x^5 - 10x^4 + 35x^3 - 50x^2 + 24x$

Therefore we have Table 5.1 giving the numbers $a_{n,k}$, for $1 \leq n \leq 5$ and $1 \leq k \leq n$.

PROBLEM 5.2
Calculate the numbers $b_{n,k}$, for $1 \leq n \leq 4$ and $1 \leq k \leq n$.

Solution
$[x]_1 = x$, and hence $b_{1,1} = 1$.
$[x]_2 = x^2 - x$, and hence $x^2 = [x]_2 + x = [x]_2 + [x]_1$. Therefore $b_{2,2} = 1$ and $b_{2,1} = 1$.
$[x]_3 = x^3 - 3x^2 + 2x$, and hence $x^3 = [x]_3 + 3x^2 - 2x = [x]_3 + 3([x]_2 + [x]_1) - 2[x]_1 = [x]_3 + 3[x]_2 + [x]_1$. Therefore $b_{3,3} = 1$, $b_{3,2} = 3$, and $b_{3,1} = 1$.
$[x]_4 = x^4 - 6x^3 + 11x^2 - 6x$, and hence $x^4 = [x]_4 + 6x^3 - 11x^2 + 6x = [x]_4 + 6([x]_3 + 3[x]_2 + [x]_1) - 11([x]_2 + [x]_1) + 6[x]_1 = [x]_4 + 6[x]_3 + 7[x]_2 + [x]_1$.
Therefore $b_{4,4} = 1$, $b_{4,3} = 6$, $b_{4,2} = 7$, and $b_{4,1} = 1$.

TABLE 5.1

			k			
$a_{n,k}$		1	2	3	4	5
n	1	1				
	2	1	1			
	3	2	3	1		
	4	6	11	6	1	
	5	24	50	35	10	1

TABLE 5.2

$b_{n,k}$	k				
n	1	2	3	4	5
1	1				
2	1	1			
3	1	3	1		
4	1	7	6	1	
5	1	15	25	10	1

Using the same method you should find that $x^5 = [x]_5 + 10[x]_4 + 25[x]_3 + 15[x]_2 + [x]_1$. Hence we have Table 5.2.

The above coefficients cry out for investigation. Do any of them ring any bells? Table 5.1 may not be familiar, but look at the entries in Table 5.2. Yes, they are the same as the numbers in Table 3.2 giving the values of $S(n,k)$. Recall that $S(n,k)$ is the number of ways of placing n distinct balls in k identical boxes, so that no box is empty. Comparing the two tables we see that for $1 \leq n \leq 5$ and $1 \leq k \leq n$, $b_{n,k} = S(n,k)$. Is this just a coincidence? As we now show, we can prove, using mathematical induction, that this equality is always true.

THEOREM 5.1

For each integer $n \geq 1$, we have

$$x^n = S(n,n)[x]_n + S(n,n-1)[x]_{n-1} + \ldots + S(n,1)[x]_1,$$

that is,

$$x^n = \sum_{k=1}^{n} S(n,k)[x]_k.$$

Proof

Base. As $S(1,1) = 1$, and $[x] = x$, we have that $x = S(1,1)[x]_1$, and so the result holds for $n = 1$.

Induction step. Suppose the result holds for $n = t$, that is, that

$$x^t = \sum_{k=1}^{t} S(t,k)[x]_k. \qquad (5.1)$$

It follows from Equation 5.1 that

$$x^{t+1} = x(x^t) = x \sum_{k=1}^{t} S(t,k)[x]_k = \sum_{k=1}^{t} S(t,k)[x]_k x = \sum_{k=1}^{t} S(t,k)[x]_k ((x-k)+k). \qquad (5.2)$$

Now $[x]_k (x-k) = x(x-1)(x-2)\ldots(x-[k-1])(x-k) = [x]_{k+1}$, and hence it follows from Equation 5.2 that

$$x^{t+1} = \sum_{k=1}^{t} S(t,k)[x]_{k+1} + \sum_{k=1}^{t} kS(t,k)[x]_k \qquad (5.3)$$

$$= S(t,1)[x]_1 + \sum_{k=2}^{t} (kS(t,k) + S(t,k-1))[x]_k + S(t,t)[x]_{t+1}.$$

Now, $S(t,1) = 1 = S(t+1,1)$, $S(t,t) = 1 = S(t+1,t+1)$, and, by Theorem 3.4, $kS(t,k) + S(t,k-1) = S(t+1,k)$. Hence, it follows from Equation 5.3 that $x^{t+1} = \sum_{k=1}^{t+1} S(t+1,k)[x]_k$. Therefore the results hold also for $n = t + 1$. It follows, by mathematical induction, that the result is true for all integers $n \geq 1$.

Problem 5.A is another instance where the numbers $S(n,k)$ arise. In calculus the operator $\theta = xD = x(x/dx)$ occurs frequently. We let, as usual, θ^n and D^n be these operators repeated n times, so that, in particular, D^n is the operator d^n/dx^n. Then, if f is a function that can be differentiated as often as required, and as $\theta f = xDf$, we see that $\theta^2 f = xD(xDf) = x(Df + xD^2f) = xDf + x^2D^2f$. (Note that $D(xDf) = Df + xD^2f$ is obtained by applying the product rule to differentiate the product xDf.) Thus the operators θ^2 and $xD + x^2D^2$ are identical.

PROBLEM 5.3

Show that $\theta^3 = xD + 3x^2D^2 + x^3D^3$, and find a similar expression for θ^4.

Solution

We have that $\theta^3 = \theta(\theta^2) = xD(xD + x^2D^2) = x([D + xD^2] + [2xD^2 + x^2D^3]) = xD + 3x^2D^2 + x^3D^3$.

Similarly, $\theta^4 = \theta(\theta^3) = xD(xD + 3x^2D^2 + x^3D^3) = x([D + xD^2] + [6xD^2 + 3x^2D^3] + [3x^2D^3 + x^3D^4]) = xD + 7x^2D^2 + 6x^3D^3 + x^4D^4$.

At this stage, you will notice that the coefficients 1, 7, 6, 1 that occur in this last expression are the values of the Stirling numbers, $S(4,k)$, for $k = 1,2,3,4$. We ask you to prove that this generalizes in Exercise 5.1.7B.

The numbers $S(n,k)$ are usually called *Stirling numbers of the second kind*. The numbers $a_{n,k}$ were also introduced by Stirling and are usually called *Stirling numbers of the first kind*. We shall use the notation $s(n,k)$ for these numbers. As the names "Stirling numbers of the second kind" and "Stirling numbers of the first kind" are rather cumbersome, we shall follow the notation $S(n,k)$ and $s(n,k)$ and call these numbers *Stirling numbers* and *stirling numbers*, respectively.

Therefore, we define the *stirling number*, $s(n,k)$, as follows.

DEFINITION 5.1

For each pair of integers k, n such that $1 \leq k \leq n$ we define the *stirling number*, $s(n,k)$, to be the modulus of the coefficient of x^k in the expansion of $[x]_n$. That is, $s(n,k) = a_{n,k}$.

Because $[x]_n = x(x-1)(x-2)\ldots(x-[n-1])$, the coefficient of x^k in $[x]_n$ is positive or negative according to whether $n - k$ is even or odd. Thus $s(n,k) = (-1)^{n-k} \times$ the coefficient of x^k in $[x]_n$, and, conversely, the coefficient of x^k in $[x]_n$ is $(-1)^{n-k}s(n,k)$.

Table 5.1 gives the values of $s(n,k)$ for $1 \leq k \leq n \leq 5$. To extend this table it would be useful to have a recurrence relation for the numbers $s(n,k)$, analogous to that one we have already given for $S(n,k)$ in Theorem 3.4. And, indeed, we have:

THEOREM 5.2

For all integers k, n such that $2 \leq k < n$, we have

$$s(n,k) = s(n-1,k-1) + (n-1)s(n-1,k).$$

Proof

We have that $[x]_n = x(x-1)(x-2)\ldots(x-[n-2])(x-[n-1]) = [x]_{n-1}(x-[n-1])$. Thus

$$[x]_n = x[x]_{n-1} - (n-1)[x]_{n-1}. \tag{5.4}$$

Now the coefficient of x^k in $x[x]_{n-1}$ is the same as the coefficient of x^{k-1} in $[x]_{n-1}$, and the coefficient of x^k in $-(n-1)[x]_{n-1}$ is $-(n-1) \times$ the coefficient of x^k in $[x]_{n-1}$. Hence $(-1)^{n-k}s(n,k) = (-1)^{(n-1)-(k-1)}s(n-1,k-1) - (n-1)(-1)^{(n-1)-k}s(n-1,k) = (-1)^{n-k}[s(n-1,k-1) + (n-1)\,s(n-1,k)]$. Hence $s(n,k) = s(n-1,k-1) + (n-1)s(n-1,k)$.

PROBLEM 5.4

Use Theorem 5.2 to calculate the values of $s(n,k)$ for $n = 6,7$ and $1 \leq k \leq n$, and thus extend Table 5.1, to give the values of $s(n,k)$ for $1 \leq k \leq n \leq 7$.

Solution

We have, for example, that, by Theorem 5.2, $s(6,4) = s(5,3) + 5s(5,4) = 35 + 5 \times 10 = 85$. In a similar way we obtain the other values shown in Table 5.3.

From the very idea of writing the polynomials $[x]_n$ in terms of the polynomials x^k, using the coefficients $(-1)^{n-k}s(n,k)$ and the polynomials x^n in terms of the polynomials $[x]_k$ using the coefficients $S(n,k)$, it is not surprising that the stirling and Stirling numbers are kinds of "inverses" of each other. In fact, if, for some n, you take the $n \times n$ matrices $\left((-1)^{i+j}s(i,j)\right)$ and $\left(S(i,j)\right)$ and multiply these matrices together, you *should* obtain something that might surprise you (see Exercises 5.1.2A and 5.1.2B). Of course, it shouldn't be *so* surprising when you recall that, given the two pairs of equations $u = ax + by, v = cx + dy$, then if these can be solved to give $x = pu + qv, y = ru + sv$,

the matrices $\begin{pmatrix} a & b \\ c & d \end{pmatrix}$ and $\begin{pmatrix} p & q \\ r & s \end{pmatrix}$ are inverses of each other.

TABLE 5.3

$s(n,k)$		1	2	3	4	5	6	7
n	6	120	274	225	85	15	1	
	7	720	1764	1624	735	175	21	1

But is that all these stirling numbers are good for? By no means! Although the stirling numbers, $s(n,k)$, do not seem to occur as prominently in combinatorics as do the Stirling numbers, $S(n,k)$, they have some intimate connections with permutations, as shown below.

Exercises

5.1.1A Write $[x]_6$ as a polynomial in x.

5.1.1B Write $[x]_7$ as a polynomial in x.

5.1.2A For each n we let P_n be the matrix $\left((-1)^{i+j} s(i,j)\right)_{1 \le i \le n, 1 \le j \le n}$ made up of the coefficients in the polynomials $[x]_k$ for $1 \le k \le n$, and let Q_n be the matrix $(S(i,j))_{1 \le i \le n, 1 \le j \le n}$ of Stirling numbers. Thus from Tables 5.1 and 5.2 we see that

$$P_2 = \begin{pmatrix} 1 & 0 \\ -1 & 1 \end{pmatrix}, P_3 = \begin{pmatrix} 1 & 0 & 0 \\ -1 & 1 & 0 \\ 2 & -3 & 1 \end{pmatrix}, Q_2 = \begin{pmatrix} 1 & 0 \\ 1 & 1 \end{pmatrix}, Q_3 = \begin{pmatrix} 1 & 0 & 0 \\ 1 & 1 & 0 \\ 1 & 3 & 1 \end{pmatrix}.$$

Check that Q_2 is the inverse of P_2 and that Q_3 is the inverse of P_3.

5.1.2B Using the same notation as in Exercise 5.1.2A, check that the matrix Q_4 is the inverse of the matrix P_4.

5.1.3A Show that, for each integer $n \ge 1$, $s(n,1) = (n-1)!$.

5.1.3B Show that, for each integer $n \ge 1$, $S(n,n-1) = [n(n-1)]/2 = s(n,n-1)$.

5.1.4A Show that, for each integer $n \ge 1$, $s(n,1) + s(n,2) + \ldots + s(n,n) = n!$.

5.1.4B Show that, for each integer $n > 1$, $s(n,1) - s(n,2) + \ldots + (-1)^{n-1} s(n,n) = 0$.

5.1.5A Show that, for all positive integers k, n, $\sum_{r=1}^{n} s(n,r) k^r = n! C(k+n-1, n)$.

5.1.5B Show that, for all positive integers k, n, $\sum_{r=1}^{n} (-1)^{n+r} r! S(n,r) C(k+r-1, r) = k^n$.

5.1.6A Show, directly from the definition, that for all integers k, n, with $1 \le k \le n$, $s(n,k)$ is the sum of all the products of $n - k$ different integers taken from the set $\{1,2,\ldots,n-1\}$.

5.1.6B Show by mathematical induction that for all integers k, n, with $1 \le k < n$, $S(n,k)$ is the sum of all products of $n - k$ integers taken from the set $\{1,2,\ldots,k\}$ with repetitions allowed.

5.1.7A Let θ be the operator of Problem 5.3. Find θ^5 in terms of the expressions $x^k D^k$ for $1 \le k \le 5$, and verify that the coefficient of $x^k D^k$ is the Stirling number $S(5,k)$.

5.1.7B Let θ be the operator of Problem 5.3. Prove that, for each positive integer n,

$$\theta^n = \sum_{k=1}^{n} S(n,k) x^k D^k.$$

5.2 PERMUTATIONS AND STIRLING NUMBERS

Recalling, from Chapter 2, how we expressed permutations as products of disjoint cycles, we begin with a simple definition.

DEFINITION 5.2

For all integers k, n with $1 \le k \le n$, we let $Perm(n,k)$ be the set of permutations of the set $\{1,2,3,\ldots,n\}$ made up of exactly k disjoint cycles, and we let $p(n,k)$ be the number of permutations in $Perm(n,k)$, that is, $p(n,k) = \#(Perm(n,k))$.

Note that in counting the number of disjoint cycles, we include the cycles of length 1, so that, for example, the permutation (1 6 2)(3)(4 5) is in the set *Perm*(6,3).

PROBLEM 5.5

Show that, for all positive integers n, $p(n,n) = 1$ and $p(n,1) = (n-1)!$

Solution

A permutation in *Perm*(*n,n*) is a permutation of $\{1,2,3,...,n\}$ made up of exactly n disjoint cycles. Such a permutation must be made up entirely of cycles of length 1, and so must be the *identity permutation* that fixes each member of the set. Thus there is just one such permutation. That is, $p(n,n) = 1$. We have shown that $p(n,1) = (n-1)!$ in the solution to Exercise 2.6.3A.

PROBLEM 5.6

Find a formula for $p(n,n-1)$.

Solution

A permutation in *Perm*(*n,n*–1) is a permutation of $\{1,2,3,...,n\}$ made up of $n-1$ disjoint cycles. Such a permutation must consist of $n-2$ cycles of length 1 and one cycle of length 2, and is thus entirely determined by the two numbers that occur in the cycle of length 2. These two numbers may be chosen in $C(n,2)$ ways. Hence $p(n,n-1) = C(n,2) = [n(n-1)]/2$.

PROBLEM 5.7

Prove that

$$p(n,2) = \begin{cases} n!\sum_{k=1}^{m}\dfrac{1}{k(n-k)}, & \text{where } n \text{ is odd, with } n = 2m+1, \\ n!\left(\dfrac{1}{2m^2} + \sum_{k=1}^{m-1}\dfrac{1}{k(n-k)}\right), & \text{where } n \text{ is even, with } n = 2m. \end{cases}$$

Solution

A permutation in *Perm*(*n*,2) consists of two disjoint cycles, of lengths, say, k and l, where $k + l = n$. There are $C(n,k)$ ways to choose the numbers making up the cycle of length k. For any k of these numbers there are $(k-1)!$ different cycles of length k made up from them, and $(n-k-1)!$ different cycles of length $n-k$ made up from the remaining numbers. So, for $k \neq l$, there are

$$C(n,k) \times (k-1)! \times (n-k-1)! = \frac{n!(k-1)!(n-k-1)!}{k!(n-k)!} = \frac{n!}{k(n-k)}$$

permutations of $\{1,2,3,...,n\}$ made up of a cycle of length k and a cycle of length $n-k$. In the case where $n = 2m$ is even, and $k = l = m$, we need to divide this number by 2, as the order in which the two permutations of length m are written is immaterial. So in this case there are

$$\frac{1}{2}\left(\frac{n!}{(2m-m)(2m-m)}\right) = \frac{n!}{2m^2}$$

permutations made up of two cycles each of length m. Thus the formula of the problem is correct.

We now show that the numbers $p(n,k)$ are the stirling numbers, $s(n,k)$, in disguise. We first show that they satisfy the same recurrence relation as given in Theorem 5.2, namely,

$$s(n,k) = s(n-1,k-1) + (n-1)s(n-1,k).$$

THEOREM 5.3

For all integers k, n, such that $2 \leq k < n$,

$$p(n,k) = p(n-1,k-1) + (n-1)p(n-1,k).$$

Proof

Suppose $2 \leq k < n$. The permutations in $Perm(n,k)$ are the permutations of $\{1,2,3,...,n\}$ made up of exactly k cycles. This set can be partitioned into two disjoint subsets, say X_1 and X_2, where X_1 consists of those permutations in $Perm(n,k)$ in which n is in a cycle by itself, and X_2 consists of all the other permutations in $Perm(n,k)$.

If we have a permutation in X_1 and delete the cycle (n) we are left with a permutation of $\{1,2,3,...,n-1\}$ made up of $k-1$ cycles, that is, a permutation in $Perm(n-1,k-1)$. Conversely, given a permutation in $Perm(n-1,k-1)$, by adding the cycle (n) we obtain a permutation in X_1. This gives a one–one correspondence between X_1 and $Perm(n-1,k-1)$. Hence $\#(X_1) = \#(Perm(n-1,k-1)) = p(n-1,k-1)$.

Now consider a permutation in X_2, namely, a permutation of $\{1,2,3,...,n\}$ made up of k cycles, and in which n is not in a cycle by itself. If we remove n from the cycle in which it occurs, then we are left with a permutation of $\{1,2,3,...,n-1\}$ that is still made up of k cycles, that is, a permutation in $Perm(n-1,k)$. We note that in this process different permutations in X_2 give rise to the same permutation in $Perm(n-1,k)$. For example, when 6 is deleted from each of the permutations

(1 6 2)(3)(4 5), (1 2 6)(3)(4 5), (1 2)(3 6)(4 5), (1 2)(3)(4 5 6), (1 2)(3)(4 6 5),

we are left with the same permutation, namely, (1 2)(3)(4 5) from $Perm(5,3)$.

Going the reverse direction, given a permutation $(a_1 a_2 ... a_r)(a_{r+1} ... a_s)...(a_{y+1} ... a_{n-1})$ in $Perm(n-1,k)$, we can obtain $n-1$ different permutations in X_2 by inserting n before each a_j, for $1 \leq j \leq n-1$ in this permutation. Hence $\#(X_2) = (n-1)\#(Perm(n-1,k)) = (n-1)p(n-1,k)$.

Now $Perm(n,k) = X_1 \cup X_2$, and the sets X_1, X_2 are disjoint. It follows that $P(n,k) = \#(Perm(n,k)) = \#(X_1) + \#(X_2) = p(n-1,k-1) + (n-1)p(n-1,k))$.

THEOREM 5.4

For all integers k, n with $1 \leq k \leq n$, $p(n,k) = s(n,k)$.

Proof

We have seen that for all $n \geq 1$, $p(n,1) = (n-1)! = s(n,1)$. (See Exercise 5.1.3A and Problem 5.5.) Also, by Theorems 5.2 and 5.3, the numbers $s(n,k)$, $p(n,k)$ satisfy the same recurrence relation. It follows that for all integers k, n with $1 \leq k < n$, $p(n,k) = s(n,k)$. Finally, for each positive integer n, $p(n,n) = 1 = s(n,n)$. Hence for all integers k, n, $1 \leq k \leq n$, $p(n,k) = s(n,k)$.

If you compare the entries in Tables 5.1 and 5.2 you will find that we always have $S(n,k) \leq s(n,k)$. Does this state of affairs persist? Theorem 5.5 shows that the answer is in the affirmative.

THEOREM 5.5

For all integers k, n, such that $1 \leq k \leq n$, we have $S(n,k) \leq s(n,k)$.

Proof

$S(n,k)$ is the number of partitions of the integers $\{1,2,3,...,n\}$ into k disjoint, nonempty subsets. To each such partition $P = \{a_1,a_2,...,a_k\} \cup \{b_1,b_2,...,b_l\} \cup ... \cup \{z_1,z_2,...,z_t\}$, we associate the permutation P^* written in cycle notation as $(a_1 \, a_2 \, ... \, a_k)(b_1 \, b_2 \, ... \, b_l)...(z_1 \, z_2 \, ... \, z_t)$. P^* is a permutation of $\{1,2,3,...,n\}$ made up of exactly k cycles, that is, $P^* \in Perm(n,k)$. The mapping $P \to P^*$ is injective. It follows that $S(n,k) \leq \#(Perm(n,k)) = s(n,k)$.

Exercises

5.2.1A Find a general formula for $p(n,n-2)$, with $n \geq 4$.

5.2.1B Find general formulas for $p(n,n-3)$.

5.2.2A Show that, for each integer $n > 1$, $0!S(n,1) - 1!S(n,2) + ... + (-1)^{n-1}(n-1)!S(n,n) = 0$.

5.2.2B Show that, for all positive integers k, n, with $k \leq n$, $k^n = C(k,1)S(n,1)1! + C(k,2)S(n,2)2! + ... + C(k,k)S(n,k)k!$

5.2.3A Show that, for all positive integers k, n, with $k \leq n$, $\sum_{r=k}^{n} C(n,r)S(r,k) = S(n+1, k+1)$.

5.2.3B Show that, for all positive integers k, n, with $k \leq n$, $\sum_{r=k}^{n} C(r,k)S(n,r) = s(n+1, k+1)$.

5.3 CATALAN NUMBERS

Applying hindsight, we begin with a definition "out of the blue."

DEFINITION 5.3

For each integer, $n \geq 0$, we define the *nth Catalan number* C_n by

$$C_n = \frac{1}{n+1} C(2n,n).$$

These numbers are singled out because they occur in a large number of combinatorial problems. It is not immediately obvious that $C(2n,n)$ must be divisible by $n+1$. However, this follows from the alternative formula for C_n given at the end of the proof of Theorem 5.6, and also from the result of Exercise 5.3.1A.

The Catalan numbers first came to prominence in the work of Euler in the late 1750s when he determined the number of ways that the area of a convex polygon could be split into triangles when joining vertices by "diagonals" no two of which intersect, except at a vertex. We discuss this problem below (see Theorem 5.9). They had previously appeared in the work of the Mongolian astronomer and mathematician Ming Antu (called Myangat in Mongolian) circa 1730. Euler's problem was fully solved, in 1758, by J. A. von Segner (1704–1777), who used a recurrence relation* of the form $T_{n+1} = T_1T_n + T_2T_{n-1} + ... + T_{n-1}T_2 + T_nT_1$ (see Theorem 5.8). Catalan first came across the numbers (subsequently named after him by Netto in 1901) in 1838 in a paper asking, "In how many ways can one evaluate the product of n different factors?" (See the discussion after Problem 5.8.) He solved the problem by use of the above recurrence relation.† (We shall solve it by a different method here.) As happens from time to time in mathematics, the name of the first investigator is not always that which is attached to a concept!

As we have already said, there are very many ways in which the Catalan numbers arise.‡ We shall start with a curious problem involving sequences of 1s and –1s since the proof we offer is especially pretty and rather clever, and several other problems involving Catalan numbers can be easily reduced to it.

THEOREM 5.6

The number of sequences $a_1, a_2, ..., a_{2n}$ that can be formed using n 1s and n –1s for which each *partial sum* $s_k = a_1 + a_2 + ... + a_k$ satisfies $s_k \geq 0$ is the nth Catalan number, C_n.

Proof

Using one ploy (but not yet the clever one!) we shall head toward the desired result by first counting the total number of sequences in the set, A, comprising n 1s and n–1s and then subtract the number of sequences in the subset, say F, of those that are "forbidden" by the conditions of the theorem.

The number of sequences of n 1s and n –1s is the number of ways of selecting the n positions in a sequence of $2n$ 1s in which to insert a minus sign. Thus $\#(A) = C(2n,n)$. So, now, how do we calculate $\#(F)$?

Let $S = a_1, a_2, ..., a_{2n}$ be a sequence in F. By the definition of F, at least one of the partial sums of this sequence is negative. Let r be the least integer such that $s_r = a_1 + a_2 + ... + a_r < 0$. It must then be that $a_r = -1$, and, if $r > 1$, $s_{r-1} = 0$ and so $r-1$ must be even. Suppose $r-1 = 2k$.

---

* See Chapter 7, Section 7.3, for a discussion of what is meant by a "recurrence relation."
† Eugene Catalan was born in Bruges on March 30, 1814. At this time Bruges, which is now in Belgium, was part of the French empire and Catalan considered himself to be French. Like several mathematicians, he initially intended to follow his father's profession and become an architect. But he soon found that his forte was mathematics. As in the famous case of Evariste Galois, Catalan's political convictions interfered greatly with his career both as a student and as a teacher, but, aged 51, he secured a professorship at the University of Liège, retiring 19 years later. He published much on the theory of numbers, including a famous conjecture that the numbers 8 and 9 provide the only example of a cube being 1 more than a square, a conjecture that was proved correct by Mihailescu only in 2003. Catalan died in Liège on February 14, 1894.
‡ R. P. Stanley has collected over 150 problems whose solution yields the Catalan numbers.

The sequence $a_1, a_2, \ldots, a_{r-1}$ includes equal numbers of 1s and −1s. Now comes the clever idea, a "reflection" method due, in the form we present it, to D. Miriamoff (1923).*

We produce, from S, a new sequence, $S' = b_1, b_2, \ldots, b_{2n}$, by changing the sign of each of the first r terms of S. We then count the number of 1s and −1s in this new sequence. In the subsequence $a_1, a_2, \ldots, a_r$ there are k 1s and $k + 1$ −1s, and hence in the subsequence $a_{r+1}, \ldots, a_{2n}$ there are $n-k$ 1s and $n-(k+1)$ −1s. Thus in the new sequence S' there are $(k+1) + (n-k) = n + 1$ 1s and $n-1$ −1s.

Conversely, suppose that we have a sequence $S' = b_1, b_2, \ldots, b_{2n}$ made up of $n + 1$ 1s and $n-1$ −1s. Then there is a least integer, $t \leq 2n$, such $b_1 + b_2 + \ldots + b_t = 1$. We let S'' be the sequence obtained from S' by changing the signs of the first t terms. Then S'' is a sequence in F. Furthermore, it is easy to see that for each sequence S in F, $S'' = S$. Thus the mapping $S \to S'$ is a bijection from F to the set G of sequences made up of $n + 1$ 1s and $n-1$ −1s. Thus $\#(F) = \#(G) = C(2n, n + 1) = C(2n, n-1)$.

We can now deduce that $\#(A) - \#(F) = C(2n, n) - C(2n, n + 1)$. Finally, we have that

$$C(2n,n) - C(2n,n+1) = \frac{(2n)!}{n!n!} - \frac{(2n)!}{(n+1)!(n-1)!} = \frac{(2n)!}{n!(n+1)!}((n+1)-n) = \frac{(2n)!}{n!(n+1)!}$$

$$= \frac{1}{n+1}\left(\frac{(2n)!}{n!n!}\right) = \frac{1}{n+1} C(2n,n) = C_n.$$

Because of this theorem we will call a sequence of 1s and −1s that meets the conditions of Theorem 5.6 a *Catalan sequence* of 1s and −1s. Note also that we have shown in the last line of the above proof that $C_n = C(2n, n) - C(2n, n + 1)$. It follows that C_n is an integer.

We can picture Catalan sequences in the way shown in the next problem, which is Problem 5B of Chapter 1.

PROBLEM 5.8

Suppose we have an $n \times n$ grid. How many paths are there, following edges of the grid, from the bottom left corner to the top right corner that may touch, but not go above, the diagonal shown in Figure 5.1? (Here by "path" we mean any route from the bottom left-hand corner to the top right corner that always moves along the lines of the grid either upward or to the right.)

Solution

We can describe each path in a grid by a sequence of 1s and −1s, with 1 corresponding to a horizontal segment and −1 to a vertical segment. For example, the path shown above in a 10×10 grid is described by the sequence 1,1,1,−1,−1,1,1,1,−1,−1,−1,−1,1,−1,1,1,−1,1,−1,−1 of 10 1s and 10 −1s. More generally, every path in an $n \times n$ grid may be represented by a sequence of n 1s and n −1s. The condition that the path does not go above the diagonal is exactly the condition that all the partial sums of the corresponding sequence

---

* The reflection principle is often attributed to Désiré André (1887). A paper *Lost (and Found) in Translation: André's Actual Method and Its Application to the Generalized Ballot Problem*, Marc Renault, American Mathematical Monthly, 115, 2008, pp. 358–363, explains why this is incorrect and how the misattribution came about.

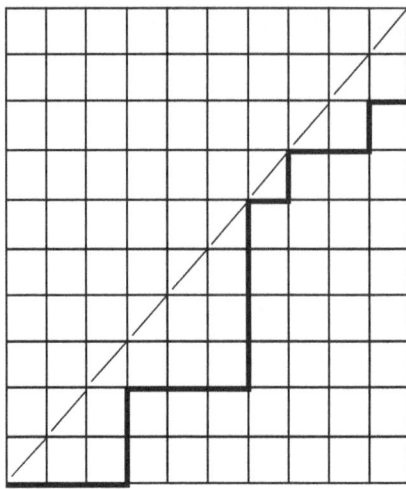

FIGURE 5.1

are nonnegative, that is, the condition that it is a Catalan sequence. Thus, by Theorem 5.6, the number of paths with the given property is the nth Catalan number C_n.

We now come to the problem that engaged Catalan. Suppose that we have three square matrices,* A,B,C, all of the same size. If we wish to multiply them together in this order, we could either first evaluate AB and then postmultiply the resulting matrix by C, or first evaluate BC and then premultiply the resulting matrix by A. That is, we could either calculate the product as $((AB)C)$ or as $(A(BC))$. If you are familiar with matrix multiplication, you will know that $((AB)C) = (A(BC))$. With four matrices A,B,C,D, there are five ways of multiplying them together in this order, given by $(((AB)C)D)$, $((AB)(CD))$, $((A(BC))D)$, $(A((BC)D))$, and $(A(B(CD)))$. This raises the question of how many different ways a product of n matrices may be evaluated.

We have seen that a given way of evaluating the product of matrices in a given order can be described by expressions of the kind used above. To tackle this matrix problem it is helpful to give a careful definition of these *expressions*. Although we have used A, B, C, and D above, to indicate that the matrices we are multiplying might be different, this is not relevant when it comes to counting the number of ways the product of n matrices may be evaluated. So in counting the number of different expressions, it makes no difference if we replace A, B, C, D, ... by a single letter that might as well be X. This leads us to the following definition.

DEFINITION 5.4

We define the set of *expressions* built up from the letter X, the left parenthesis, (, and the right parenthesis,), as follows.

a. The letter X by itself is an expression.
b. If Γ, Δ are expressions, then so also is $(\Gamma\Delta)$.
c. Nothing is an expression unless this can be deduced from (a) and (b).

---

* Do not worry if you have not met matrices before. The combinatorial problem is the same if we regard A,B,C as, for example, real numbers.

The next problem shows how Definition 5.4 can be used to prove facts about expressions.

PROBLEM 5.9

Prove that in any expression containing n letters, there are $n-1$ left-hand parentheses and $n-1$ right-hand parentheses.

Solution

We prove this by mathematical induction. An expression containing just one letter must be X by itself. So it contains one letter and zero parentheses. Hence the result is true in the case $n = 1$. Now suppose that the result is true for all expressions with fewer than n letters, and let Σ be an expression containing n letters. By Definition 5.4, Σ must be of the form $(\Gamma\Delta)$ where Γ and Δ are expressions. Suppose Γ contains c letters. Then Δ contains $n-c$ letters. As $c, n-c < n$, it follows from the induction hypothesis that Γ contains $c-1$ pairs of parentheses and Δ contains $(n-c)-1$ pairs of parentheses. Thus $(\Gamma\Delta)$ contains $(c-1) + ((n-c)-1) + 1 = n-1$ pairs of parentheses. So the result holds also for expressions containing n letters. Hence, by mathematical induction, the result is true for all expressions.

THEOREM 5.7

For each integer $n > 1$, there are C_{n-1} expressions containing n Xs.

Comment

The idea behind the proof is to find a one–one correspondence between the set of expressions and the set of Catalan sequences of 1s and –1s. The first idea that comes to mind is to associate 1 with a left-hand parenthesis, (, and –1 with a right-hand parenthesis,). However, this natural idea doesn't work, as, for example, the different expressions $((XX)X)$ and $(X(XX))$ both correspond in this way to the same Catalan sequence 1,1,–1,–1.

Another idea would be to associate 1 with a left-hand parenthesis and –1 with an X. At first glance it seems that this is no good either since, for example, the expression $((XX)(XX))$ corresponds to the sequence 1,1,–1,–1,1,1,–1,–1, which is not a Catalan sequence, as it consists of one more –1 than + 1. What can we do? In fact, we can "forget" the last X. For, in each expression, as we read from left to right, there must be at least one X after the final left-hand parenthesis, (, and hence the associated sequence of 1s and –1s must end with a –1. (Why is this true? It can be proved using mathematical induction, in a similar way to Problem 5.9 – see Exercise 5.3.4A.)

Proof

We associate with each expression, say Γ, a sequence, Γ^*, of 1s and –1s obtained by replacing each left-hand parenthesis by 1 and each letter, *except the final letter*, with a –1. Γ^* will be a Catalan sequence. (See Exercise 5.3 4A.)

We need to show that the mapping $\Gamma \to \Gamma^*$ is a bijection between the set of expressions containing n letters and the set of Catalan sequences of length $2n-2$. We tackle this by dealing with the following issue. Given a Catalan sequence, say C, can we find a unique expression Γ, such that $\Gamma^* = C$?

We describe a general process for doing this, but to help fix ideas, we explain this with reference to the following the Catalan sequence made up of seven 1s and seven −1s.

$$C = 1,1,1,-1,-1,1,1,-1,1,-1,1,-1,-1,-1 \qquad (5.5)$$

We first replace each term 1 in C by a left-hand parenthesis, and each of the −1 terms by the letter X, with an additional X at the end. In this way the Catalan sequence C becomes the sequence C^+

$$(((XX((X(X(XXXX. \qquad (5.6)$$

As C is a Catalan sequence, the sequence C^+ includes one more letter than "(." Since the sequence must start with a "(," at some point there are three successive symbols consisting of "(" followed by two letters. We consider the first place where this occurs. In C^+ these symbols are $(XX$. We let Γ_1 be this sequence followed by ")" and we replace the given three successive symbols by Γ_1. So $\Gamma_1 = (XX)$, and the new sequence is, say Σ_1, where

$$\Sigma_1 = ((\Gamma_1((X(X(XXXX. \qquad (5.7)$$

In going from Equation 5.6 through Equation 5.7 we have replaced the two letters XX by the single symbol Γ_1, and we have deleted one left-hand parenthesis, "(." So if we count Γ_1 as a letter, we still have a sequence beginning with "(" and with one more letter than "(." Therefore, continuing to use *letter* in an extended sense to include symbols of the form Γ_i, where i is an integer, there is again a first place from the left in Equation 5.7 where the symbol "(" is followed by two letters. We let Γ_2 be the sequence consisting of "(" and the two letters followed by ")" and substitute Γ_2 for them in Equation 5.7. In this way we obtain:

$$\Gamma_2 = (XX) \quad \text{and} \quad \Sigma_2 = ((\Gamma_1((X(X\Gamma_2XX.$$

We can now continue in this way, as follows:

$$\Gamma_3 = (X\Gamma_2) \quad \text{and} \quad \Sigma_3 = ((\Gamma_1((X\Gamma_3XX,$$

$$\Gamma_4 = (X\Gamma_3) \quad \text{and} \quad \Sigma_4 = ((\Gamma_1(\Gamma_4XX,$$

$$\Gamma_5 = (\Gamma_4X) \quad \text{and} \quad \Sigma_5 = ((\Gamma_1\Gamma_5X,$$

$$\Gamma_6 = (\Gamma_1\Gamma_5) \quad \text{and} \quad \Sigma_6 = (\Gamma_6X, \text{and}$$

$$\Gamma_7 = (\Gamma_6X) \quad \text{and} \quad \Sigma_7 = \Gamma_7.$$

Since there are seven 1s in the Catalan sequence C, this process ends after seven steps, and substituting backwards we see that

$$\Sigma_7 = (((XX)((X(X(XX)))X))X).$$

It should be clear that this process converts a Catalan sequence, *C*, into a uniquely determined expression, say C^+. Furthermore, if we start with an expression Γ, then $Γ^{+^*} = Γ$. We leave it to you to check this.

Thus there is a one–one correspondence between the set of expressions containing *n* Xs and the set of Catalan sequences made up of *n* 1s and *n* –1s. Since there are C_{n-1} such sequences, Theorem 5.7 is proved. It is convenient to specify that $C_0 = 1$, so that the result of Theorem 5.7 holds also for the case where *n* = 1.

We can use Theorem 5.7 to prove the following relationship between the Catalan numbers.

THEOREM 5.8

For each integer $n \geq 1$, $C_{n+1} = \sum_{k=0}^{n} C_k C_{n-k}$.

Proof

By Theorem 5.7 there are C_{n+1} expressions involving *n* + 2 letters. Suppose that Σ is one such expression. It follows from Exercise 5.3.4B that there are unique expressions $Σ_1, Σ_2$ such that Σ is the expression $(Σ_1 Σ_2)$. Then $Σ_1$ contains *k* + 1 Xs for some integer *k*, where $0 \leq k \leq n$. Therefore $Σ_2$ contains $(n + 2) - (k + 1) = (n - k) + 1$ Xs. Thus, for $0 \leq k \leq n$ there are C_k choices for $Σ_1$ and there are C_{n-k} choices for $Σ_2$, and hence $C_k C_{n-k}$ choices for the pair $Σ_1, Σ_2$. It follows that, by taking together all the cases as *k* runs from 0 to *n*, $C_{n+1} = \sum_{k=0}^{n} C_k C_{n-k}$.

In Chapter 7, Section 7.8, we show how the recurrence relation of Theorem 5.8 may be used to find the formula for C_n as given in Definition 5.3.

We now look at Problem 5C, first considered by Euler, to which the answer is given by the Catalan numbers. By a *triangulation* of a polygon we mean a way of dividing the polygon into triangles by nonintersecting diagonals, that is, lines joining two vertices. For example, Figure 5.2 shows two different triangulations of a square.

Note that we count the two triangulations of the square as different even though one can be obtained from the other by a rotation or reflection of the square. In fact, a square has just two different triangulations, namely, the two shown in Figure 5.2.

For each integer $n \geq 1$, we let T_n be the number of different triangulations of a polygon with *n* + 2 edges. Note that *n*–1 diagonals are needed to triangulate a polygon with *n* + 2 edges. Thus $T_1 = 1$ and $T_2 = 2$. It turns out to be a useful convention to put $T_0 = 1$, as this simplifies some of the formulas.

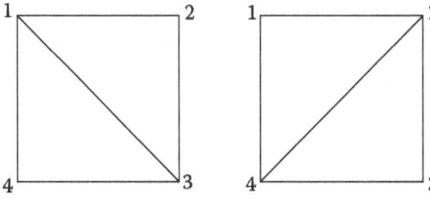

FIGURE 5.2

PROBLEM 5.10

How many different triangulations does a pentagon have?

Solution

There are five different triangulations, as shown in Figure 5.3, and thus $T_3 = 5$.

You may have noticed that for $1 \leq n \leq 3$, $T_n = C_n$. The next result shows that this is not a coincidence.

THEOREM 5.9

For each integer $n \geq 1$, $T_n = C_n$.

Proof

Suppose $n \geq 1$, and consider the triangulations of the polygon with $n + 3$ vertices labeled $1, 2, \ldots, n + 3$. We denote the edge joining the vertices labeled i and j by (i, j).

In each triangulation the edge $(n + 2, n + 3)$ will be an edge of just one triangle (see Figure 5.4). We classify the triangulations according to the third vertex, say $k + 1$, of this triangle, where $0 \leq k \leq n$. The complete triangulation will include a triangulation of the polygon, say P, with the $k + 2$ vertices $1, 2, \ldots, k + 1, n + 3$, and a triangulation of the polygon, say Q, with the $n-k + 2$ vertices $k + 1, \ldots, n, n + 1, n + 2$. These triangulations of P and Q can be chosen in T_k and T_{n-k} ways. Thus there are $T_k T_{n-k}$ triangulations of which the vertices $n + 2$, $n + 3$, and $k + 1$ form a triangle. Hence the total number of triangulations of the given polygon is $\sum_{k=0}^{n} T_k T_{n-k}$. That is, $T_{n+1} = \sum_{k=0}^{n} T_k T_{n-k}$. Thus, we see from Theorem 5.8 that the numbers, T_n, satisfy the same relationship as do the Catalan numbers. Hence, as $T_0 = C_0$, we deduce that for all integers $n \geq 0$, $T_n = C_n$.

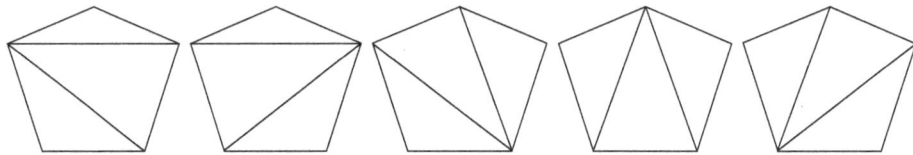

FIGURE 5.3

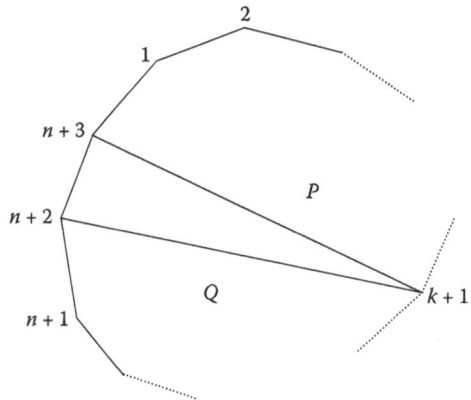

FIGURE 5.4

Exercises

5.3.1A i. Prove that for each positive integer n, $C_n = C(2n-1,n-1) - C(2n-1,n+1)$.
ii. Deduce that each Catalan number is an integer.

5.3.1B Calculate the value of C_n, for $1 \le n \le 8$.

5.3.2A Prove that the Catalan number C_n is odd if and only if n is one less than a power of 2, that is, for some positive integer k, $n = 2^k - 1$.

5.3.2B We have that $C_2 = 2$ and $C_3 = 5$. Prove that these are the only Catalan numbers that are prime numbers.

5.3.3A Show that in the situation described in Problem 5.8, the number of paths from the bottom left corner to the top right corner that, except for the endpoints, are always below the diagonal, is C_{n-1}.

5.3.3B Show that in the situation described in Problem 5.8, but with a $p \times q$ grid, where $p \ge q$ (that is, with p columns and q rows), the number of paths from the bottom left corner to the top right corner that do not go above the diagonal sloping at 45° from the bottom left hand corner

$$\frac{p-q+1}{p+1} C(p+q,p).$$

5.3.4A i. Prove that in any expression (see Definition 5.4) there must be at least one X after the final left-hand parenthesis (counting from left to right).
ii. Prove that, if we associate with each expression, Γ, a sequence, Γ^*, of 1s and -1s obtained by replacing each (by 1 and each X, *except the final X*, with a -1, then Γ^* will be a Catalan sequence.

5.3.4B Suppose that the brackets in an expression are assigned numbers according to the following rule. The first left-hand bracket is assigned the number 1. Each subsequent left-hand bracket is assigned the number that is 1 more than the number of the previous bracket. Each subsequent right-hand bracket is assigned the number that is one less than the number of the previous bracket. For example the numbers assigned to the brackets in the expression $(((XX)((X(X(XX)))X))X)$ are as shown:

```
( ( ( X X ) ( ( X ( X ( X X ) ) ) X ) ) X )
1 2 3       2 3 4   5   6     5 4 3   2 1   0
```

i. Assign numbers according to the above rule to the brackets in the expression $(((XX)X)(X(XX)))$.

ii. Prove that for every expression the numbers assigned to the brackets according to the above rule satisfy the following property: All the numbers are positive, except the number associated with the final bracket, which is zero.

iii. Prove that, for every expression, there are at most two right-hand brackets that are assigned the number 1.

iv. Deduce that, given any expression, Σ, containing more than one letter, there are uniquely determined expressions Σ_1, Σ_2 such that Σ is the expression $(\Sigma_1 \Sigma_2)$.

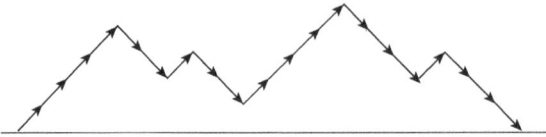

FIGURE 5.5

5.3.5A Prove that there are C_n sequences of n nonnegative integers $a_1, a_2, \ldots, a_n$ such that $a_1 + a_2 + \ldots + a_n = n$ and for all integers k, with $1 \le k < n$, $a_1 + a_2 + \ldots + a_k \le k$.

5.3.5B Consider those sequences, $a_1, a_2, \ldots, a_n$ of n integers such that $1 \le a_1 \le a_2 \le \ldots \le a_n$, and for $1 \le k \le n$, $a_k \le k$. Determine the number of such sequences. (Note that for $n = 3$ we have five sequences 1,1,1; 1,1,2; 1,1,3; 1,2,2; and 1,2,3.)

5.3.6A Look at the picture in Figure 5.5 of a "mountain path," where each step is either up or down by the same distance, and the path ends at the same level as the starting position.

Show that there are C_n such "mountain paths" made up of n up-sloping arrows and n down-sloping arrows.

5.3.6B Consider the number of sequences $a_1, a_2, \ldots, a_{2n+1}$ of $2n + 1$ nonnegative integers such that $a_1 = a_{2n+1} = 0$ and for all integers k, with $1 \le k \le 2n$, we have $|a_k - a_{k+1}| = 1$. Show that there are C_n such sequences.

5.3.7A The sequence HHTHHTTTHT represents a sequence of 10 tosses of a coin, using "H" for head and "T" for tail, in which there are equal numbers of heads and tails, and at each stage, the number of heads is at least as large as the number of tails. How many sequences of 10 tosses are there with this property?

5.3.7B Suppose that we toss a fair coin. What is the probability that we obtain an equal number of heads and tails *for the first time* after $2n$ tosses?

5.3.8A In an election between two candidates, A and B, the votes are counted one by one. Show that if both candidates end up with a votes, the probability that at no stage of the count B is ahead of A is $1/(a + 1)$.

5.3.8B In an election between two candidates, A and B, the votes are counted one by one. Show that if A ends up with a votes, and B with b votes, where $a \ge b$, the probability that at no stage of the count B is ahead of A is $(a-b + 1)/(a + 1)$.

CHAPTER 6

Partitions and Dot Diagrams

6.1 PARTITIONS

In Section 3.3 of Chapter 3 we saw how the problem of placing n identical balls in k identical boxes could be looked at in a different way by thinking of it as being about expressing n as the sum of at most k positive integers, when the order of the terms does not matter. For example, we wrote 6 as the sum of at most three positive integers in the following seven ways:

$$6, 5+1, 4+2, 4+1+1, 3+3, 3+2+1, 2+2+2,$$

and this corresponds to the fact that six identical balls may be placed in three identical boxes, with some of these boxes possibly remaining empty, in seven ways. Note that we count 6, *by itself*, as one way of writing 6 as the sum of positive integers.

We have introduced the notation $p_k(n)$ for the number of ways expressing n as the sum of at most k positive integers. Using this notation, we have that $p_3(6) = 7$. We call the above representations of the number 6 *partitions* of n into at most three *parts*. The functions p_k are not too difficult to deal with. However, much more attention has been paid to the function p, where $p(n)$ is the total number of ways of expressing n as the sum of *any* number of positive integers. For example, $p(6) = 11$, since if we drop the restriction of having at most three parts, we obtain the following four other partitions of 6,

$$3+1+1+1, 2+2+1+1, 2+1+1+1+1, 1+1+1+1+1+1,$$

making 11 partitions altogether. The problem of finding a succinct formula for $p(n)$ is very hard, but in this chapter we shall be able to use some nice theorems about the p_k functions to get some kind of grip on the values of $p(n)$. We now record the definitions of the functions p_k and p formally.

DEFINITION 6.1

For each positive integer n, $p(n)$ is the number of ways in which n can be expressed as a sum of positive integers, with the order not counting. Each such sum is called a *partition* of n. For all positive integers k, n, $p_k(n)$ is the number of ways in which n can be expressed as the sum of at most k positive integers. By convention (and for reasons that emerge later), we specify that, for all integers $n > 0$, $p_0(n) = 0$ and, for all integers $k \geq 0$, $p_k(0) = 1$ and $p(0) = 1$.

We emphasize that in these definitions the *order* of the positive integers in the sum is not relevant. For example, $2 + 1 + 3$ and $3 + 1 + 2$ count as the same partition of 6. The numbers that occur in these partitions are called *parts*. The standard convention is to write the parts in decreasing order. So the standard way to write this partition of 6 is $3 + 2 + 1$. Partitions, as well as being interesting in their own right, have a number of applications to other problems. We hope you will read far enough to see that we make a good deal of use of them in Chapters 13 and 14.

PROBLEM 6.1
Find the value of $p(5)$.

Solution
We have the following seven partitions of 5, and hence $p(5) = 7$.

$$5, 4+1, 3+2, 3+1+1, 2+2+1, 2+1+1+1, 1+1+1+1+1.$$

PROBLEM 6.2
Find the values of $p_k(6)$ for all integers $k \geq 1$.

Solution
From the partitions of 6 given above, we see that $p_1(6) = 1$, $p_2(6) = 4$, $p_3(6) = 7$, $p_4(6) = 9$, $p_5(6) = 10$, and $p_6(6) = 11$ (and, of course, $p_k(6) = 11$ for all integers $k \geq 6$).

As well as considering partitions in which the number of parts is limited, it is also natural, interesting, and useful to consider partitions in which we limit the *size* of the individual parts.

DEFINITION 6.2
For all positive integers k and n, $q_k(n)$ is the number of partitions of n in which no part is greater than k.

PROBLEM 6.3
Find the values of $q_k(6)$ for all integers $k \geq 1$.

Solution
From the partitions of 6 given above, we see that $q_1(6) = 1$, $q_2(6) = 4$, $q_3(6) = 7$, $q_4(6) = 9$, $q_5(6) = 10$, and $q_6(6) = 11$. In any partition of 6, the largest part must be at most 6 and hence, for all integers $k \geq 6$, $q_k(6) = q_6(6) = 11$.

You may have noticed that with $n = 6$, we have that, for all integers $k \geq 1$, $p_k(n) = q_k(n)$. Is this a mere fluke? Theorem 6.1 gives the answer: "No." But not only that. The ability to swap between $p_k(n)$ and $q_k(n)$ proves to be rather useful. So we should waste no time in proving that equality.

THEOREM 6.1

For all positive integers k and n, $p_k(n) = q_k(n)$.

We can offer a very convincing pictorial "proof" of Theorem 6.1, via a specific example, if we use a clever pictorial idea ascribed to Norman Macleod Ferrers (1829–1903).[*] Not only can the pictures help us see *immediately* why Theorem 6.1 is true, but this particular (re)interpretation also suggests methods of finding other interesting relationships between certain restricted partitions of any given integer n that are not immediately apparent from the basic definition (Definition 6.1). These pictorial representations are variously called Ferrers diagrams (or shapes or graphs). We shall call them *dot diagrams*.

6.2 DOT DIAGRAMS

Since "a picture is worth a thousand words"[†] we begin with two pictures.

The dot diagram of the partition $18 = 7 + 4 + 3 + 3 + 1$ is that drawn in Figure 6.1, comprising five rows of dots containing seven, four, three, three, and one dot, respectively. When we interchange the rows and columns, we obtain the so-called *dual dot diagram* (Figure 6.2), from which we obtain the *dual partition* $18 = 5 + 4 + 4 + 2 + 1 + 1 + 1$.

FIGURE 6.1

FIGURE 6.2

---

[*] Norman Macleod Ferrers was educated at Eton and at Gonville and Caius College, Cambridge. He was senior wrangler in 1851 and became a fellow of his college in 1852 and its master in 1880. He was the author of *Trilinear Coordinates* and *Spherical Harmonics*. After his death he was described as follows: "As a lecturer he was extremely successful. He was probably the best lecturer, in his subject, in the university of his days; besides great natural powers in mathematics, he possessed an unusual capacity for vivid exposition." It is interesting that this comment, quoted from the *Dictionary of National Biography*, Oxford University Press, Oxford, 2004, Vol. 14, pp. 430–431, was written by another mathematician who is remembered for a method of representing a mathematical idea by a diagram, namely, John Venn.

[†] We quote from *Brewer's Quotations* by Nigel Rees, Cassell, London, 1994, p.23: "This famous saying which occurs, for example, in the song 'If' popularized by Bread in 1971, is sometimes said to be a Chinese proverb. *Bartlett's Familiar Quotations* (15th ed, 1980) listed it as such in the form 'One picture is worth more than ten thousand words' and compared what Turgenev says in *Father and Sons*: 'A picture shows me at a glance what it takes dozens of pages of a book to expound.' But the *Concise Oxford Dictionary of Proverbs*, second edition, edited by John Simpson, Oxford University Press, Oxford, 1992, p. 201. points out that it originated in an American paper *Printer's Ink* (8 December 1921) in the form 'One look is worth a thousand words.' It was later reprinted in the same paper (10 March 1927) and there ascribed by its actual author, Frederick R. Barnard, to a Chinese source ('so that people would take it seriously', he told Burton Stevenson in 1948)."

Given any dot diagram D, we use D^* for its *dual dot diagram* formed from D by writing the rows (columns) of dots in D as the corresponding columns (rows) of D^*. Similarly, for each partition π of n, we let π^* be its dual. In transposing the rows and columns, the number of dots does not change, so π^* is also a partition of n. Usually, but not always, π^* will be different from π. Note also that, for each partition π, the partition π^{**}, that is, the dual of its dual, is identical with π.

PROOF OF THEOREM 6.1

Let S be the set of all partitions of n into at most k parts, and let T be the set of all partitions of n into parts with size at most k. Now suppose π is a partition in S. Since π is a partition of n into at most k parts, there are at most k rows in its dot diagram, D. Hence in the dual dot diagram, D^*, each row contains at most k dots. Therefore the dual partition π^* is a partition of n into parts of size at most k. Thus $\pi^* \in T$. Clearly, the converse also holds, so $\pi \in S$ if and only if $\pi^* \in T$. Thus $\pi \mapsto \pi^*$ is a mapping from S to T. This mapping is surjective (onto), since for $\pi \in T$, we have $\pi^* \in S$ and $\pi^{**} = \pi$. Also, it is injective (one–one) as $\pi^* = \rho^*$ implies $\pi^{**} = \rho^{**}$, that is, that $\pi = \rho$. Therefore, the mapping $\pi \mapsto \pi^*$ is a bijection between S and T. Hence $\#(S) = \#(T)$, that is, $p_k(n) = q_k(n)$.

The next result, making typical use of the dot diagram method, plays a role in eventually suggesting a rough lower bound for the size of $p_k(n)$. Note that conventions that $p_0(n) = 0$ and $p_k(0) = 1$ mean that we do not need to exclude the cases where $k = 1$ and $k = n$.

THEOREM 6.2

If k, n are positive integers with $k \leq 1$, then $p_k(n) = p_{k-1}(n) + p_k(n-k)$.

The proof can be followed without the need of a diagram. Nevertheless, there is no harm in offering a helpful picture, so we have included Figure 6.3.

Proof

We divide the set, say S, of those partitions of n that have at most k parts into two subsets. We let S_1 be the set of partitions of n into fewer than k parts, and we let S_2 be the set of the partitions of n into exactly k parts. Clearly, each partition in S is in either S_1 or S_2 but not both. Therefore, $\#(S) = \#(S_1) + \#(S_2)$. Now, S_1 is the set of partitions of n into at most $k-1$ parts. Hence $\#(S_1) = p_{k-1}(n)$.

Each partition in S_2 has a dot diagram with at least one dot in each of the k rows. So if we remove one dot from each row, we obtain the dot diagram corresponding to a partition of $n-k$ into at most k parts. Conversely, suppose we are given a dot diagram for a partition of $n-k$ into at most k parts, then, by placing a column of k dots on the left, we obtain a partition of n into exactly k parts. This is illustrated in Figure 6.3. It follows that $\#(S_2) = p_k \cdot (n-k)$.

We therefore have that $p_k(n) = \#(S) = \#(S_1) + \#(S_2) = p_{k-1}(n) + p_k(n-k)$.

We can extend Theorem 6.2 in the following way.

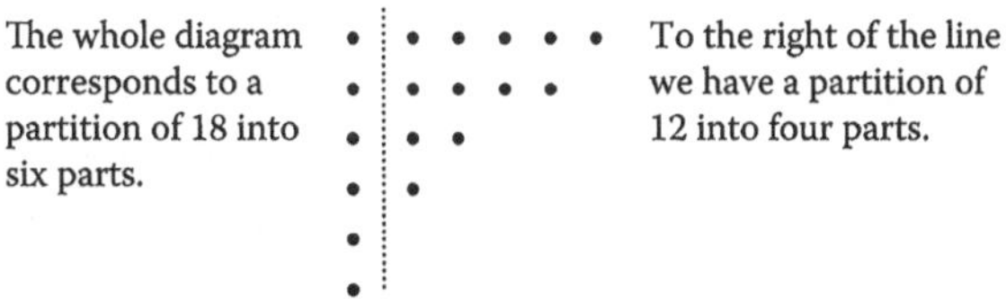

The whole diagram corresponds to a partition of 18 into six parts.

To the right of the line we have a partition of 12 into four parts.

FIGURE 6.3

THEOREM 6.3

If k, n are integers with $1 \leq k \leq n$, then

$$p_k(n) = p_{k-1}(n) + p_{k-1}(n-k) + p_{k-1}(n-2k) + \ldots + p_{k-1}(n-[s-1]k) + p_k(n-sk), \quad (6.1)$$

where $s = \lfloor n/k \rfloor$ (this notation means that s is the integer part of n/k). [Note that, as $n-sk < k$, we have that $p_k(n - sk) = p_{k-1}(n - sk)$, and we could have replaced $p_k(n - sk)$ by $p_{k-1}(n - sk)$ in this theorem.]

Proof

A very formal proof might proceed by induction. A somewhat more informal (and understandable) one can go as follows.

We use Theorem 6.2 repeatedly. First, we obtain

$$p_k(n) = p_{k-1}(n) + p_k(n-k). \quad (6.2)$$

Then, replacing n by $n-k$ in Theorem 6.2,

$$p_k(n-k) = p_{k-1}(n-k) + p_k(n-2k). \quad (6.3)$$

Next, replacing n by $n-2k$ in Theorem 6.2,

$$p_k(n-2k) = p_{k-1}(n-2k) + p_k(n-3k), \quad (6.4)$$

and so on.

Then, substituting from Equation 6.4 into Equation 6.3 and then from Equation 6.3 into Equation 6.2, we deduce that

$$p_k(n) = p_{k-1}(n) + p_{k-1}(n-k) + p_{k-1}(n-2k) + p_k(n-3k).$$

In this way we obtain $p_k(n)$ as the sum of $p_{k-1}(n), p_{k-1}(n-k), p_{k-1}(n-2k), p_k(n-3k)$, ... Where do we stop? We stop when we arrive at $p_k(n - sk)$, where s is an integer such that $0 \leq n - sk < k$ so that we cannot pass on to $p_k(n - [s+1]k)$. Now, $n - sk < k$ when $(n/k) - s < 1$. And the largest s for which this holds is given by $s = \lfloor n/k \rfloor$. It follows that Equation 6.1 is correct.

Comment

There is a small point we have skated over. We mention it to satisfy those who have spotted it and to remind those who haven't of the care that is often taken in mathematics, which adds to its reputation as an excellent subject to study if you are keen to develop your logical thinking powers.

The "small" point is what happens if k is a divisor of n and hence $n - ks = 0$. In this case $p_k(0)$ occurs in the formula of Theorem 6.3. We have said that "by convention" $p_k(0) = 1$, but does this value for $p_k(0)$ make sense? We could argue that you can draw a dot diagram of the integer 0 using at most k rows *in just one way*, namely, *by doing nothing!* This gives $p_k(0) = 1$. This agrees nicely with Theorem 6.2, as by this theorem, $p_k(k) - p_{k-1}(k) = p_k(0)$. There is just one partition of k into at most k parts that is not a partition of k into at most $k - 1$ parts (namely, $1 + 1 + 1 + \ldots + 1$ with k parts), and this gives $p_k(k) - p_{k-1}(k) = 1$. So our convention that, for all integers $k \geq 0$, $p_k(0) = 1$ makes good sense.

Theorem 6.2 can be used to calculate the value of $p_k(n)$ for small values of n provided we have some "starting" values, just as we were able to calculate values of $S(n,k)$ in Chapter 3. The following result gives us some starting values.

THEOREM 6.4
The Values of $p_k(n)$, for $k = 1, 2, 3$

For all positive integers n,

i. $p_1(n) = 1$;   ii. $p_2(n) = \lfloor \tfrac{1}{2}n + 1 \rfloor = \lfloor \tfrac{1}{2}n \rfloor + 1$;   iii. $p_3(n) = \lfloor \tfrac{1}{12}n^2 + \tfrac{1}{2}n + 1 \rfloor$.

Proof

i. This is obvious, as there is just one partition of n into at most one part, namely, n by itself.
ii. The partitions of n into at most two parts are shown in Figure 6.4. We need to consider separately the case where n is even, say $n = 2q$ for some integer q, and the case where n is odd, say $n = 2r + 1$, where r is an integer.

    We see that when $n = 2q$, $p_2(n) = q + 1 = \tfrac{1}{2}n + 1$, and when $n = 2r + 1$, $p_2(n) = r + 1 = \tfrac{1}{2}n + \tfrac{1}{2}$. As these values are integers, the formula $p_2(n) = \lfloor \tfrac{1}{2}n + 1 \rfloor$ holds in both cases.
iii. Careful analysis of the cases $n = 1,2,3,4,5,6,7,8,\ldots$ leads to the suggestion* that

$$p_3(n) = \tfrac{1}{12}(n^2 + 6n + t), \text{ where } t = 5, 8, 9, 8, 5, 12,$$

according to whether $n \equiv 1, 2, 3, 4, 5, 0 \pmod 6$. (6.5)

$2q$	$(2r+1)$
$(2q-1)+1$	$(2r)+1$
$(2q-2)+2$	$(2r-1)+2$
$\vdots$	$\vdots$
$q+q$	$(r+1)+r$
n even, $n = 2q$	n odd, $n = 2r+1$

FIGURE 6.4

---

* You should not regard this as being wise after the event. Not infrequently in mathematics a theorem emerges only after sufficient (numerical) evidence has accumulated to make an attempt to find a proof worthwhile.

We prove Equation 6.5 by mathematical induction.

Base. We will suppose that you can confirm that Equation 6.5 holds by direct calculation, for $n = 1,2,3,4,5,6$ (see Exercise 6.2.1A).

Induction step. We assume that $n \geq 6$ and that Equation 6.5 is true for all positive integers $k < n$. Using Theorem 6.2 twice, we have that $p_3(n) = p_2(n) + p_3(n-3) = p_2(n) + p_2(n-3) + p_3(n-6)$. Hence, by (ii), $p_3(n) = \lfloor \frac{1}{2}n+1 \rfloor + \lfloor \frac{1}{2}(n-3)+1 \rfloor + p_3(n-6)$. Now, it can be shown that $\lfloor \frac{1}{2}n+1 \rfloor + \lfloor \frac{1}{2}(n-3)+1 \rfloor = n$ (we ask you check this in Exercise 6.2.6A), and hence, using the induction hypothesis,

$$p_3(n) = n + \frac{1}{12}((n-6)^2 + 6(n-6) + t) = \frac{1}{12}(n^2 + 6n + t),$$

where $t = 5,8,9,8,5,12$ according to whether $n-6 \equiv 1,2,3,4,5,0 \pmod{6}$. Since $n \equiv n-6 \pmod{6}$, it follows that the result is also true for n.

This completes the proof, by mathematical induction, that Equation 6.5 is true for all positive integers n.

We now need to show that it follows that the formula given in (iii) is correct. To do this, we first note that if $n = 6m + r$, then $\frac{1}{12}(n^2 + 6n + t) = 3m^2 + mr + 3m + \frac{1}{12}r^2 + \frac{1}{2}r + \frac{1}{12}t$, whereas $\lfloor \frac{1}{12}n^2 + \frac{1}{2}n + 1 \rfloor = \lfloor 3m^2 + mr + 3m + \frac{1}{12}r^2 + \frac{1}{2}r + 1 \rfloor = 3m^2 + mr + 3m + \lfloor \frac{1}{12}r^2 + \frac{1}{2}r + 1 \rfloor$, as $3m^2 + mr + 3m$ is an integer. So we need only check that $\frac{1}{12}r^2 + \frac{1}{2}r + \frac{1}{12}t = \lfloor \frac{1}{12}r^2 + \frac{1}{2}r + 1 \rfloor$ for the cases where $(r,t) = (1,5), (2,8), (3,9), (4,8), (5,5)$, and $(6,12)$. This you can do easily.

Theorems 6.2, 6.3, and 6.4 are useful in determining the values of $p_k(n)$ for small values of k and n. We offer some of these values in Table 6.1. Because $p(n) = p_n(n)$ the table also gives us the values of $p(n)$ for small values of n. These are, of course, the diagonal entries 1, 2,3,5,7,11,15,... in the table.

You may notice that, in each column, the numbers immediately *below* this diagonal are equal to the numbers *on* the diagonal. (Why?) Now, to find the values in any new column you can use the formula $p_k(n) = p_{k-1}(n) + p_k(n-k)$, remembering to take $p_k(0) = 1$ for each k. So, for example, $p_{10}(10) = p_9(10) + p_{10}(0) = 41 + 1 = 42$.

It was not *too* hard to find $p(n)$ for $n = 1,2,...,10$. However you can see that something better will be needed if we are to find the value of $p(n)$ for all values of n up to, say, 200—as was accomplished (using only pencil and paper) by Major Percy MacMahon*—(obviously!) without the need for listing all the partitions for each n, which, as you will see from the size of the numbers, would be an impossible task. Table 6.2 gives some of the values of $p(n)$ beyond those given in Table 6.1.

---

* Percy Alexander Macmahon (1854–1929) was born in Malta. He became a cadet at the Royal Military College in 1871 and subsequently served in the army in Malta and India, becoming a major in 1889. He taught at the Royal Military College, ultimately becoming professor of physics. Despite this title his main work was in combinatorics. He published *Combinatory Analysis*, Cambridge University Press, Cambridge, in two volumes in 1915 and 1916.

TABLE 6.1

	$p_k(n)$	1	2	3	4	5	6	7	8	9	10
	1	1	1	1	1	1	1	1	1	1	1
	2	1	2	2	3	3	4	4	5	5	6
	3	1	2	3	4	5	7	8	10	12	14
k	4	1	2	3	5	6	9	11	15	18	23
	5	1	2	3	5	7	10	13	18	23	30
	6	1	2	3	5	7	11	14	20	26	35
	7	1	2	3	5	7	11	15	21	28	38
	8	1	2	3	5	7	11	15	22	29	40
	9	1	2	3	5	7	11	15	22	30	41
	10	1	2	3	5	7	11	15	22	30	42

TABLE 6.2

n	$p(n)$
20	627
30	5,604
40	37,338
50	204,226
60	966,467
70	4,087,968
80	15,796,476
90	56,634,173
100	190,569,292
200	3,972,999,029,388

A remarkable result giving superb approximations for the values of $p(n)$ for *all* n was given in 1918 by two of the most powerful mathematicians of the first half of the twentieth century, G. H. Hardy and Srinivasa Ramanujan. We describe their work in Chapter 8.

Exercises

6.2.1A For $1 \leq n \leq 8$, list the partitions of n into at most three parts, and hence verify that the values of $p_3(n)$ given in Equation 6.5 of Theorem 6.4 and in Table 6.1 are correct.

6.2.1B Verify that for $1 \leq n \leq 8$, $q_3(n) = p_3(n)$, by listing the partitions of n into parts of size at most three.

6.2.2A Write down all the partitions of 7. Hence confirm that $p_4(7) = q_4(7)$.

6.2.2B Write down all the partitions of 8 that involve only *odd* parts and those that involve only *distinct* parts. How many are there of each?

6.2.3A Find the number of partitions of n into k parts *in which order matters*. (So, for example, there are six such partitions of 5 into three parts, namely, $3+1+1, 1+3+1, 1+1+3, 2+2+1, 2+1+2, 1+2+2$.)

6.2.3B Show that the number of partitions of n into exactly k parts is at least $(1/k!)\, C(n-1, k-1)$.

6.2.4A Let $u_k(n)$ be the number of ways that the integer n can be expressed as a sum of *distinct* positive integers *of unequal size* whose *size is at most k*. Prove that, for all positive integers k, n with $k < n$, we have $u_k(n) = u_{k-1}(n) + u_{k-1}(n-k)$.

6.2.4B Let $s_k(n)$ be the number of partitions of n into distinct parts of which the smallest is of size k. Prove that, for all positive integers k, n with $k < n$, we have $s_k(n) = s_{k+1}(n+1) + s_k(n-k)$. [*Hint:* Split the partitions of n according to whether they (i) do have or (ii) do not have just one part of size k. For partitions of type (i) form a partition of $n+1$ by adding one dot to the smallest part.]

6.2.5A Show, using Theorem 6.4(iii) that $p_3(n)$ is the integer nearest to $\frac{1}{12}(n+3)^2$.

6.2.5B Use Theorem 6.4(iii) to show that
  i. For all positive integers n, $\frac{1}{12}n^2 < p_3(n)$, and
  ii. There is an integer n_0 such that for all integers $n \geq n_0$, $p_3(n) \leq \frac{1}{11}n^2$.

6.2.6A Show that, for all integers $n \geq 3$, $\lfloor \frac{1}{2}n+1 \rfloor + \lfloor \frac{1}{2}(n-3)+1 \rfloor = n$. (*Hint:* Treat the cases where n is even and n is odd separately.)

6.2.6B Show that, for all positive integers m, $\lfloor \frac{1}{3}m \rfloor + \lfloor \frac{1}{3}(m+1) \rfloor + \lfloor \frac{1}{3}(m+2) \rfloor = m$.

6.3 A BIT OF SPECULATION

It follows from Theorem 6.4(ii) and (iii) that $p_2(n) \geq n/2 = n/(1!2!)$ and $p_3(n) \geq n^2/12 = n^2/(2!3!)$. This is scant basis for making a conjecture, but we shall be bold!

Conjecture

For all positive integers n, k with $k \leq n$, we have

$$p_k(n) \geq \frac{n^{k-1}}{(k-1)!k!}.$$

Comment

It is a consequence of the logical order of presentation of mathematical arguments that the lemma must be presented before it can be used in the following theorem. However, let no one persuade you that some clever person thought up the lemma and then, hey presto!, found a theorem needing precisely that lemma. No! The main aim is to prove the conjecture, and the proof needs a bit of technical argument in the middle, which we have extracted as a lemma so that the proof of the theorem doesn't get clogged up with technicalities.

Lemma 6.5

Let t, r, and n be real numbers such that $t \geq 2$, $r \geq 0$, and $(r+1)t \geq n$. Then,

$$(n-rt)^{t-2} > \left[\frac{(n-rt)^{t-1} - (n-(r+1)t)^{t-1}}{t(t-1)} \right]. \tag{6.6}$$

Proof

Suppose $0 < a < b$, and k is an integer with $k \geq 1$, then

$$b^k - a^k = (b-a)(b^{k-1} + ab^{k-2} + \ldots + a^{k-2}b + a^{k-1})$$
$$< (b-a)(b^{k-1} + bb^{k-2} + \ldots + b^{k-2}b + b^{k-1}) = k(b-a)b^{k-1},$$

and hence, as $b - a > 0$, we can deduce that

$$kb^{k-1} > \frac{b^k - a^k}{b-a}. \tag{6.7}$$

We now put $k = t - 1$, $b = n - rt$, $a = n - (r+1)t$ in inequality 6.7. This gives

$$(t-1)(n-rt)^{t-2} > \frac{(n-rt)^{t-1} - (n-(r+1)t)^{t-1}}{t},$$

and, as $t - 1 > 0$, the inequality 6.6 follows.

We can now establish the conjecture.

Proof of the Conjecture

We prove that our conjecture holds for all integers $k \geq 1$ using mathematical induction. The base of the induction is the case $k = 1$. In this case we have that, for all $n \geq 1$, $p_k(n) = p_1(n) = 1 = n^0/(0!1!)$, and so the conjecture is true in this case.

For the induction step we assume that the result holds for $k = t - 1$. That is, we assume that

$$\text{for all } n \geq t-1, \quad p_{t-1}(n) \geq \frac{n^{t-2}}{(t-2)!(t-1)!}. \tag{6.8}$$

By Theorem 6.3, $p_t(n) = p_{t-1}(n) + p_{t-1}(n-t) + \ldots + p_{t-1}(n-st)$, where $s = \lfloor n/t \rfloor$. Hence by our induction assumption (inequality 6.8),

$$p_t(n) \geq \frac{1}{(t-1)!(t-2)!}\left[n^{t-2} + (n-t)^{t-2} + \ldots + (n-(s-1)t)^{t-2} + (n-st)^{t-2}\right].$$

Using Lemma 6.5 on the first s terms on the right-hand side above, we obtain

$$p_t(n) \geq \frac{1}{(t-2)!(t-1)!}$$
$$\times \left[\frac{n^{t-1} - (n-t)^{t-1}}{t(t-1)} + \frac{(n-t)^{t-1} - (n-2t)^{t-1}}{t(t-1)} + \ldots + \frac{(n-(s-1)t)^{t-1} - (n-st)^{t-1}}{t(t-1)} + (n-st)^{t-2}\right].$$

Most of the terms cancel, leaving just

$$p_t(n) \geq \frac{1}{(t-2)!(t-1)!}\left[\frac{n^{t-1}}{t(t-1)} - \frac{(n-st)^{t-1}}{t(t-1)} + (n-st)^{t-2}\right]$$

$$= \frac{1}{(t-2)!(t-1)!}\left[\frac{n^{t-1}}{t(t-1)} + (n-st)^{t-2}\left(1 - \frac{(n-st)}{t(t-1)}\right)\right]. \tag{6.9}$$

Since $s = \lfloor n/t \rfloor$, $n - st \geq 0$, and hence $(n-st)^{t-2} \geq 0$. Also, $n - st < t$, and $t - 1 \geq 1$, and hence $(n-st)/[t(t-1)] < 1$, so that $1 - (n-st)/t(t-1) > 0$. Consequently, it follows from inequality 6.9 that

$$p_t(n) \geq \frac{1}{(t-2)!(t-1)!}\left[\frac{n^{t-1}}{t(t-1)}\right] = \frac{n^{t-1}}{(t-1)!t!}$$

and therefore the conjecture holds also for $k = t$.

This completes the proof by mathematical induction that the conjecture holds for all integers $k \geq 1$.

Since we have proved the conjecture, we can now dignify it by calling it a theorem, which we restate for the record.

THEOREM 6.6

For all positive integers n, k with $k \leq n$, we have

$$p_k(n) \geq \frac{n^{k-1}}{(k-1)!k!}.$$

We can deduce from this that the values of the function p grow faster than any polynomial function, in the sense given by our next result.

THEOREM 6.7

For all positive integers k and all positive constants A, there is a positive integer n_0 such that for all integers $n > n_0$, $p(n) > An^k$.

Proof

Suppose k is a positive integer and A is a positive constant. We let n_0 be the larger of $k + 2$ and $A(k + 1)! (k + 2)!$ [Of course, unless A were very small, $A(k + 1)! (k + 2)!$ would be much the larger of the two, but we need to cover all possible cases.] Replacing k by $k + 2$ in Theorem 6.6, we deduce that, if $n \geq k + 2$, then

$$\frac{n^{k+1}}{(k+1)!(k+2)!} \leq p_{k+2}(n) \leq p(n). \tag{6.10}$$

Hence, if $n > n_0$, $n \geq k + 2$, and so inequality 6.10 holds, and also $n > A(k + 1)!\,(k + 2)!$, and so

$$\frac{n^{k+1}}{(k+1)!(k+2)!} > A(k+1)!(k+2)!\left(\frac{n^k}{(k+1)!(k+2)!}\right) = An^k. \tag{6.11}$$

It follows from inequalities 6.10 and 6.11 that if $n > n_0$, then $p(n) > An^k$. This completes the proof.

Theorem 6.7 tells us that the values of p must grow very rapidly. For example, it tells us that there are infinitely many values of n for which

$$p(n) > n^{1,000,000},$$

and we could replace the right-hand side of this inequality by any power of n we choose. In Chapter 8, we say more about how the values of p grow.

6.4 MORE PROOFS USING DOT DIAGRAMS

We have claimed that dot diagrams are very useful for suggesting relationships between different sets of partitions that would not be apparent from the basic definition. Here, to whet your appetite, we present just four somewhat simpler ones, leaving some to the exercises.

PROBLEM 6.4

Show that, for each positive integer n, the number of partitions of n into three nonzero parts is equal to the number of partitions of $2n$ into three parts each no larger than $n - 1$.

Solution

We illustrate the general solution by looking at a particular example. Consider the dot diagram in Figure 6.5, which corresponds to the partition $6 + 3 + 2$ of 11, to which we have added squares so that there are 11 symbols in each row.

The squares make up the partition $5 + 8 + 9$ of 22 into three parts. Because there is at least one dot in each row, there are at most 10 squares in each row.

In general, we can, in this way, establish a one–one correspondence between the partitions of 11 into three nonzero parts and partitions of 22 into three nonzero parts of size at most 10. The same method defines a one–one correspondence between partitions of n into three nonzero parts and partitions of $2n$ into three nonzero parts of size at most $n - 1$. Hence there are equal numbers of partitions of the two types.

Another pretty result coming readily from dot diagrams is:

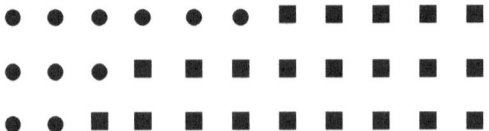

FIGURE 6.5

THEOREM 6.8

For all positive integers n, $p(2n) > p(1) + p(2) + \ldots + p(n)$.

Proof

Clearly, if $k \leq n$, none of the dot diagrams for partitions of k can have a row of dots with more than n dots in it. Also, $2n - k \geq n$, and so, if we take $2n - k$ more dots, we can form a dot diagram for $2n$ by placing a row of $2n - k$ dots above the top row of the original dot diagram for k. This procedure associates, with each dot diagram for each k, where $k \leq n$, a dot diagram for $2n$. It is not difficult to check that all dot diagrams for $2n$ so obtained are different from one another, and we therefore see that there are at least $p(1) + p(2) + \ldots + p(n)$ of them. In addition, we have a dot diagram that consists of a single row of $2n$ dots that does not arise in this way. Hence $p(2n) > p(1) + p(2) + \ldots + p(n)$.

As a variant of Theorem 6.8, you might speculate from Tables 6.1 and 6.2 that, for all positive integers n, we have (i) $p(2n) \geq 2p(n)$ and (ii) $p(n+2) \leq p(n+1) + p(n)$. You are asked to show this in Exercise 6.4.4A. Furthermore, (ii) follows immediately from Theorem 8.6. Nevertheless, a simpler proof of a lesser result is often of interest in mathematics, and here we can give such a proof using dot diagrams and a bit of ingenuity (which some might call "fiddling"). We have:

THEOREM 6.9

For each positive integer n, $p(n+2) \leq p(n+1) + p(n)$.

Proof

Let D be the dot diagram for the partition P of $n + 2$. We consider three distinct cases.

a. D contains at least one row containing just one dot. In this case we let D' be the dot diagram obtained by deleting the bottom row of D containing just one dot. Then D' corresponds to a partition of $n + 1$.
b. D contains no rows containing just one dot but at least one row containing two dots. Suppose that the next smallest row contains k dots, so that $k \geq 2$. We let D' be the dot diagram obtained from D by deleting the bottom row containing two dots and replacing the lowest row of k dots by k rows each containing one dot. Since $k \geq 2$, D' corresponds to a partition of n with more than one part of size 1.
c. The smallest row of D contains at least three dots. In this case we let D' be the partition obtained from D by deleting two dots from the bottom row of D. Then D' corresponds to a partition of n with at most one part of size 1.

Thus, in each case D' corresponds to either a partition of $n + 1$ or of n. Also, if $D_1 \neq D_2$, then $D_1' \neq D_2'$. This shows that there is a one–one correspondence between the partitions of $n + 2$ and some of the partitions of either n or $n + 1$. Hence $p(n + 2) \leq p(n) + p(n + 1)$.

Note: Exercise 6.4.5B asks you to obtain, for all positive n, the lower bound $p(n + 2) \geq p(n + 1) + p(n) - p(n-1)$. This and the intriguing "averaging" result in Exercise

6.4.5A, namely, $p(n) \leq \frac{1}{2}(p(n+1)+p(n-1))$, can be established using the following theorem.

THEOREM 6.10

The number of partitions of n which have no parts of size 1 is $p(n) - p(n-1)$.

Proof

Given a dot diagram, say D, corresponding to a partition of n that has at least one part of size 1, we let D' be the dot diagram obtained by removing the bottom row consisting of a single dot. So D' corresponds to a partition of $n-1$. Conversely, given a dot diagram corresponding to a partition of $n-1$, by adding one row of just a single dot, we obtain a partition of n with at least one part of size 1. This shows that the partitions of $n-1$ can be put in a one–one correspondence with the partitions of n that have at least one part of size 1. Thus, $p(n) - p(n-1)$ is the number of partitions of n with no parts of size 1.

As announced above, we shall return to partitions in Chapter 8 where we shall also describe a diagram-free way of discovering relationships between various partitions of integers.

Exercises

6.4.1A Prove that for all positive integers k and n with $1 < k \leq n$, $p(n) - p(n-1) \leq p_k(n) + p(n-k)$.

6.4.1B Prove that for $n \geq 2$, $p(n) - p(n-2)$ is the number of partitions of n with at most one part of size 1. How may this result be generalized?

6.4.2A Prove that the number of partitions of n into *exactly* k parts is equal to the number of partitions of n whose maximum part has size *exactly* k.

6.4.2B Prove that the number of partitions of n into *exactly* k parts is equal to the number of partitions of $n + C(k,2)$ into k *distinct* parts.

6.4.3A Show that the number of partitions of n into at most k parts is equal to the number of partitions of $n + k$ into *exactly* k parts.

6.4.3B Show that, for all positive integers n, k, the number of partitions of $2n + k$ into exactly $n + k$ parts is the same for each k, namely, $p(n)$.

6.4.4A Prove that, for every integer $n \geq 2$, $p(2n) > 2p(n)$.

6.4.4B Is $p(3n) > 3p(n)$ for all $n > 1$?

6.4.5A Use Theorem 6.10 to prove that, for every positive integer n, $p(n) \leq \frac{1}{2}(p(n+1)+p(n-1))$.

6.4.5B From numerical evidence it appears that, for all $n \geq 2$, $p(n + 2) \geq p(n + 1) + p(n) - p(n - 1)$. Use Theorem 6.10 to confirm this.

6.4.6A Prove that, for $n \geq 2$, $p(n) - p(n - 1)$ is the number of partitions of n whose two largest parts are equal.

6.4.6B Find (and prove) a similar formula for the number of partitions whose three largest parts are equal.

CHAPTER 7

Generating Functions and Recurrence Relations

7.1 FUNCTIONS AND POWER SERIES

In this chapter we describe a powerful algebraic technique that can be used to solve many combinatorial problems. We apply this technique to the solution of recurrence relations and especially to finding a succinct formula for the Fibonacci numbers (see Problem 7 of Chapter 1). In Chapter 8 this same technique is applied to investigate partitions of n.

We saw a connection between algebra and counting in Chapter 2, where we gave a combinatorial proof of the binomial theorem. Recall that we did this by considering the terms that are obtained when the product

$$(a + b)(a + b)\ldots(a + b)$$

with n terms is multiplied out in full. We can apply a similar idea in other contexts, as is shown by the following problem, which, though rather simple, will set us off in the right direction.

PROBLEM 7.1

Consider the following product:

$$(1 + x + x^2 + x^3 + \ldots)(1 + x^2 + x^4 + x^6 + \ldots). \tag{7.1}$$

What is the coefficient of x^7 in this product?

Solution

The coefficient of x^7 is 4 because when the terms in the first bracket are multiplied by those in the second bracket we obtain just four terms of degree seven, namely, $x \cdot x^6$, $x^3 \cdot x^4$, $x^5 \cdot x^2$, and $x^7 \cdot 1$, or, as we now prefer to write them,

$$x \cdot x^6, \; x^3 \cdot x^4, \; x^5 \cdot x^2, \; x^7 \cdot x^0.$$

These four terms correspond to the four different ways of writing 7 as the sum of two nonnegative integers, where the second is a multiple of 2, namely, $1+6, 3+4, 5+2$, and $7+0$. We can see, in general, that the coefficient of x^n in the product 7.1 is the number of ways of representing n as the sum of two nonnegative integers of which the second is even. Thus the coefficients give us an answer to a combinatorial problem. In fact, this problem is similar to the partition problems we have already discussed in Chapter 6. We return to this type of problem in Chapter 8. Meanwhile, the interest for us is the way we can use an infinite series to represent a sequence of numbers.

The above solution indicates that if a_n is the number of ways of writing n as the sum of two nonnegative integers of which the second is even, then

$$(1+x+x^2+x^3+\ldots)(1+x^2+x^4+x^6+\ldots) = \sum_{n=0}^{\infty} a_n x^n. \tag{7.2}$$

Equation 7.2 involves the sums of infinite series, and we have not yet said how these are to be interpreted. In fact, we can view the matter in two ways, *algebraically* and *functionally*. We could regard equations such as Equation 7.2 as dealing with formal algebraic expressions. The awkwardness of this approach arises when it comes to saying exactly what we mean by a *formal algebraic expression*. The coefficients in the series look like numbers, but what sort of animal is x?

It is tempting to say that x is a symbol, but then how can we combine a symbol with an abstract object like a number to form such things as $3x$? In earlier days writers used to talk about "an indeterminate x" without ever really saying what this means. The modern approach, which aims to explain all mathematical concepts ultimately in terms of set theory, is to regard an algebraic expression such as

$$a_0 + a_1 x + a_2 x + \ldots$$

as simply a convenient way of representing the sequence of numbers

$$(a_0, a_1, a_2, \ldots),$$

which is to be manipulated according to certain rules that arise from the algebraic background to this approach. The sequences are infinite if we are dealing with power series and finite if we are dealing with polynomials. There is, of course, no difficulty in defining what we mean by an infinite sequence in terms of sets. If X is a set, an infinite sequence of elements of X can be regarded as a function $f: N \to X$.

In this approach the rule for addition of sequences of numbers becomes

$$(a_0, a_1, a_2, \ldots) + (b_0, b_1, b_2, \ldots) = (a_0 + b_0, a_1 + b_1, a_2 + b_2, \ldots), \tag{7.3}$$

corresponding exactly to term-by-term addition of power series and polynomials. The multiplication rule for sequences is a little more complicated. If we multiply the terms in the series $a_0 + a_1 x + a_2 x^2 + \ldots$ by those in the series $b_0 + b_1 x + b_2 x^2 + \ldots$ and gather together terms of the same degree, we see that the coefficient of x^n is

$$a_0 b_n + a_1 b_{n-1} + \ldots + a_n b_0,$$

which we can write as

$$\sum_{i=0}^{n} a_i b_{n-i}.$$

Thus the multiplication rule is

$$(a_0,a_1,a_2,\ldots)\times(b_0,b_1,b_2,\ldots)=(c_0,c_1,c_2,\ldots), \quad \text{where} \quad c_n = \sum_{i=0}^{n} a_i b_{n-i}. \tag{7.4}$$

The Equations 7.3 and 7.4 define the algebra of formal power series without any need to mention x. Nonetheless, we shall continue to manipulate power series using the traditional notation involving powers of x, as this is more convenient, and certainly more familiar.

The second approach to Equation 7.2 is to regard it as an equation between *functions*. From this point of view it says that for a certain range of values of x, the function defined by the formula on the left-hand side of the equation has the same values as the function defined by the formula on the right-hand side. Since, in the cases of equations such as Equation 7.2, the functions are defined by infinite series, this approach involves knowing something about their convergence.* Fortunately, it is usually not important for us to know the exact range of values of x for which the series converge. What matters is that they should converge for at least some nonzero values of x. We will use standard theorems that provide us with this information.

There is no difficulty with the particular series in Equation 7.2. The series on the left-hand side are geometric series, and standard theorems tell us that they converge for $|x|<1$, and hence that the series on the right-hand side also converges in this range.

These two approaches, algebraic and functional, are closely related. A standard theorem tells us that if the series $\sum_{n=0}^{\infty} a_n x^n$ and $\sum_{n=0}^{\infty} b_n x^n$ both converge for $|x|<R$, for some $R > 0$, then

$$\text{for all } x, \text{ with } |x|<R, \sum_{n=0}^{\infty} a_n x^n = \sum_{n=0}^{\infty} b_n x^n \Leftrightarrow \text{for all } n \in \mathbb{N}, a_n = b_n. \tag{7.5}$$

This tells us that an equation such as Equation 7.2 is true when regarded as a functional equation if and only if it is also true when regarded as an algebraic equation between formal power series. It is this interplay that enables us to use both algebraic and analytic methods as appropriate and that turns out to be so fruitful. For example, we often use Equation 7.5 to argue that if two functions are the same, then the coefficients in the

---

* If you have not yet met *convergence* of series, don't worry if you don't fully understand these remarks. It will be sufficient to interpret "the series converges" as meaning "the series defines a specific number." If you have struggled with the $\varepsilon - n$ definition of convergence in an analysis course, we hope that you now appreciate how it relates to some fairly concrete mathematical problems.

power series that represent the functions must match exactly. We call this the *method of equating coefficients*.

You may think that rather too much fuss has been made about the two different interpretations of Equation 7.2, especially now that we have seen that they are equivalent. However, it is important to be aware that the equivalence in Equation 7.5 does depend on both series converging for some positive values of x. It also depends on properties of the field of real numbers. In other cases the analogous result need not be true. For example, if you are familiar with the field Z_3 made up of the numbers 0, 1, and 2 with addition and multiplication modulo 3, you will be able to check that although $1 + x^2 + x^3$ and $1 + x + x^2$ are distinct as polynomials, they define exactly the same functions on this field Z_3.

It is convenient at this point to introduce some abbreviated notation for sequences. We write $\{a_n\}$ to represent the infinite sequence $(a_0, a_1, a_2, \ldots)$. In some cases the infinite sequences we are considering will begin with a_1 rather than a_0. Normally this will be clear from the context, and so we sometimes write $\{a_n\}$ for the infinite sequence $(a_1, a_2, a_3, \ldots)$.

Exercises

7.1.1A Find the coefficient of x^n in the series

$$(1 + x + x^2 + x^3 + \ldots)^2.$$

7.1.1B Find the coefficient of x^n in the series

$$(1 + x + x^2 + x^3 + \ldots)^3.$$

7.1.2A Use the method of equating coefficients to find the terms of the sequence $\{a_n\}$ such that

$$(a_0 + a_1 x + a_2 x^2 + a_2 x^3 + \ldots)(1 + x^3 + x^6 + x^9 + \ldots) = (1 + x + x^2 + x^3 + \ldots).$$

7.1.2B Find a polynomial $p(x)$ such that

$$p(x)(x + 4x^2 + 9x^3 + \ldots + n^2 x^n + \ldots) = x(1 + x).$$

7.2 GENERATING FUNCTIONS

We begin with a definition.

DEFINITION 7.1

The *generating function* of the sequence $\{a_n\}$ is the function

$$x \mapsto \sum_{n=0}^{\infty} a_n x^n.$$

We should use this terminology only when the power series has a positive radius of convergence, so that the method of equating coefficients is applicable. In the combinatorial applications that we consider, $\{a_n\}$ will almost always be a sequence of natural numbers, since a_n will be the number of arrangements of some kind, but the notion of a generating function has wider applications. We could equally consider generating functions corresponding to sequences of real numbers or complex numbers.

We begin our discussion of generating functions by giving some standard examples.

PROBLEM 7.2

Find the generating function of the sequence $\{C(n,r)\}_{r=0}^{n}$.

Solution

By the binomial theorem, $\sum_{r=0}^{n} C(n,r)x^r = (1+x)^n$. Hence $x \mapsto (1+x)^n$ is the generating function of the sequence $\{C(n,r)\}_{r=0}^{n}$. Since the sequence is finite, the question of convergence does not arise.

PROBLEM 7.3

a. Show that the function $x \mapsto 1/(1-x)$ is the generating function for the constant sequence $\{1\}$.
b. Determine the generating function for the sequence $\{n\}$.

Solution

a. The geometric series $1 + x + x^2 + x^3 + \ldots$, that is, the series $\sum_{n=0}^{\infty} x^n$, all of whose coefficients equal 1, converges for $|x|<1$, and for this range of values of x, $\sum_{n=0}^{\infty} x^n = 1/(1-x)$. Thus $x \mapsto 1/(1-x)$ is the generating function for the sequence $\{1\}$ all of whose terms are equal to 1.
b. A theorem of analysis tells us that a power series may be differentiated term by term within the range of values of x for which it converges. Thus, as

$$\frac{1}{1-x} = \sum_{n=0}^{\infty} x^n \text{ for } |x|<1, \qquad (7.6)$$

we deduce by differentiating both sides of Equation 7.6 that $1/(1-x)^2 = \sum_{n=0}^{\infty} nx^{n-1}$ and hence

$$\frac{x}{(1-x)^2} = \sum_{n=0}^{\infty} nx^n, \text{for } |x|<1. \qquad (7.7)$$

Therefore $x \mapsto x/(1-x)^2$ is the generating function for the sequence $\{n\}$.

In Exercise 7.2.1A you are asked to find the generating function for the sequence $\{n^2\}$. This, together with the sequence for $\{n\}$ that we have just found, enables us to write down the generating function for the sequence $\{\frac{1}{2}n(n+1)\}$ by using the fact that it can be obtained by adding the generating functions for the sequences $\{n^2\}$ and $\{n\}$ and then dividing by 2. You are asked to do this in

Exercise 7.2.1A, but there is also a more direct method that exploits the fact that $\frac{1}{2}n(n+1)$ is the sum of the first n natural numbers. That is, the sequence $\{\frac{1}{2}n(n+1)\}$ is the sequence $\{a_n\}$ that is given by $a_0 = 0$, and

$$a_{n+1} = a_n + n + 1. \tag{7.8}$$

Let f be the generating function for this sequence. Thus,

$$f(x) = \sum_{n=0}^{\infty} a_n x^n. \tag{7.9}$$

If we multiply Equation 7.8 by x^n and sum the resulting terms, we obtain

$$\sum_{n=0}^{\infty} a_{n+1} x^n = \sum_{n=0}^{\infty} a_n x^n + \sum_{n=0}^{\infty} n x^n + \sum_{n=0}^{\infty} x^n. \tag{7.10}$$

Now

$$\sum_{n=0}^{\infty} a_{n+1} x^n = \frac{1}{x} \sum_{n=0}^{\infty} a_{n+1} x^{n+1}$$

$$= \frac{1}{x} \sum_{n=0}^{\infty} a_n x^n, \text{because } a_0 = 0,$$

$$= \frac{1}{x} f(x) \tag{7.11}$$

Hence, from Equations 7.7, 7.9, 7.10, and 7.11, we deduce that

$$\frac{1}{x} f(x) = f(x) + \frac{x}{(1-x)^2} + \frac{1}{1-x}$$

$$= f(x) + \frac{1}{(1-x)^2},$$

from which it follows that

$$f(x) = \frac{x}{(1-x)^3}. \tag{7.12}$$

Thus Equation 7.12 gives the generating function for the sequence $\{\frac{1}{2}n(n+1)\}$.

Exercises

7.2.1A i. Find the generating functions for the following sequences.

  a. $\{n(n-1)\}$ b. $\{n^2\}$ c. $\{n^3\}$

  ii. Deduce the generating function for the sequence $\{\frac{1}{2}n(n+1)\}$, and verify that your answer agrees with Equation 7.12.

7.2.1B For each positive integer k, let $x \mapsto F_k(x)$ be the generating function for the sequence $\{n^k\}$.

  i. Prove that, for each positive integer k, $F_{k+1}(x) = xF_k'(x)$, where $F_k'(x)$ is the derivative of $F_k(x)$ with respect to x.

  ii. Hence deduce that, for each positive integer k,

$$F_k(x) = \frac{P_k(x)}{(1-x)^{k+1}},$$

where $P_k(x)$ is a polynomial of degree k in which the coefficient of x^k is 1.

7.2.2A Find the generating function for the sequence $\{a_n\}$, where a_n is the sum of the first n squares. That is, the sequence $\{a_n\}$ is defined by $a_0 = 0$, and $a_{n+1} = a_n + (n+1)^2$.

7.2.2B Find the generating function for the sequence $\{a_n\}$, where a_n is the sum of the first n cubes.

7.3 WHAT IS A RECURRENCE RELATION?

We have seen in the previous section that the sequence $\{a_n\}$, where a_n is the sum of the first n natural numbers, can be defined by $a_0 = 0$ and $a_{n+1} = a_n + n + 1$. Thus the sequence is specified by giving the value of the first term in the sequence, and a formula telling us how to calculate each subsequent term from the previous term. This is an example of what is called a *recurrence relation*. Before explaining exactly what we mean by this, we give another example of a combinatorial problem that gives rise to a relation of this form.

PROBLEM 7.4

How many strings are there made up of 10 of the digits 0 and 1, in which there are no two successive 0s?

Solution

We let A_n be the set of strings made up of n of the digits 0 and 1 that do not contain consecutive 0s. We put $a_n = \#(A_n)$. We are asked to find the value of a_{10}, and we can do this by obtaining a general method for generating the numbers in the sequence $\{a_n\}$ by considering how strings in A_n can be built up from shorter strings that do not contain consecutive zeros.

Clearly each of the two strings consisting of just one digit meets the required condition. The two-digit strings 11, 10, and 01, but not 00, meet this condition. Hence

$$a_1 = 2 \text{ and } a_2 = 3. \tag{7.13}$$

Now suppose $n \geq 3$. The strings in A_n can be divided into those that begin with a 0 and those that begin with 1. Those that begin with 0 must have 1 as their second digit and so must have the form 01\$, where \$ is a string of $n-2$ digits that do not contain consecutive zeros, that is, a string in A_{n-2}. So there are a_{n-2} strings of this form. The strings in A_n that begin with 1 must have the form 1\$, where \$ is a string in A_{n-1}, and hence, there are a_{n-1} strings of this form. It therefore follows that for each integer $n \geq 3$,

$$a_n = a_{n-2} + a_{n-1}. \tag{7.14}$$

It is straightforward, using Equations 7.13 and 7.14, to calculate a_n for any particular value of n. For example, we have that $a_3 = a_1 + a_2 = 2 + 3 = 5$ and so $a_4 = a_2 + a_3 = 3 + 5 = 8$ and so on. In this way you can check that $a_{10} = 144$.

Although these calculations could be extended as far as we like, this provides a rather long-winded method for calculating the values of a_n for large values of n. In later sections of this chapter we discuss ways of converting Equations 7.13 and 7.14 into a succinct formula that enables us to calculate values of a_n very quickly. Before we do this, we want to use this example to help us to make clear what, in general, is meant by a recurrence relation for the terms of a sequence $\{a_n\}$.

It is characteristic of a recurrence relation that it enables us to calculate each term of a sequence from the values of earlier terms in the sequence. As a first shot we could say that a recurrence relation has the form

$$a_n = f(a_{n-1}, a_{n-2}, \ldots), \tag{7.15}$$

where f is a given function. In our Equation 7.14, f is the function given by $f:(x,y) \mapsto x+y$. In order to make the definition as general as possible, our notation in Equation 7.15 avoids specifying how many of the preceding terms are needed to calculate a_n. Nonetheless, Equation 7.15 is not general enough, as it does not even cover Equation 7.8 of the previous section. We repeat this for convenience, but, to fit in with Equation 7.15, we rewrite it to give a_n in terms of a_{n-1}, as follows:

$$a_n = a_{n-1} + n. \tag{7.16}$$

Notice that the definition of a_n in Equation 7.16 involves, on the right-hand side, not only the preceding term in the sequence, a_{n-1}, but also the number n itself. So we need to revise Equation 7.15 so as to include cases where the value of n enters explicitly into the formula for a_n. Thus a general recurrence relation has the form

$$a_n = f(n, a_{n-1}, a_{n-2}, \ldots), \tag{7.17}$$

where f is some given function. Again, our notation is deliberately vague about exactly which terms of the sequence are involved on the right-hand side of Equation 7.17 as we wish to allow for such cases as

$$a_n = \sum_{k=1}^{n-1} a_k,$$

where the number of terms of the sequence that are involved in the definition of a_n can vary with the value of n.

The recurrence relation does not, by itself, define the sequence. For example, Equation 7.14 does not enable us to calculate the terms of the sequence unless we are also given the values of a_1 and a_2 to start with. In general, a recurrence relation needs to be accompanied by one or more *initial conditions* that specify the first few terms of the sequence. A recurrence relation together with the appropriate initial conditions enables us to calculate all the terms of the sequence, assuming, of course, that we have a way to calculate the values of the function f that enters into the recurrence relation. By *solving* a recurrence relation we mean finding an explicit formula for a_n.

It is worth remarking here that what we have called a *recurrence relation* is, in other contexts, called a *recursive definition*. Recursive definitions are usually thought of as defining functions rather than sequences, but as we have noted earlier, a sequence $\{a_n\}$ is really a function $n \mapsto a_n$ in disguise. Recursive functions are studied in the branch of mathematical logic known as *recursive function theory* or *computability theory*. Recursive definitions are also allowed in several programming languages.

Recurrence relations are classified according to the form of the function f that occurs in the relation. In this chapter we discuss recurrence relations that can be solved using the device of generating functions.

Exercises

7.3.1A Find a recurrence relation and initial conditions for the sequence $\{a_n\}$, where a_n is the number of strings of n digits (that is, 0, 1, 2, 3, 4, 5, 6, 7, 8, 9) that do not contain consecutive even digits (where 0 counts as an even digit).

7.3.1B Find a recurrence relation and initial conditions for the sequence $\{a_n\}$, where n is the number of strings of n digits that contain no consecutive 0s, no consecutive 1s, and no consecutive 2s.

7.4 FIBONACCI NUMBERS

We have already seen in Section 7.2 how to work out the generating function of the sequence defined by the recurrence relation given by Equation 7.8. We are now going to take this idea further. If we have an explicit formula for the generating function, we may be able to use this to derive a formula for the coefficients in its power series. These coefficients are, of course, just the terms of the sequence in which we are interested. Before discussing the general method we give another illustration in relation to, perhaps, the best known of all sequences defined by a recurrence relation, namely, the *Fibonacci numbers*. These numbers are named after the Italian mathematician Leonardo of Pisa,[*] who introduced them in connection with the following problem.

---

[*] Leonardo of Pisa lived during the late twelfth and early thirteenth century. His alternative name *Fibonacci* comes from *filius Bonacci*, meaning "the son of Bonacci." His biggest contribution to mathematics was his book *Liber Abaci*, in which, among other things, he advocated the use of the Indian place-value system for numbers, which we now regard as standard. He also described the problem about rabbits that gives rise to what we now call Fibonacci numbers.

Rabbits take one month to reach maturity. A mature pair of rabbits produces another pair of rabbits each month. If you start with one pair of newly born rabbits, how many pairs will you have after n months?

We let f_n, for $n \geq 1$, be the number of pairs of rabbits alive after n months. Thus f_n is the number of pairs of rabbits, f_{n-2}, born to the rabbits who have reached maturity, plus the number, f_{n-1}, who were alive the previous month. Thus, for $n \geq 1$,

$$f_n = f_{n-2} + f_{n-1}. \qquad (7.18)$$

We obtain the *Fibonacci numbers* when we solve this recurrence relation subject to the initial conditions

$$f_1 = 1 \text{ and } f_2 = 1. \qquad (7.19)$$

Note that, except for a change of notation, the recurrence relation in Equation 7.18 is the same as in Equation 7.14. The only difference is that the initial conditions given by Equation 7.19 are different from those given by Equation 7.13.

The numbers in the sequence defined by Equations 7.18 and 7.19 are the Fibonacci numbers. It is easy to calculate the first few terms in the sequence:

$$1, 1, 2, 3, 5, 8, 13, 21, 34, 55, 89, 144, 233, 377, 610, 987, 1597, \ldots,$$

but can we find a simple formula for the numbers in this sequence? This was Problem 7 of Chapter 1. We now show how we can answer this question by finding the generating function for the sequence.

We let F be the generating function for this sequence. Thus, $F(x) = \sum_{n=1}^{\infty} f_n x^n$. If we multiply both sides of the recurrence relation in Equation 7.18 by x^n and sum for all integers $n \geq 3$, we obtain

$$\sum_{n=3}^{\infty} f_n x^n = \sum_{n=3}^{\infty} f_{n-1} x^n + \sum_{n=3}^{\infty} f_{n-2} x^n. \qquad (7.20)$$

The sum on the left-hand side of Equation 7.20 is just the series for $F(x)$ without the first two terms, that is, $\sum_{n=3}^{\infty} f_n x^n = F(x) - f_1 x - f_2 x^2 = F(x) - x - x^2$. In a similar way, $\sum_{n=3}^{\infty} f_{n-1} x^n = x \sum_{n=3}^{\infty} f_{n-1} x^{n-1} = x(F(x) - x)$ and $\sum_{n=3}^{\infty} f_{n-2} x^n = x^2 \sum_{n=3}^{\infty} f_{n-2} x^{n-2} = x^2 F(x)$. Therefore, it follows from Equation 7.20 that

$$F(x) - x - x^2 = x(F(x) - x) + x^2 F(x). \qquad (7.21)$$

Equation 7.21 can be rearranged to give

$$F(x) = \frac{x}{1 - x - x^2}. \qquad (7.22)$$

We have now achieved the first stage of our objective. We have obtained an explicit formula for the generating function of the sequence $\{f_n\}$. More than one method can now be used to derive from Equation 7.22 a formula for the coefficients in the power series for $F(x)$. One method is to rewrite the formula for $F(x)$ using the technique of *partial fractions** and then using the standard power series for $1/(1-x)$, namely, $1/(1-x) = 1 + x + x^2 + \ldots$, which generalizes to give

$$\frac{1}{1-\alpha x} = \sum_{n=0}^{\infty}(\alpha x)^n \quad \text{and} \quad \frac{1}{1+\beta x} = \frac{1}{1-(-\beta x)} = \sum_{n=0}^{\infty}(-\beta x)^n.$$

The first move is to factorize the denominator $1 - x - x^2$ in the form $(1 - \alpha x)(1 - \beta x)$. If $1 - x - x^2 = (1 - \alpha x)(1 - \beta x)$, then equating the coefficients of x and x^2 gives $\alpha + \beta = 1$ and $\alpha\beta = -1$. Hence $\alpha(1 - \alpha) = -1$, and so $\alpha^2 - \alpha - 1 = 0$. It is readily seen that β satisfies the same quadratic equation. Thus α, β are the roots of $x^2 - x - 1 = 0$, so

$$\alpha, \beta = \frac{1}{2}(1 \pm \sqrt{5}). \tag{7.23}$$

We now use partial fractions. The process is in principle straightforward, but the calculations are a little messy because α and β are the irrational numbers given by Equation 7.23. This gives you, the reader, three choices. If you are already familiar with the technique of partial fractions, you can take the details of our calculations on trust, and skip directly to Equation 7.26 or even Equation 7.28. If you would like to see an example that, though more artificial, is numerically easier, look ahead to Problem 7.5, and then come back to the Fibonacci numbers. The third choice is to work through the following calculations with us.

We have that

$$\frac{1}{1-x-x^2} = \frac{1}{(1-\alpha x)(1-\beta x)}.$$

We now seek to find constants A and B so that

$$\frac{x}{(1-\alpha x)(1-\beta x)} \equiv \frac{A}{1-\alpha x} + \frac{B}{1-\beta x}. \tag{7.24}$$

The symbol $\equiv$ in Equation 7.24 is intended to mean that it is an *identity*, that is, that the left-hand side is equal to the right-hand side *for all values of x for which they are defined*. We first multiply both sides of Equation 7.24 by $(1 - \alpha x)(1 - \beta x)$ to give

$$x \equiv A(1 - \beta x) + (1 - \alpha x). \tag{7.25}$$

---

* You will probably have used the method of partial fractions for evaluating integrals of rational functions. We hope that our methods for solving problems using this method will explain it to those who have not seen it before.

Because Equation 7.25 is an identity we can now find the constants A and B either by equating the constant terms and the coefficients of x, or by substituting particular values for x. The second method is more straightforward here. If we put $x = 1/\alpha$ in Equation 7.25, we obtain $1/\alpha = A(1 - (\beta/\alpha))$, from which it follows that $A = 1/(\alpha - \beta)$. Similarly, putting $x = 1/\beta$ in Equation 7.25 gives $B = 1/(\beta - \alpha)$. Using the values of α and β given by Equation 7.23, we thus see that

$$A = \frac{1}{\sqrt{5}} \text{ and } B = \frac{-1}{\sqrt{5}}. \tag{7.26}$$

Hence, from Equations 7.22 and 7.24,

$$F(x) = \frac{x}{1-x-x^2} = \left(\frac{1}{\sqrt{5}}\right)\left(\frac{1}{1-\alpha x}\right) + \left(\frac{-1}{\sqrt{5}}\right)\left(\frac{1}{1-\beta x}\right)$$

$$= \left(\frac{1}{\sqrt{5}}\right)\sum_{n=0}^{\infty}(\alpha x)^n + \left(\frac{-1}{\sqrt{5}}\right)\sum_{n=0}^{\infty}(\beta x)^n.$$

If we now substitute in the values of α and β as given in Equation 7.23, we obtain

$$F(x) = \frac{1}{\sqrt{5}}\sum_{n=0}^{\infty}\left[\left(\frac{1+\sqrt{5}}{2}\right)^n - \left(\frac{1-\sqrt{5}}{2}\right)^n\right]x^n, \tag{7.27}$$

and, as f_n is the coefficient of x^n in the power series in Equation 7.27, we can deduce that

$$f_n = \frac{1}{\sqrt{5}}\left[\left(\frac{1+\sqrt{5}}{2}\right)^n - \left(\frac{1-\sqrt{5}}{2}\right)^n\right]. \tag{7.28}$$

Here, at last, we have our formula for the Fibonacci numbers. Since these numbers are all integers, it is rather surprising to find the irrational number $\sqrt{5}$ in the formula for them. However, if in the formula in Equation 7.28 we expand $(1+\sqrt{5})^n$ and $(1-\sqrt{5})^n$, we find that all the terms involving $\sqrt{5}$ cancel, and we are left with an expression in which irrational numbers do not occur, albeit a rather more complicated expression than that given in Equation 7.28.

The next problem can be solved using a similar method, but there are no irrational numbers involved in the calculations, which are therefore less complicated.

PROBLEM 7.5

Sequences are formed using the letters A,B,C and the digits $1,2,3,4$. Find a formula for the number of sequences made up of n of these symbols in which there are not two letters in succession.

Solution

We let A_n, for $n \geq 1$, be the set of sequences made up of n of the symbols $A,B,C,1,2,3,4$ in which there are not two successive letters, and we put $a_n = \#(A_n)$. It is easy to see that $a_1 = 7$ and $a_2 = 40$. Now suppose $n \geq 3$. A sequence in A_n begins with either a letter or a digit. If a sequence in A_n begins with a letter, the second symbol must be a digit and so the sequence must have the form $Ld\$$, where $L \in \{A,B,C\}$, $d \in \{1,2,3,4\}$, and $\$ \in A_{n-2}$. Hence there are $3 \times 4 \times a_{n-2} = 12a_{n-2}$ sequences beginning with a letter. A sequence in A_n that begins with a digit has the form $d\$$, where $d \in \{1,2,3,4\}$ and $\$ \in A_{n-1}$, and hence there are $4a_{n-1}$ sequences beginning with a digit. It follows that, for $n \geq 3$,

$$a_n = 12a_{n-2} + 4a_{n-1}, \text{ with } a_1 = 7 \text{ and } a_2 = 40. \tag{7.29}$$

We now let $A(x) = \sum_{n=1}^{\infty} a_n x^n$. Now, by multiplying Equation 7.29 by x^n and summing, we obtain

$$\sum_{n=3}^{\infty} a_n x^n = 12 \sum_{n=3}^{\infty} a_{n-2} x^n + 4 \sum_{n=3}^{\infty} a_{n-1} x^n,$$

that is,

$$\sum_{n=3}^{\infty} a_n x^n = 12x^2 \sum_{n=3}^{\infty} a_{n-2} x^{n-2} + 4x \sum_{n=3}^{\infty} a_{n-1} x^{n-1},$$

and thus

$$A(x) - 7x - 40x^2 = 12x^2 A(x) + 4x(A(x) - 7x),$$

from which it follows that

$$A(x) = \frac{7x + 12x^2}{1 - 4x - 12x^2}. \tag{7.30}$$

Since $1 - 4x - 12x^2 = (1 - 6x)(1 + 2x)$, irrational numbers are not involved in the partial-fraction calculation that now ensues. The other difference from the case of the Fibonacci numbers is that in the quotient on the right-hand side of Equation 7.30, both the numerator and the denominator are polynomials of the same degree. So we first divide the denominator into the numerator to obtain

$$A(x) = \frac{7x + 12x^2}{1 - 4x - 12x^2} = -1 + \frac{3x + 1}{1 - 4x - 12x^2} = -1 + \frac{3x + 1}{(1 - 6x)(1 + 2x)}. \tag{7.31}$$

We now seek to find constants B and C such that

$$\frac{3x + 1}{(1 - 6x)(1 + 2x)} = \frac{B}{1 - 6x} + \frac{C}{1 + 2x}. \tag{7.32}$$

Multiplying both sides of Equation 7.32 by $(1 - 6x)(1 + 2x)$, we obtain

$$3x + 1 \equiv B(1 + 2x) + C(1 - 6x). \tag{7.33}$$

Putting, successively, $x = \frac{1}{6}$ and $x = -\frac{1}{2}$ in Equation 7.33, we obtain $\frac{3}{2} = \frac{4}{3}B$ and $-\frac{1}{2} = 4C$, from which it follows that $B = \frac{9}{8}$ and $C = -\frac{1}{8}$. It follows from Equations 7.31 and 7.32 that

$$A(x) = -1 + \frac{9}{8}\left(\frac{1}{1-6x}\right) - \frac{1}{8}\left(\frac{1}{1+2x}\right) = -1 + \frac{9}{8}\sum_{n=0}^{\infty}(6x)^n - \frac{1}{8}\sum_{n=0}^{\infty}(-2x)^n. \tag{7.34}$$

Consequently, $a_n = \frac{9}{8}6^n - \frac{1}{8}(-2)^n$.

We have solved Problem 7.5 by first finding the generating function for the sequence $\{a_n\}$ and then using the generating function to find a formula for a_n. You may think that this is rather a cumbersome method as it involves first finding the generating function, and only then a formula for a_n. There is a more direct method for solving simple recurrence relations, which we describe in the next section. We can get a clue to this method by looking at the solutions to the two recurrence relations we have solved, namely, that for the Fibonacci numbers and that of Problem 7.5. We encourage you to think about what these have in common before you begin reading the next section.

There are lots of well-known relationships between the Fibonacci numbers. We deduce one of them by rewriting the generating function as follows. We have that

$$F(x) = \frac{x}{1-x-x^2} = \frac{x}{1-x(1+x)} = x\sum_{k=0}^{\infty}(x(1+x))^k,$$

and it follows that f_n equals the coefficient of x^{n-1} in $\sum_{k=0}^{\infty}(x(1+x))^k$. The product $(x(1+x))^k = x^k(1+x)^k$ includes a term x^r for all integers r such that $k \leq r \leq 2k$, and hence includes a term involving x^{n-1}, provided that $k \leq n-1 \leq 2k$, that is, for $\frac{1}{2}(n-1) \leq k \leq n-1$, when the coefficient is the same as the coefficient of x^{n-1-k} in $(1+x)^k$, that is, $C(k, n-1-k)$. It follows that $f_n = \sum_{(1/2)(n-1) \leq k \leq n-1} C(k, n-1-k)$.

The numbers $C(k, n-1-k)$ for $\frac{1}{2}(n-1) \leq k \leq n-1$ occur on the diagonals of Pascal's triangle, such as the one shown by the large bold numbers in Figure 7.1, corresponding to the case $n = 8$, when our formula gives

```
                        1
                     1     1
                  1     2     1
               1     3     3     1
            1     4     6     4     1
         1     5    10    10     5     1
      1     6    15    20    15     6     1
   1     7    21    35    35    21     7     1
1
```

FIGURE 7.1

$$f_8 = \sum_{(7/2) \le k \le 7} C(k, 7-k) = C(4,3) + C(5,2) + C(6,1) + C(7,0) = 4 + 10 + 6 + 1 = 21.$$

For other relationships involving the Fibonacci numbers, see Exercise 7.4.3B.

Exercises

7.4.1A Find the generating function for the sequence $\{a_n\}$ defined by

$$a_1 = 2, a_2 = 16, \text{ and, for } n \ge 3, a_n = 8a_{n-1} - 15a_{n-2}.$$

Hence find a formula for a_n.

7.4.1B Find the generating function for the sequence $\{a_n\}$ defined by

$$a_1 = 5, a_2 = 15, \text{ and, for } n \ge 3, a_n = 3a_{n-1} + 4a_{n-2}.$$

Hence find a formula for a_n.

7.4.2A We have seen that the solution to Problem 7.4 is given by the recurrence relation $a_n = a_{n-2} + a_{n-1}$, for $n \ge 3$, with $a_1 = 2$ and $a_2 = 3$. Find the generating function for the sequence $\{a_n\}$.

7.4.2B Find the generating function for the sequence defined by the same recurrence relation as arises in Problem 7.5, namely, $a_n = 12a_{n-1} + 4a_{n-2}$, for $n \ge 3$, but with the general initial conditions, $a_1 = \alpha$, $a_2 = \beta$.

7.4.3A Consider the sequence $\{g_n\}$, which is defined by the same recurrence relation as the Fibonacci numbers, that is, $g_n = g_{n-2} + g_{n-1}$, but with the initial conditions $g_1 = \alpha$ and $g_2 = \beta$. Show that for $n \ge 3$, $g_n = \alpha f_{n-2} + \beta f_{n-1}$, where f_n is the nth Fibonacci number.

7.4.3B Prove that the Fibonacci numbers, f_n, have the following properties.
  i. For all $n \ge 3$, $f_n f_{n-2} - f_{n-1}^2 = (-1)^{n-1}$.
  ii. For all $n \ge 1$, $f_1^2 + f_2^2 + \ldots + f_n^2 = f_n f_{n+1}$.
  iii. For all $m, n \ge 1$, $f_{m+n} = f_{m-1} f_n + f_m f_{n+1}$.
  iv. For all $m, n \ge 1$, if n is divisible by m, then f_n is divisible by f_m.

7.5 SOLVING HOMOGENEOUS LINEAR RECURRENCE RELATIONS

In this section, we describe a more practical method for solving the type of recurrence relation that we looked at in the previous section. We postpone the theory that justifies this method until later. We need first to be precise about the type of recurrence relations to which our method applies.

DEFINITION 7.2

A *linear function* is a function of the form

$$f:(x_1, x_2, \ldots x_k) \to b_1 x_1 + b_2 x_2 + \ldots + b_k x_k,$$

where $b_1, b_2, \ldots, b_k$ are constants. (To be really precise we should specify the domain of the function and the possible values of the constants. In combinatorial applications,

$x_1, x_2, \ldots, x_k$ will usually be natural numbers, and the constants will normally be rational numbers. However, to allow for all possibilities, we could say that they are all real numbers, or even complex numbers, so that the domain of f is $\mathbb{R}^n$ or $\mathbb{C}^n$.)

A *homogeneous linear recurrence relation* is a recurrence relation of the form

$$a_n = f(a_{n-1}, a_{n-2}, \ldots, a_{n-k}), \tag{7.35}$$

where f is a linear function, with $1 \leq k < n$.

The Equation 7.35 brings out the fact that we can use it to calculate a_n in terms of the previous k terms of the sequence, but in some ways it is more natural to write the equation in the form

$$a_n - f(a_{n-1}, a_{n-2}, \ldots, a_{n-k}) = 0, \tag{7.36}$$

and thus we can give the following alternative definition.

DEFINITION 7.2

A *homogeneous linear recurrence relation* is a recurrence relation of the form

$$c_0 a_n + c_1 a_{n-1} + \ldots + c_k a_{n-k} = 0 \tag{7.37}$$

for some integer k, where $1 \leq k < n$, and where $c_0, c_1, \ldots, c_k$ are constants with $c_k \neq 0$. (Of course, if Equation 7.37 is identical with Equation 7.36, c_0 is 1, but we allow for other values of c_0.)

The term *homogeneous* is rather a mouthful. It is used to indicate that the right-hand side of Equation 7.36 is 0. We leave the more complicated case of nonhomogeneous linear recurrence relations until later. It will be readily seen that the Fibonacci sequence is defined by a homogeneous linear recurrence relation and that Problem 7.5 also leads to a homogeneous linear recurrence relation.

We illustrate our direct method by returning to this example. We begin by rewriting Equation 7.29 as

$$a_n - 4a_{n-1} - 12a_{n-2} = 0, \tag{7.38}$$

$$\text{with } a_1 = 7 \text{ and } a_2 = 40. \tag{7.39}$$

The idea of the method is to try to find a solution of Equation 7.38 of the form $a_n = x^n$. This may seem a bold idea, but it does not come completely out of the blue, as we have seen that the formula for the Fibonacci numbers involves expressions of the form x^n with $x = \frac{1}{2}(1 \pm \sqrt{5})$, and that the solution of Equation 7.38 involves 6^n and $(-2)^n$. However, it remains to see whether it works.

If we put $a_n = x^n$ in Equation 7.38, we obtain

$$x^n - 4x^{n-1} - 12x^{n-2} = 0, \tag{7.40}$$

which, if we divide by x^{n-2} (we hope you will pause to satisfy yourself that we can assume that $x^{n-2} \neq 0$ before dividing by it), becomes

$$x^2 - 4x - 12 = 0. \tag{7.41}$$

Since $x^2 - 4x - 12 = (x - 6)(x - 2)$, the solutions of Equation 7.41 are $x = 6, -2$. We thus obtain two different solutions of the recurrence relation in Equation 7.38, that is, $a_n = 6^n$ and $a_n = (-2)^n$. As the subsequent theory will show, it follows that the general solution of Equation 7.38 is given by

$$a_n = A6^n + B(-2)^n, \tag{7.42}$$

where A and B are constants that we now choose so that the initial conditions in Equation 7.39 are satisfied. If we successively put $n = 1$ and $n = 2$ in Equation 7.42, we see that for the initial conditions in Equation 7.39 to hold, we need to have

$$6A - 2B = 7 \text{ and } 36A + 4B = 40. \tag{7.43}$$

We can solve this pair of simultaneous equations to give $A = \frac{9}{8}$, $B = -\frac{1}{8}$, giving $a_n = \frac{9}{8}6^n - \frac{1}{8}(-2)^n$, which agrees with the solution to Problem 7.5 that we found in the previous section. (Just as well – because if our new "method" did not give the correct answer we would have been wasting our time.)

Looking back at the above solution we see that we can cut out one step, as we can go directly from the recurrence relation in Equation 7.38 to the quadratic Equation 7.41. Note that the coefficients in Equations 7.38 and 7.41 are the same. We adopt this shortcut in tackling the next problem, which we encourage you to try before reading our solution. It does not arise from a natural combinatorial problem but has been chosen to illustrate the method, with numbers chosen to make the calculation straightforward.

PROBLEM 7.6

Find the solution of the recurrence relation

$$a_n - 10a_{n-1} + 21a_{n-2} = 0, \quad \text{for} \quad n \geq 3, \tag{7.44}$$

subject to the initial conditions

$$a_1 = 2 \text{ and } a_2 = 146. \tag{7.45}$$

Solution

The quadratic equation associated with the recurrence relation given by Equation 7.44 is $x^2 - 10x + 21 = 0$, which has the solutions $x = 3, 7$. So the general solution of Equation 7.44 is $a_n = A3^n + B7^n$. The initial conditions in Equation 7.45 give $3A + 7B = 2$ and $9A + 49B = 146$. These equations have the solution $A = -11$ and $B = 5$. Hence the solution is $a_n = -11(3^n) + 5(7^n)$.

The quadratic equation corresponding to the recurrence relation is called the *auxiliary equation* associated with the recurrence relation. For example, in Problem 7.6 the auxiliary equation is the quadratic equation $x^2 - 10x + 21$.

The same method also works for homogeneous linear recurrence relations where the formula for a_n involves more than the two previous terms of the sequence. However, these cases are a little more complicated as the auxiliary equation will not be a quadratic equation but, instead, a polynomial equation of higher degree. This is illustrated by the next problem.

PROBLEM 7.7

Find the solution of the recurrence relation

$$a_n - 4a_{n-1} - 11a_{n-2} + 30a_{n-3} = 0, \quad \text{for} \quad n \geq 4, \tag{7.46}$$

subject to the initial conditions

$$a_1 = 5, a_2 = -5, \text{ and } a_3 = 155. \tag{7.47}$$

Solution

The auxiliary equation associated with the recurrence relation in Equation 7.46 is the cubic equation $x^3 - 4x^2 - 11x + 30 = 0$, that is, $(x - 5)(x + 3)(x - 2) = 0$, with solutions $x = 5, -3, 2$. Hence the general solution of Equation 7.46 is $a_n = A5^n + B(-3)^n + C2^n = 0$, where A, B, and C are constants. The initial conditions give

$$5A - 3B + 2C = 5, \; 25A + 9B + 4C = -5, \; 125A - 27B + 8C = 155,$$

with the solution $A = 1$, $B = -2$, $C = -3$ (of course, the recurrence relation and the initial conditions have been carefully chosen to give these integer values). Hence $a_n = 5^n - 2(-3)^n - 3(2^n)$.

We can sum up the method we have used as follows:

SOLVING HOMOGENEOUS LINEAR RECURRENCE RELATIONS

To solve the homogeneous linear recurrence relation

$$c_0 a_n + c_1 a_{n-1} + \ldots + c_k a_{n-k} = 0$$

subject to the initial conditions

$$a_1 = \alpha_1, \; a_2 = \alpha_2, \ldots, a_k = \alpha_k,$$

1. Write down the *auxiliary equation* associated with the recurrence relation. This is the polynomial equation of degree k

$$c_0 x^k + c_1 x^{k-1} + \ldots + c_k = 0.$$

2. Solve the auxiliary equation.
3. If the auxiliary equation has *distinct* roots $x_1,\ldots,x_k$, then the general solution of the recurrence relation is

$$a_n = A_1 x_1^n + A_2 x_2^n + \ldots + A_k x_k^n.$$

4. Find the constants $A_1, A_2, \ldots A_k$ by solving the system of linear equations

$$x_1 A_1 + x_2 A_2 + \ldots + x_k A_k = \alpha_1$$
$$\vdots$$
$$x_1^k A_1 + x_2^k A_2 + \ldots + x_k^k A_k = \alpha_k$$

given by the initial conditions.

The case where the auxiliary equation has repeated roots is a little more complicated. We explain what to do in this case in Section 7.7 where we give the theoretical justification of the method. If you have already met the "try $y = e^{mx}$" method for solving linear differential equations with constant coefficients, that is, equations such as

$$\frac{d^2y}{dx^2} - 4\frac{dy}{dx} - 12y = 0, \tag{7.48}$$

you will notice a resemblance with the method we have just used to solve linear recurrence relations. Indeed, the two theories are almost exactly parallel, as they both concern linear operators on vector spaces. (The general solution of the differential equation in Equation 7.48 is $y = Ae^{6x} + Be^{-2x}$, which you should compare with the general solution $a_n = A6^n + B(-2)^n$ of Equation 7.40.) Although it is not necessary to know about vector spaces in order to understand the piece of theory that is needed to justify the method we have just used, we discuss the vector space approach briefly in Section 7.7.

Exercises

7.5.1A Solve the recurrence relation $a_n - 3a_{n-1} - 4a_{n-2} = 0$, for $n \geq 3$, subject to the initial conditions $a_1 = 1$ and $a_2 = 3$.

7.5.1B Solve the recurrence relation $a_n - a_{n-1} - 6a_{n-2} = 0$, for $n \geq 3$, subject to the initial conditions $a_1 = 9$ and $a_2 = 57$.

7.5.2A Let a_n be the number of sequences of 0s, 1s, and 2s of length n in which a 0 may only be followed by a 1. Find a recurrence relation satisfied by the sequence $\{a_n\}$ and hence find a formula for a_n.

7.5.2B Let a_n be the number of sequences of n letters of the alphabet in which a vowel may only be followed by a consonant. Find a recurrence relation satisfied by the sequence $\{a_n\}$, and hence find a formula for a_n.

7.5.3A Let A_n be the $n \times n$ matrix that has 1s on the leading diagonal, and on the diagonals above and below the leading diagonal, and 0s everywhere else. Thus A_n is the $n \times n$ matrix (a_{ij}) where

$$a_{ij} = \begin{cases} 1, & \text{if } |i-j| \leq 1, \\ 0, & \text{otherwise.} \end{cases} \quad \text{For example, } A_5 = \begin{bmatrix} 1 & 1 & 0 & 0 & 0 \\ 1 & 1 & 1 & 0 & 0 \\ 0 & 1 & 1 & 1 & 0 \\ 0 & 0 & 1 & 1 & 1 \\ 0 & 0 & 0 & 1 & 1 \end{bmatrix}.$$

Let $d_n = \det(A_n)$. Find a recurrence relation for the sequence $\{d_n\}$ and hence find a formula for d_n.

7.5.3B Richard Richardson,* known to his friends as "Rich-Rich," opens a savings account with his bank by paying in £1000. At the end of the year he pays a further £2000, and in general, at the end of n years he pays in an additional amount of £1000(n + 1). At the end of each year the bank adds to his account 4% of everything in his account at the start of the year, with an additional 2% of everything that has been in his account for at least two years. How much money does he have in his account after n years? How long does it take Rich-Rich to accumulate £100,000 in his savings account?

7.6 NONHOMOGENEOUS LINEAR RECURRENCE RELATIONS

To illustrate what we mean by an nonhomogeneous linear recurrence relation we begin with a problem that leads to a recurrence relation of this type. The problem concerns the evaluation of determinants. If this is something you are not familiar with, you should skip the discussion of the problem, and go straight to Equation 7.49 at the end of the solution, where the recurrence relation is given.

PROBLEM 7.8

Find a formula for the number, r_n, of arithmetic operations that are needed to evaluate the determinant of an $n \times n$ matrix, by row-reducing the matrix to upper triangular form, and then multiplying together the diagonal entries.

Solution

The first stage of the row-reduction process applied to an $n \times n$ matrix, as shown on the left below, is to row-reduce it to the form shown on the right by subtracting multiples of the top row from the remaining rows.

$$\begin{bmatrix} a_{11} & \cdots & a_{1n} \\ \vdots & & \vdots \\ \vdots & & \vdots \\ a_{n1} & & a_{nn} \end{bmatrix} \quad \begin{bmatrix} a_{11} & \cdots & a_{1n} \\ 0 & a'_{22} & \cdots & a'_{2n} \\ 0 & \vdots & \vdots & \vdots \\ 0 & a'_{n2} & \cdots & a'_{nn} \end{bmatrix}.$$

---

* Not, we hasten to add, the West Indian cricketer Richard Benjamin Richardson.

(This assumes that $a_{11} \neq 0$. If $a_{11} = 0$, we need first to put a nonzero number in the first row and first column by interchanging the top row with the ith row, for some i for which $a_{1i} \neq 0$. Of course, if all the entries in the first column are 0, the determinant of the matrix is 0, and no arithmetic is needed to calculate it.)

The reduction process involves the following steps. First, for $2 \leq i \leq n$, we calculate the appropriate multiple of the first row that we need to subtract from the ith row. This means that we need to calculate, for $2 \leq i \leq n$,

$$c_i = \frac{a_{i1}}{a_{11}}.$$

Thus, this involves $n - 1$ divisions. Then, for $2 \leq i \leq n$, we subtract c_i times row 1 from row i. Thus, for $2 \leq i \leq n$ and $2 \leq j \leq n$, we need to calculate

$$a'_{ij} = a_{ij} - c_i \times a_{1j}.$$

It takes one multiplication and one subtraction to calculate each a'_{ij}, and hence $2(n - 1)^2$ arithmetic operations altogether.

Having carried out all these row operations, we then need to calculate the determinant of the $(n - 1) \times (n - 1)$ matrix

$$\begin{bmatrix} a'_{22} & \cdots & a'_{2n} \\ \vdots & & \vdots \\ a'_{n2} & \cdots & a'_{nn} \end{bmatrix}$$

and then multiply this determinant by a_{11}.

It follows that

$$r_n = (n-1) + 2(n-1)^2 + r_{n-1} + 1.$$

Clearly, the initial condition is $r_1 = 0$, as no arithmetic is needed to calculate the determinant of a 1×1 matrix. Thus, r_n is given by

$$r_n - r_{n-1} = 2n^2 - 3n + 2, \quad \text{for} \quad n \geq 2 \text{ with } r_1 = 0. \tag{7.49}$$

The recurrence relation in Equation 7.49 is linear, but it is not homogeneous because the right-hand side is not zero but is instead a function of n. The general definition is as follows.

DEFINITION 7.4

A *nonhomogeneous linear recurrence relation* is a recurrence relation of the form

$$c_0 a_n + c_1 a_{n-1} + \ldots + c_k a_{n-k} = f(n) \tag{7.50}$$

for some integer $k \geq 1$, where $c_0, c_1, \ldots, c_k$ are constants, and f is some non-zero function.

Our method for solving such nonhomogeneous recurrence relations is as follows. The theoretical justification for this method is given in Section 7.7.

SOLVING NONHOMOGENEOUS RECURRENCE RELATIONS

To solve the nonhomogeneous recurrence relation

$$c_0 a_n + c_1 a_{n-1} + \ldots + c_k a_{n-k} = f(n), \tag{7.51}$$

1. Find the general solution, say $\{y_n\}$, of the associated homogeneous recurrence relation

$$c_0 a_n + c_1 a_{n-1} + \ldots + c_k a_{n-k} = 0. \tag{7.52}$$

2. Find any particular solution, say $\{z_n\}$, of the nonhomogeneous recurrence relation in Equation 7.51.
3. Then the general solution of Equation 7.51 is given by

$$a_n = y_n + z_n,$$

and the constants occurring in this solution can be found so as to satisfy any given initial conditions.

As we already know how to solve homogeneous recurrence relations, all we need to be able to solve nonhomogeneous equations is to find some way to come across a particular solution $\{z_n\}$. We illustrate one way this can be done by returning to the recurrence relation given in Equation 7.49, which we repeat for convenience:

$$r_n - r_{n-1} = 2n^2 - 3n + 2 \text{ for } n \geq 2 \text{ with } r_1 = 0. \tag{7.49}$$

We first need to solve the associated homogenous recurrence relation

$$r_n - r_{n-1} = 0. \tag{7.53}$$

We couldn't ask for anything easier! By Equation 7.53 all the terms, r_n, are the same, so the general solution of Equation 7.53 is given by

$$r_n = A, \tag{7.54}$$

where A is a constant.

Next, we seek a particular solution of Equation 7.49. The idea here is that if the difference between r_n and r_{n-1} is a polynomial of degree 2, we should look for a polynomial of degree 3 for r_n. The reason will emerge from the solution. (However, this is a special case because of the particular form of the left-hand side of Equation 7.49. In general, with a polynomial of degree k on the right-hand side, we try to find a particular solution that is also a polynomial of degree k; see also Exercise 7.6.5A.) So we put

$$r_n = \alpha n^3 + \beta n^2 + \gamma n + \delta. \tag{7.55}$$

and aim to find values of $\alpha, \beta, \gamma,$ and δ that satisfy Equation 7.49.
Substituting from Equation 7.55 into Equation 7.49 gives

$$(\alpha n^3 + \beta n^2 + \gamma n + \delta) - (\alpha(n-1)^3 + \beta(n-1)^2 + \gamma(n-1) + \delta) = 2n^2 - 3n + 2,$$

that is,

$$3\alpha n^2 + (-3\alpha + 2\beta)n + (\alpha - \beta + \gamma) = 2n^2 - 3n + 2. \tag{7.56}$$

The polynomials on the left- and right-hand sides of Equation 7.56 must be equal for all integers $n \geq 2$, and hence their coefficients must be equal. This gives

$$3\alpha = 2, -3\alpha + 2\beta = -3, \alpha - \beta + \gamma = 2, \tag{7.57}$$

from which it follows that $\alpha = \frac{2}{3}$, $\beta = -\frac{1}{2}$, and $\gamma = \frac{5}{6}$. We cannot determine the value of δ from Equation 7.56, but this should not be a surprise because our general solution, Equation 7.54, of the associated homogeneous recurrence relation already involves an undetermined constant.

The general solution of the recurrence relation is the sum of the right hand sides of Equations 7.54 and 7.55, giving

$$r_n = A + \left(\frac{2}{3}n^3 - \frac{1}{2}n^2 + \frac{5}{6}n + C\right).$$

From the initial condition that $r_1 = 0$, we deduce that $A + 1 + C = 0$, giving $A + C = -1$. Hence the solution of Equation 7.49 is

$$r_n = \frac{2}{3}n^3 - \frac{1}{2}n^2 + \frac{5}{6}n - 1. \tag{7.58}$$

This result shows that although the determinant of an $n \times n$ matrix involves $n!$ terms when expanded in full, and $n!$ increases extremely rapidly, the number of arithmetic operations needed to evaluate it by the row-reduction method only grows in proportion to n^3. This makes it possible to carry out calculations in linear algebra with very large matrices in a feasible amount of time.

As it happens, because the recurrence relation in Equation 7.49 is particularly simple, there is a more direct way to find this solution (see Exercise 7.6.6A). The next problem covers a more typical case.

PROBLEM 7.9

Solve the recurrence relation $a_n - 5a_{n-1} + 6a_{n-2} = 2n^2$, subject to the initial conditions $a_1 = 28$ and $a_2 = 34$.

Solution

The auxiliary equation is $x^2 - 5x + 6 = 0$ with solutions $x = 2$ and $x = 3$. So the general solution of the associated homogeneous equation is $a_n = A2^n + B3^n$. Because n^2 is a polynomial of degree 2 we try to find a particular solution of the same form, that is, with $a_n = \alpha n^2 + \beta n + \gamma$. Substituting this into the recurrence relation gives

$$(\alpha n^2 + \beta n + \gamma) - 5(\alpha(n-1)^2 + \beta(n-1) + \gamma) + 6(\alpha(n-2)^2 + \beta(n-2) + \gamma) = n^2.$$

That is,

$$2\alpha n^2 + (-14\alpha + 2\beta)n + (19\alpha - 7\beta + 2\gamma) = 2n^2.$$

Equating the coefficients of n^2 and n and the constant terms gives $2\alpha = 2$, $-14\alpha + 2\beta = 0$, and $19\alpha - 7\beta + 2\gamma = 0$. Hence $\alpha = 1$, $\beta = 7$, and $\gamma = 15$. Therefore, the general solution of the nonhomogeneous recurrence relation is given by

$$a_n = A2^n + B3^n + n^2 + 7n + 15.$$

The initial conditions give $2A + 3B + 23 = 28$ and $4A + 9B + 33 = 34$, that is, $2A + 3B = 5$ and $4A + 9B = 1$. Hence $A = 7$ and $B = -3$. Therefore the solution is

$$a_n = 7(2^n) - 3(3^n) + n^2 + 7n + 15.$$

Exercises

7.6.1A Find a formula for a_n given that $a_n - 2a_{n-1} - 8a_{n-2} = 18 - 9n$, with $a_1 = 1$ and $a_2 = 3$.

7.6.1B Find a formula for a_n given that $a_n + 5a_{n-1} + 6a_{n-2} = 2n^3 - 21n^2 + 57n - 43$ for $n \geq 3$ with $a_1 = 1$ and $a_2 = 14$.

7.6.2A Let a_n be the number of sequences of length n of letters of the English alphabet in which between them the five vowels, A, E, I, O, U, occur an even number of times.
  i. Show that $a_1 = 21$ and that, for $n \geq 2$, $a_n = 16a_{n-1} + 5(26^{n-1})$.
  ii. Find a formula for a_n.

7.6.2B Let a_n be the number of n-digit sequences formed using only the digits 0, 1, 2, and 3 in which 0 occurs an odd number of times.
  i. Show that $a_n = 2a_{n-1} + 4^{n-1}$.
  ii. Find a formula for a_n.

7.6.3A Let m and n be positive integers with $m < n$. How many arithmetic operations are needed to row-reduce an $m \times n$ matrix to echelon form?

7.6.3B Let $\mathbf{A}$ be an invertible (that is, nonsingular) $n \times n$ matrix. The inverse of $\mathbf{A}$ may be calculated by forming the $n \times 2n$ matrix $(\mathbf{A}\mathbf{I}_n)$ by writing the $n \times n$ identity matrix $\mathbf{I}_n$ alongside $\mathbf{A}$ and then row-reducing this matrix to echelon form, thus obtaining a matrix of the form $(\mathbf{I}_n\mathbf{B})$, where $\mathbf{B}$ is the inverse of $\mathbf{A}$. How many arithmetic operations are needed to calculate the inverse of $\mathbf{A}$ by this method?

7.6.4A *The Tower of Hanoi*
  Suppose that we have three pegs. On one of these pegs there are n disks of different sizes arranged in order of size with the largest disk at the bottom. The task is to transfer these n disks to the third peg, by moving one disk at

a time to another peg, subject to the rule that no disk may ever be placed on top of a smaller disk.

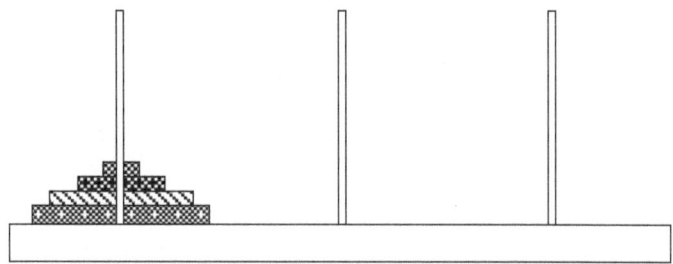

(This is an old toy known, for no good reason as the Tower of Hanoi.) If, as the legend says,* in the great temple of Benares, the priests of Bramah are carrying out this task with 64 disks, and they move one disk each second, without making any mistakes, how long will it take before all the disks have been transferred and "tower, temple and Brahmins alike will crumble into dust, and with a thunder clap the world will vanish"?

7.6.4B In this exercise we generalize the Tower of Hanoi problem to one with n disks and r (> 3) pegs.

We let $T(n,r)$ denote the minimum number of moves needed to transfer the n disks from one peg to another with, as usual, the restriction that no disk is can be placed above a smaller one on the same peg. The Frame Stewart algorithm is as follows.

For each k, $1 \le k \le n-1$:
  i. Move the top k disks (legally) to another peg;
  ii. Move the remaining $n-k$ disks (legally) to one of the remaining pegs;
  iii. Move the k disks (legally) to the peg containing the $n-k$ disks.

Deduce that, for $1 \le k \le n-1$, we have $T(n,r) \le 2T(k,r) + T(n-k,r-1)$. Using this inequality find upper bounds for the values of $T(n,4)$ for $1 \le n \le 8$.

7.6.5A Show that
  i. It is not possible to find a particular solution of Equation 7.49 of the form $r_n = En^2 + Dn + C$, where C, D, and E are constants.
  ii. The recurrence relation $r_n - 2r_{n-1} = 2n^2 - 3n + 2$ has a solution of the form $r_n = En^2 + Dn + C$, where C, D, and E are constants.

7.6.5B Show that the recurrence relation $r_n - r_{n-1} = n^3$ has solutions of the form $r_n = Gn^4 + Fn^3 + En^2 + Dn + C$.

7.6.6A i. Show that the solution of the recurrence relation

$$r_n - r_{n-1} = f(n)$$

is given by $r_n = r_1 + \sum_{k=2}^{n} f(k)$, for $n \ge 2$.

---

* This is quoted in *Mathematical Recreations and Essays* by W. Rouse Ball, Macmillan & Co., London, 1892, but is almost certainly a modern invention.

ii. Use the formula of (i) to solve the recurrence relation in Equation 7.49 of Problem 7.8.

7.6.6B Show that the solution of the recurrence relation

$$r_n - r_{n-1} - r_{n-2} = g(n)$$

is given by $r_n = f_{n-1}r_1 + f_{n-2}r_2 + \sum_{k=3}^{n} f_{n+1-k}g(k)$, where f_n is the nth Fibonacci number.

7.7 THE THEORY OF LINEAR RECURRENCE RELATIONS

In this section we describe briefly the theory that underlies the methods for solving homogeneous and nonhomogeneous recurrence relations covered in the previous two sections.

Throughout the first part of this discussion, we assume that we are dealing with the recurrence relation

$$c_0 a_n + c_1 a_{n-1} + \ldots + c_k a_{n-k} = 0, \text{ where } c_k \neq 0 \qquad (7.59)$$

subject to the initial conditions

$$a_1 = \alpha_1, a_2 = \alpha_2, \ldots, a_k = \alpha_k. \qquad (7.60)$$

The first theorem justifies our use of a solution of the form $a_n = x^n$.

THEOREM 7.1

A solution of the recurrence relation corresponding to Equation 7.59 is given by $a_n = x^n$ if and only if either $x = 0$ or x is a solution of the polynomial equation

$$c_0 x^k + c_1 x^{k-1} + \ldots + c_{k-1} x + c_k = 0. \qquad (7.61)$$

Proof

The sequence $\{x^n\}$ is a solution of the recurrence relation corresponding to Equation 7.59 if and only if

$$c_0 x^n + c_1 x^{n-1} + \ldots + c_{k-1} x^{n-k+1} + c_k x^{n-k} = 0 \Leftrightarrow x^{n-k}(c_0 x^k + c_1 x^{k-1} + \ldots + c_{k-1} x + c_k) = 0$$

$$\Leftrightarrow x = 0 \text{ or } c_0 x^k + c_1 x^{k-1} + \ldots + c_{k-1} x + c_k = 0.$$

Of course, $x = 0$ leads to a "trivial" solution in which $\{x^n\}$ is the constant sequence all of whose terms are 0.

The next theorem shows how we may combine solutions of the recurrence relation corresponding to Equation 7.59 to obtain other solutions of the same recurrence relation.

THEOREM 7.2

If the sequences $\{x_n\}$ and $\{y_n\}$ are both solutions of the recurrence relation in Equation 7.59, then for all choices of the constants A and B the sequence

$$\{Ax_n + By_n\} \tag{7.62}$$

is also a solution.

Proof

If the sequences $\{x_n\}$, $\{y_n\}$ are solutions of Equation 7.59, we have that

$$c_0 x_n + c_1 x_{n-1} + \ldots + c_k x_{n-k} = 0. \tag{7.63}$$

and

$$c_0 y_n + c_1 y_{n-1} + \ldots + c_k y_{n-k} = 0. \tag{7.64}$$

Multiplying Equation 7.63 by A and adding this to Equation 7.64 multiplied by B, we deduce that

$$c_0 (Ax_n + By_n) + c_1 (Ax_{n-1} + By_{n-1}) + \ldots + c_k (Ax_{n-k} + By_{n-k}) = 0, \tag{7.65}$$

which shows that the sequence $\{Ax_n + By_n\}$ is also a solution of Equation (7.59).

In the language of vector spaces we have shown that the solution set of Equation 7.59 is *closed* under linear combinations; that is, it is a *linear subspace* of the vector space of all sequences.

Theorems 7.1 and 7.2 show us how to obtain some of the solutions of the recurrence relation in Equation 7.59, but does this method yield *all* the solutions, and will this method enable us to find a solution compatible with whatever initial conditions are specified?

The easiest case to deal with is where the polynomial equation in Equation 7.61 of Theorem 7.1 has k distinct solutions. Our assumption that $c_k \neq 0$ implies that these must be nonzero solutions. Suppose that these k distinct solutions of Equation 7.61 are $x_1, x_2, \ldots, x_k$. Then, for each choice of constants, $A_1, A_2, \ldots, A_k$, the sequence

$$\{A_1 x_1^n + A_2 x_2^n + \ldots + A_k x_k^n\} \tag{7.66}$$

provides a solution of Equation 7.59. It is then straightforward to check that, whatever initial conditions are specified, there are always uniquely determined values of the constants $A_1, A_2, \ldots, A_k$ so that the sequence in expression 7.66 is a solution of the recurrence relation satisfying the specified initial conditions. In terms of vector spaces, this amounts to saying that the solution space of Equation 7.59 has k dimensions, and the sequences $\{x_i^n\}$, for $1 \leq i \leq k$, form a basis for the solution space (see Exercises 7.7.1A for the details). Notice that in this discussion we have not assumed that $x_1, x_2, \ldots, x_k$ are real numbers, so it also covers the case where some or all of the roots of the polynomial equation in Equation 7.61 are complex numbers.

We now come to the theory of nonhomogeneous linear recurrence relations, namely, recurrence relations of the form

$$c_0 a_n + c_1 a_{n-1} + \ldots + c_k a_{n-k} = f(n), \tag{7.67}$$

where $c_0, c_1, \ldots, c_k$ are constants, with $c_k \neq 0$, and where f is some function. The *associated homogeneous recurrence relation* is

$$c_0 a_n + c_1 a_{n-1} + \ldots + c_k a_{n-k} = 0. \tag{7.68}$$

The method we described and used in Section 7.6 can now be justified.

THEOREM 7.3

The general solution of the nonhomogeneous linear recurrence relation in Equation 7.67 has the form

$$\{a_n + b_n^0\},$$

where $\{b_n^0\}$ is one particular solution of Equation 7.67 and $\{a_n\}$ is any solution of the associated homogeneous recurrence relation in Equation 7.68.

Proof

Let $\{b_n^0\}$ be one particular solution of Equation 7.67, and let $\{a_n\}$ be any solution of Equation 7.68. Thus,

$$c_0 b_n^0 + c_1 b_{n-1}^0 + \ldots + c_k b_{n-k}^0 = f(n) \tag{7.69}$$

and

$$c_0 a_n + c_1 a_{n-1} + \ldots + c_k a_{n-k} = 0. \tag{7.70}$$

Adding Equations 7.69 and 7.70 gives

$$c_0(a_n + b_n^0) + c_{n-1}(a_{n-1} + b_{n-1}^0) + \ldots + c_{n-k}(a_{n-k} + b_{n-k}^0) = f(n).$$

It follows that $\{a_n + b_n^0\}$ is a solution of the recurrence relation in Equation 7.67. Conversely, suppose that $\{d_n\}$ is a solution of Equation 7.67. Then,

$$c_0 d_n + c_1 d_{n-1} + \ldots + c_k d_{n-k} = f(n). \tag{7.71}$$

Subtracting Equation 7.69 from Equation 7.71 gives

$$c_0(d_n - b_n^0) + c_1(d_{n-1} - b_{n-1}^0) + \ldots + c_k(d_{n-k} - b_{n-k}^0) = 0,$$

from which it follows that the sequence $\{d_n - b_n^0\}$ is a solution of the homogeneous Equation 7.70. Since $d_n = (d_n - b_n^0) + b_n^0$, this shows that every solution of Equation 7.67 has the form stated in the theorem.

The case where Equation 7.61 has multiple roots is only a little more complicated. It can be shown (see Exercise 7.7.1B) that if x_0 is a root of Equation 7.61 of multiplicity r, then the sequences $\{x_0^n\}$, $\{nx_0^n\}$, ... , $\{n^{r-1}x_0^n\}$ form a set of r linearly independent solutions of Equation 7.59. This is illustrated by the solution to the following problem.

PROBLEM 7.10

Find the solution to the recurrence relation

$$a_n - 6a_{n-1} + 9a_n = 0$$

subject to the initial conditions $a_1 = 9$ and $a_2 = 9$.

Solution

The auxiliary equation is $x^2 - 6x + 9 = 0$, that is, $(x-3)^2 = 0$. Hence $x = 3$ is a root of multiplicity 2. It follows that the general solution of the recurrence relation is $a_n = A3^n + Bn3^n$, where A and B are constants. The initial conditions give $3A + 3B = 9$ and $9A + 18B = 9$. Hence $A = 5$ and $B = -2$, and hence the solution is $a_n = 5(3^n) - 2n3^n = (5-2n)3^n$.

Exercises

7.7.1A *Vector spaces and recurrence relations*

We let R^∞ be the set of all infinite sequences, $\{u_n\}$, of real numbers. This set becomes a vector space if we define vector addition by $\{u_n\} + \{v_n\} = \{u_n + v_n\}$ and scalar multiplication by $c\{u_n\} = \{cu_n\}$. Theorem 7.2 shows that the solution set, S, of the linear recurrence relation

$$\alpha_0 a_n + \alpha_1 a_{n-1} + \ldots + \alpha_k a_{n-k} = 0 \qquad (7.72)$$

is a subspace of R^∞. The aim of this exercise is to show that if the polynomial equation

$$\alpha_0 x^k + \alpha_1 x^{k-1} + \ldots + \alpha_{k-1} x + \alpha_k = 0$$

has the k distinct roots, $x_1, x_2, \ldots, x_k$, then the general solution of Equation 7.72 is given by

$$a_n = A_1 x_1^n + A_2 x_2^n + \ldots + A_k x_k^n,$$

where $A_1, A_2, \ldots, A_k$ are constants.

i. Prove that if the real numbers $r_1, r_2, \ldots, r_k$ are all different, then the sequences $\{r_1^n\}, \{r_2^n\}, \ldots, \{r_k^n\}$ form a linearly independent subset of R^∞.

ii. For $1 \le i \le k$, let $\{y(i)_n\}$ be the solution of the above recurrence relation that satisfies the initial conditions

$$y(i)_n = \begin{cases} 1, & \text{if } n=i, \\ 0, & \text{otherwise.} \end{cases}$$

Prove that the set of sequences $\{y(1)_n\}, \{y(2)_n\}, \ldots, \{y(k)_n\}$ forms a spanning set for the solution space, S, of the recurrence relation in Equation 7.72.

iii. Prove that, if $x_1, x_2, \ldots, x_k$ are k distinct solutions of the polynomial equation

$$\alpha_0 x^k + \alpha_1 x^{k-1} + \ldots + \alpha_{k-1} x + \alpha_k = 0,$$

then every solution of the above recurrence relation has the form

$$\{A_1 x_1^n + A_2 x_2^n + \ldots + A_k x_k^n\}$$

for some appropriate choice of the constants $A_1, A_2, \ldots, A_k$.

7.7.1B *Repeated roots of the auxiliary equation*
Prove that if x_0 is a nonzero solution of the polynomial equation $b_0 x^k + b_1 x^{k-1} + \ldots + b_{k-1} x + b_k = 0$ of multiplicity r, then the sequences $\{x_0^n\}, \{n x_0^n\}, \ldots, \{n^{r-1} x_0^n\}$ form a set of r linearly independent solutions of the recurrence relation $b_0 a_n + b_1 a_{n-1} + \ldots + b_k a_{n-k} = 0$.

7.7.2A Find the solution of the following recurrence relations.
i. $a_n - 10 a_{n-1} + 25 a_{n-2} = 0$, subject to the initial conditions $a_1 = 15$ and $a_2 = 325$.
ii. $a_n + 4 a_{n-1} + 4 a_{n-2} = 0$, subject to the initial conditions $a_1 = 4$ and $a_2 = 4$.
iii. $a_n - 7 a_{n-1} + 16 a_{n-2} - 12 a_{n-3} = 0$, subject to the initial conditions $a_1 = 8$, $a_2 = 42$, and $a_3 = 142$.
iv. $a_n - 6 a_{n-1} + 12 a_{n-2} - 8 a_{n-3} = 0$, subject to the initial conditions $a_1 = 0$, $a_2 = 8$, and $a_3 = 16$.

7.7.2B Find the solution of the following recurrence relations.
i. $a_n + 10 a_{n-1} + 25 a_{n-2} = 0$, subject to the initial conditions $a_1 = 15$ and $a_2 = 325$.
ii. $a_n - 4 a_{n-1} + 4 a_{n-2} = 0$, subject to the initial conditions $a_1 = 0$ and $a_2 = 4$.
iii. $a_n - 9 a_{n-1} + 27 a_{n-2} - 27 a_{n-3} = 0$, subject to the initial conditions $a_1 = 0$, $a_2 = 0$, and $a_3 = 54$.
iv. $a_n - 4 a_{n-1} - 3 a_{n-2} + 18 a_{n-3} = 0$, subject to the initial conditions $a_1 = 4$, $a_2 = -29$, and $a_3 = 40$.

7.8 SOME NONLINEAR RECURRENCE RELATIONS

When it comes to nonlinear recurrence relations, life becomes much more difficult. In this section we consider nonlinear recurrence relations of just one type. This type has been chosen because it occurs naturally in some counting problems and, by good fortune, also succumbs to the generating function method.

We illustrate this idea by finding the formula for the generating function for the Catalan numbers, C_n, as described in Chapter 5. Recall that we have specified that $C_0 = 1$ and that, by Theorem 5.8, we have that for each integer $n \geq 1$, $C_{n+1} = \sum_{k=0}^{n} C_k C_{n-k}$. This formula should remind you of the formula in Equation 7.4 for multiplying power series. If we let C be the generating function for the Catalan numbers, that is, $C(x) = \sum_{n=0}^{\infty} C_n x^n$, then, by Equation 7.4, $(C(x))^2 = \sum_{n=0}^{\infty} d_n x^n$, where $d_n = \sum_{i=0}^{n} C_i C_{n-i}$ and thus $d_n = C_{n+1}$. Hence $(C(x))^2 = \sum_{n=0}^{\infty} C_{n+1} x^n$. Therefore $x(C(x))^2 = \sum_{n=0}^{\infty} C_{n+1} x^{n+1} = C(x) - 1$. Therefore, writing C for $C(x)$ we have that

$$xC^2 - C + 1 = 0. \tag{7.73}$$

Solving Equation 7.73 by the standard formula, we have that, for $x \neq 0$,

$$C = \frac{1 \pm \sqrt{1-4x}}{2x}. \tag{7.74}$$

Using the power series expansion for $\sqrt{1-4x}$, Equation 7.74 may be used to derive the formula $C_n = [1/(n+1)]C(2n,n)$ for the Catalan numbers. You are asked to do this in Exercise 7.8.1A.

Another example of this type is given in Exercise 7.8.1B.

Exercises

7.8.1A Deduce from Equation 7.73 that, for all positive integers n, $C_n = [1/(n+1)]C(2n,n)$.

Hint: The power series expansion for $\sqrt{1-4x}$, that is, $(1-4x)^{1/2}$ may be derived either from the general binomial theorem

$$(1+x)^\alpha = 1 + \alpha x + \frac{\alpha(\alpha-1)}{2!}x^2 + \frac{\alpha(\alpha-1)(\alpha-2)}{3!}x^3 + \dots$$

in the case where x is replaced by $-4x$ and α by $\frac{1}{2}$, or from Taylor's theorem, namely, that, for an infinitely differentiable function f,

$$f(x) = f(0) + f'(0)x + \frac{f''(0)}{2!}x^2 + \frac{f'''(0)}{3!}x^3 + \dots.$$

7.8.1B Find a formula for the generating function of the sequence $\{a_n\}$ defined by $a_0 = a_1 = 1$, and for $n \geq 2$, $a_n = a_{n-1} + \sum_{i=0}^{n-2} a_i a_{n-i-2}$.

CHAPTER 8

Partitions and Generating Functions

8.1 THE GENERATING FUNCTION FOR THE PARTITION NUMBERS

Chapter 7 showed the usefulness of using a generating function to determine an explicit formula for the nth term of the Fibonacci sequence. This chapter further demonstrates that usefulness in establishing results about partitions. In Chapter 1 we raised the following question.

PROBLEM 8

Is it true that for each integer n the number of ways of writing n as the sum of odd positive integers is the same as the number of ways of writing n as the sum of positive integers that are all different?

In this chapter we show how this and similar questions can be answered.

PROBLEM 8.1

Check that the number of ways of writing 7 as the sum of odd positive numbers is the same as the number of ways of writing 7 as the sum of different positive integers.

Solution

There are five partitions of 7 into parts that are all odd, as follows:

$$7, \quad 5+1+1, \quad 3+3+1, \quad 3+1+1+1+1, \quad 1+1+1+1+1+1+1.$$

There are also five partitions of 7 into distinct parts:

$$7, \quad 6+1, \quad 5+2, \quad 4+3, \quad 4+2+1.$$

We will solve Problem 8 by showing that the generating functions for the sequences of these two types of numbers are identical. Furthermore, by tampering only slightly with this particular generating function, we can obtain a recurrence relation for the number, $p(n)$, of partitions of n that enabled Major P.A. MacMahon, using pencil and paper before the days of electronic computers, to determine the exact value of $p(n)$ for each $n \leq 200$ without having to obtain a precise list of partitions in each case. [When you see the size of $p(200)$ you should be mightily impressed—especially if you have already tried to find, for example, the value of $p(20)$ by listing the partitions of 20.]

The following simple problem leads us in the right direction.

PROBLEM 8.2

What is the coefficient of x^{89} in the following product?

$$(1+x^3+x^6+x^9+....)(1+x^8+x^{16}+x^{24}+....). \tag{8.1}$$

Solution

It is fairly readily seen that when terms in the first paranthesis are multiplied by terms in the second bracket we obtain the term x^{89} in just four ways:

$$x^9 \cdot x^{80}, \quad x^{33} \cdot x^{56}, \quad x^{57} \cdot x^{32}, \text{ and } x^{81} \cdot x^8,$$

and hence the coefficient of x^{89} in the product is 4.

We have seen a problem of this type before—Problem 7.1 in Chapter 7. Looking at the solution of Problem 7.1, we can see that the coefficient of x^n in the product in expression 8.1 is the same as the number of ways of writing 89 in the form $3m_1 + 8m_2$, where m_1 and m_2 are nonnegative integers. In other words, the coefficient of x^n in the product of expression 8.1 is the number of partitions of n using only parts of sizes 3 and 8.

We can readily generalize this. Since a partition of the integer n may use parts of any size ($\leq n$), to obtain a series whose coefficients count *all* the partitions of n, we will have to multiply together (infinitely many) infinite series corresponding to parts of size 1, 2, 3, ... Indeed, we have the following result, which gives us the generating function for the sequence $\{p(n)\}$ of partition numbers.

THEOREM 8.1

$$(1+x+x^2+...)(1+x^2+x^4+...)(1+x^3+x^6+...)... = 1 + p(1)x + p(2)x^2 + p(3)x^3 + ...$$

In other words, if P is the generating function for the sequence $\{p(n)\}$, then

$$P(x) = \prod_{k=1}^{\infty}(1+x^k+x^{2k}+x^{3k}+...). \tag{8.2}$$

Since, for $|x| < 1$, $1 + x^k + x^{2k} + x^{3k} + ... = 1/(1-x^k)$, we may write this last equation even more succinctly as

$$P(x) = \prod_{k=1}^{\infty}\frac{1}{1-x^k}. \tag{8.3}$$

Proof

The terms in the infinite product in Equation 8.2 are obtained by taking, in all possible ways, terms of positive degree from a *finite* number of the infinite series and multiplying these together (see Problem 8.3). Clearly, in seeking the coefficient x^n, we need not consider terms $1 + x^k + x^{2k} + \ldots$, of the product in Equation 8.2, where $k > n$. Thus the terms of degree n have the form

$$[x^1]^{m_1} \times [x^2]^{m_2} \times \ldots \times [x^k]^{m_k},$$

where k is a positive integer and $m_1, m_2, \ldots, m_k$ are nonnegative integers such that

$$m_1.1 + m_2.2 + \ldots + m_k.k = n. \tag{8.4}$$

As we have seen, Equation 8.4 corresponds to a partition of n. Thus the coefficient of x^n in the product of Equation 8.2 counts the number of partitions of n.

PROBLEM 8.3

Use the generating function, P, to evaluate $p(n)$ for $1 \leq n \leq 5$.

Solution

Since we are counting the partitions of the positive integers ≤ 5, we may restrict our attention to the terms in Equation 8.2 of degree at most 5, that is, to the product:

$$(1 + x + x^2 + x^3 + x^4 + x^5)(1 + x^2 + x^4)(1 + x^3)(1 + x^4)(1 + x^5).$$

The first six terms of this product are $1 + x + 2x^2 + 3x^3 + 5x^4 + 7x^5$, giving $p(1) = 1$, $p(2) = 2$, $p(3) = 3$, $p(4) = 5$, and $p(5) = 7$.

We are now ready to tackle Problem 8 using generating functions. We let $p_o(n)$ be the number of partitions n into odd parts and $p_d(n)$ be the number of partitions of n into distinct parts. We let P_o and P_d be the generating functions for the sequences $\{p_o(n)\}$ and $\{p_d(n)\}$, respectively. The proof of the following lemma is straightforward:

LEMMA 8.2

a. $P_o(x) = \dfrac{1}{(1-x)(1-x^3)(1-x^5)(1-x^7)(1-x^9)\ldots}$, that is, $P_o(x) = \prod\limits_{t=1}^{\infty} \dfrac{1}{1 - x^{2t-1}}$.

b. $P_d(x) = (1+x)(1+x^2)(1+x^3)(1+x^4)(1+x^5)\ldots$, that is, $P_d(x) = \prod\limits_{k=1}^{\infty} (1 + x^k)$.

Proof

Since $p_o(n)$ counts the partitions of n into odd parts, the generating function for the sequence $\{p_o(n)\}$ is obtained by just taking the terms in the product in Equation 8.3 corresponding to parts of odd size. Since $p_d(n)$ counts the partitions of n into distinct parts, that is, partitions of n in which there is at most one part of any given size, the generating function for the sequence $\{p_d(n)\}$ is obtained by just taking, from each bracket $(1 + x^k + x^{2k} + x^{3k} + \ldots)$ of Equation 8.2, the terms 1 and x^k, that is, those that correspond to having either no part of size k or just one part of size k.

So to solve Problem 8 we only have to prove that the two generating functions of Lemma 8.2 are identical. Can that really be true? Yes! And the proof is as follows.

THEOREM 8.3

For all positive integers n, $p_d(n) = p_o(n)$.

Proof

Since $(1 - x^{2k}) = (1 + x^k)(1 - x^k)$, it follows that $(1 + x^k) = (1 - x^{2k})/(1 - x^k)$. Hence

$$P_d(x) = (1+x)(1+x^2)(1+x^3)(1+x^4)(1+x^5)\ldots$$
$$= \frac{(1-x^2)(1-x^4)(1-x^6)(1-x^8)(1-x^{10})\ldots}{(1-x)(1-x^2)(1-x^3)(1-x^4)(1-x^5)\ldots}.$$

Now, canceling each term in the numerator of this last expression with the identical term in the denominator leaves us just the terms $(1 - x)$, $(1 - x^3)$, $(1 - x^5)$, ... in the denominator, that is, the terms of the form $(1 - x^{2t-1})$, where t is a positive integer t. Thus

$$P_d(x) = \frac{1}{(1-x)(1-x^3)(1-x^5)\ldots} = \prod_{t=1}^{\infty} \frac{1}{(1-x^{2t-1})} = P_o(x). \tag{8.5}$$

Since the generating functions for the two sequences $\{p_d(n)\}$ and $\{p_o(n)\}$ are identical, their coefficients must be equal. That is, for each positive integer n, we have $p_d(n) = p_o(n)$.

Note that we have proved Theorem 8.3 by establishing the algebraic identity given in Equation 8.5, which could be written as

$$(1+x)(1+x^2)(1+x^3)\ldots = \frac{1}{(1-x)(1-x^3)(1-x^5)\ldots}.$$

In the next section we come to a work saver!

Exercises

8.1.1A Find the coefficient of x^{100} in the following products.
  i. $(1 + x^7 + x^{14} + x^{21} + \ldots)(1 + x^{11} + x^{22} + x^{33} + \ldots)$
  ii. $(1 + x^3 + x^6 + x^9 + \ldots)(1 + x^{14} + x^{28} + x^{42} + \ldots)$

8.1.1B What is the coefficient of x^{24} in the product

$$(1 + x^2)(1 + x^3)(1 + x^5)(1 + x^7)(1 + x^{11})(1 + x^{13})(1 + x^{17})(1 + x^{19})?$$

Interpret the answer in terms of the partitions of 24.

8.1.2A Write down the generating function for the sequence $\{a(n)\}$, where,
  i. For each positive integer n, $a(n)$ is the number of partitions of n in which no even part occurs more than once, and
  ii. For each positive integer n, $a(n)$ is the number of partitions of n in which no odd part occurs more than once.

8.1.2B For each positive integer k, let $q_{k,e}(n)$ be the number of partitions of n into even parts, all of size at most k. Write down the generating function, say $Q_{k,e}$, for the sequence $\{q_{k,e}(n)\}$.

8.1.3A Let $p_{d,o}(n)$ be the number of partitions of n into distinct odd parts.
  i. Calculate $p_{d,o}(n)$ for $1 \le n \le 10$.
  ii. Write down the generating function for the sequence $\{p_{d,o}(n)\}$.

8.1.3B Let $a(n)$ be the number of partitions of n into parts that are not multiples of 3, and let $b(n)$ be the number of partitions of n in which there are not more than two parts of the same size.
  i. Calculate the values of $a(n)$ and $b(n)$ for $1 \le n \le 8$.
  ii. Find the generating functions for the sequences $\{a(n)\}$ and $\{b(n)\}$.
  iii. Prove, by the method of Theorem 8.3, that for each positive integer n, $a(n) = b(n)$.
  iv. How may the result of (c) be generalized?

8.1.4A Let $t(n)$ be the total number of partitions of all integers from 1 to n. Show that the generating function for the sequence $\{t(n)\}$ is $x \mapsto P(x)/(1-x)$ (where P is the generating function for the sequence $\{p(n)\}$).

8.1.4B Prove that $p_2(n) = \lfloor \tfrac{1}{2} n \rfloor + 1$ [this is the result of Theorem 6.4(ii)] by using the identity

$$\frac{1}{(1-x)(1-x^2)} = \frac{1}{2}\left(\frac{1}{(1-x)^2}\right) + \frac{1}{2}\left(\frac{1}{1-x^2}\right).$$

8.1.5A Let $a(n)$ be the number of partitions of n in which only even parts can be repeated, and let $b(n)$ be the number of partitions of n in which the only even parts are multiples of 4. Use the generating function method to show that, for all positive integers n, $a(n) = b(n)$.

8.1.5B Show that the number of partitions of n in which only odd parts may be repeated is equal to the number of partitions of n in which no part occurs more than three times.

8.1.6A i. Show that, for each positive integer n,

$$(1+x^{2^0})(1+x^{2^1})(1+x^{2^2})\ldots(1+x^{2^n}) = 1 + x + x^2 + x^3 + \ldots + x^{2^{n+1}-1},$$

that is, that

$$\prod_{t=0}^{n} 1 + x^{2^t} = \sum_{s=0}^{2^{n+1}-1} x^s.$$

ii. Deduce that

$$1-x = \frac{1}{(1+x^{2^0})(1+x^{2^1})(1+x^{2^2})\ldots}.$$

iii. Comparing the expression in (ii) with the generating function for the sequence $\{c(n)\}$, where $c(n)$ is the number of ways of expressing n as a sum of powers of 2, prove that, for $n \geq 2$, the number of ways of expressing n as a sum of an *even* number of powers of 2 is equal to the number of ways of expressing n as an *odd* number of powers of 2.

8.1.6B Use the generating function method to show that the number of partitions of n into even parts is equal to the number of partitions of n with each part occurring an even number of times.

8.2 A QUICK(ISH) WAY OF FINDING p(n)

The following is a little involved but, like much good mathematics, a huge labor-saving idea in the long run, the pièce de résistance being the formula that is given in Theorem 8.6. We establish the formula via some calculations with generating functions and a wonderful idea, due to Fabian Franklin* (see Theorem 8.5), who described an ingenious way of redistributing the dots in each dot diagram that, initially, has unequal parts.

We begin with some generating function calculations and see how our interest in partitions into distinct parts arises. To ease the way, we look, first, at a particular example.

We have seen that $P_d(x) = (1 + x)(1 + x^2)(1 + x^3)\ldots$ is the generating function for the sequence $\{p_d(n)\}$. Now consider the series

$$P_d(a,x) = (1+ax)(1+ax^2)(1+ax^3)\ldots,$$

which we may think of as a power series in a and so write as $1 + u_1(x)a + u_2(x)a^2 + \ldots$, where the coefficients, $u_i(x)$, are functions of x, also given as power series. For example, $u_3(x)$, the coefficient of a^3, is the sum of all those powers of x that can be obtained by multiplying together exactly three of the powers $x, x^2, x^3, \ldots$. Hence, in expanding $P_d(a, x)$, the coefficients of (for example) those powers of x of degree at most 10 that are involved in $u_3(x)$ arise from the individual contributions as follows:

$$x^{1+2+3} + x^{1+2+4} + x^{1+2+5} + x^{1+2+6} + x^{1+2+7} + x^{1+3+4} + x^{1+3+5} + x^{1+3+6}$$
$$+ x^{1+4+5} + x^{2+3+4} + x^{2+3+5}$$

---

* Fabian Franklin was born in Hungary in 1853 and emigrated to the United States in 1855. He was trained as a civil engineer but achieved a PhD in mathematics in 1880. He taught at Johns Hopkins University until 1895, when he was appointed editor of the *New York Evening Post*. In 1882 he married Christine Ladd, who was also a strong mathematician but who had to struggle in an era that did not give the same opportunities to women as to men. In 1869, she graduated from Vassar, a women's college where there was little mathematics in the curriculum. She sought admission to Johns Hopkins, then an all-male university, in 1878. She had the support of James Sylvester, but although she was allowed to attend his lectures, she was not granted formal admission to the university. Franklin died in 1939.

so that $u_3(x) = x^6 + x^7 + 2x^8 + 3x^9 + 4x^{10} + \dots$.

Clearly, the coefficient of x^n in $u_3(x)$ gives us the number of partitions of n into three *distinct*, that is, *unequal*, parts. In particular, the number of partitions of 10 into three *distinct* parts is the coefficient, namely 4, of x^{10} corresponding to the four partitions

$$10 = 1+2+7 = 1+3+6 = 1+4+5 = 2+3+5.$$

Even more generally, the number of partitions of n into any number of distinct parts is the coefficient of x^n in the infinite sum $u_1(x) + u_2(x) + u_3(x) + \dots$, which we may write as $(u_1(x) + u_3(x) + u_5(x) + \dots) + (u_2(x) + u_4(x) + u_6(x) + \dots)$, where the coefficient of x^n in the first paranthesis counts the number of partitions of n into an odd number of distinct parts and that in the second bracket the number of partitions of n into an even number of distinct parts.

If we now put $a = -1$ in $P_d(a, x)$, we obtain

$$(1-x)(1-x^2)(1-x^3)\dots = 1 - u_1(x) + u_2(x) - u_3(x) + \dots \\ 1 + (u_2(x) + u_4(x) + u_6(x) + \dots) - (u_1(x) + u_3(x) + u_5(x) + \dots). \tag{8.6}$$

This shows us that the coefficient of x^n in Equation 8.6 is the coefficient of x^n in $1 + (u_2(x) + u_4(x) + u_6(x) + \dots)$ minus the coefficient of x^n in $(u_1(x) + u_3(x) + u_5(x) + \dots)$. So, if we let $u_{e,d}(n)$ be the number of partitions of n into an even number of distinct parts ("u" for "*unequal*") and $u_{o,d}(n)$ be the number of partitions of n into odd number of distinct parts, we have just proved the following theorem.

THEOREM 8.4

The coefficient of x^n in the infinite product $(1-x)(1-x^2)(1-x^3)\dots$ is $u_{e,d}(n) - u_{o,d}(n)$.

Our next task is to try to get more to grips with this quantity $u_{e,d}(n) - u_{o,d}(n)$. This clearly requires that we look at decompositions of n into distinct parts. We shall see that the above quantity is, quite often, zero!

THEOREM 8.5

For all positive integers n, $u_{e,d}(n) - u_{o,d}(n) = 0$ except when n is of the form $\frac{1}{2}(3k^2 \pm k)$, in which case $u_{e,d}(n) - u_{o,d}(n) = (-1)^k$.

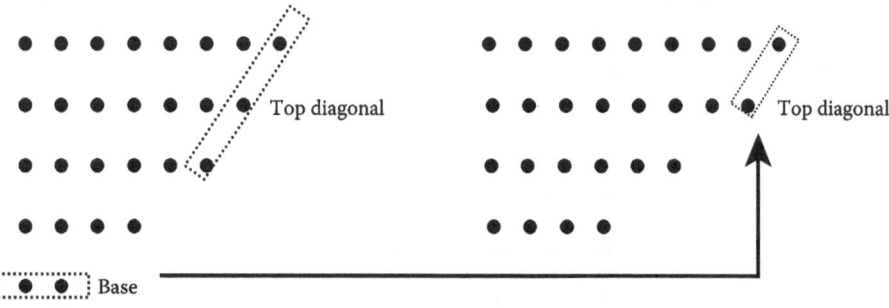

FIGURE 8.1

Setting the scene. To introduce the proof gently, we begin with an example that indicates the general method. Consider, then, a typical partition into distinct parts, say $27 = 8 + 7 + 6 + 4 + 2$, whose dot diagram is shown on the left-hand side of Figure 8.1. This has five distinct parts, the smallest of which, 2, we call the *base*.

Notice that the final dots in the first three (but not four) rows lie on a line at 45° to the horizontal. Such a line of maximal length, and starting on the top row, is called the *top diagonal*. To change the given partition into one with four distinct parts, we move the dots of the base, adding one dot to the end of each of the first two original rows, both of which end with a dot on the top diagonal. The new partition, $27 = 9 + 8 + 6 + 4$, is shown on the right-hand side of Figure 8.1.

This process converts one partition into distinct parts into another partition of the same number into distinct parts, but with one fewer part. The process can be reversed by taking the two dots that form the top diagonal in the new partition and returning them to their former base position. Since the first partition has one more part than the second, it has an even number of parts if and only if the second partition has an odd number of parts.

At first sight this process sets up a one–one correspondence between the partitions of n into an even number of distinct parts, and the partitions of n into an odd number of distinct parts. If this always worked, we would have a proof that $u_{e,d}(n) - u_{o,d}(n) = 0$, for all positive integers n.

But does this argument ever break down? Well, yes! For example, the (forward) process breaks down when the number of dots in the base exceeds the number of dots in the top diagonal! Fortunately, this situation can be kept under control. Here is the proof.

PROOF OF THEOREM 8.5

Let D be any dot diagram corresponding to a partition of n into distinct parts and let b, t be the number of dots in the base and the top diagonal of D, respectively. We deal with the three cases (i) $b < t$, (ii) $b = t$, and (iii) $b > t$ separately.

Case (i), $b < t$. Here the b dots of the base are moved to sit at the right-hand end of the first b rows of the dot diagram, as described above. In this way, a diagram with an even number of distinct parts has associated with it a diagram with an odd number of distinct parts, and vice versa.

Case (ii), $b = t$. Here the same argument applies *except* where the final dot of the base is also the lowest dot of the top diagonal, so that removing all of the base would, at the same time, also remove the last element of the top diagonal—as in Figure 8.2, which corresponds to the partition $22 = 7 + 6 + 5 + 4$.

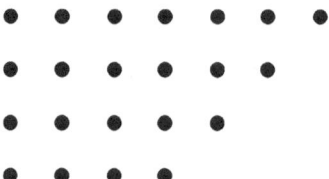

FIGURE 8.2

In general (with $b = t$), this case can arise only when we have a partition in which the number of parts equals the number of dots in the base, and each part has one fewer dot than the part above it. Thus it occurs only when we have a partition of the form

$$n = (2b - 1) + (2b - 2) + \ldots + (b + 1) + b$$

with b dots in the base and b parts. Consequently, in this case we have that $n = \frac{1}{2}b(3b-1) = \frac{1}{2}(3b^2 - b)$.

Thus this case can arise only when n has the form $\frac{1}{2}(3b^2 - b)$, and in this case there is just one partition of n that cannot be matched with another partition in the above manner. In our example b is even, and the partition that is the only one we cannot match up in the above manner has an even number of parts. All the other partitions of 22 into an even number of distinct parts can be matched one-one with the partitions of 22 into a distinct number of odd parts. Thus $u_{e,d}(22) - u_{o,d}(22) = 1$.

Generally, when we have the case of $b = t$ with b even, then $u_{e,d}(n) - u_{o,d}(n) = 1$. Likewise, if b had been odd, then the unmatchable partition would have been one into an odd number of distinct parts so that $u_{d,e}(n) - u_{d,o}(n) = -1$. We can thus summarize these two cases by saying that when n has the form $\frac{1}{2}(3b^2 - b)$, then $u_{e,d}(n) - u_{o,d}(n) = (-1)^b$.

Finally, case (iii), $b > t$. This time the matching between partitions into an even and an odd number of distinct parts is achieved by *removing* the dots from the top diagonal and forming a new base with them. This could fail only when there is a dot that is both in the base and on the top diagonal. However, where $t \leq b - 2$, the former base still has $b - 1$ dots when the final dot, which forms part of the top diagonal, is removed from it, whereas the new base has $b - 2$ dots in it. This case is illustrated in Figure 8.3, where the partition $34 = 10 + 9 + 8 + 7$ of 34 into four distinct parts on the left is converted into the partition $34 = 9 + 8 + 7 + 6 + 4$ of 34 into five distinct parts on the right.

However, a transformation of this kind is not possible in the case where $t = b - 1$, and we have a partition into t parts. This is illustrated in Figure 8.4, which corresponds to the partition $15 = 6 + 5 + 4$.

FIGURE 8.3

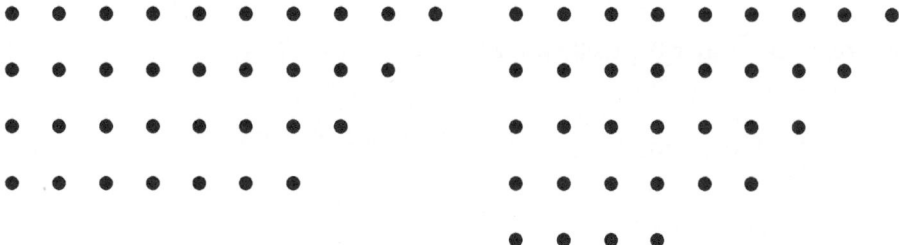

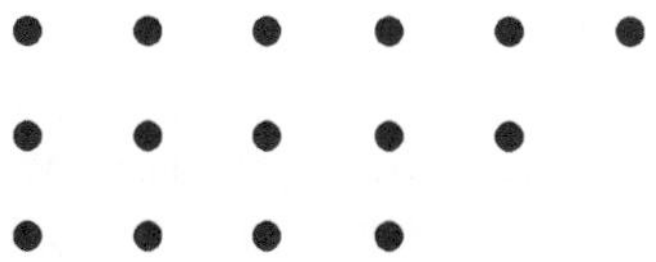

FIGURE 8.4

In general, this case arises only when we have a partition of the form

$$n = 2t + (2t - 1) + (2t - 2) + \ldots + (t + 2) + (t + 1),$$

and so $n = \frac{1}{2}t(3t+1) = \frac{1}{2}(3t^2 + t)$. Further, if t is even, then the unmatched partition has an even number of parts, and if t is odd, it has an odd number of parts. Therefore, in these cases $u_{e,d}(n) - u_{o,d}(n) = (-1)^t$.

This completes the proof of Theorem 8.5.

Now we employ Theorem 8.5 to find our long-awaited recurrence relation for $p(n)$.

THEOREM 8.6

For any positive integer n, we have

$$p(n) = p(n-1) + p(n-2) - p(n-5) - p(n-7) + \ldots$$
$$+ (-1)^{k+1} p(n - \tfrac{1}{2}(3k^2 - k)) + (-1)^{k+1} p(n - \tfrac{1}{2}(3k^2 + k)) + \ldots, \quad (8.7)$$

where the expression on the right-hand side is continued until we reach a value of k for which either $n - \tfrac{1}{2}(3k^2 - k)$ or $n - \tfrac{1}{2}(3k^2 + k)$ is negative.

PROOF OF THEOREM 8.6: FINALE

Theorem 8.5 shows that $u_{e,d}(n) - u_{o,d}(n)$ has the value 0 if n is not of the form $\tfrac{1}{2}(3k^2 \pm k)$ and the value $(-1)^k$ if $n = \tfrac{1}{2}(3k^2 \pm k)$, for some positive integer k. It follows that

$$u_{e,d}(n) - u_{o,d}(n) = \begin{cases} -1, & \text{for } n = 1, 2, \ 12, 15, \ 35, 40, \ldots; \\ +1, & \text{for } n = \ 5, 7, \ 22, 26, \ 51, 57, \ldots; \\ 0, & \text{otherwise.} \end{cases}$$

Hence, from Theorem 8.4 it follows that

$$(1-x)(1-x^2)(1-x^3)\ldots = 1 - x - x^2 + x^5 + x^7 - x^{12} - x^{15} + x^{22} + x^{26} - \ldots, \quad (8.8)$$

an identity due to Euler.*

Finally, by Theorem 8.1 we have that

---

* Leonhard Euler, born in Basel on April 15, 1707, was a powerful and prolific mathematician who made valuable contributions to most areas of mathematics, far too many to list here. We owe the symbols π, e, and i to Euler, as well as the remarkable formula $e^{i\pi} = -1$ that relates them. In 1766 he was called to St. Petersburg by the Empress Catherine to become director of the Academy of Sciences. Despite losing his sight he continued to carry out his duties and to contribute to mathematics until his death on September 18, 1783.

$$1 + p(1)x + p(2)x^2 + p(3)x^3 + \ldots = \frac{1}{(1-x)(1-x^2)(1-x^3)\ldots}. \quad (8.9)$$

It follows from Equation 8.9 that

$$(1 + p(1)x + p(2)x^2 + p(3)x^3 + \ldots)(1-x)(1-x^2)(1-x^3)\ldots = 1.$$

And hence, by Equation 8.8,

$$(1 + p(1)x + p(2)x^2 + p(3)x^3 + \ldots)(1 - x - x^2 + x^5 + x^7 - x^{12} - x^{15} + x^{22} + x^{26} - \ldots) = 1. \quad (8.10)$$

Now, equating the coefficients of x^n in Equation 8.10, we have

$$p(n) - p(n-1) - p(n-2) + p(n-5) + p(n-7) - p(n-12) - p(n-15)$$
$$+ p(n-22) + p(n-26) - \ldots = 0,$$

and hence we obtain the formula of Theorem 8.6.

The recurrence relation given by Theorem 8.6 can be used to determine, reasonably quickly and by pencil and paper, the values of $p(n)$, at least for modest-sized values of n.

PROBLEM 8.4

In Chapter 6, we obtained the values of $p(n)$ for $0 \leq n \leq 10$ as given in Table 8.1.*
Use Theorem 8.6 to evaluate $p(n)$ for $11 \leq n \leq 15$.

TABLE 8.1

n	$p(n)$
0	1
1	1
2	2
3	3
4	5
5	7
6	11
7	15
8	22
9	30
10	42

---

* The comment after Theorem 6.3 explains why we take the value of $p(0)$ to be 1. The other values are taken from Table 6.1.

Solution

By the formula in Equation 8.7 of Theorem 8.6, we have

$$p(11) = p(10) + p(9) - p(6) - p(4) = 42 + 30 - 11 - 5 = 56,$$

$$p(12) = p(11) + p(10) - p(7) - p(5) + p(0) = 56 + 42 - 15 - 7 + 1 = 77,$$

$$p(13) = p(12) + p(11) - p(8) - p(6) + p(1) = 77 + 56 - 22 - 11 + 1 = 101,$$

$$p(14) = p(13) + p(12) - p(9) - p(7) + p(2) = 101 + 77 - 30 - 15 + 2 = 135,$$

$$p(15) = p(14) + p(13) - p(10) - p(8) + p(3) + p(0) = 135 + 101 - 42 - 22 + 3 + 1 = 176.$$

We next give another demonstration of the power (no pun intended!) of generating functions to prove results quickly. We present it in a manner commonly used in mathematics in which the requirements of logical order transcend any thought of detailed explanation to the reader as to how the proof was discovered. It may look like magic!

We let $p_{e,e}(n)$ be the number of partitions of n in which there is an *even* number of *even* parts, and $p_{o,e}(n)$ be the number of partitions of n in which there is an *odd* number of *even* parts. As in Exercise 8.1.3A, $p_{d,o}(n)$ is the number of partitions of n into distinct odd parts.

PROBLEM 8.5

Evaluate $p_{e,e}(n)$, $p_{o,e}(n)$, and $p_{d,o}(n)$ for $n = 5$ and 6.

Solution

There are seven partitions of 5, namely, 5, 4 + 1, 3 + 2, 3 + 1 + 1, 2 + 2 + 1, 2 + 1 + 1 + 1, and 1 + 1 + 1 + 1 + 1. Of these, 5, 3 + 1 + 1, 2 + 2 + 1, and 1 + 1 + 1 + 1 + 1 have an even number of even parts (we include those partitions with zero even parts), and so $p_{e,e}(5) = 4$. The other three partitions have an odd number of even parts, and so $p_{o,e}(5) = 3$. The only partition of 5 into distinct odd parts is 5. Hence $p_{d,o}(5) = 1$.

The partitions of 6 in which there is an even number of even parts are 5 + 1, 4 + 2, 3 + 3, 3 + 1 + 1 + 1, 2 + 2 + 1 + 1, and 1 + 1 + 1 + 1 + 1 + 1. Hence $p_{e,e}(6) = 6$. The partitions of 6 in which there is an odd number of even parts are 6, 4 + 1 + 1, 3 + 2 + 1, 2 + 2 + 2, and 2 + 1 + 1 + 1 + 1. Hence $p_{o,e}(6) = 5$. There is just one partition of 6 into distinct odd parts, namely, 5 + 1. Hence $p_{d,o}(6) = 1$.

Note that in both these cases $p_{e,e}(n) - p_{o,e}(n) = p_{d,o}(n)$. It would be very bold to conjecture, on the basis of just these two cases, that this is always true. But this is just what Derrick Henry Lehmer* proved.

THEOREM 8.7

For each positive integer n, $p_{e,e}(n) - p_{o,e}(n) = p_{d,o}(n)$.

---

* Derrick (Dick) Henry Lehmer was born in Berkeley, California, on February 23, 1905, and died there on May 22, 1991. His father, Derrick Norman Lehmer, was also a mathematician. In 1928 Dick Lehmer married Emma Markovna Trotskaia, who was also a distinguished mathematician and who lived to be 100. Dick Lehmer was a pioneer of the application of mechanical devices and, later, electronic computers to problems in number theory, and he wrote papers on a wide range of topics in number theory and combinatorics.

We promised you a bit of magic, but we offer a hint now to help you see what is afoot. The move we describe is quite clever!

As we have seen, the total number, $p(n)$, of partitions of n is the coefficient of x^n in

$$(1+x^2+x^4+\ldots)(1+x^4+x^8+\ldots)(1+x^6+x^{12}+\ldots)\ldots$$
$$\times (1+x+x^2+\ldots)(1+x^3+x^6+\ldots)(1+x^5+x^{10}+\ldots)\ldots, \quad (8.11)$$

where, in the expression 8.11, we have now put the terms corresponding to the *even* parts of a partition in the first line and the terms corresponding to the *odd* parts in the second line.

We now consider a particular partition of n, which for $1 \leq i \leq s$ has m_i parts of size $2k_i$, where $k_1 < k_2 < \ldots < k_s$, and whose odd parts add up to q. Thus

$$n = m_1(2k_1) + m_2(2k_2) + \ldots + m_s(2m_s) + q. \quad (8.12)$$

The even parts of this partition are obtained by multiplying together the terms $x^{2k_1 m_1}$, $x^{2k_2 m_2}$, ..., $x^{2k_s m_s}$ from the first line of expression 8.11.

Now consider the same terms in the expression

$$(1-x^2+x^4-\ldots)(1-x^4+x^8-\ldots)(1-x^6+x^{12}-\ldots)\ldots$$
$$\times (1+x+x^2+\ldots)(1+x^3+x^6+\ldots)(1+x^5+x^{10}+\ldots)\ldots, \quad (8.13)$$

which is the same as expression 8.11 except for the alternating + and − signs in the first line of expression 8.13. It follows that the multiplier associated with the expression $x^{2k_1 m_1 + 2k_2 m_2 + \ldots + k_s m_s}$ in expression 8.13 is $(-1)^{m_1 + m_2 + \ldots + m_s}$, which is +1 or −1 according as whether $m_1 + m_2 + \ldots + m_s$ is even or odd.

Thus, if in the partition in Equation 8.12 there is an even number of even parts, the multiplier associated with the term $x^{2k_1 m_1 + 2k_2 m_2 + \ldots + 2k_s m_s + q}$ is +1, and if there is an odd number of even parts, the multiplier is −1. It follows that the coefficient of x^n in expression 8.13 counts

> the number of partitions of n with an even number of even parts
> —the number of partitions of n with an odd number of even parts;

that is, the coefficient of x^n in expression 8.13 is $p_{e,e}(n) - p_{o,e}(n)$.

We can rewrite the product in expression 8.13 as

$$\frac{1}{(1+x^2)(1+x^4)(1+x^6)\ldots(1-x)(1-x^3)(1-x^5)\ldots}. \quad (8.14)$$

Thus to prove Theorem 8.7 it is sufficient to prove that the generating function in expression 8.14 for the sequence $\{p_{e,e}(n) - p_{o,e}(n)\}$ is identical to the generating function for the sequence $\{p_{d,o}(n)\}$, which by Exercise 8.1.3A is

$$(1+x)(1+x^3)(1+x^5)\ldots. \quad (8.15)$$

So, here we go!

PROOF OF THEOREM 8.7

It follows from Theorem 8.3 that

$$(1+x)(1+x^2)(1+x^3)\ldots = \frac{1}{(1-x)(1-x^3)(1-x^5)\ldots}.$$

Hence

$$\frac{1}{(1+x^2)(1+x^4)(1+x^6)\ldots(1-x)(1-x^3)(1-x^5)\ldots}$$

$$= \frac{1}{(1+x^2)(1+x^4)(1+x^6)\ldots} \times (1+x)(1+x^2)(1+x^3)\ldots$$

$$= (1+x)(1+x^3)(1+x^5)\ldots,$$

which is the desired result.

We conclude this section with an identity due to Euler, who gave an algebraic proof of Theorem 8.8, for which we give a combinatorial proof. Exercise 8.1.3A(ii) shows that the generating function for the sequence $\{p_{d,\,o}(n)\}$ giving the number of partitions of n into distinct odd parts is $(1+x)(1+x^3)(1+x^5)\ldots$. We can obtain another expression for this generating function by analyzing the dot diagram for a partition of n into distinct odd parts. We illustrate this by looking at the partition $26 = 13 + 7 + 5 + 1$. We represent this partition by a dot diagram where, instead of using rows of dots to represent the parts, we use symmetrical L-shaped arrays of dots, as shown on the left of Figure 8.5. The L shapes appear because we are considering partitions into *odd* parts.

We can split up the dot diagram on the left in the way shown in the middle diagram. In the bottom left-hand corner there is a 4×4 array of dots. The dots to the right of this square, read in columns, form a partition of 5 into parts of size at most 4. The dots above the square, read in rows, form an identical partition, because of the symmetry. If we put these two identical partitions of 5 alongside one another, we get a partition of 10 into even parts of size at most 8, as shown on the right of Figure 8.5.

We now generalize this. Suppose we are given a partition of n into unequal odd parts. We can represent this by an L-shaped array of dots. The bottom left-hand corner will form a square array of dots. If k is the largest integer for which there is a square

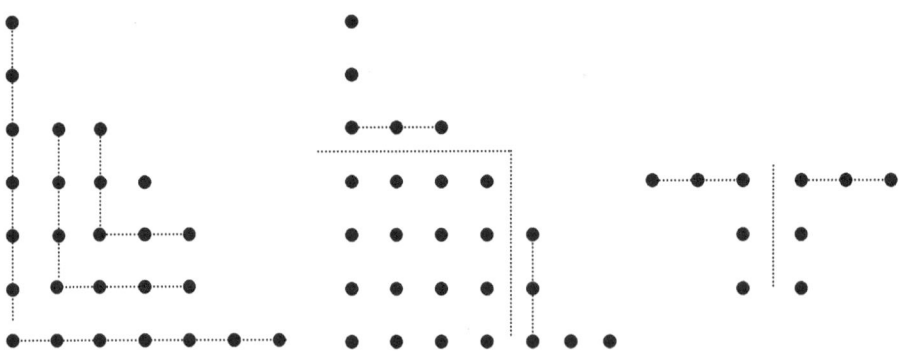

FIGURE 8.5

$k \times k$ array of dots in the dot diagram, then clearly we have $1 \leq k^2 \leq n$, and so $k \leq \lfloor \sqrt{n} \rfloor$. There will be $n - k^2$ dots that are not in this square, and these may be arranged to form a partition of $n - k^2$ into parts that are even and of size at most $2k$.

In this way we obtain a one–one correspondence between the partitions of n into unequal odd parts and partitions of $n - k^2$ into even parts of size at most $2k$, for some integer k with $1 \leq k \leq \lfloor \sqrt{n} \rfloor$. Thus, if $q_{2k,e}(n)$ is the number of partitions of n into even parts of size at most $2k$, we have that

$$P_{d,o}(n) = \sum_{k=0}^{\lfloor \sqrt{n} \rfloor} q_{2k,e}(n - k^2). \tag{8.16}$$

It follows from Equation 8.16 that the generating function $P_{d,o}$ for the sequence $\{p_{d,o}(n)\}$ is given by

$$P_{d,o}(x) = \sum_{n=0}^{\infty} \left(\sum_{k=0}^{\lfloor \sqrt{n} \rfloor} q_{2k,e}(n - k^2) \right) x^n. \tag{8.17}$$

For a given value of k, the sum in Equation 8.17 includes a term $q_{2k,e}(n - k^2) x^n$, provided that $k^2 \leq n$. Hence, when we change the order of summation we get

$$P_{d,o}(x) = \sum_{k=0}^{\infty} \left(\sum_{n=k^2}^{\infty} q_{2k,e}(n - k^2) x^n \right). \tag{8.18}$$

Now we let $Q_{2k,e}$ be the generating function for the sequence $\{q_{2k,e}(n)\}$. Thus,

$$Q_{2k,e}(x) = \sum_{n=0}^{\infty} q_{2k,e}(n) x^n$$

and hence multiplying through by x^{k^2} and then replacing n by $n - k^2$ throughout, we have

$$x^{k^2} Q_{2k,e}(x) = \sum_{n=0}^{\infty} q_{2k,e}(n) x^{n+k^2} = \sum_{n=k^2}^{\infty} q_{2k,e}(n - k^2) x^n. \tag{8.19}$$

It follows from Equations 8.18 and 8.19 that

$$P_{d,o}(x) = \sum_{k=0}^{\infty} x^{k^2} Q_{2k,e}(x). \tag{8.20}$$

Hence, using the result of Exercises 8.1.2B and 8.1.3A, we can deduce from Equation 8.20 the following result.

THEOREM 8.8

$$(1+x)(1+x^3)(1+x^5)\ldots = 1 + \frac{x}{1-x^2} + \frac{x^4}{(1-x^2)(1-x^4)} + \frac{x^9}{(1-x^2)(1-x^4)(1-x^6)} + \ldots \tag{8.21}$$

Note that the identity in Equation 8.21 could also be written more economically as

$$\prod_{k=0}^{\infty}(1+x^{2k+1}) = 1 + \sum_{n=1}^{\infty} x^{n^2} \prod_{k=1}^{n} \frac{1}{1-x^{2k}}.$$

Exercises

8.2.1A Use Theorem 8.6 to calculate $p(n)$ for $16 \leq n \leq 20$.
8.2.1B Use Theorem 8.6 to calculate $p(25)$.
8.2.2A Find the values of $u_{e,d}(n)$ and $u_{o,d}$ for $1 \leq n \leq 10$. Hence, verify that the formula of Theorem 8.5 holds for this range of values of n.
8.2.2B Prove the following analogue of Theorem 8.8:

$$(1+x^2)(1+x^4)(1+x^6)\ldots = 1 + \frac{x^2}{(1-x^2)} + \frac{x^6}{(1-x^2)(1-x^4)} + \frac{x^{12}}{(1-x^2)(1-x^4)(1-x^6)} + \ldots,$$

that is,

$$\prod_{n=1}^{\infty}(1+x^{2n}) = 1 + \sum_{n=1}^{\infty} x^{n(n+1)} \prod_{k=1}^{n} \frac{1}{1-x^{2k}}.$$

This is another result due to Euler.

8.3 AN UPPER BOUND FOR THE PARTITION NUMBERS

In this section we show how the generating function for the partition numbers, $p(n)$, may be used to obtain an upper bound for them. More precisely, we show that $p(n)$ does not grow as fast as the exponential function, e^n. We use only elementary methods, and, as a result, our upper bound is very crude. In the next section we discuss a method that is too sophisticated to be given in this book but provides an asymptotic estimate for $p(n)$.

We have seen in Section 8.1 that the generating function, P, for the sequence $\{p(n)\}$ of partition numbers is given by the formula in Equation 8.3, which we repeat for convenience:

$$P(x) = \prod_{k=1}^{\infty} \frac{1}{1-x^k}. \qquad [(8.3)]$$

It can be shown that the product on the right-hand side of Equation 8.3 converges for $0 < x < 1$.

From Equation 8.3, by taking the natural logarithm of each side,* we have that

$$\log(P(x)) = \sum_{k=1}^{\infty} \log\left(\frac{1}{1-x^k}\right). \qquad (8.22)$$

---

* We use here the "logarithm of a product is the sum of the logarithms" formula, that is, $\log(ab) = \log(a) + \log(b)$, extended to infinite products. This requires that we are dealing with convergent products of positive numbers. The proof uses the fact that the natural logarithm function is continuous.

We have the standard series $\log(1 - x) = - x - (x^2/2) - (x^3/3) - \ldots$, that is, $\log(1-x) = -\sum_{m=1}^{\infty}(x^m/m)$, which is valid for $0 < x < 1$, and so, substituting x^k for x,

$$\log\left(\frac{1}{1-x^k}\right) = -\log(1-x^k) = \sum_{m=1}^{\infty} \frac{(x^k)^m}{m}, \tag{8.23}$$

and hence, by Equations 8.22 and 8.23,

$$\log(P(x)) = \sum_{k=1}^{\infty}\sum_{m=1}^{\infty} \frac{x^{km}}{m}. \tag{8.24}$$

All the terms in the double series in Equation 8.24 are positive. Hence we can interchange the order of summation to give

$$\log(P(x)) = \sum_{m=1}^{\infty}\sum_{k=1}^{\infty} \frac{x^{km}}{m}$$

$$= \sum_{m=1}^{\infty}\frac{1}{m}\sum_{k=1}^{\infty}(x^m)^k. \tag{8.25}$$

Now, $\sum_{k=1}^{\infty}(x^m)^k$ is a geometric series with common ratio x^m that for $0 \le x \le 1$ converges with sum $x^m/(1 - x^m)$. Thus we can deduce from Equation 8.25 that

$$\log(P(x)) = \sum_{m=1}^{\infty}\frac{1}{m}\left(\frac{x^m}{1-x^m}\right). \tag{8.26}$$

Now, because $0 < x < 1$,

$$\frac{1-x^m}{1-x} = 1 + x + x^2 + \ldots + x^{m-1} > mx^{m-1},$$

and hence

$$\frac{x^m}{1-x^m} < \frac{1}{m}\left(\frac{x}{1-x}\right).$$

Therefore it follows from Equation 8.26 that

$$\log(P(x)) < \left(\frac{x}{1-x}\right)\sum_{m=1}^{\infty}\frac{1}{m^2}$$

$$= \frac{\pi^2}{6}\left(\frac{x}{1-x}\right), \tag{8.27}$$

using the well-known fact that the series $\sum_{m=1}^{\infty}(1/m^2)$ converges with sum $\pi^2/6$.

Each term in the series for $P(x)$ is positive. Hence $P(x)$ is larger than any single term in the series. That is, for each positive integer n,

$$p(n)x^n < P(x),$$

and hence

$$p(n) < \frac{1}{x^n} P(x). \tag{8.28}$$

It follows that

$$\log(p(n)) < n\log\left(\frac{1}{x}\right) + \log(P(x)). \tag{8.29}$$

It is a standard inequality that, for $0 < h$, $\log(1 + h) < h$. Putting $h = (1/x) - 1$ in this inequality gives

$$\log\left(\frac{1}{x}\right) < \frac{1}{x} - 1. \tag{8.30}$$

By Equations 8.27, 8.29, and 8.30,

$$\log(p(n)) < n\left(\frac{1}{x} - 1\right) + \frac{\pi^2}{6}\left(\frac{x}{1-x}\right). \tag{8.31}$$

We now apply calculus to find the minimum value of the right-hand side of inequality 8.31. We put

$$f(x) = n\left(\frac{1}{x} - 1\right) + \frac{\pi^2}{6}\left(\frac{x}{1-x}\right).$$

Then

$$f'(x) = -\frac{n}{x^2} + \frac{\pi^2}{6(1-x)^2} \quad \text{and} \quad f''(x) = \frac{2n}{x^3} + \frac{\pi^2}{3(1-x)^3}.$$

Thus

$$f'(x) = 0 \Leftrightarrow 6n(1-x)^2 = \pi^2 x^2 \Leftrightarrow \sqrt{6n}(1-x) = \pm \pi x \Leftrightarrow x = \frac{\sqrt{6n}}{\sqrt{6n} \pm \pi}.$$

It follows that when $x = \sqrt{6n}/(\sqrt{6n} + \pi)$, we have $0 < x < 1$, $f'(x) = 0$, and $f''(x) > 0$. Thus $x = \sqrt{6n}/(\sqrt{6n} + \pi)$ gives the minimum value of $f(x)$, for $0 < x < 1$, namely,

$$f\left(\frac{\sqrt{6n}}{\sqrt{6n} + \pi}\right) = \pi\sqrt{\frac{2n}{3}}.$$

Hence, from inequality 8.31 we can deduce that $\log(p(n)) < \pi\sqrt{2n/3}$ and therefore that

$$p(n) < e^{\pi\sqrt{2n/3}}.$$

Thus, we have proven the following theorem.

THEOREM 8.9

For each positive integer n,

$$p(n) < e^{\pi\sqrt{2n/3}}. \qquad (8.32)$$

The upper bound given for $p(n)$ in Theorem 8.9 is very crude. For example, it gives

$$p(100) < 1.4 \times 10^{11},$$

whereas in fact

$$p(100) = 190{,}569{,}292.$$

This is only to be expected because the inequality $p(n) < (1/x^n)P(x)$, used in the proof, came from neglecting all but one of the terms in the series for $P(x)$. Also, we have used only elementary methods. However, Theorem 8.9 is good enough to show that $p(n)$ does not grow as fast as the exponential function. Indeed, for $n > 2\pi^2/3$, $e^{\pi\sqrt{2n/3}} < e^n$. This should be contrasted with Theorem 6.7, which shows that $p(n)$ grows faster than any power of n.

In the next section we describe the Hardy–Ramanujan formula, which approximates $p(n)$ very accurately and enables us to provide an asymptotic formula that describes how fast $p(n)$ grows.

8.4 THE HARDY–RAMANUJAN FORMULA

Most of the applications of generating functions in this chapter have used algebraic or combinatorial methods. The Hardy–Ramanujan asymptotic formula for $p(n)$, which we alluded to in Chapter 6, arises from a sophisticated use of complex analysis. We give a brief sketch of this work without going into any of the technical details.

So far we have thought of generating functions as having as their domains intervals of real numbers. Since these functions are defined by power series, we can equally well regard them as functions defined for complex numbers. In this way the powerful methods of complex analysis become available. For example, the generating function for the sequence $\{p(n)\}$ of partition numbers, regarded as a complex function, is the function given by

$$P(z) = \prod_{k=1}^{\infty} \frac{1}{1-z^k} \text{ for } z \in \mathbb{C} \text{ with } |z| < 1,$$

since the infinite product converges for $|z|<1$. The coefficients in the power series $\sum_{n=0}^{\infty} a_n z^n$ for a complex function, f, are given by Cauchy's formula

$$a_n = \frac{1}{2\pi i} \int_C \frac{f(z)}{z^{n+1}} dz,$$

where C is a suitably chosen contour. Since $p(n)$ is the coefficient of z^n in the power series for the generating function P, it is given by

$$p(n) = \frac{1}{2\pi i} \int_C \frac{P(z)}{z^{n+1}} dz, \qquad (8.33)$$

where C is any simple closed contour that surrounds the origin and is wholly inside the unit disk, $D_1 = \{z \in \mathbb{C}: |z|<1\}$.

It follows from this that the values of $p(n)$ can be approximated by estimating the contour integral in Equation 8.33. This is, however, more easily said than done. The biggest contributions to the values of the integral arise from parts of the contour nearest to the singularities of the integrand. The function P has a singularity at each point $z \in \mathbb{C}$, where $z^k = 1$ for some positive integer k. Thus P has singularities at all the complex roots of 1, and these are distributed densely on the boundary of the unit disk, D_1. Thus every point of D_1 is an essential singularity of P. This leads to tremendous technical difficulties.

In a celebrated paper published in 1918, G.H. Hardy and S. Ramanujan[*] overcame these difficulties and derived a remarkable formula that approximates $p(n)$ very accurately. It is not possible to describe their method here. However, it is possible to give some idea of the formula that they derived.

The first two terms of their formula are

$$\frac{1}{2\sqrt{2}\pi} \frac{d}{dn}\left(\frac{e^{(1/3)\sqrt{2(n-(1/24))}\pi}}{\sqrt{n-(1/24)}}\right) + \frac{(-1)^n}{2\pi} \frac{d}{dn}\left(\frac{e^{(1/6)\sqrt{2(n-(1/24))}\pi}}{\sqrt{n-(1/24)}}\right). \qquad (8.34)$$

Here d/dn denotes differentiation with respect to n. We use $HR_2(n)$ for the formula in expression 8.34. Hardy and Ramanujan were able to use the first term of their series to derive the following asymptotic formula for $p(n)$:

$$p(n) \sim \frac{1}{4\sqrt{3}n} e^{\pi\sqrt{(2/3)n}}. \qquad (8.35)$$

---

[*] G.H. Hardy and S. Ramanujan, Asymptotic Formulae in Combinatory Analysis, *Proceedings of the London Mathematical Society*, (2), 17 (1918), pp 75–115.

TABLE 8.2

n	p(n)	$HR_2(n)$
1	1	1.04038
2	2	2.06478
3	3	2.93888
4	5	5.02656
5	7	7.02937
6	11	10.93145
7	15	15.04265
8	22	22.05504
9	30	29.86557
10	42	42.06952
20	627	627.05738
30	5,604	5,603.54788
40	37,338	37,338.04032
50	204,226	204,226.7970
60	966,467	966,464.8362
70	4,087,968	4,087,969.043
80	15,796,476	15,796,477.76
90	56,634,173	56,634,167.93
100	190,569,292	190,569,293.7
200	3,972,999,029,388	3.97300×10^{12}
300	9,253,082,936,723,602	9.25308×10^{15}
400	6,727,090,051,741,041,926	6.72708×10^{18}

The great accuracy of the Hardy–Ramanujan formula can be seen from the values in Table 8.2. The values of $p(n)$ in this table have been calculated using Theorem 8.6.

8.5 THE STORY OF HARDY AND RAMANUJAN

The story of Hardy and Ramanujan has often been told, but it is well worth retelling. Hardy was christened Godfrey Harold, but he was known to most people by his initials G.H. He was born in Cranleigh, Surrey, in England on February 7, 1877. His father was art master and bursar at Cranleigh School. After attending this school, Hardy went as a scholar to Winchester and then to Trinity College, Cambridge. Although his family was not wealthy, they had good connections in the English educational world, and after Hardy's mathematical talent had been noticed at an early age, there was no difficulty in obtaining for him the best mathematical education then available in England.

In fact, this was not at a very high level. When Hardy arrived at Trinity College in 1896, pure mathematics at Cambridge was rather in the doldrums. Although England had produced some good mathematicians during the nineteenth century, they were cut off from continental European developments in mathematics, particularly developments in analysis associated with such mathematicians as Cauchy, Dedekind, and Weierstrass, among others. The reason for this isolation goes back to the seventeenth century, when a dispute broke out between English and continental mathematicians as to whether Newton or Leibniz should be given the credit for having been the first to develop the basic ideas of differential and integral calculus. To demonstrate their loyalty, English mathematicians continued to use Newton's notation, instead of the superior notation due to Leibniz—the dy/dx notation that is still common today.

Hardy's introduction to modern mathematical analysis came from the applied mathematician A.E.H. Love,[*] who advised him to read Jordan's *Course d'Analyse*. Hardy's first book, *A Course of Pure Mathematics*, first published in 1908, was one of the earliest books published in England to give a rigorous presentation of the basic concepts of mathematical analysis in the modern style. Unusually for a text at this level, Hardy's book was still in print 100 years after it was first published.

Hardy remained as a lecturer in Cambridge until 1919, when he succeeded to the Savilian Chair of geometry in Oxford. He returned to Cambridge in 1931, when he became the Sadlerian professor of pure mathematics. He retired from this Chair in 1942 and died in 1947.

Hardy's mathematical work is remarkable for two great collaborations. In 1912 he wrote the first of a large number of papers with J.E. Littlewood. Altogether Hardy and Littlewood published 93 joint papers (as well as two papers written together with G. Pólya and one with E. Landau). Hardy, Littlewood, and Pólya also jointly wrote the book *Inequalities*.[†] Hardy and Littlewood's collaboration continued for the rest of Hardy's life, with their final joint paper being published in 1948 after Hardy's death. In the long history of mathematics there is no other example of such a fruitful partnership lasting so long.

Early in 1913, while Hardy was still in Cambridge, he received a letter from India. Thus began his association with the Indian mathematician Srinivasa Ramanujan Aiyangar.[‡] Ramanujan was born to a poor Brahmin family in Erode, a town 175 km south of Bangalore, in southern India, on December 22, 1887. His mathematical ability appeared early. After leaving school he entered the Government College at Kumbakonam in 1904 with a scholarship. Due to his neglect of nonmathematical subjects he failed an examination and lost his scholarship. For the same reason he failed another examination in 1907 and had to abandon the hope of further education, but he continued to work at mathematics. In 1909 he married and needed to obtain a job to support his wife, Janaki. Eventually in 1912 he became a clerk with the Madras Port Trust.

In his search for employment in 1909, Ramanujan visited V. Ramaswamy Iyer, the founder of the Indian Mathematical Society. With the encouragement of Ramasawamy Iyer and Sheshu Aiyar, one of his professors at the Government College, Ramanujan wrote on January 16, 1913, to Hardy, enclosing over 100 mathematical formulas that he had obtained. Hardy recognized some of Ramanujan's formulas. Of others he wrote,

---

[*] Augustus Edward Hough Love was born in 1863. He became Sedleian professor of natural philosophy at the University of Oxford in 1899, where he remained until his death in 1940. He is renowned especially for his books on the mathematical theory of elasticity and on geodynamics.

[†] G.H. Hardy, *A Course of Pure Mathematics*, Cambridge University Press, Cambridge, first edition 1908.

[‡] There is an excellent biography of Ramanujan: Robert Kangel, *The Man Who Knew Infinity*, Scribners, New York and London, 1991. This biography also includes a good deal of information about Hardy.

The formulae (1.10)–(1.13) are on a different level and obviously both difficult and deep. An expert in elliptic functions can see at once that (1.13) is derived somehow from the theory of "complex multiplication", but (1.10)–(1.12) defeated me completely; I had never seen anything the least like them before. A single look at them is enough to show that they could only be written down by a mathematician of the highest class. They must be true because, if they were not true, no one would have the imagination to invent them. Finally (you must remember that I knew nothing of Ramanujan, and had to think of every possibility), the writer must be completely honest, because great mathematicians are commoner than thieves or humbugs of such incredible skill.*

Hardy made strenuous efforts to bring Ramanujan to England, initially against Ramanujan's own wishes. However, by April 1914 Ramanujan was in Cambridge, with financial support from the Government of Madras and Trinity College. Although Ramanujan was now free from financial worries and could devote himself full-time to mathematics, his health soon deteriorated. He was elected as fellow of the Royal Society and a fellow of Trinity College in 1918. In 1919 his health had sufficiently recovered to enable him to travel home to India, but his health again failed and he died at a tragically early age on April 26, 1920, in Madras.

Hardy and Ramanujan are associated not only by their mathematical work but also in one of the most often told stories about mathematicians. The story merits repetition, since, apart from anything else, it is marked out from many of the anecdotes that litter the history of mathematics by being directly attested by one of the participants. We tell the story in Hardy's own words:

> I remember once going to see him when he was lying ill at Putney. I had ridden in taxi-cab No. 1729, and remarked that the number (7·13·19) seemed to me rather a dull one, and I hoped it was not an unfavourable omen. "No", he replied "it is a very interesting number; it is the smallest number expressible as a sum of two cubes in two different ways."†

Hardy summed up Ramanujan's mathematical achievement as follows:

> It was his insight into algebraic formulae, transformations of infinite series, and so forth, that was most amazing. On this side most certainly I have never met his equal, and I can compare him only with Euler or Jacobi. He worked, far more than the majority of modern mathematicians, by induction from numerical examples; all of his congruence properties of partitions, for example, were discovered in this way. But with his memory, his patience, and his power of calculation, he combined the power of generalization, a feeling for form, and a capacity for rapid modification of

---

* G.H. Hardy, *Ramanujan, Twelve Lectures on Subjects Suggested by His Life and Work*, Cambridge University Press, Cambridge, 1940, p. 9.
† G.H. Hardy, P. V. Seshu Aiyar, and B. M. Wilson (editors), *Collected Papers of Srinivasa Ramanujan*, Cambridge University Press, Cambridge, 1927, p. xxxv. Note that $1729 = 1^3 + 12^3 = 9^3 + 10^3$.

his hypotheses, that were often really startling, and made him, in his own peculiar field, without a rival in his day.

It is often said that it is much more difficult now for a mathematician to be original than it was in the great days when the foundation of modern analysis was laid; and no doubt in a measure it is true. Opinions may differ as to the importance of Ramanujan's work, the kind of standard by which it should be judged, and the influence which it is likely to have on the mathematics of the future. It has not the simplicity and the inevitableness of the very greatest work; it would be greater if it were less strange. One gift it has which no one can deny, profound and invincible originality. He would probably have been a greater mathematician if he had been caught and tamed a little in his youth; he would have discovered more that was new, and that, no doubt, of greater importance. On the other hand he would have been less of a Ramanujan, and more of a European professor, and the loss might have been greater than the gain.*

Writing 13 years later, Hardy endorsed this judgment except for the last sentence, which he described as "ridiculous sentimentalism."

---

* Hardy et al., *Collected Papers of Ramanujan*, pp. xxxv–xxxvi.

CHAPTER 9

Introduction to Graphs

9.1 GRAPHS AND PICTURES

As we remarked in Chapter 1, graphs (which comprise points—usually called *vertices*—some pairs of which are joined by arcs—usually called *edges*) occur in many practical situations, for example, in representing, constructing, and improving on road, rail, electrical, and oil-pipeline networks and in scheduling problems, to name but a few. In some situations (for example, in one-way streets on a road map) a *direction* may also be indicated; in the theory of *directed graphs*, also called *digraphs*, some, maybe all, edges have an associated direction—indicated by an arrow. (In that case there could well be an edge from vertex a to vertex b and a distinct edge from b to a.) We shall, however, have no occasion to consider digraphs in this book.

Graphs also feature in recreational problems, as Problems 9A and 9B from Chapter 1 indicate. We discuss the *bridges of Königsberg* in Section 9.6 and the *utilities problem* in Section 9.5. The *four-color Problem* is discussed in Section 9.8.

Labeling vertices with letters or numbers, which is just another form of "coloring" them, is also employed in the determination of the number of different *isomers* of various chemical compounds, as we shall see in the next chapter.

While it is more natural, convenient, instructive, and entertaining to study graphs by using pictures, these pictures should be regarded as only informal representations. Pictures can give rise to ambiguities, even to false assumptions, since it is not always obvious to the eye that different pictures represent the same graph. Indeed, we have not yet even made precise what we mean by this. Roughly speaking, we mean that the two pictures show that the vertices are joined by edges in the same way. We make this more precise when we give the definition of an *isomorphism* between graphs in Section 9.3.

It turns out that the pictures in Figure 9.1 all represent the same graph called the *Petersen graph*,* as we (and you!) will confirm in Exercise 9.3.1B. The vertices are labeled for later convenience.

---

* This graph is named after Julius Peter Christian Petersen (born June 16, 1839; died August 5, 1910). He did not introduce the graph merely because it looks so pretty. Indeed it was invented to show that several plausible conjectures about certain types of graphs were actually false. Furthermore Alfred Bray Kempe seems to have beaten Petersen to it. He studied the same graph in 1886. There is more information about Kempe in a footnote in Section 9.8.

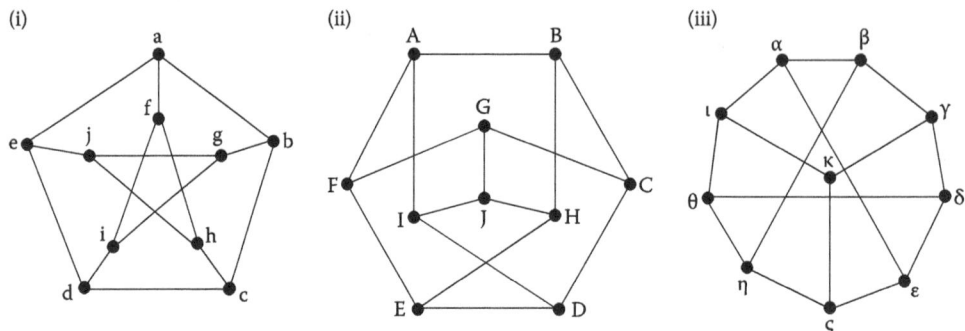

FIGURE 9.1

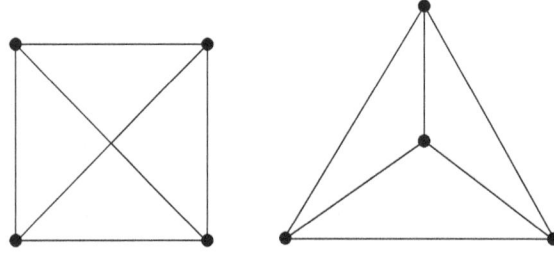

FIGURE 9.2

In these pictures the vertices are indicated by the solid dots (•). One point to notice is that we have allowed edges to cross each other at points that are not vertices. In general this is unavoidable. Indeed it turns out that only relatively few graphs, naturally called *planar graphs* (see Definition 9.8), can be exhibited in two dimensions without having to draw edges crossing other than at a vertex.

Sometimes a graph that is drawn so that two edges cross at a point that is not a vertex can be redrawn so as to avoid this. You can see an example of this in Figure 9.2.

You will naturally now ask if the Petersen graph is planar. That is, is there yet another way to draw the graph shown in Figure 9.1 so that the edges cross only at vertices? This is a good question that we ask you to answer in Exercises 9.5.2B and 9.5.3A.

Exercises

9.1.1A Try to formulate a precise definition of what is meant by *two pictures represent the same graph*.

9.1.1B Redraw the graph shown in Figure 9.3 so that edges meet only at vertices.

9.2 GRAPHS: A PICTURE-FREE DEFINITION

Since, in all but the study of planar graphs, the only purpose of drawing an edge is to indicate that its endpoints are joined, it is immaterial how the edges are drawn. The following formal picture-free definition seems to capture our ideas.

DEFINITION 9.1

A *graph* G is a set V of elements, called *vertices* (or *nodes* or *points*), together with a set E of *pairs* of distinct elements of V, called the *edges* (or *arcs*) of G. We write this as $G = (V,E)$.

For example, we first consider the graph $G = (V,E)$, where $V = \{a,b,c,d\}$ and $E = \{\{a,b\},\{a,c\},\{a,d\}\}$. We see that G has four vertices, one of which is joined by an edge to the other three. We can picture this graph as in Figure 9.4.

It is rather cumbersome to write an edge as a set, say $\{x, y\}$, of two vertices. So, where it is unlikely to lead to confusion or ambiguity, we will instead write xy for the edge joining the vertex x to the vertex y.

PROBLEM 9.1

Draw pictures for each of the following graphs:
  i. $V = \{e, f, g, h\}$, $E = \{ef, fg, fh\}$
  ii. $V = \{i, j, k, l\}$, $E = \{ij, jk, kl\}$
  iii. $V = \{m, n, p, q\}$, $E = \{mp, mq, np\}$
  iv. $V = \{r, s, t, u\}$, $E = \{rs, ru, st, su\}$
  v. $V = \{v, w, x, y\}$, $E = \{vw, vx, vy, wx, wy, xy\}$

Solution

The solution is given in Figure 9.5. (Of course, there are alternative ways of drawing these pictures.)

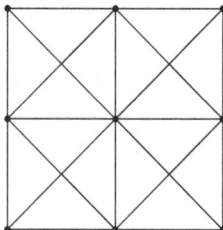

FIGURE 9.3

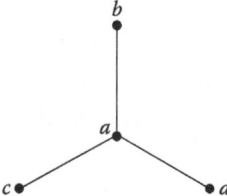

FIGURE 9.4

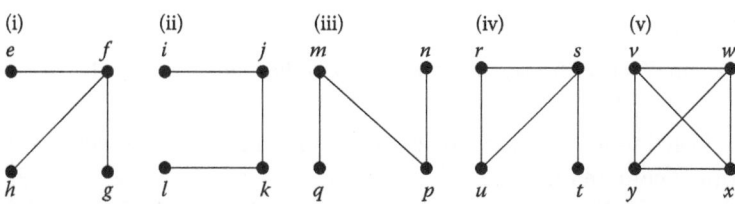

FIGURE 9.5

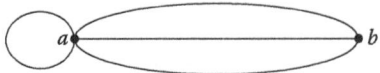

FIGURE 9.6

There are several consequences of Definition 9.1 that we need to note. Because *edges* are defined as *sets* of pairs of vertices, we can have at most one edge joining any pair of vertices. This rules out, for example, the diagram that Euler used in 1736 to represent the Königsberg bridges problem, which was the origin of graph theory and which we discuss in Section 9.6. We could modify our definition to allow for more than one edge to join two vertices. Sometimes *multigraph* is used to describe such "graphs." Also, because an edge is defined as a pair of *distinct* vertices, we have ruled out *loops*, that is, edges that join a vertex to itself.

In the multigraph shown in Figure 9.6, the vertices a and b are joined by three edges, and there is a loop joining a to itself.

Those people who allow graphs to have multiple edges and loops then call those graphs with no multiple edges and no loops *simple graphs*. Except for the Königsberg "graph," just about all of our graphs will be simple graphs in this sense, and yet many of our theorems will be applicable to multigraphs. In what follows the reader should take "graphs" in the statement of theorems to include the case of multigraphs unless we add something to exclude this.

Note also that in Definition 9.1 we have not said that V must be a *finite* set, so, in general, infinite graphs are allowed, but in this book *all the graphs we consider will be finite*; that is, both the set of vertices and the set of edges will be finite. At the other extreme, we have not said that V must be nonempty, so the *empty graph*, $(\emptyset, \emptyset)$, is covered by the definition. The empty graph is not very interesting; however, it is more pertinent that we allow graphs where V is not empty, but $E = \emptyset$, that is, there are no edges. In the solution to Problem 9.5 below, graph (i) is a graph of this type. It has four vertices and no edges.

9.3 ISOMORPHISM OF GRAPHS

We have reached the stage where we need to be more precise about the relationship between a graph, as defined abstractly in Definition 9.1, and the pictures we have used to illustrate them. We also need to specify exactly what we mean by two graphs "being the same."

It is not difficult to explain what we mean by a "picture" of a graph, or, to use more mathematical language, a *representation of a graph in the plane*, $\mathbb{R}^2$.

DEFINITION 9.2

Given a graph $G = (V, E)$, a *representation of G in the plane* consists of

  i. An injection (one–one mapping) $\phi : V \to \mathbb{R}^2$ that associates with each vertex a distinct point in the plane
  ii. A mapping Γ that associates with each edge $\{a, b\}$ a curve in the plane from $\phi(a)$ to $\phi(b)$ that, for each vertex v, other than a and b, does not go through the point $\phi(v)$

To make this definition even more precise, we should really say what we mean by a "curve in the plane from ϕ(a) to ϕ(b)" rather than just relying on geometric intuition. For those with a taste for this level of precision, we could add that, by a curve in the plane from a point p to a point q, we mean a continuous injective (one–one) function $\gamma:[0,1]\to \mathbb{R}^2$, such that $\gamma(0)=p$ and $\gamma(1)=q$. Note the requirement that these curves should not go through (the point corresponding to) any other vertex, but we do allow curves to cross.*

In most of the pictures we have drawn above, the curves joining the points are all straight lines. However, our definition allows for edges to be represented by curves of any shape. For example, Figure 9.7 is a second representation in the plane of the graph that is represented by Figure 9.4.

We are now ready to give the definition of two graphs "being the same." If we look at the graphs represented by the pictures (ii) and (iii) in Figure 9.5 we see that each graph consists of four vertices connected, as it were, in a string, so we can redraw the pictures so that they look the same, as shown in Figure 9.8.

In other words, we can match up the vertices of the graphs in such a way that two vertices of the first graph are connected by an edge if and only if the corresponding vertices of the second graph are also connected by an edge. Here is the formal definition.

DEFINITION 9.3
Isomorphism of Graphs

The two graphs $G_1 = (V_1, E_1)$ and $G_2 = (V_2, E_2)$ are said to be isomorphic† if there is a bijection $\phi: V_1 \to V_2$ such that for all $a, b \in V_1$, $\{a,b\} \in E_1 \Leftrightarrow \{\phi(a),\phi(b)\} \in E_2$. Such a mapping ϕ is said to be an isomorphism between the two graphs.

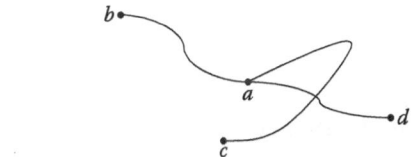

FIGURE 9.7

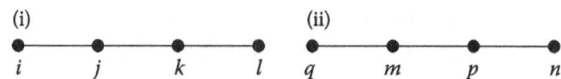

FIGURE 9.8

---

* For gluttons for formality, we require that there is mapping, say Γ, such that for each edge $\{a, b\} \in E$, $\Gamma(\{a, b\}) = \gamma$ is a curve with the property that $\gamma(0) = a$, $\gamma(1) = b$, and for each vertex $v \notin \{a, b\}$, and all $t \in \mathbb{R}$ with $0 \le t \le 1$, $\gamma(t) \ne \phi(v)$.

† From the Greek ισοσ + μορϕη (isos-morphe), meaning "same form." *Isomorphic* is used throughout mathematics to mean "having the same structure," the exact definition depending on which mathematical structures are being considered.

For example, with $V_1 = \{i, j, k, l\}$ and $V_2 = \{m, n, p, q\}$, the mapping $\phi: V_1 \to V_2$ defined by $\phi(i) = q$, $\phi(j) = m$, $\phi(k) = p$, and $\phi(l) = n$ is an isomorphism between the graphs shown in Figure 9.8.

PROBLEM 9.2

a. Give another example of an isomorphism between the graphs shown in Figure 9.8.
b. Explain why the mapping $\theta: V_1 \to V_2$, where $\theta(i) = m$, $\theta(j) = n$, $\theta(k) = p$, and $\theta(l) = q$, is not an isomorphism between the graphs shown in Figure 9.8.

Solution

a. The only other isomorphism between these graphs is the mapping $\psi: V_1 \to V_2$, where $\psi(i) = n$, $\psi(j) = p$, $\psi(k) = m$, and $\psi(l) = q$.
b. For example, $\{i, j\}$ is an edge of the first graph, but $\{\theta(i), \theta(j)\} = \{m, n\}$ is not an edge of the second graph.

We shall normally not operate at this level of formality, and we shall be content to say that two graphs are isomorphic if they can be represented by pictures that can be seen to be the same.

PROBLEM 9.3

Which pairs of the six graphs pictured in Figures 9.4 and 9.5 are isomorphic?

Solution

We have already seen that the graphs (ii) and (iii) of Figure 9.5 are isomorphic. You can easily see that the graph of Figure 9.4 is isomorphic to graph (i) of Figure 9.5. However, (i) is not isomorphic to (ii). (Why not? See a precise answer below.) It should also be fairly clear that graphs (iv) and (v) of Figure 9.5 are not isomorphic, nor are they isomorphic to any of the other graphs. Indeed, the graph (iv) is the only one with four edges, and (v) is the only graph with six edges.

We can prove that two graphs *are* isomorphic by giving an isomorphism between them. To prove that two graphs are *not* isomorphic, we need to show that no isomorphism between them can exist. Merely being *unable to find* an isomorphism between them is not enough. If two graphs each have n vertices, there are $n!$ bijections between their vertex sets, and thus even for quite small values of n it is not practical to list them all and check that none of them is an isomorphism. So normally when we are trying to show that two graphs are not isomorphic we seek a property that one graph has but the other does not.

We adopted this strategy in our solution to Problem 9.3 when, for example, we said that graph (v) of Figure 9.5 is not isomorphic to the other graphs as it is the only graph with six edges. We now need to justify this strategy. Our first theorem is an immediate consequence of Definition 9.1, which says that when we have a pair of isomorphic graphs there is a bijection between their sets of vertices and a bijection between their sets of edges.

THEOREM 9.1

If the graph $G_1 = (V_1, E_1)$ is isomorphic to the graph $G_2 = (V_2, E_2)$, then $\#(V_1) = \#(V_2)$ and $\#(E_1) = \#(E_2)$.

We have already used Theorem 9.1 in Problem 9.3 to show that graphs with different numbers of edges are not isomorphic. Of course, the converse of Theorem 9.1 is not true. For example, the graphs (i) and (ii) of Figure 9.5 are not isomorphic even though they are both graphs with four vertices and three edges.

Up to now we have relied on geometric pictures to justify this last statement, but we now introduce an important idea that enables us to make the argument more precise.

DEFINITION 9.4

We say that the edge $\{a, b\}$ of a graph is *adjacent* to the vertex v if either $v = a$ or $v = b$. The *degree* of a vertex v is the number of edges that are adjacent to v. We use $\delta(v)$ for the degree of the vertex v. We also say that the vertices a and b are *adjacent* if $\{a, b\}$ is an edge.

In terms of pictures, the degree of a vertex is just the number of edges that meet at the vertex. If a graph has n vertices, each vertex can be joined to all, some, or none of the other $n - 1$ vertices, so the degree of a vertex v in a graph with n vertices is an integer, $\delta(v)$, in the range $0 \leq \delta(v) \leq n-1$. (In a multigraph where multiple edges and loops are allowed there are no restrictions on the degrees of individual vertices.) Note that a minority of authors uses *valence* instead of *degree*.

PROBLEM 9.4

Calculate the degrees of the vertices of the graphs pictured in Figure 9.5.

Solution

(i)		(ii)		(iii)		(iv)		(v)	
v	$\delta(v)$	v	$\delta(v)$	v	$\delta(v)$	v	$\delta(v)$	v	$\delta(v)$
e	1	i	1	m	2	r	2	v	3
f	3	j	2	n	1	s	3	w	3
g	1	k	2	p	2	t	1	x	3
h	1	l	1	q	1	u	2	y	3

Because of the way an isomorphism between graphs matches the vertices and the edges of two graphs, in the following theorem the first statement is evident, and the second is an immediate consequence.

THEOREM 9.2

a. If $\phi: V_1 \to V_2$ is an isomorphism between the graphs $G_1 = (V_1, E_1)$ and $G_2 = (V_2, E_2)$, then for each $v \in V_1$, $\delta(v) = \delta(\phi(v))$.
b. If two graphs are isomorphic, they have the same number of vertices of each degree.

It is immediate from Theorem 9.2 that the graphs (i) and (ii) in Figure 9.5 are not isomorphic even though they both have four vertices and three edges. We see from the tables in the solution to Problem 9.4 that graph (i) has three vertices of degree 1, and one vertex of degree 3, whereas graph (ii) has two vertices of degree 1 and two of degree 2. So, by Theorem 9.2, these graphs are not isomorphic.

Unfortunately, the converse of Theorem 9.2(b) is not true. We ask you to find a counterexample in Exercise 9.3.2A. So although Theorem 9.2 can often be used to prove that two graphs are *not* isomorphic, it cannot be used to determine in general whether or not two graphs *are* isomorphic. Of course, in principle, the question can always be answered by trying each potential isomorphism in turn. However, in practice this is not feasible for graphs with more than a small number of vertices.

PROBLEM 9.5

Give a full list of all the different graphs with four vertices. (Note that by "different" we mean that no two of the graphs should be isomorphic, and by "full list" we mean that every graph with four vertices should be isomorphic to one of the graphs in the list.)

Solution

There are 11 different graphs with four vertices, as shown in Figure 9.9.

Here we can be fairly sure that we have included all the graphs with four vertices. But what if we were to want to know how many different graphs there are with, say, 10 or 20 vertices? Could we be confident we could draw them all—and repeat none? Pólya's theorem, in Chapter 14, provides a method for calculating, for each positive integer n, how many different graphs there are with n vertices.

We have seen that we can associate with each graph a sequence of nonnegative integers giving the degrees of its vertices. We now turn this around and ask how we can decide, given a finite sequence of nonnegative integers, whether there is a *simple* graph whose vertices have these numbers as its degrees. We emphasize that we are considering only simple graphs at this stage, as the corresponding problem for multigraphs has a rather different solution.

It is convenient to assume the list of degrees are arranged in nondecreasing order. Thus we frame the following definition.

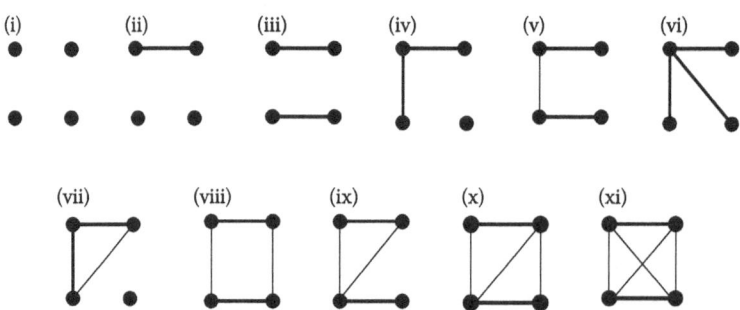

FIGURE 9.9

DEFINITION 9.5

A sequence $d_1, d_2, \ldots, d_n$ of nonnegative integers is said to be a *simple graph degree sequence* if $d_1 \leq d_2 \leq \ldots \leq d_n$ and there is a simple graph with n vertices $v_1, v_2, \ldots, v_n$ such that for $1 \leq i \leq n$, $\delta(v_i) = d_i$.

For example, from the graph in Figure 9.2 we see that 3,3,3,3 is a simple graph degree sequence, and from the graph in Figure 9.4 so also is 1,1,1,3. In contrast, 1,1,3,3 is not a simple graph degree sequence because if we had a graph with four vertices, two of which had degree 3, these two vertices would have to be joined by edges to both the other two vertices, which would therefore have degrees at least 2. In these cases it was easy to tell whether or not the sequence was a simple graph degree sequence, but what about, for example, the sequence

$$1,2,3,4,5,7,8,9,9,9,9\ ?$$

It would take rather a lot of work to try to find a graph with 11 vertices having these degrees. However, the next theorem provides a straightforward way to answer the question.

THEOREM 9.3
The Havel–Hakimi Theorem*

The sequence $d_1, d_2, \ldots, d_n$ is a simple graph degree sequence if and only if so also is the sequence $d_1, \ldots, d_{n-k-1}, d_{n-k}-1, d_{n-(k-1)}-1, \ldots, d_{n-1}-1$ (reordered if necessary), where $k = d_n$.

Comment

We illustrate what this theorem says in the examples given after the proof. Note that the second sequence is obtained from the first by deleting the final term d_n and subtracting 1 from the preceding d_n numbers. Of course this will be possible only if $d_n < n$. If $d_n \geq n$, there cannot be a simple graph with only n vertices but a vertex of degree d_n. In subtracting 1 from the preceding terms we may end up with a sequence that is no longer nondecreasing, which is why we have added "reordered if necessary." Also, some of the numbers may become 0. These are usually dropped from the sequence. The second sequence contains one fewer term than the first sequence. The original sequence will be a simple graph degree sequence if and only if it reduces to a sequence of 0s (which is the degree sequence of a graph with no edges).

Proof

We deal with the easier "if" part of the theorem first. So suppose

$$d_1, d_2, \ldots, d_{n-k-1}, d_{n-k}-1, \ldots, d_{n-1}-1$$

---

* A proof of this theorem was first published by Václav Havel, Poznáma o existenci konečných grafů, *Časopis pro pěstování matematiky*, 80, 1955, pp. 477–480. It seems that this paper, published in Czech with a German summary, was not widely read, as in 1963 another proof was published independently by S. L. Hakimi, On Realizability of a Set of Integers as Degrees of the Vertices of a Linear Graph I, *Journal of the Society for Industrial and Applied Mathematics*, 10, 1962, pp. 496–506.

is a simple graph degree sequence, and $d_n = k$. Then there is a simple graph, say G, with $n-1$ vertices, say $v_1,\ldots,v_{n-1}$, with degrees $d_1, \ldots, d_{n-k-1}, d_{n-k}-1,\ldots, d_{n-1}-1$, respectively.

We obtain a new simple graph, G', by adding a vertex, v_n, and joining it by an edge to the vertices $v_{n-k}, \ldots, v_{n-1}$. The degrees of the vertices $v_1, \ldots, v_n$ in G' are then $d_1,\ldots, d_{n-k-1}, d_{n-k}, \ldots, d_{n-1}, d_n$, respectively, and therefore this latter sequence is a simple graph degree sequence.

For the "only if" converse we suppose that $d_1, \ldots, d_n$ is a simple graph degree sequence, with $d_1 \le d_2 \le \ldots \le d_n = k$. It follows that there is at least one simple graph whose n vertices, $v_1, v_2, \ldots, v_n$, have these degrees. Among all such graphs let G be one in which the sum of the degrees of the k vertices adjacent to v_n is as large as possible. We show that, in this case, the degrees of the vertices adjacent to v_n are $d_{n-k}, \ldots, d_{n-1}$.

Suppose not. That is, suppose that v_n is *not* adjacent to k vertices whose degrees are $d_{n-k}, \ldots, d_{n-1}$. Then, since v_n is adjacent to k vertices, there must be vertices v_r, v_s such that

i. The vertex v_r is adjacent to v_n and $\delta(v_r) < d_{n-k}$, and
ii. The vertex v_s is not adjacent to v_n and $\delta(v_s) \ge d_{n-k}$.

Now, since $\delta(v_r) < \delta(v_s)$, there must be a vertex, say v_t, that is adjacent to v_s but that is not adjacent to v_r. We now change G to a new graph G' by replacing the edges $\{v_r, v_n\}$ and $\{v_s, v_t\}$ by edges $\{v_r, v_t\}$ and $\{v_n, v_s\}$ (see Figure 9.10).

The degrees of the vertices v_n, v_r, v_s, v_t are unchanged, and so G' has the same degree sequence as G. However, as in G' the vertex v_n is adjacent to v_s rather than v_r and $\delta(v_s) > \delta(v_r)$, we see that the sum of the degrees of the vertices of G' to which v_n is adjacent is greater than the sum of the degrees of the vertices of G to which v_n is adjacent. This contradicts our choice of G.

We can therefore deduce that the k vertices to which v_n is adjacent in G have degrees $d_{n-k},\ldots,d_{n-1}$. Thus if we delete the vertex v_n and the edges joining it to these k vertices, the result will be a graph with $n-1$ vertices with degrees $d_1,\ldots, d_{n-k-1}, d_{n-k}-1,\ldots, d_{n-1}-1$, and so this sequence of degrees, reordered if necessary, is also a degree sequence. This completes the proof.

We can illustrate this by the examples we have previously mentioned. We use $S \Rightarrow S'$ to indicate the transition from the sequence $d_1,\ldots,d_n$ to the sequence $d_1,\ldots, d_{n-k-1}, d_{n-k}-1,\ldots, d_{n-1}-1$ and $S' \sim S''$ to indicate that S'' is obtained from the sequence S' by reordering where necessary.

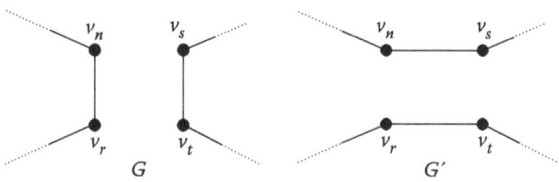

FIGURE 9.10

We have seen that 3,3,3,3 is a simple graph degree sequence. We can confirm this using the reductions the Havel–Hakimi theorem. We have

$$3,3,3,3 \Rightarrow 2,2,2 \Rightarrow 1,1 \Rightarrow 0.$$

Now, 0 is the degree sequence of the simple graph with one vertex. So from the Havel–Hakimi theorem, we know that 3,3,3,3 is a simple graph degree sequence.

We also saw that 1,1,3,3 is not a simple graph degree sequence. Here the Havel–Hakimi method gives

$$1,1,3,3 \Rightarrow 0,0,2,$$

and no further reduction is possible. Indeed, it is easy to see that we cannot have a simple graph with one vertex that has degree 2 and two with degree 0. Hence 0,0,2 is not a simple graph degree sequence. Consequently by the Havel–Hakimi theorem we deduce that 1,1,3,3 is not a simple graph degree sequence.

The next problem deals with a more complicated example in which we show how to construct a graph corresponding to any given simple graph degree sequence.

PROBLEM 9.6

Determine whether the following are simple graph degree sequences. In the case of simple graph degree sequences, construct a simple graph with the degree sequence.

a. 1, 3, 3, 3, 4, 4
b. 1, 2, 3, 4, 5, 7, 8, 9, 9, 9, 9

Solution

a. $1,3,3,3,4,4 \Rightarrow 1,2,2,2,3 \Rightarrow 1,1,1,1 \Rightarrow 1,1,0 \sim 0,1,1 \Rightarrow 0,0$. As 0,0 is the degree sequence of a graph with two vertices of degree 0, it follows that 1,3,3,3,4,4 is a simple graph degree sequence. Starting with the graph with two vertices of degree 0, we can construct a graph following the method of the "if" part of the Havel–Hakimi theorem, adding one vertex at a time; see Figure 9.11.
b. $1,2,3,4,5,7,8,9,9,9,9 \Rightarrow 1,1,2,3,4,6,7,8,8,8 \Rightarrow 1,0,1,2,3,5,6,7,7 \sim 0,1,1,2,3,5,6,7,7 \Rightarrow 0,0,0,1,2,4,5,6$, and no further reduction is possible, as 6 is larger than the number of nonzero terms in the sequence. Consequently, 0,0,0,1,2,4,5,6 is not a simple graph degree sequence and hence neither is the original sequence.

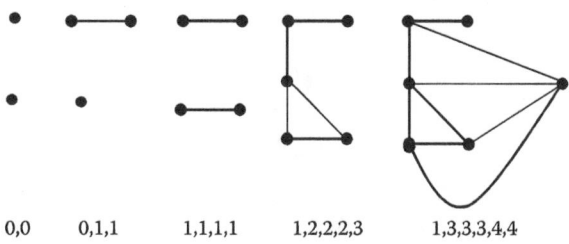

0,0　　0,1,1　　1,1,1,1　　1,2,2,2,3　　1,3,3,3,4,4

FIGURE 9.11

Exercises

9.3.1A Determine which pairs of the graphs in Figure 9.12 are isomorphic. We have, for convenience, in each case labeled the vertices of the graphs. Where a pair of graphs is isomorphic you should indicate, using these labels, a bijection between the vertices of the two graphs, which is an isomorphism between them. Where the graphs are not isomorphic, you should give a reason why no such isomorphism exists.

9.3.1B Show that each pair of the graphs in Figure 9.1 is isomorphic.

9.3.2A Give an example of two graphs with the same number of vertices of each degree, but that are not isomorphic. Aim to find an example of a pair of such graphs with the smallest possible number of vertices and edges.

9.3.2B The *dual* of a graph $G = (V, E)$ is defined to be the graph $G^* = (V, E^*)$, where for all $u,v \in V$, we have $\{u,v\} \in E^* \Leftrightarrow \{u,v\} \notin E$. Thus G^* has the same set of vertices as G, and two vertices are joined by an edge in G^* if and only if they are not joined by an edge in G. G is said to be *self-dual* if it is isomorphic to G^*. For example, in Figure 9.13, graph (ii) is the dual of graph (i). Since these graphs are isomorphic, each graph is self-dual.

  i. Find all the graphs with four vertices that are self-dual.

  ii. Give an example of a graph with five vertices that is self-dual.

  iii. Show that there is no graph with six vertices that is self-dual.

9.3.3A Determine whether the following are simple graph degree sequences. Where they are, construct a simple graph whose vertices have these degrees.

  i. 2, 2, 2, 3, 3, 4   ii. 1, 1, 3, 3, 5, 5   iii. 3, 3, 3, 3, 3, 3, 3, 3

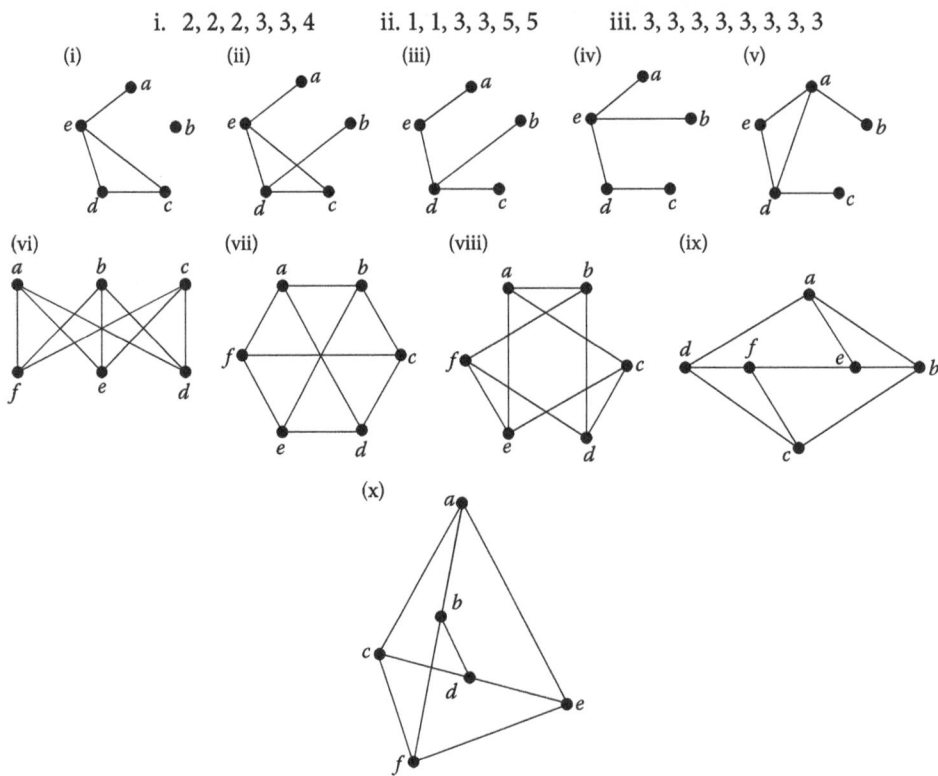

FIGURE 9.12

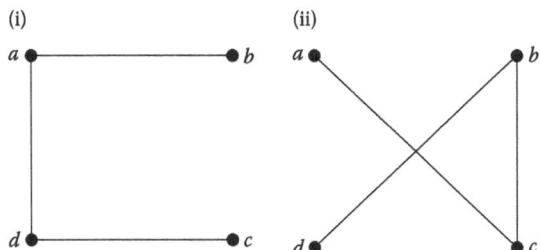

FIGURE 9.13

9.3.3B i. Determine whether the following are simple graph degree sequences. Where they are construct a simple graph whose vertices have these degrees.

a. 1, 1, 3, 3, 5, 5, 7, 7   b. 0, 1, 2, 3, 4, 5, 6

ii. Is it possible to have, for any positive integer n, a simple graph with n vertices with degrees 0, 1, 2, …, $n-1$? [*Hint:* Part (b) of (i) provides a clue.]

9.3.4A Prove that, for each positive integer n, two simple graphs with n vertices and $\frac{1}{2}n(n-1)$ edges are isomorphic.

9.3.4B i. Let G_1 and G_2 be two graphs with 10 vertices of which eight have degree 9 and two have degree 8. Prove that these graphs must be isomorphic.

ii. Generalize this result.

9.3.5A We say that a nondecreasing sequence $d_1,\ldots,d_n$ of positive integers is a *multigraph degree sequence* if there is a graph that may have multiple edges, *but not loops*, and that has n vertices with degrees $d_1,\ldots,d_n$, respectively. Which of the following is a multigraph degree sequence?

i. 1, 2, 3, 4, 5   ii. 1, 2, 2, 3, 5, 5   iii. 2, 2, 2, 8

9.3.5B Prove that a nondecreasing sequence of nonnegative integers $d_1,\ldots,d_n$ is a multigraph degree sequence if and only if (i) $\sum_{i=1}^{n} d_i$ is even, and (ii) $d_n \leq \sum_{i=1}^{n-1} d_i$. This result is also given in the paper by Hakimi cited in the footnote to Theorem 9.3.

9.4 PATHS AND CONNECTED GRAPHS

Since this is a book on counting, we begin with an amusing—and not *completely* obvious—result that introduces a simple but typical graph theoretical procedure.

PROBLEM 9.7

You hold a party at your house with 20 guests. Those pairs of guests who know each other shake hands (once!), those who do not … don't! Show that when all this bonhomie is completed, the *number* of guests who have shaken hands an odd number of times must be even.

Solution

Consider a graph with vertices $g_1, g_2, \ldots, g_{20}$ representing the 20 guests, with an edge connecting g_i and g_j if and only if the corresponding guests shook hands. Then the degree, $\delta(g_i)$, of g_i is just the number of people with whom g_i shook hands. Since each handshake involves two people, the sum

$$D = \delta(g_1) + \delta(g_2) + \ldots + \delta(g_{20}), \tag{9.1}$$

is twice the number of handshakes (or, equivalently, twice the number of edges in the graph). So D is an even number. Hence the number of terms on the right-hand side of Equation 9.1 that are odd numbers must itself be even.

Note that the argument used in the above solution applies to all graphs (and also multigraphs), as it always true that each edge contributes one to the degree of each of the two vertices that it joins (and in a multigraph a loop contributes two to the degree of its vertex). Thus we have the following result.

THEOREM 9.4
The Handshaking Lemma*

In each graph the sum of the degrees of the vertices is twice the number of edges.

[If you like equations, you can express this by saying that, given a graph $G = (V, E)$, we have

$$\sum_{v \in V} \delta(v) = 2 \#(E), \tag{9.2}$$

where, you will recall, $\#(E)$ is the number of elements in the set E.]

Although this is a very simple result, it turns out to be extremely useful. For example, it immediately tells us that we cannot have a graph with seven vertices each with degree 3, as this would imply that the sum of the vertex degrees is the odd number 21.

When a graph is used to represent a road map, and in many other applications, it is natural to consider the ways you can travel from one vertex to another by going along the edges, and other related problems. To discuss these problems it is helpful to introduce some more definitions, and to make some subtle distinctions. There is no absolutely standard terminology in this part of graph theory. So if you look at books by other authors (which we are not *completely* against!) you need to check this. Our own review of a good number of books of graph theory suggests our terminology mostly follows that of the majority.

---

* The term *lemma* is usually used in mathematics for a subsidiary result, used to prove a "theorem" that is of greater significance. By historical accidents a number of results significant enough to count as theorems have *lemma* as part of their standard names. The "handshaking lemma" is a theorem of this kind. The definitive discussion of the difference between what a thing is and what it is called is to be found in Chapter 8 of *Through the Looking Glass* by Lewis Carroll, Macmillan, London, 1871.

DEFINITION 9.6

Let G be a graph.

i. A sequence $v_0, v_1, \ldots, v_{k-1}, v_k$ of vertices, not necessarily distinct, of G such that for $0 \leq i \leq k$, $\{v_i, v_{i+1}\}$ is an edge of G, is called a *walk*. The length of the walk is the number, k, of edges in it. We will often use the suggestive notation:

$$v_0 \to v_1 \to \ldots \to v_{k-1} \to v_k \qquad (9.3)$$

to indicate a walk. Not surprisingly, v_0, v_k are called the *ends* of the walk, with v_0 being called the *start* of the walk and v_k the *finish*.

ii. If the edges $\{v_i, v_{i+1}\}$ are all different, the walk is called a *trail*.
iii. If the vertices are all different, the walk is called a *path*. (Thus a path is necessarily a trail.)
iv. If $v_0 = v_k$ the walk given by the expression 9.3 is said to be *closed*. Likewise a *closed trail* is a trail in which the start is the same as the finish.
v. We have to be rather more careful with what we mean by a closed path, as we have said that in a path all the vertices are distinct, and this rules out the case $v_0 = v_k$. So we can't just say that a closed path is a path in which the start is the same as the finish. Instead, we define a *closed path* to be a trail (so the edges are all distinct) and $v_0 = v_k$, but otherwise all the vertices are distinct.

Notes: Many authors use *circuit* to mean a closed trail, and *cycle* to mean a closed path, but we will avoid introducing these additional terms. Definition 9.6 can be extended to multigraphs, in a fairly obvious way. However, since given two vertices v_i, v_{i+1} there may be more than one edge joining them, to specify a walk (or trail or path) in a multigraph we need to list not just the vertices in a specified order but also which edge is used to get from one vertex to the next.

PROBLEM 9.8

Let G be the graph illustrated in Figure 9.14. (This is the Petersen graph, which first appeared in Figure 9.1 and which we have repeated here for convenience.)

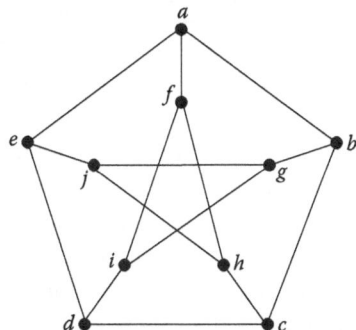

FIGURE 9.14

For each of the following sequence of vertices, determine whether it is (i) a walk, (ii) a trail, and/or (iii) a path. In each case where the sequence is a walk, determine whether it is closed, and also its length.

a. $a \to b \to g \to j \to e \to a \to f \to h \to c$
b. $a \to b \to c \to h \to i \to d \to e \to a$
c. $a \to b \to g \to j \to e \to d \to c \to b \to a$
d. $a \to f \to i \to d \to e \to a$
e. $a \to e \to j \to h \to c$
f. $a \to b \to g \to j \to e \to d \to c \to b \to g \to i \to f \to a$

Solution
a. This is a walk and a trail but not a path, as the vertex a occurs twice. Since $a \neq c$, it is not closed. It has length 8.
b. This is not even a walk, because hi is not an edge. Hence it is neither a trail nor a path.
c. This is a walk but not a trail, as the edge ab occurs twice, and hence it is not a path. It is closed and has length 8.
d. This is a walk, a trail, and a path. It is closed and has length 5.
e. This is a walk, a trail, and a path. It is not closed and has length 4.
f. This is a walk but not a trail, as the edge bg occurs twice. So it is also not a path. It is closed and has length 11.

The following useful result gives a sufficient condition for a graph (or multigraph) to have a closed path.

THEOREM 9.5

If in a graph (or a multigraph), G, each vertex has degree at least 2, then there is a closed path in G.

Proof

If in the graph there is a loop, that is, an edge joining a vertex v to itself, then $v \to v$ is a closed path and the result holds. If there are vertices v and w joined by more than one edge, then by following one edge from v to w and a second edge from w to v we obtain a closed path in G. So from now on we assume G is a simple graph.

Select some vertex v_0 of G. Since v_0 has degree at least 2, there is some vertex v_1 different from v_0 that is joined to v_0 by an edge. Now suppose we have constructed a path $v_0 \to v_1 \to ... \to v_{k-1} \to v_k$ in G, where all the vertices, other than possibly v_0 and v_k, are different. If $v_k = v_0$ we have a closed path. Otherwise, as v_k has degree at least 2, there is a vertex, say v_{k+1}, which is different from v_{k-1} and joined to v_k by an edge. If v_{k+1} is different from all of $v_0, v_1, ..., v_{k-2}$, then $v_0 \to v_1 \to ... \to v_{k-1} \to v_k \to v_{k+1}$ is also a path. However, as there is only a finite number of vertices, we must eventually reach a stage where for some j, $0 \leq j \leq k-2$, $v_{k+1} = v_j$. But then, when this first occurs, $v_j \to v_{j+1} \to ... \to v_k \to v_{k+1}$ is a closed path in G and our theorem is proved.

DEFINITION 9.7

i. Two vertices in a graph are said to be *connected* if there is a walk in the graph with the given vertices as its ends.
ii. A graph is said to be *connected* if each pair of vertices is connected.
iii. If a graph is not connected it can be split up into separate connected parts, which are called the *connected components* of the graph.

For example, in the graph of Figure 9.15, the pairs of vertices *a* and *b*; *a* and *c*; *b* and *c*; and *d* and *e* are connected, but none of *a*, *b*, and *c* is connected to either *d* or *e*. Since not all the pairs of vertices are connected, the graph is not connected. It has two connected components, one with vertex set {*a,b,c*} and edges *ab* and *bc*; and the other with vertex set {*d,e*} and edge *de*.

PROBLEM 9.9

How many of the graphs with four vertices, as shown in Figure 9.9, are connected?

Solution

Six of these graphs are connected, namely, (v), (vi), (viii), (ix), (x), and (xi).

Exercises

9.4.1A Prove that if G is a connected graph with n vertices and no closed paths, then G has $n-1$ edges.

9.4.1B Prove that if G is a simple graph with n vertices such that, for all pairs of vertices u, v that are not joined by an edge, $\delta(u) + \delta(v) \geq n-1$, then G is connected.

9.4.2A Prove that, given a connected graph G, the removal of a particular edge results in a graph that is still connected if and only if the edge that is removed occurs in a closed path in G.

9.4.2B Suppose that G is a connected graph with 16 edges in which each vertex has degree at least three. Find the maximum number of vertices it can have. Draw a picture of such a graph.

9.4.3A Show that the Petersen graph has a closed path of length 6 but no closed path of length 7.

9.4.3B Does that the Petersen graph have a closed path of length 9? Does it have a closed path of length 8?

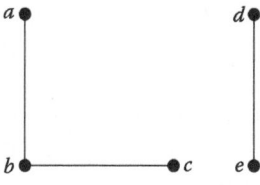

FIGURE 9.15

9.4.4A Show that a simple graph with n vertices and $\frac{1}{2}(n-1)(n-2)+1$ edges is connected.

9.4.4B Show that for all integers $n \geq 2$, there is a simple graph with n vertices and $\frac{1}{2}(n-1)(n-2)$ edges that is not connected.

9.4.5A i. Devise an algorithm for determining whether a graph given by listings of its vertices and edges is connected.

ii. Apply the algorithm of part (i) to determine whether the following graphs are connected:

a. $G = (V, E)$, where $V = \{a, b, c, d, e, f, g, h\}$ and $E = \{ad, ae, af, bc, bg, bh, cg, de, ef, gh\}$

b. $G = (V, E)$, where $V = \{a, b, c, d, e, f, g, h\}$ and $E = \{ab, ac, ad, ae, af, bf, cg, ch, ef, gh\}$

c. $G = (V, E)$, where $V = \{a, b, c, d, e, f, g, h\}$ and $E = \{ae, af, bc, ce, ch, de, gh, hj\}$

9.4.5B i. Devise an algorithm for deciding whether a closed path can be formed from the edges in a given list.

ii. Apply the algorithm in the following cases.

a. The list of edges is *ac,ag,ai,bd,be,bh,cf*.

b. The list of edges is *ae,ag,bk,ce,dk,eh,fg,gj,hj,ik*.

9.5 PLANAR GRAPHS

Do you recognize the graph in Figure 9.12vi? It would be the graph to solve the *utilities problem* stated in Chapter 1—if only it could be redrawn in the plane without any of the edges crossing one another! Well, can it? Let us now turn our attention to that problem.

We have already seen one example of a graph with edges crossing at a nonvertex that can be redrawn to get rid of this property. In Figure 9.2, the graph on the right is a redrawing of the one on the left, arranged so that edges meet only at vertices. We have already used the term *planar* for graphs that can be redrawn in such a way. It is appropriate now to introduce this term in a formal definition.

DEFINITION 9.8

Any graph that can be represented in the plane without any pair of edges meeting, except at a vertex, is called a *planar graph*—and any representation of it in the plane a *plane graph*.

There are some topological aspects of this definition that we mention briefly but do not pursue here because our chief focus is on the combinatorial aspects of graphs. What happens if instead of considering representing graphs in the plane, R^2, we consider representations of graphs in other spaces, S? We can explain what this means by taking Definition 9.2 and replacing the reference to the plane by a reference to the space S. It is not too difficult to see that every graph can be represented in R^3. The vertices can be represented by distinct points on, say, the x-axis. The edges joining the vertices can then be represented by arcs, for example, semicircles, in planes each making a different angle with the xy-plane. In fact, with a little more care, we can show that every simple graph can be represented in R^3 with the edges represented by straight line segments (see Exercise 9.5.8A).

However, there is some interest in considering which graphs can be represented in two-dimensional surfaces other than R². Because the surface of a sphere, with one point deleted, is topologically equivalent to the plane R² (the stereographic projection of a sphere with one point deleted onto the plane shows this), any graph that can be represented in the plane can also be represented on the surface of a sphere and vice versa. However, there are some nonplanar graphs that can be represented on two-dimensional surfaces that are not topologically equivalent to R². For example, we will see that the graph, called K_5, consisting of five vertices, each joined to each of the other four vertices, is not planar. However it can be represented on a torus. For this see Exercise 9.5.6A.

It is also interesting to note that each planar graph can be represented in the plane in such a way that the curve corresponding to each edge is a straight line segment. We prove this result in Section 9.8, as Theorem 9.13, after we have established some relevant facts about planar graphs. Note that this result clearly does not hold for multigraphs, as we cannot connect two points in the plane by two straight line segments having no points, except their endpoints, in common.

We let K_n be the simple graph with n vertices in which each vertex is joined to each other vertex by an edge. We call K_n the *complete graph with n vertices*. These graphs, for $1 \leq n \leq 5$, are shown in Figure 9.16.

It can be seen from Figure 9.16 that K_1, K_2, K_3, and K_4 are all planar graphs, but what about K_5? We have not succeeded in drawing it in the plane with edges meeting only at vertices, but maybe that is just a sign of our incompetence, and with more skill we could represent this graph in the plane.

To get useful information about all planar multigraphs that enables us to deal with this problem, we begin with a lovely result due to Euler. [It is lovely because (i) the result is a nice surprise, being far from obvious, and yet (ii) the proof is not hard.] Euler's formula, discovered by him in around 1750 and proved by Legendre in 1794, establishes a relationship between the number of regions (or faces), edges, and vertices in a planar graph. To avoid an overlengthy (and possibly boring!) discussion we shall show you, by means of the diagram in Figure 9.17, what is meant by the term *region*.

It is easily seen that in the graph represented in Figure 9.17 there are 13 vertices and 17 edges. These edges divide the plane into six regions, which we have labeled A, B, C, D, E, and F. Note that we have included in this count the unbounded region F that surrounds the figure in the diagram. Note also that the region E is bounded by two curves that represent edges joining the same pair of vertices. Thus the diagram represents a multigraph.

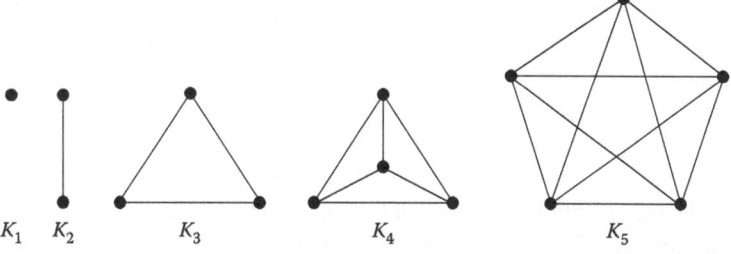

FIGURE 9.16

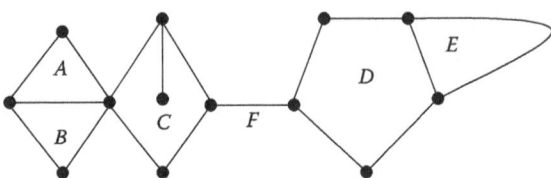

FIGURE 9.17

THEOREM 9.6
Euler's Formula for Planar Graphs

Let G be a connected plane graph with v vertices and e edges and drawn so as to divide the plane into f regions. Then $v - e + f = 2$.

Comment on the notation

We have used "f" for the number of regions for historical reasons. The problem Euler considered was the relationship between the number of vertices, edges, and *faces* of a polyhedron. For example a cube has 8 vertices, 12 edges, and 6 faces, as shown on the left in Figure 9.18.

Imagine the cube sitting on a plane, and squashing the cube onto this plane, while slightly reducing the size of the top face. We would end up with a plane graph as shown on the right of Figure 9.18. This graph also has 8 vertices and 12 edges, and it divides the plane into 6 regions. Thus the formula $v - e + f = 2$ holds both for the cube and for the plane graph resulting from it. So a standard way to prove Euler's formula for polyhedra is to reduce it to the case of plane graphs in the way we have indicated. So "f" comes from "face."

Proof

We give a proof by mathematical induction on the number of edges in the graph. The base case is that of a connected graph with zero edges. Such a graph must have only one vertex, and so when drawn in the plane it consists of a single point. Consequently, there is just one region in this graph. So we have $v = 1$, $e = 0$, and $f = 1$ and hence in this case the formula $v - e + f = 2$ certainly holds.

We now come to the induction step. We assume, as our induction hypothesis, that the formula of the theorem holds for all connected plane graphs with k edges. Let G be a connected plane graph with $k + 1$ edges.

We now consider two separate cases. The first case is where G has a vertex of degree one, say a. In this case we let b be the one vertex joined to a by an edge.

Let G' be the graph that results from G by deleting the vertex a and the edge joining it to b. As G is connected, any two vertices of G' are joined by a path in G. Because such a path cannot include the edge joining a to b, this path is also a path in G'. Hence G' is also connected. Since G' has k edges, by the induction hypothesis, Euler's formula holds for G'.

Since a has degree 1, the edge from a to b lies in the boundary of just one region (either a bounded region, as shown on the left in Figure 9.19, or the unbounded region, as shown on the right). Thus the removal of the edge from a to b does not affect the number of regions. However, G' has one fewer edge and one fewer vertex than G. Thus the value of "$v - e + f$" is the same for G as for G'. Hence Euler's formula holds also for G.

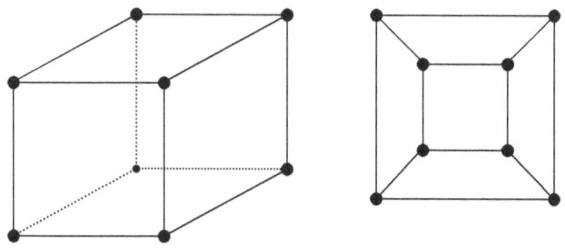

FIGURE 9.18

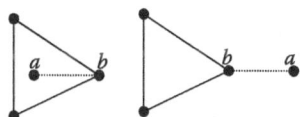

FIGURE 9.19

The second case is when each vertex of G has degree at least 2. Then, by Theorem 9.5, there is a closed path in G. Let a and b be vertices of G such that an edge from a to b occurs in a closed path in G, and let G' be the graph that results when this edge (but neither of the vertices a and b) is deleted. (Note that we have said "*an* edge" as we want our argument to cover a multigraph with more than one edge from a to b. In addition, note that the subsequent argument works also in the case where $a = b$ so that the edge from a to b is a loop.)

G' is also a connected graph (by Exercise 9.4.2A) with one fewer edge than G and so, by the induction hypothesis, Euler's formula holds for G'.

Since the edge ab is part of a closed path in G, it forms part of the boundary between two regions (see Figure 9.20). These two regions become one region when the edge is deleted. Hence G' has one fewer face than does G. It has the same number of vertices as G but one fewer edge. So "$v - e + f$" is the same for G' as for G. Therefore Euler's formula holds also for G.

These two cases cover all the possibilities. Therefore, this completes the proof by mathematical induction that Euler's formula holds for all connected plane graphs.

One consequence of this theorem is that although a planar graph may, in general, be drawn in the plane in more than one way (compare, for example, the two graphs in Figure 9.19), whenever it is drawn as a plane graph the number of regions is always the same, as this number is determined by the number of vertices and edges of the graph using the formula $f = e - v + 2$.

Theorem 9.6, and a little bit of thought, lead immediately to some nontrivial information about planar maps in general. When a graph is drawn in the plane, each region is surrounded by a closed walk of edges. The edges in this walk form the *boundary* of the region. For example, in the graph of Figure 9.17, the region A has a boundary made up of three edges, region C has a boundary with five edges, and region E has a boundary made up of two edges. In the case of region A the boundary makes up a closed path (that is, a closed walk in which there are no repeated vertices other than the endpoints), but

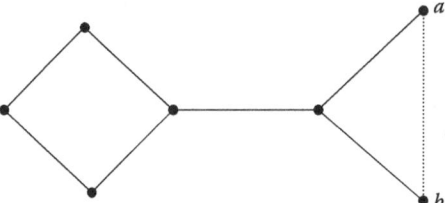

FIGURE 9.20

this not the case for region *C*. Because the graph of Figure 9.17 has two vertices joined by two edges, there is one region, namely *E*, which has a boundary made up of just two edges. This cannot happen in a simple graph, where the boundary of a region must be made up of at least three edges. This simple observation has a useful consequence.

THEOREM 9.7

If *G* is a connected planar simple graph with v vertices and e edges, then $e \leq 3v - 6$.

Proof

Suppose that *G* is drawn in the plane with f regions. Then, by Euler's formula,

$$f = e - v + 2. \tag{9.4}$$

Since each region has a boundary made up of at least three edges, and each edge can be part of the boundary of at most two regions, we have

$$3f \leq 2e. \tag{9.5}$$

From Equation 9.4 and inequality 9.5 we obtain

$$3(e - v + 2) \leq 2e,$$

from which it immediately follows that

$$e \leq 3v - 6.$$

This inequality enables us to answer the question as to whether the complete graph K_5 is planar. In this graph there are five vertices, and, as each pair of vertices is joined by an edge, there are 10 edges. Thus $e(= 10) > 3v-6(= 9)$, and hence, by Theorem 9.7, K_5 is not planar.

And now, what about our utilities problem? This was Problem 9B of Chapter 1. Recall that the problem is whether the graph shown in Figure 9.21 in nonplanar form is in fact planar.

We now reveal that the utilities problem *cannot be solved*; that is, the graph of Figure 9.21 is not planar. We don't mean that, so far, no one has been clever enough to find a way to do it. We *do* mean that someone *has* been clever enough to *prove* that no one will ever be able to do it because *it can't be done!*

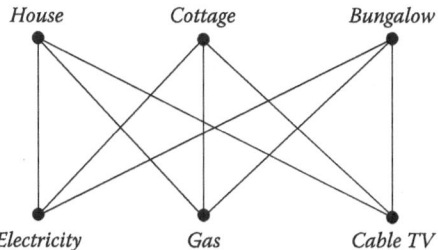

FIGURE 9.21

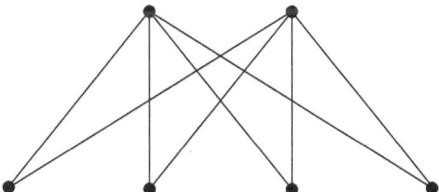

FIGURE 9.22

In this graph $v = 6$ and $e = 9$ and thus $e < 3v-6$. It follows that we cannot use Theorem 9.7 to show that this graph is not planar. However, by noting that in this graph all the closed paths have length at least 4, we can derive a sharper inequality, namely, $e \leq 2(v - 2)$, that the graph does not satisfy, and in this way show that it is not planar. This is Exercise 9.5.2A. Note that, as this graph satisfies $e \leq 3v - 6$ but is not a planar graph, the condition $e \leq 3v - 6$ is a necessary but not a sufficient condition for a graph to be planar.

In the utilities graph, the vertices can be divided into two sets of three {house, cottage, bungalow} and {electricity, gas, cable TV} such that no two vertices in the same set are joined by an edge, but every vertex in one set is joined to every vertex in the other. We use the notation $K_{3,3}$ for this graph.

In general, if the vertices of a graph G can be divided into disjoint sets V_1 and V_2 in such a way that edges only join a vertex in V_1 to a vertex in V_2 (but not vertices that are both in V_1 or both in V_2), then we say that G is a *bipartite graph*. For all positive integers m, n, $K_{m,n}$ is the bipartite graph with $m + n$ vertices that can be partitioned into two sets V_1, V_2 containing m and n vertices, respectively, such that each vertex in V_1 is joined by an edge to each vertex in V_2, but with no edges joining two vertices both from V_1 or both from V_2. For example, the graph $K_{2,4}$ is shown in Figure 9.22.

Of course, the solution to the utilities problem is of no practical consequence whatsoever since the various utility pipelines can be laid under or over each other in the ground. However, planarity is not without significance in the production of printed circuits where the thickness (or, rather, the thinness!) is improved if no conducting parts of the circuit are forced to bridge over one another.

The graphs K_5 and $K_{3,3}$ play a very special role in the theory of planar graphs because in 1930 the Polish mathematician Kazimierz Kuratowski* proved that a graph is not planar if and only if it "contains" either K_5 or $K_{3,3}$. We need first to explain what "contains" means in this context.

By a *subgraph* of a graph, G, we mean a graph obtained from G by deleting vertices and edges. Of course, if we delete a vertex, we must delete all the edges adjacent to it, but we can obtain a subgraph by deleting edges but no vertices. By the usual slight abuse of language in mathematics, a graph counts as a subgraph of itself. More formally, we have:

DEFINITION 9.9

Let $G = (V, E)$ be a graph. A *subgraph* of G is a graph $G' = (V', E')$ such that $V' \subseteq V$ and $E' \subseteq E$.

For example, in Figure 9.23, graph (ii) is a subgraph of graph (i).

We say that a graph G' has been obtain from a graph G by *dropping vertices of degree 2* if we obtain G' by carrying out the following process a finite number of times. Let v be a vertex of degree 2, joined to the vertices u and w. Then we delete v and the edges joining it to u and w and add an edge joining u and w.

For example, in the graph, G', of Figure 9.23(ii), the vertices c, d, and e have degree 2. If we drop these vertices we obtain the graph shown in Figure 9.24.

Note that although the graph we started with was a simple graph, by dropping the vertices c and e and adding an edge from b to g, we have ended up with a multigraph.

We say that the graph G *contains* the graph G' if G' is isomorphic to a graph obtained from a subgraph of G by dropping vertices of degree 2. For example, the graph in Figure 9.24 is contained in the graph of Figure 9.23(i).

It is clear that if G is a planar graph, then so also is any graph that it contains. Hence if a graph contains either K_5 or $K_{3,3}$, then it cannot be planar. This gives a sufficient condition for a graph not being planar. Kuratowski's theorem, whose topological proof is beyond the scope of this book, asserts that this condition is also necessary.

THEOREM 9.8
Kuratowski's Theorem

A graph is planar if and only if it does not contain either K_5 or $K_{3,3}$.

In Exercise 9.5.3A we ask you to show that the Petersen graph is not planar by showing that it contains $K_{3,3}$.

---

* Kazimierz (Casimir) Kuratowski was born in Warsaw on February 2, 1896, and died there on June 18, 1980. In 1896 Poland was occupied by Russia, and it was not possible to proceed to university in Poland without passing an examination in Russian. For this reason many Poles went abroad for their university education. Kuratowski entered the University of Glasgow in 1913 as an engineering student, but his studies were interrupted by World War I. In the course of this war, the Russians withdrew from Poland, and the University of Warsaw became a Polish university, where Kuratowski enrolled as one of the first mathematics students. He had a distinguished career and became one of the founders of a very strong Polish school of mathematicians. His paper in which his theorem was first published is Sur le problème des courbes gauches en Topologie, *Fundamenta Mathematicae*, 15, 1930, pp. 271–283. In this paper the graph $K_{3,3}$ (not so called) is drawn to look like graph (x) in Figure 9.12.

We end this section with an old problem to which Euler's formula provides a ready answer.

PROBLEM 9.10

If n points are placed on the circumference of a circle in such way that when all the lines joining them are drawn, no three lines meet at a point inside the circle, into how many regions do these lines divide the circle?

Solution

In Figure 9.25 we have pictured the cases for $n = 1,2,3,4$, and 5. You can check that in these cases the circle is divided into 1,2,4,8, and 16 regions, respectively. This suggests an obvious conjecture, namely, that with n points there are 2^{n-1} regions. This problem is famous* because appearances here are very deceptive, as this formula breaks down for $n = 6$, in which case there are only 31 regions.

Euler's formula for planar graphs, $v - e + f = 2$, readily yields the correct result. To use this formula we need to add vertices at the points where the lines meet inside the circle so that the graph is planar. Each such, new, vertex inside the circle lies at the intersection of two chords and so corresponds to a choice of four points on the circumference of the circle. So, including the n points on the circle, we have that

$$v = C(n,4) + n.$$

As we don't count the region outside the circle, the number, say r, of regions inside the circle is related to the number of faces of the graph by the equation $r = f - 1$ and so

$$r = (e - v + 2) - 1 = e - v + 1.$$

We need to use the handshaking lemma to work out the number of edges. Each of the vertices on the circle has $n - 1$ straight lines coming from it, but also two curved arcs joining it to the adjacent vertices on the circle. So each vertex on the circle has degree $n + 1$. Since only two lines meet at each point inside the circle, regarded as vertices of the graph, each of these points has degree 4, and we have already seen that there are $C(n,4)$ of these points. So by the handshaking lemma, $2e = n \times (n + 1) + C(n,4) \times 4$, and therefore

$$e = \frac{n(n+1)}{2} + 2C(n,4).$$

It follows that

$$r = \left(\frac{n(n+1)}{2} + 2C(n,4)\right) - \left(n + C(n,4)\right) + 1,$$

and therefore

---

* We do not know the origin of this problem. It was attributed to Leo Moser by Martin Gardner in his Mathematical Games article, *Scientific American*, Vol. 221, No. 2, August 1969, p. 121. Torsten Sillke cites *Puzzles and Curious Problems* by Henry Dudeney, T. Nelson & Sons, London, 1931, as the earliest reference to a closely related problem.

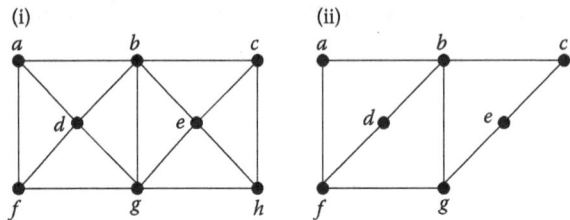

FIGURE 9.23

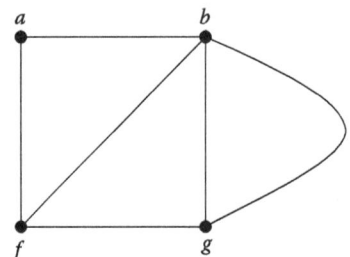

FIGURE 9.24

FIGURE 9.25

$$r = 1 + \frac{n(n+1)}{2} - n + C(n,4).$$

As

$$\frac{n(n+1)}{2} - n = \frac{n(n-1)}{2} = C(n,2), \text{ and } C(n,4) = \frac{n(n-1)(n-2)(n-3)}{4!},$$

we can rewrite this in these equivalent ways:

$$r = 1 + C(n,2) + C(n,4) = 1 + \frac{n(n-1)}{2!} + \frac{n(n-1)(n-2)(n-3)}{4!}$$

$$= \frac{1}{24}n^4 - \frac{1}{4}n^3 + \frac{23}{24}n^2 - \frac{3}{4}n + 1.$$

Note that $C(n,2)$ is the number of straight lines joining the points on the circle. So the formula for the number of regions can be expressed as

Number of regions = 1 + Number of lines + Number of crossing points inside the circle, and this suggests an alternative proof, which we leave to the reader.

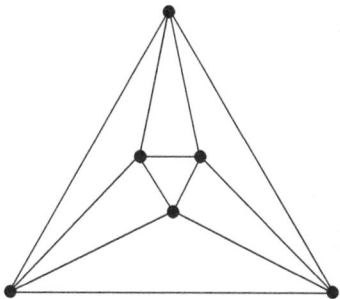

FIGURE 9.26

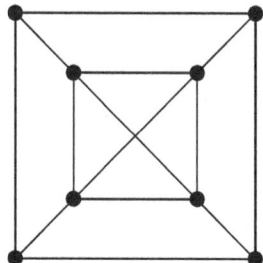

FIGURE 9.27

Exercises

9.5.1A Check that Euler's formula holds for the graphs in Figure 9.23.

9.5.1B Check that Euler's formula holds for the *octahedron graph*, that is, the planar graph obtained from a regular octahedron in the same way that the graph on the right of Figure 9.18 is obtained from a cube. This graph is shown in Figure 9.26.

9.5.2A i. Let G be a connected planar graph with v vertices and e edges, in which each closed path contains at least four edges. Show that $e \leq 2v - 4$.
ii. Deduce that the graph $K_{3,3}$ is not planar.
iii. Show that if any edge is removed from $K_{3,3}$, then the resulting graph is planar.

9.5.2B i. Find an inequality of the kind $e \leq av - b$, where a,b are constants, which holds for all connected planar graphs with v vertices and e edges, with $v \geq 6$, and in which all closed paths contain at least five edges.
ii. Deduce that the Petersen graph (see Figure 9.14) is not planar.

9.5.3A Prove that the Petersen graph (see Figure 9.14) contains the graph $K_{3,3}$ and hence is not planar.

9.5.3B Determine whether the graph in Figure 9.27 is planar.

9.5.4A Prove that a connected graph is bipartite if and only if all the closed paths in the graph have even length.

9.5.4B For which values of m and n is the bipartite graph $K_{m,n}$ planar?

9.5.5A Determine whether the graph shown in Figure 9.28(i) is planar.

9.5.5B Determine whether the graph shown in Figure 9.28(ii) is planar.

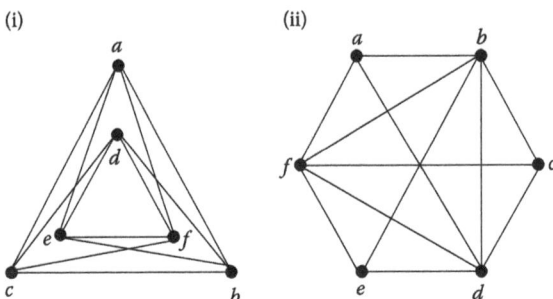

FIGURE 9.28

9.5.6A Show that the complete graph K_7 may be drawn on a torus, so that edges meet only at vertices. It follows that K_5 and K_6 can also be drawn on a torus in a similar way.

9.5.6B Show that the bipartite graph $K_{3,3}$ may be drawn on a torus, so that edges meet only at vertices.

Note: If you don't have a convenient torus to hand, you can make one from a rectangular piece of paper $ABCD$, as in Figure 9.29, by first gluing together the edges AB and CD to make a tube and then gluing the resulting circular edges. Of course, you can draw the figures on the rectangle, and just imagine the gluing.

9.5.7A Show that if six points, A, B, C, D, E, and F, are placed around a circle in this order in such a way that the lines AD, BE, and CF meet at a single point inside the circle, and all the other lines joining these points are drawn, then the circle is divided into only 30 regions. (This could be done by drawing a picture, but we encourage you to answer this question by using Euler's formula.)

9.5.7B i. Show that in Problem 9.10, in the case where $n = 10$, the number of regions is a power of 2.

ii. Does this ever happen for $n > 10$? (This question is, we believe, unsolved and is probably very hard.)

9.5.8A Show that, for all positive integers n, it is possible to find a sequence of points $P_1, P_2, \ldots, P_n$ on the surface of the sphere, such that for $4 \leq m \leq n$, the point P_m is not in the plane defined by any three points P_i, P_j, P_k with $1 \leq i < j < k < m$. Deduce that every simple graph can be drawn in $\mathbb{R}^3$ with the edges represented by straight line segments.

9.5.8B What is the maximum number of points that you can place on the *edges* of a tetrahedron so that the line segments joining these points in pairs meet only at their endpoints?

9.6 EULERIAN GRAPHS

We now return to the classic problem of the Königsberg bridges that we mentioned as Problem 9A in Chapter 1.

In Figure 9.30 we have reproduced from Chapter 1 a schematic map of Königsberg in the eighteenth century, showing the seven bridges over the river Pregel. The problem for the citizens of Königsberg was:

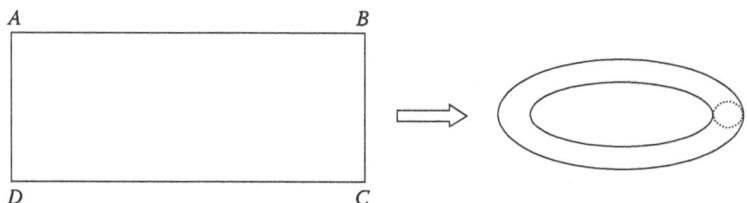

FIGURE 9.29

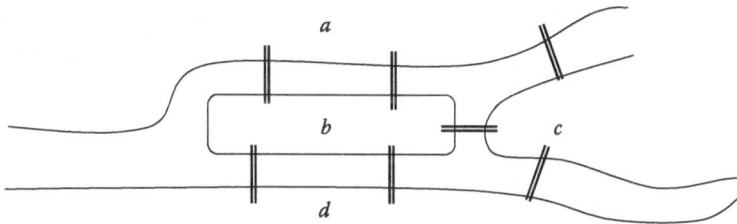

FIGURE 9.30

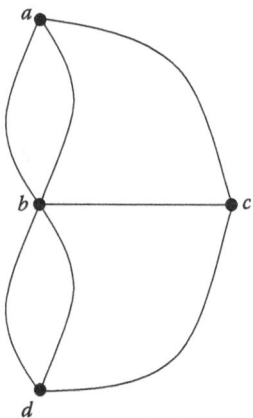

FIGURE 9.31

PROBLEM 9.11
The Bridges of Königsberg

Is there a route that would have taken the burghers of Königsberg over each of the bridges exactly once?

In Figure 9.31 we have represented this map by a graph in which the vertices correspond to the land areas and the bridges are represented by edges. It is a multigraph. Since there are two bridges between areas a and b, in the graph there are two edges joining the vertices a and b. Similarly, there are two edges joining the vertices b and d.

Solution

In terms of the graph this problem is equivalent to that of finding a trail that contains each edge of the graph. Recall that, in a trail, no edge occurs more than once, so the problem is to find a trail using each edge exactly once. Euler noted that in following such a trail, for each vertex other than the first and last vertices, we would have to reach the vertex along one edge and leave along another. Hence such a trail would use an even number of edges adjacent to each of these vertices. Therefore, as it would use each edge exactly once, the degrees of these vertices much be even numbers. However, we see in the graph of Figure 9.31 that there are more than two vertices with odd degrees. So the graph has no trail containing each edge exactly once. Hence the answer to Problem 9.11 is no.

Euler's argument, as applied to the particular graph of Figure 9.31, is not difficult, and it is hard to believe that it had not previously occurred to some smart citizen of Königsberg. Euler's real contribution was to realize that his argument applied in general, and also that his necessary condition for the existence of a trail using each edge once is also sufficient. This we now explain.

We first introduce some terminology, for graphs where there is a closed trail using each edge once, and for those where there is such a trail that need not be closed. These graphs are named in honor of Euler. We again remind you that here by "graph" we mean both simple graphs and multigraphs.

DEFINITION 9.10
A graph that is connected and has a trail that includes each of its edges is said to be *semi-Eulerian*. It is said to be *Eulerian* if it is connected and has a closed trail including each edge. We specify that the graph consisting of just one vertex and no edges is not Eulerian.

THEOREM 9.9
(EULER 1735*)

a. If G is a graph with more than one vertex, then G is Eulerian if and only if G is a connected graph in which the degree of each vertex is an even number.
b. If G is a graph with more than one vertex, then G is semi-Eulerian if and only if G is a connected graph in which there are at most two vertices whose degrees are odd numbers.

Proof

As we have already seen, the "only if" parts of this theorem are straightforward.

a. If G is Eulerian there is a closed trail containing each edge exactly once, and hence in which each vertex occurs. So there is a trail and hence a path connecting each pair of vertices. So G is connected. If a vertex, say a, occurs k times in the closed trail, there must be $2k$ edges adjacent to a (namely, k "in" edges and k "out" edges) and so a has even degree.

---

* Euler stated only the "if" part of his theorem. The first published proof is attributed to C. Hierholzer, Über die Möglichkeit einen Linienzug ohne Wiederholung und ohne Unterbrechnung zu umfahren, *Mathematische Annalen*, 6, 1873, pp. 30–32. Hierholzer, who lived a short life (1840–1871), died before his paper was published.

b. Similarly, if G is semi-Eulerian, it has a trail that includes each edge exactly once. It follows that G is connected, and every vertex, except possibly the first and last vertices in the trail, has even degree.

We now come to the harder "if" part of the theorem. We prove by mathematical induction on the number of edges that if G is a connected graph with more than one vertex in which each vertex has even degree, then G is Eulerian. Such a graph must have at least two edges if it is a multigraph and at least three edges if it is simple graph. It is easy to see that a connected multigraph with two edges and a connected simple graph with three edges must both be Eulerian. This provides the base for the proof by induction.

Now suppose that the result is true for all graphs with fewer than k edges and that G is a connected graph with k edges, with $k \geq 3$, in which each vertex has even degree. So each vertex has degree at least 2 and hence, by Theorem 9.5, G has a closed path, say C.

If C includes all the edges of G, our proof is complete. If not, then we continue the proof by first removing from G the edges (but none of the vertices) included in C. This produces a new graph, which may not be connected but in which each vertex still has even degree, since by deleting the edges in C we remove two edges adjacent to each vertex that occurs in C. The new graph has fewer than k edges, and so by our induction hypothesis each connected part of this graph has a closed trail involving all its edges. Let these closed trails be $T_1, T_2, \ldots, T_n$. Thus between them these closed trails together with the closed path C include all the edges of G. Since G is connected, each of the closed trails has at least one vertex that also occurs in C. It follows that we can obtain a closed trail that includes all the edges of G by following the closed path C and, for $1 \leq i \leq n$, the first time we come to a vertex that also occurs in T_i, going round this trail before continuing around G.

This completes the proof of the "if" part of (a). The "if" part of (b) can easily be deduced from this. We ask you to do this in Exercise 9.6.2A. We have deliberately not included a diagram to illustrate this proof, but you are strongly encouraged to draw your own diagrams to help you to understand the proof.

Euler's paper also contains some useful mathematical advice. He observes that one could solve the Königsberg problem merely by trying all cases. He dismisses this as "too difficult and laborious." (And, he might have added, such a trial by cases would not expose the real reason why the same result is true in general.) He also says (essentially), "Before I attempt to solve the problem I shall try to determine if it *can* be solved, for if it can't, then attempts to find a solution would be a waste of time."

Euler's theorem could be of some interest to people who deliver the mail, or newspapers, and to street cleaners, indeed everyone whose job requires them to visit a number of streets in a certain area. It can tell them, in advance, if it is possible to avoid traveling down any street more than once before returning to base: Namely, if there is a point at which an odd number of streets meet, then Euler's theorem tells you that they will be wasting their time trying to discover such a route. If you are delivering the mail or newspapers, you may need to deliver to both sides of a street, and this involves walking along the street twice. This corresponds to connecting the

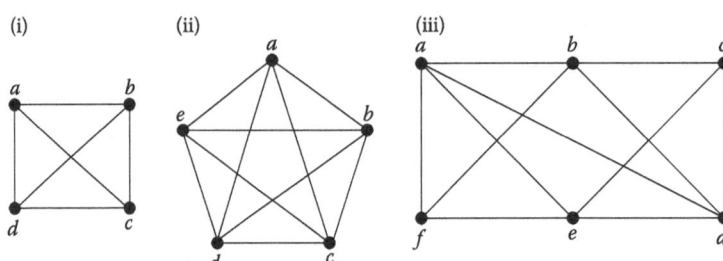

FIGURE 9.32

points where streets meet by two edges, and, if this is the case for every street, the resulting graph must be Eulerian.

We have seen that we can tell, by looking at the degrees of its vertices, whether a connected graph has a closed trail including each *edge* exactly once. It is natural to ask a related question: When will a graph have a closed trail that includes each *vertex* once and once only? To achieve this we do not insist that we use all the edges, and since the required closed trail does not visit any vertex more than once, it will be a closed path. This question does not yet have as satisfactory an answer as the analogous question about Eulerian graphs. We discuss this problem in the next section.

Exercises

9.6.1A Determine which of the graphs in Figure 9.32 are Eulerian and which are semi-Eulerian. Where the graph is semi-Eulerian, find a trail that includes all its edges. Where it is Eulerian, find a closed trail that includes all its edges.

9.6.1B For which positive integers n is the complete graph with n vertices, K_n, Eulerian?

9.6.2A Show that if G is a connected graph in which there are exactly two vertices whose degrees are odd numbers, then G is semi-Eulerian. (In proving this result you will be completing the proof of Theorem 9.9.)

9.6.2B For which values of m, n is $K_{m,n}$ semi-Eulerian, and for which values of m, n is it Eulerian?

9.6.3A What is the least number of bridges that needed to be built in Königsberg so that burghers could have walked around their city crossing each bridge once and ending up where they had started?

9.6.3B How many of the simple graphs with four vertices (see Problem 9.5) are Eulerian?

9.7 HAMILTONIAN GRAPHS

In an Eulerian graph there is a trail that uses each edge exactly once. It is natural also to consider the related question as to whether in a given graph there is a path that includes each vertex exactly once. This question came to attention in 1856 when the distinguished

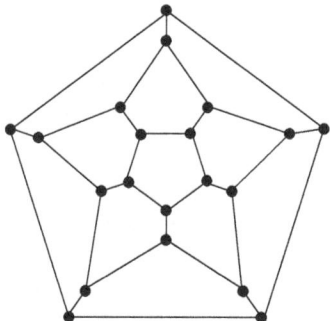

FIGURE 9.33

mathematician William Rowan Hamilton* invented a game, the object of which was to take a dodecahedron, choose one of its 20 vertices, and then try to trace a path around (some of) the edges of the dodecahedron, visit every vertex once, and finish up at the original vertex. The 20 vertices and 30 edges of the dodecahedron may be represented by a graph, as shown in Figure 9.33.

We leave you the pleasure of showing that this graph is *Hamiltonian* according to the following definition.

DEFINITION 9.11

A graph is said to be *Hamiltonian* if it has a closed path in which each vertex occurs. Such a path is called a *Hamiltonian path*. A graph is said to be *semi-Hamiltonian* if it has a path in which each vertex occurs. We specify that the graph consisting of just one vertex and no edges is not Hamiltonian.

Since each vertex can occur just once in a path, in a Hamiltonian path each vertex of the graph occurs exactly once. It follows that a Hamiltonian path cannot involve more than one edge between any pair of vertices. Thus the problem of the existence of Hamiltonian paths in multigraphs is essentially the same as for simple graphs. Consequently, in this section, we assume that all the graphs we are dealing with are simple graphs.

Hamilton's name is used for these graphs because he promoted the dodecahedron problem. However, other mathematicians had discussed similar problems earlier than Hamilton. Indeed, Euler gave his attention in the middle of the eighteenth century to the related problem of finding a *knight's tour* on a chessboard.

On a chessboard, a knight can move to the opposite corner of any 2×3 rectangle of which the square it is on forms one corner. Thus a knight near the middle of a

---

* Hamilton was born in Dublin in 1805. He was a child prodigy, who at the age of only 22 was appointed professor of astronomy at Trinity College, Dublin, where he had been an undergraduate. However, Hamilton devoted himself to mathematics rather than astronomy. In his lifetime Hamilton was best known for his invention of *quaternions*. These form a four-dimensional number system, extending the complex numbers, but where multiplication is not commutative. In the end quaternions did not play the important role in physics that Hamiltonian had hoped, as they have been superseded by vectors. Hamilton is today best remembered for *Hamilton's principle* (of least action) and for *Hamiltonian operators*, which are used in quantum mechanics. Hamilton died of gout, it is said, in 1865.

chessboard has eight possible moves, but it has fewer moves the nearer it is to a corner of the board. Some of the possibilities are shown in Figure 9.34. A *knight's tour* is a sequence of 64 moves by a knight in which it visits each square of the board once and returns to its original square. One example of a knight's tour is shown in Figure 9.35. It is not quite trivial to find a knight's tour, but equally there are known to be a large number of them. Euler studied heuristic methods for finding them, and using his methods, which essentially give a way to include squares that have been left out, it is not difficult to find them.* The method used in the proof of Theorem 9.10 to turn a semi-Hamiltonian path into a Hamiltonian path is the same as one that Euler used.

But what has this to do with Hamiltonian graphs? The graph corresponding to the problem of a knight's tour has the 64 squares of the chessboard as its vertices, with two vertices being joined by an edge if a knight can go from one square to the other in one move. A knight's tour then corresponds to a Hamiltonian path in this graph.

So far, our examples of Hamiltonian paths have been drawn from recreational problems. They are, however, also related to some practical problems. If you need to visit a number of different towns, it would seem most efficient not to have to return to the same town. Thus it seems best to follow a Hamiltonian path in the graph, which has the towns as the vertices and the connecting roads as its edges. In this practical problem, you may need to take account of the distances between the towns, or the time or cost involved in traveling between them. This kind of problem is modeled by a graph in which we assign a number (distance, time, cost, etc.) to each edge. In general these numbers are called *weights*, and the resulting structure is called a *weighted graph*. The

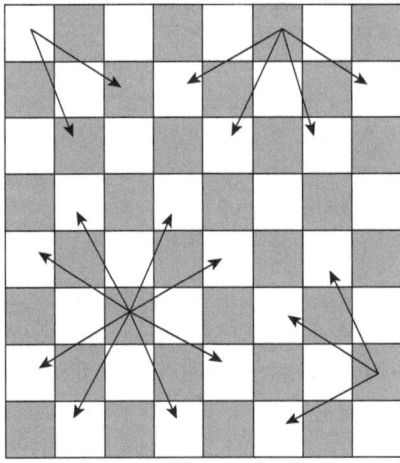

A knight's move

FIGURE 9.34

---

* There is a good discussion of this problem in *Mathematical Recreations and Problems* by W. W. Rouse Ball. First published in 1892 by Macmillan, London, this book has gone through many editions, with a revised edition with the title *Mathematical Recreations and Essays* by H. S. M. Coxeter in 1939 by Macmillan, London. According to David Hooper and Ken Whyld, *The Oxford Companion to Chess*, new edition, 1984, Oxford University Press, Oxford, there are more than 122,000,000 knight's tours, and "almost an infinity" (!) if you drop the restriction that the path is closed.

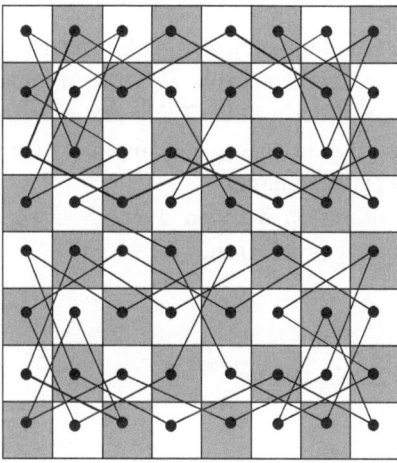
A knight's tour

FIGURE 9.35

problem is then to find a Hamiltonian path with minimal total weight. Because of its origins this problem is often called the *traveling salesman problem*. This is a hard problem since at present no better method for finding its best solution is known other than to enumerate all possible Hamiltonian paths in the graph in question. There are, however, efficient methods for obtaining "good" solutions, which are not necessarily the best possible.

A detailed discussion of the traveling salesman problem and related problems is beyond the scope of this book. We just consider the problem of deciding whether or not a graph has a Hamiltonian path. We have already asked you to consider this problem for the dodecahedron graph of Figure 9.33. You may also like to consider whether the Petersen graph shown in Figure 9.14 is Hamiltonian. This may help you to appreciate the nature of this problem.

Of course, in principle, it is a straightforward question to answer. Given a graph with n vertices, just list all possible ways of arranging these vertices in order and for any given ordering, say $v_1, v_2, \ldots, v_n$, check whether all of $\{v_1, v_2\}, \{v_2, v_3\}, \ldots, \{v_{n-1}, v_n\}$ and $\{v_n, v_1\}$ are edges of the graph. If they are, you have found a Hamiltonian path. If you don't find a Hamiltonian path in this way, you can be sure that the graph is not Hamiltonian. However, this is not a practical method for graphs with more than a small number of vertices. With n vertices they can be ordered in $n!$ ways, and this number grows very rapidly with n. For example, if a computer could check 10^{12} arrangements in one second, it would still take more than a year to check all possible orderings of the vertices of a graph with 21 vertices.

Thus it would be good if we had a simple criterion for whether or not a graph is Hamiltonian, analogous to the criterion given by Theorem 9.9 for Eulerian graphs. Unfortunately, no such criterion is known. There are a number of theorems that give either a necessary or a sufficient condition, but so far no one has found a simple condition that is both necessary and sufficient. It is easily seen that for a graph to be Hamiltonian each vertex must have degree at least 2. A connected graph in which each vertex has degree 2 must be Hamiltonian as it must consist of n vertices in a single path. For each $n \geq 3$ there is a Hamiltonian graph with n vertices and only n edges. At the other extreme, for $n \geq 3$ the complete graph , K_n, is Hamiltonian as we can follow a path through each vertex in any order we like. These examples indicate that there cannot be a straightforward condition in terms of the number of edges or the degrees of the vertices for a graph to be Hamiltonian.

Also, there is no relationship between a graph being Eulerian and its being Hamiltonian, as the following problem shows.

PROBLEM 9.12

Give examples of simple graphs that are

a. Eulerian and Hamiltonian
b. Eulerian but not Hamiltonian
c. Hamiltonian but not Eulerian
d. Neither Eulerian nor Hamiltonian but with more than one vetex

Solution

See Figure 9.36. There are, of course, lots of possible solutions. We have given examples with the smallest possible number of vertices and, for this number of vertices, the smallest number of edges.

We content ourselves with just one theorem, which gives a straightforward sufficient condition for a graph to be Hamiltonian.

THEOREM 9.10
(ORE 1960*)

If G is a simple graph with n vertices with $n \geq 3$, such that for each pair of vertices u, v not joined by an edge, $\delta(u) + \delta(v) \geq n$, then G is Hamiltonian.

Proof

Suppose that G satisfies the condition of the theorem but is not Hamiltonian. Since the complete graph K_n is Hamiltonian, we may assume that G is not complete. However, adding to G one "missing" edge at a time, we shall eventually reach a graph, G_1, say, such that G_1 is not Hamiltonian and yet adding one more edge, say $\{x, y\}$, to G_1 makes

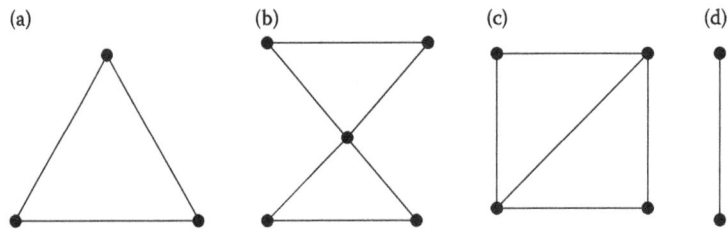

FIGURE 9.36

---

* O. Ore, Note on Hamiltonian Circuits, *American Mathematical Monthly*, 67, 1960, p. 55.

the new graph, which we will write as $G_1 + \{x, y\}$, Hamiltonian. Since G_1 is obtained from G by adding edges, it also satisfies the condition that for each pair of vertices u,v not joined by an edge, $\delta(u) + \delta(v) \geq n$.

As G_1 is not Hamiltonian but $G_1 + \{x, y\}$ is Hamiltonian, each Hamiltonian path in $G_1 + \{x, y\}$ must include the edge $\{x, y\}$. If we delete this edge from one such Hamiltonian path, we obtain a path from x to y that includes every vertex of G_1. Let

$$x = v_1 \to v_2 \to \ldots \to v_n = y$$

be such a path.

The vertices v_1 and v_n are not joined by an edge in G_1, and hence $\delta(v_1) + \delta(v_n) \geq n$. Suppose $\delta(v_1) = k$ and that v_1 is joined by an edge to the vertices $v_{i_1}, \ldots, v_{i_k}$. Then $\delta(v_n) \geq n-k$, and hence v_n must be joined by an edge to at least one of the vertices $v_{i_1-1}, \ldots, v_{i_k-1}$ since otherwise there would be at most $n - k - 1$ vertices it could be joined to. Thus for some i, both $\{v_1, v_i\}$ and $\{v_{i-1}, v_n\}$ are edges of G_1. Here $3 \leq i \leq n-1$, as v_1 is not joined to v_n.

$$x = v_1 \to v_2 \to \ldots \to v_{i-1} \to v_i \to \ldots \to v_n = y$$

It follows that

$$v_1 \to v_i \to v_{i+1} \to \ldots \to v_n \to v_{i-1} \to v_{i-2} \to \ldots \to v_2 \to v_1$$

is a Hamiltonian path in G_1, contradicting our assumption that G is not Hamiltonian. Hence G_1 is also not Hamiltonian. This contradiction shows that G is Hamiltonian, thus completing the proof.

Exercises

9.7.1A Determine whether the graphs in Figure 9.37 are Hamiltonian.

9.7.1B Determine whether the graphs in Figure 9.38 are Hamiltonian.

9.7.2A Determine whether the Petersen graph (as shown in Figure 9.14) is Hamiltonian.

9.7.2B For which values of m and n is the bipartite graph $K_{m,n}$ Hamiltonian?

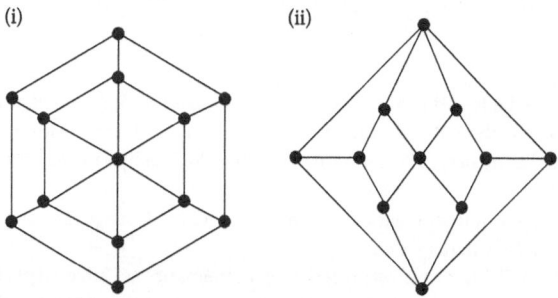

FIGURE 9.37

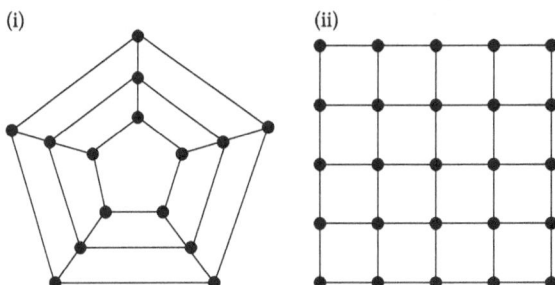

FIGURE 9.38

9.7.3A Prove that for $n \geq 3$ a graph with n vertices and $\frac{1}{2}(n-1)(n-2)+2$ edges must be Hamiltonian, but that for each integer $n \geq 3$, there is a simple graph with n vertices and $\frac{1}{2}(n-1)(n-2)+1$ vertices that is not Hamiltonian.

9.7.3B Prove that if G is a simple graph with n vertices, with $n \geq 3$, in which each vertex has degree at least $\frac{1}{2}n$, then G is Hamiltonian.

9.7.4A Determine whether the graph shown in Figure 9.39 is Eulerian, and whether it is Hamiltonian.

9.7.4B i. Determine whether the graph in Figure 9.40 is Hamiltonian.

ii. Generalize (ii) by determining whether the graph with $\frac{1}{2}n(n+1)$ vertices arranged in triangular array analogous to the graph in Figure 9.40 (which is the case where $n = 6$) is Hamiltonian.

9.8 THE FOUR-COLOR THEOREM

We now give a brief discussion of the famous *four-color theorem*.* The question was first raised in terms of coloring a map showing different countries. Can the countries always be colored using just four colors, so that countries with common boundaries are colored differently?

Although this question is described as being about maps, it seems never to have been of interest to cartographers. It was first posed in 1852 by a student, Francis Guthrie,† eventually reaching Augustus De Morgan.‡ We can turn the problem into one about graphs in the way indicated in Figure 9.41. We represent each country by a vertex, with an edge between a pair of vertices if they correspond to countries with a common boundary. As shown in Figure 9.41, this means a common boundary of nonzero length. Just having a point in common doesn't count. So, for example, there is no edge joining b and d. Likewise, there is no edge joining a and e.

---

* A good popular account may be found in Robin J.Wilson, *Four Colours Suffice*, Allen Lane, The Penguin Press, London, 2002. There is more technical coverage in Rudolf Fritsch and Gerda Fritsch, *The Four-Color Theorem*, Springer, New York, Berlin, Heidelberg, 1998 (a translation into English by Julie Peschke of a book first published as *Der Vierfarbensatz* in 1994).

† Francis Guthrie (1831–1899) was a mathematics student at University College, London, who became a professor of mathematics in South Africa, where he died.

‡ Augustus De Morgan (1806–1871) was the first professor of mathematics at University College, London, and is best remembered for *De Morgan's laws* in logic. He was a founder and the first president of the London Mathematical Society, whose headquarters are named De Morgan House in his memory.

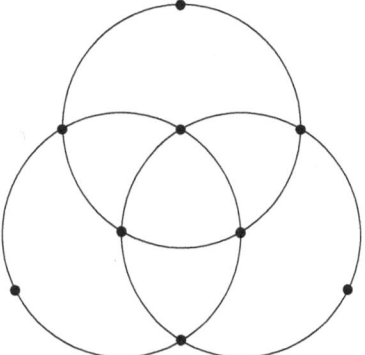

FIGURE 9.39

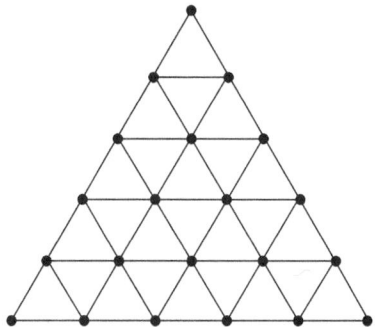

FIGURE 9.40

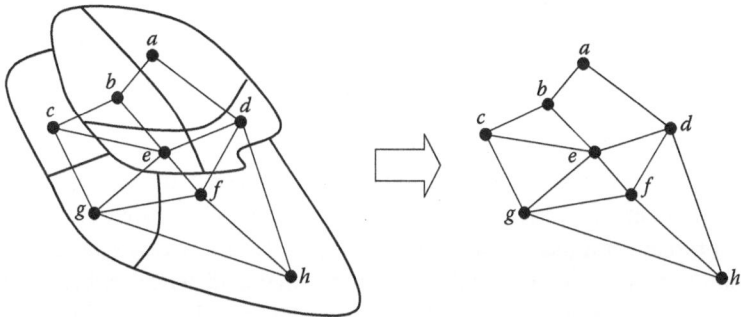

FIGURE 9.41

In this way, provided we put some restrictions on what counts as a "country," we end up with a planar graph. For example, we need to avoid countries whose territory is made up of disconnected regions (as, for example, Russia, which today includes the region round Königsberg, now called Kaliningrad, which is cut off from the rest of the country by Poland, Belarus, and Lithuania). We won't bother to specify exactly what sort of regions can count as countries, other than to say that the resulting graph must be planar. In terms of the resulting graph the question becomes: Can we assign four colors to the vertices of the graph so that vertices joined by an edge are assigned different colors? This question makes sense even for graphs that are not planar. So we adopt the following definition.

190 ■ How to Count: An Introduction to Combinatorics, Second Edition

DEFINITION 9.12

A graph, $G = (V, E)$, is said to be *k-colorable* if there is a mapping $f : V \to C$, where C is a set of k colors, such that whenever $\{u, v\}$ is an edge, $f(u) \neq f(v)$. The *chromatic number* of the graph is the least number k such that G is k-colorable but not $(k-1)$-colorable.

We call such a mapping f a *coloring of the graph*. We say that f is a *k-coloring* of G if $\#(C) = k$, that is, if f uses at most k colors.

PROBLEM 9.13

What is the chromatic number of the graph shown in Figure 9.41?

Solution

Since each pair of the vertices b, c, and e is joined by an edge, they must be assigned different colors. So the graph is not two-colorable. However, it is three-colorable since, for example, we could color a, c, and f red; b, d, and g green; and e and h yellow. So the chromatic number of the graph is 3.

PROBLEM 9.14

What is the chromatic number of the graph shown in Figure 9.42?

Solution

The vertices a, b, and f must be assigned different colors. Suppose, say, that we color a red, b green, and f yellow. Then, if we wish to avoid a fourth color we would need to color c red and e green. But then, as d is joined to c, e, and f, and these vertices are colored with three different colors, we are forced to use a fourth color for d. Thus the graph is four-colorable but not three-colorable. So its chromatic number is 4.

The question that Guthrie asked can now be formulated as: Is every planar graph four-colorable? This question turned out to be much harder to answer than most mathematicians thought when they first heard the question. There is an easy trap to fall into. The chromatic number of the complete graph K_5 is 5 since each pair of vertices is joined by an edge. We have proved that K_5 is not planar, and hence neither is any graph with a subgraph that is isomorphic to K_5. It is tempting to believe that this is enough to

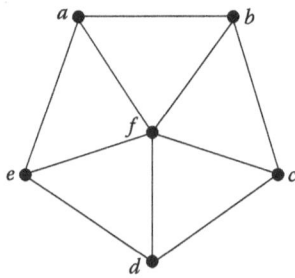

FIGURE 9.42

prove that every planar graph is four-colorable, but examples such as that of the graph in Figure 9.42 show that planar graphs can fail to be three-colorable even though they have no subgraphs isomorphic to K_4, and this immediately raises the possibility that there is a planar graph that is not four-colorable even though it has no subgraph isomorphic to K_5.

In 1879 Alfred Kempe published an argument that claimed to prove that every map can be colored with at most four colors.* His argument was built on a clever idea but included a subtle mistake that was not noticed for 11 years. It then took a further 87 years before the problem was solved, though as the solution involved a large amount of computer checking, not every mathematician accepts that it has been finally resolved. Although Kempe's argument did not establish that every planar graph is four-colorable, it does establish the next best result, namely, that every planar graph is five-colorable. We now explain the proof of this, the *five-color theorem*.

DEFINITION 9.13

We say that a plane graph is *triangulated* if every face, including the unbounded face, contains just three edges.

It is easily seen that if we have a plane graph that is not triangulated, we can extend it to a triangulated graph by adding edges, as is illustrated in Figure 9.43.

As we have illustrated, we may need to join vertices by curves that are not segments of straight lines. However by Wagner's theorem, which we prove at the end of this section, the resulting plane graph can always be redrawn so that the edges are segments of straight lines.

Since, when we triangulate a graph, we add edges but not vertices, it follows that, if the triangulated graph is k-colorable, so too will be the original graph. Thus, to prove the five-color theorem, it is enough to prove this theorem for connected triangulated plane graphs. It is clear that, in such a graph with more than three vertices, there are no vertices

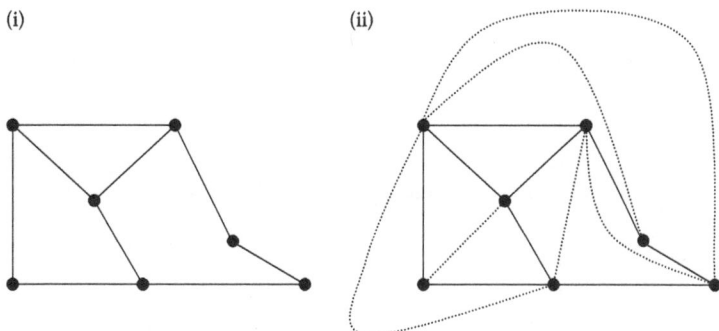

FIGURE 9.43

---

* Alfred Bray Kempe, On the Geographical Problem of Four Colours, *American Journal of Mathematics*, 2, 1879, pp. 193–200. Alfred Kempe was born in London in 1831. After obtaining a mathematics degree at Trinity College, Cambridge, he turned his attention to law and had a successful career as a barrister. He maintained his interest in mathematics, publishing work on linkages and becoming a fellow of the Royal Society of London in 1881. He died in London in 1922.

of degrees 1 and 2. A key idea behind the proof is that, however, such a graph must contain some vertices whose degrees are quite low, in the sense given by our next result.

THEOREM 9.11

A connected triangulated plane graph has at least four vertices whose degrees are at most 5.

Proof

Let $G = (V, E)$ be a connected triangulated plane graph with v vertices and e edges. By Theorem 9.7 we have that $e \leq 3v - 6$. This is good enough for our purposes, but it is worth noting that, as G is triangulated, each face is bounded by exactly three edges, and so the inequality $3f \leq 2e$ in the proof of Theorem 9.7 may be replaced by the equation $3f = 2e$, and hence in this case we can deduce that

$$e = 3v - 6. \tag{9.6}$$

Now suppose that the highest degree of any vertex of G is r and that for $k = 3,4,5,\ldots,r$, the graph G has a_k vertices of degree k. Then

$$v = a_3 + a_4 + \ldots + a_r \tag{9.7}$$

and

$$\sum_{x \in V} \delta(x) = 3a_3 + 4a_4 + 5a_5 + \ldots + ra_r. \tag{9.8}$$

Hence, by the handshaking lemma,

$$2e = 3a_3 + 4a_4 + 5a_5 + \ldots + ra_r. \tag{9.9}$$

From Equations 9.6, 9.7, and 9.9 it follows that

$$3a_3 + 4a_4 + 5a_5 + 6a_6 + \ldots + ra_r = 6(a_3 + a_4 + a_5 + a_6 + \ldots + a_r) - 12$$

and hence that

$$3a_3 + 2a_4 + a_5 = a_7 + 2a_8 + \ldots + (r-6)a_r + 12. \tag{9.10}$$

It follows that

$$3(a_3 + a_4 + a_5) \geq 3a_3 + 2a_4 + a_5 \geq 12$$

and hence that

$$a_3 + a_4 + a_5 \geq 4. \tag{9.11}$$

The inequality 9.11 is what we were aiming to prove.

Note: The inequality 9.11 plays an important role in the proof of the four-color theorem as it imposes a significant constraint on the graphs that need to be considered. We hope it is clear that we didn't really need to include the condition that the graph be triangulated in the statement of this theorem.

We are now ready to give Kempe's proof of the five-color theorem.

THEOREM 9.12
The Five-Color Theorem

Every planar graph is five-colorable.

Proof

As we have already noted, it is sufficient to prove the result for triangulated graphs. Thus we prove, by induction on the number, v, of vertices that for $v \geq 3$ a triangulated planar graph with v vertices is five-colorable. The result is obvious for $v = 3$ (and, indeed, for $v = 4$ and $v = 5$). Suppose that the result holds for $v = k$, $k \geq 3$, and that G is a triangulated planar graph with $k + 1$ vertices.

By Theorem 9.11, G has a vertex, say v_0, whose degree, say d, is at most 5. Let $v_1, v_2, \ldots, v_d$ be the vertices to which v_0 is joined by an edge. Now let G' be the graph obtained from G by deleting the vertex v_0 and the edges adjacent to it. If we triangulate G', then by our induction hypothesis the resulting graph is five-colorable and hence G' itself is five-colorable. Let f be a five-coloring of G'. If f uses at most four colors for the vertices $v_1, v_2, \ldots, v_d$, then there is at least one color available for v_0, and hence f can be extended to a five-coloring of G.

Thus the only case that gives us trouble is where $d = 5$ and the coloring f of G' uses all five colors for the vertices $v_1, \ldots, v_5$. This is where Kempe's idea can be used to show that nonetheless f can be extended to v_0 without the need for a sixth color.

In Figures 9.44 and 9.45, we have indicated the coloring of the other vertices by shading. Since G' is planar it does not have a subgraph that is isomorphic to K_5. Therefore at least two of the vertices $v_1, \ldots, v_5$ are not joined by an edge. Suppose, for example, it is v_1 and v_3 that are not joined. We suppose that the colors assigned to v_1, v_2, v_3, v_4, and v_5 are *red, blue, green, yellow,* and *purple*, respectively. We now consider the subgraph, say H, of G' consisting of the vertices that are colored red or green, and the edges joining them.

There are two cases. The easier one is where v_1 and v_3 are in different connected components of H. This case is illustrated in Figure 9.44. In this case we can swap round the colors red and green just in the component of H to which v_1 belongs. In particular, v_1 is now colored green. It is still the case that joined vertices are assigned different colors, but now there is no vertex joined to v_0 that is colored red. So we can color v_0 red, and we have found a five-coloring of G.

The second case is where v_1 and v_3 are in the same connected component of H, as shown in Figure 9.45. In this case there is a path from v_1 to v_3 made up of vertices that are colored red and green. (Such a path is shown by dashed edges in Figure 9.45.) Either

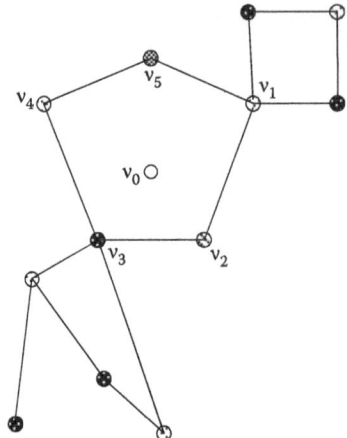

FIGURE 9.44

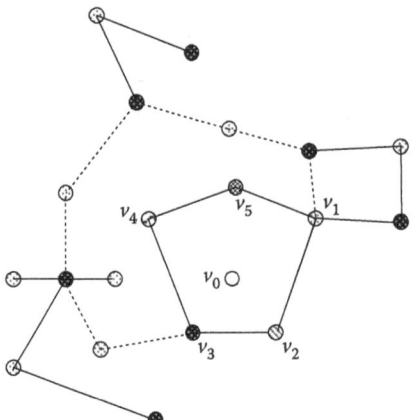

FIGURE 9.45

this path together with the edges $\{v_1, v_2\}$ and $\{v_2, v_3\}$ forms a closed path that surrounds the vertices v_4 and v_5, or this path together with the edges $\{v_1, v_5\}$, $\{v_5, v_4\}$, and $\{v_4, v_3\}$ forms a closed path that surrounds the vertex v_2. In Figure 9.45 the first of these cases is illustrated. In this case v_2 and v_4 must be in different connected components of the subgraph, say K, made up of the vertices with the same colors as v_2 and v_4, that is, blue and yellow, and the edges joining them. So we can interchange the colors blue and yellow assigned to the vertices in, say, the connected component of K that includes v_2. So v_2 becomes yellow, and this frees up blue as a color, which we can now assign to v_0. So we have again found a five-coloring of G. The second case is similar.

Thus in any case G is five-colorable, and this completes the proof by induction that all planar graphs are five-colorable.

The argument used in the above proof is often called a *Kempe chain argument*. Kempe's error was to suppose that he could carry out two of these arguments simultaneously, and thus, given a five-coloring of G, produce a coloring of G' that used only three colors for the vertices $v_1,\ldots,v_5$. However, in 1890 Percy John Heawood* published an example that showed that one color interchange could interfere with the other, and thus where Kempe's argument for the four-color theorem did not work.

It took until 1977 before a proof of the four-color theorem was published by Kenneth Appel and Wolfgang Haken.† Their proof made a great deal of use of a computer program to generate and check a large number of cases. For the details see the first footnote in Section 9.8. A proof that could be completely checked by a single person is still awaited.

Graphs that are not planar need not be four-colorable. Indeed, the complete graph K_n has the chromatic number n, and so there are graphs with chromatic numbers as large as we like. We will not discuss colorings of nonplanar graphs in any detail. We just point out that they have a practical application in scheduling, as shown by the following problem.

PROBLEM 9.15

Twelve students are studying the subjects shown in the following table. Their college wants to arrange tutorials for them, so that each student can attend all the tutorials for their subjects. How many different timetable slots are needed so that tutorials for subjects taken by the same student are at different times?

Amrit	History, Italian, Mathematics
Ben	Knitting, Latin, Mathematics
Chris	Geography, Knitting, Mathematics
David	Japanese, Latin, Mathematics
Eion	Geography, Latin, Mathematics
Fahana	History, Latin, Mathematics
Graham	French, Japanese, Latin
Hussain	History, Latin, Mathematics
Indira	French, Geography, Mathematics
Jon	French, Latin, Mathematics
Kieran	French, Japanese, Mathematics
Linda	History, Italian, Mathematics
Mary	French, Geography, Mathematics

Solution

We can represent the situation with a graph where the vertices f,g,h,i,j,k,l,m correspond to the subjects French, Geography, History, Italian, Japanese, Knitting, Latin, and Mathematics. Two vertices are joined by an edge if at least one student is studying both of them. This results in the graph in Figure 9.46.

From Figure 9.26, we see that this graph requires at least five colors, as we need five different timetable slots for French, Geography, Knitting, Latin, and Mathematics. However, History and Japanese can be in, for example, the same slot as Geography, and

---

* P. J. Heawood, Map Colour Theorem, *Quarterly Journal of Mathematics*, 24, 1890, pp. 332–338. Heawood was born in Shropshire in 1861 and studied mathematics at Exeter College, Oxford. He became a lecturer in mathematics and, subsequently, professor at the University of Durham. He was vice chancellor of the university from 1926 to 1928, and he died in Durham in 1955. He had a lifelong interest in the four-color problem.
† Kenneth Appel and Wolfgang Haken, Every Planar Map Is Four Colorable, *Illinois Journal of Mathematics*, 21, 1977, pp. 429–490.

Italian in the same slot as Latin. Thus five timetable slots, but no more, are needed. Another way to put this is that the graph in Figure 9.46 has the chromatic number 5.

You may have noticed that in proving the five-color theorem we did not make use of the full force of Theorem 9.11, as we used only the fact that a planar graph has *one* vertex of degree at most 5. However, we need the full result of Theorem 9.11 for the proof that planar graphs can always be drawn with straight line edges.

THEOREM 9.13
Wagner's Theorem*

Every planar simple graph can be drawn in the plane with the edges represented by straight lines.

Proof

We prove this by mathematical induction on the number of vertices in the graph. Clearly the result holds for a graph with one, two, or three vertices. Now suppose the result holds for all graphs with k vertices, $k \geq 3$, and that G is a planar graph with $k + 1$ vertices. Extend G to a triangulated graph G^*. Clearly if G^* can be drawn in the plane with only straight line edges, so too can G. By Theorem 9.11, G^* has at least four vertices with degrees at most 5. Since the boundary of the external face of G^* contains just three edges and hence three vertices, there is a vertex, say v_0, that is not part of this boundary and that has degree at most 5. Let G' be the graph obtained from G^* by deleting the vertex v_0 and the edges adjacent to it. As G' has k vertices, by our induction hypothesis it can be drawn in the plane using only straight line edges. The points representing the vertices adjacent to v_0, and the lines representing the edges joining them, will form a polygon that is either a triangle, a quadrilateral, or a pentagon, according to the degree of v_0. As shown in Figure 9.47, even where the quadrilateral or pentagon has a reflex angle, we can find a point inside the polygon that can be joined to all its vertices by straight line edges. Thus we can extend the drawing of G' with only straight line edges to a drawing of G^* and hence of G with just straight line edges.

This completes the proof by mathematical induction of the theorem.

Exercises

9.8.1A   Determine the chromatic number of the graph shown in Figure 9.48.

9.8.1B   Determine the chromatic number of the graph shown in Figure 9.49.

9.8.2A   i. Prove that a connected simple planar graph with no vertices of degree less than 5 must have at least 12 vertices with degree 5.

  ii. Give an example to show that there is a connected simple planar graph with exactly 12 vertices all with degree 5.

9.8.2B   i. Prove that a connected simple planar graph with no vertices of degree less than 4, and no vertices of degree 5, must have at least six vertices of degree 4.

---

* K. Wagner, Bemerkungen zum Vierfarbenproblem, *Jahrberichte der Deutsche Mathematiker-Vereinigung*, 46, 1936, pp. 26–32.

Introduction to Graphs ■ 197

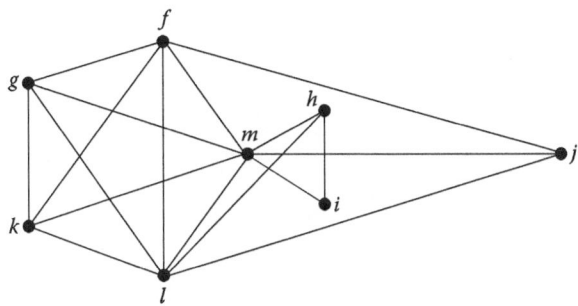

FIGURE 9.46

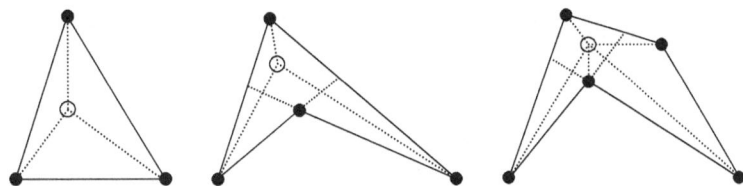

FIGURE 9.47

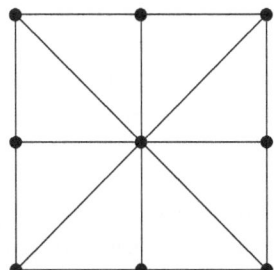

FIGURE 9.48

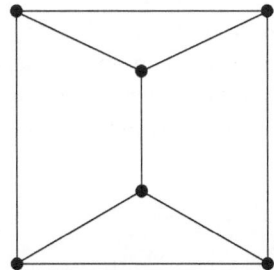

FIGURE 9.49

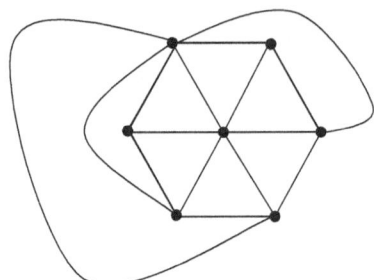

FIGURE 9.50

ii. Give an example to show that there is a connected planar graph with exactly six vertices all of degree 4.

9.8.3A Students have to take one-hour exams in different mathematical topics as shown in the following table. How many different hours are needed for the exams, so that no student has two exams in the same hour?

Michael	Arithmetic, binomial theorem
Nadia	Determinants, ellipses, functions
Oliver	Binomial theorem, calculus, ellipses
Payam	Arithmetic, determinants
Qin	Determinants, functions, hyperbolas
Rachel	Functions, graphs, hyperbolas
Satnam	Binomial theorem, calculus
Thornton	Calculus, ellipses, graphs
Ushbah	Ellipses, functions
Vanessa	Binomial theorem, ellipses
Wendy	Determinants, hyperbolas
Xiao	Ellipses, graphs

9.8.3B A zoo has to transport seven animals. To prevent one animal fighting another animal, the aardvark must travel in a different van from the cheetah; the buffalo cannot travel with either the cheetah, the giraffe, or the dormouse; the cheetah must be apart from the elephant, the flamingo, and the dormouse; the elephant must be in a different van from the aardvark, the giraffe, and the buffalo; and the flamingo cannot go in the same van as either the elephant or the buffalo. What is the smallest number of vans that the zoo needs to transport all seven animals?

9.8.4A Show that the graph (ii) in Figure 9.43 may be drawn in the plane with all the edges represented by straight line segments.

9.8.4B Show how the graph in Figure 9.50 may be redrawn in the plane so that all the edges are represented by straight line segments.

CHAPTER 10

Trees

10.1 WHAT IS A TREE?

In this chapter we look at a special class of graphs. They are singled out for particular attention because they occur in a wide range of practical and mathematical contexts. These special graphs are called *trees*, for pictorial reasons. We begin with the definition, which at first sight may look as though it was plucked from thin air. However, after we have given some examples and some explanation of how *trees* arose and why they are of mathematical and practical interest, the importance of the idea should be clear.

DEFINITION 10.1

A *tree* is a connected graph in which there are no closed paths.

A graph with no closed paths cannot have multiple edges or loops, so it must be a simple graph. Throughout this chapter we will use "graph" to mean "simple graph." Figure 10.1 gives some examples of trees, some of which are making repeat appearances, as they have already shown up in Chapter 9.

From Figure 10.1i we see that the graph consisting of a single isolated vertex counts as a tree. Graphs iv and v look more like trees in the botanical sense, though, of course, this is just because of the way we have chosen to represent them in the plane.

The *Oxford English Dictionary*[*] quotes Robert Southey as writing in 1807, "An English Esquire would as soon walk abroad in his grandfather's wedding suit, as suffer the family Tree to be seen in his hall,"[†] so the use of *tree* in a metaphorical sense is older than its use by mathematicians. In Figure 10.2 we have given an example of part of a family tree, that of Charles Darwin.

It should be noted that a family tree such as that shown in Figure 10.2 is more like a *directed tree* than a tree in which the edges have no direction. Also, a family tree need not be a tree in the mathematical sense at all, as it may contain closed paths.

---

[*] *The Oxford English Dictionary, Second Edition*, prepared by J. A. Simpson and E. S. C. Weiner, Volume 5, Clarendon Press, Oxford, 1989, p. 708.
[†] Robert Southey (1774–1843) was an English poet. The quotation is from his *Letters from England*, Longman Hurst, London, 1807.

199

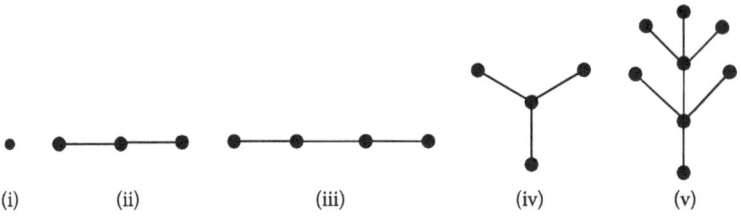

(i)   (ii)   (iii)   (iv)   (v)

FIGURE 10.1

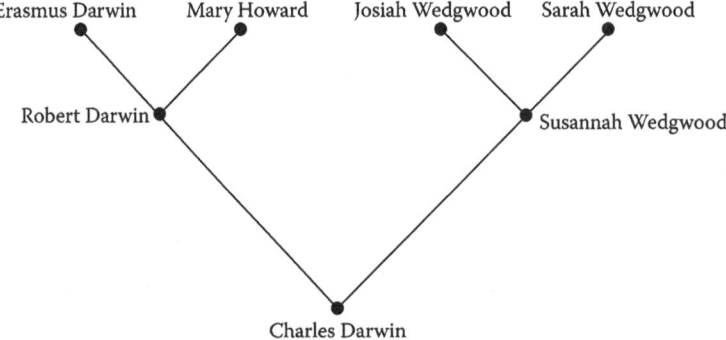

FIGURE 10.2

For example, Charles Darwin married his cousin, Emma Wedgwood. Josiah and Sarah Wedgwood were grandparents of them both, so the family trees of Charles and Emma's own children, taken back as far as their great-grandparents, include closed paths.

Another example of a "tree diagram," as its author described it, is taken from linguistics* and is shown in Figure 10.3. A sentence is analyzed with **NP** standing for "noun phrase," **VP** for "verb phrase," **N** for "noun," **V** for "verb," **Aux** for "auxiliary," **M** for "modal," and **Det** for "determiner."

One of the earliest uses of treelike structures in science may be found in the work of Gustav Robert Kirchhoff,[†] when he was investigating the flow of electrical currents in networks. He introduced the concept of what is now called a *spanning tree* (see Section 10.3) in order to determine the number of independent equations among the systems of equations derived for the circuit from Kirchhoff's own laws.

---

* Adapted from Noam Chomsky, *Aspects of the Theory of Syntax*, The M.I.T. Press Cambridge, MA, 1965.
† Gustav Kirchhoff was a German physicist who was born on March 12, 1824, in Königsberg. We don't know whether his birthplace contributed to his studying a graph-theoretic problem! He derived what are now called Kirchhoff's laws, which specify the currents in an electrical network, while he was still a student. His other great contributions were the foundation of spectroscopy and its use to identify elements in the sun's spectrum, Kirchhoff's laws for radiating bodies, and the concept of a perfect black body. He died in Berlin on October 17, 1887.

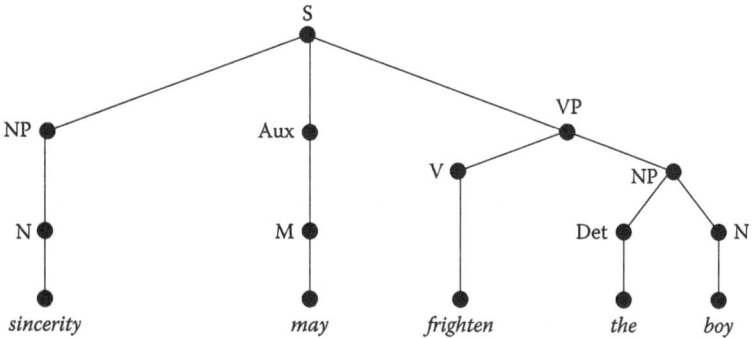

FIGURE 10.3

Trees were first employed in a mathematical context, and the word *tree* was introduced in 1857 by the English mathematician Arthur Cayley* in a paper called "On the Theory of the Analytical Forms Called Trees"† while researching some point in the differential calculus. He, somewhat later, also used trees when he was trying to determine the number of possible chemical compounds with formulas of the form C_kH_{2k+2}, the existence of some of which was only confirmed later. How trees arise in this context is described in Section 10.2, where we give Cayley's formula for the number of *labeled trees* with a given number of vertices. Clearly, trees may be used to model a large number of different situations. We give some further applications later, after we have discussed some theoretical results about trees.

Our first result gives a number of properties that, though superficially different, all turn out to be equivalent to the defining property of a tree given in Definition 10.1, and to be very useful in later work.

THEOREM 10.1

For each positive integer n, with $n \geq 3$, the following properties of a graph, G, with n vertices are equivalent.

a. G is a tree.
b. G is connected and has $n - 1$ edges.
c. G is connected, but the removal of any one edge disconnects G.
d. Each pair of vertices is connected by just one path.
e. G has no closed paths but if any new edge is added, the resulting graph has just one closed path.
f. G has no closed paths and $n - 1$ edges.

---

* Arthur Cayley was one of the most distinguished English mathematicians of the nineteenth century. He was born in Richmond, Surrey, on August 16, 1821. He graduated in mathematics from the University of Cambridge but chose not to pursue an academic career there because of the then need to become an Anglican priest. Instead, he took up the law, but after religious tests at Cambridge were relaxed he returned there in 1863 to become the first Sadlerian professor of mathematics. He died on January 26, 1895, in Cambridge. He was a prolific mathematical author, with over 900 publications. He devoted a lot of his energy to the theory of invariants, but undoubtedly his work of greatest and most long-lasting value was that on the theory of matrices, where he was a pioneer. His work on groups is mentioned in Chapter 11.

† A. Cayley, *On the theory of Analytical Forms Called Trees*, Philosophical Magazine, series 4, volume 13, 1857, pp. 172–176.

Proof

We prove that these six properties are equivalent by proving the chain of implications

$$(a) \Rightarrow (b) \Rightarrow (c) \Rightarrow (d) \Rightarrow (e) \Rightarrow (f) \Rightarrow (a).$$

(a) $\Rightarrow$ (b). For a proof of this implication see the solution to Exercise 9.4.1A.

(b) $\Rightarrow$ (c). We prove this by induction on n. The only connected graph with three vertices and two edges is that of Figure 10.1 (ii), and clearly the removal of any edge disconnects the graph. So the result is true for $n = 3$.

Now suppose that the result holds for $n = k$ and that G is a connected graph with $k + 1$ vertices and k edges. By the handshaking lemma (Theorem 9.4) the sum of the degrees of the vertices of G is $2k$, and hence G must have at least one vertex of degree 1. Suppose a is a vertex of degree 1 and that it is joined just to the vertex b. The removal of the edge $\{a,b\}$ disconnects G. By the induction hypothesis the removal of any of the remaining $k - 1$ edges disconnects the graph G' obtained from G by deleting the vertex a and the edge $\{a,b\}$. Thus, if any other edge is deleted, there is a pair of vertices, say c, d, in G' not connected by a path in G'. As $c \neq a$ and $d \neq a$, and a has degree 1, there is no path in G from c to d that includes the edge $\{a,b\}$. Thus c and d are not connected by a path in G. Hence the removal of any edge disconnects G. So (c) holds also for graphs with $k + 1$ vertices. This completes the proof by induction.

(c) $\Rightarrow$ (d). Let a, b be any two vertices of G. Since G is connected, there is at least one path from a to b. If there were more than one path, the removal of an edge that was in one path but not the other would leave G connected. Hence there must be just one path from a to b.

(d) $\Rightarrow$ (e). Suppose G is a graph in which each pair of vertices is joined by just one path. If there were a closed path, say $v_0 \to v_1 \to ... \to v_n = v_0$, in G, then there would be two paths $v_0 \to v_1$ and $v_0 = v_n \to v_{n-1} \to ... \to v_1$ from v_0 to v_1 in G, contrary to the hypothesis. So there are no closed paths in G. However, if an edge, say $\{a,b\}$, were added, then as there is already a path from a to b, say $a = v_0 \to ... \to v_n = b$, in G, there is now a closed path $a = v_0 \to ... \to v_n = b \to a$. If with the addition of this edge there were now two closed paths in G, these must both include the edge $\{a,b\}$ (Why?). Suppose these paths are $a \to v_1 \to ... \to v_m \to b \to a$ and $a \to w_1 \to ... \to w_n \to b \to a$. Then $a \to v_1 \to ... \to v_m \to b \to w_n \to ... \to w_1 \to a$ is a closed walk in the original graph G, and hence there would a closed path in G. So the addition of the edge $\{a,b\}$ creates just one closed path.

(e) $\Rightarrow$ (f). We prove the result by induction. It is obvious for $n = 3$. Suppose it holds for graphs with k vertices and that G is a graph with $k + 1$ vertices with no closed paths, but the addition of any edge creates a closed path. Since G has no closed paths it has, by Theorem 9.5, a vertex, say a, of degree 1. Let b be the vertex to which a is joined by an edge, and let G' be the graph obtained from G by deleting the vertex a and the edge $\{a,b\}$. G' has k vertices and no closed paths. The addition of any edge to G', and hence to G, creates a closed path in G, and as a closed path in G cannot include the edge $\{a,b\}$, this path must also be a closed path in G'. So, by the induction hypothesis, G' has $k - 1$ edges. So G has k edges. This completes the proof by induction that (f) follows from (e).

(f) ⇒(a). This implication can be proved by induction in a similar way. You are asked to do this in Exercise 10.1.4A.

Exercises

10.1.1A How many different (that is, nonisomorphic) trees are there with n vertices, for $n = 2,3,4$ and 5? In each case draw pictures of the different trees.

10.1.1B How many different (that is, nonisomorphic) trees are there with six vertices? Draw pictures of the different trees.

10.1.2A Prove that a tree with at least two vertices has at least two vertices with degree 1.

10.1.2B Let T be a tree with k vertices of degree 1, where $k \geq 2$, and in which all the other vertices have degree 3. How many vertices of degree 3 does T have?

10.1.3A Give an example of two trees with the same degree sequences that are not isomorphic.

10.1.3B Let T be a tree with no vertices of degree $> k$, and for $2 \leq d \leq k$ just one vertex of degree d. How many vertices of degree 1 does T have?

10.1.4A Prove that, for all positive integers $n \geq 3$, if G is a graph with n vertices, $n - 1$ edges, and no closed paths, then G is connected and hence is a tree.

10.1.4B Let T be a tree. For each pair of vertices of T let $d(x,y)$ be the number of edges in the one path from x to y. Prove that for all three vertices x, y, and z, $d(x,y) + d(y,z) \geq d(x,z)$. (If you have met the notion of a *metric* in the context of topology, you should notice that the function d is a metric, and this gives rise to a topology on the tree.)

10.1.5A Prove that, for positive integers $n \geq 2$, if $a_1,a_2,\ldots,a_n$ is a sequence of positive integers with $a_1 \leq a_2 \leq \ldots \leq a_n$, then $a_1,a_2,\ldots,a_n$ is a degree sequence for a tree if and only if $a_1 + a_2 + \ldots + a_n = 2n - 2$.

10.1.5B Prove that, for all integers $n \geq 2$, there are at least $p(n - 2)$ different (that is, nonisomorphic) trees with n vertices. [Here, $p(n - 2)$ is the number of partitions of $n - 2$, as defined in Chapter 6.]

10.1.6A Prove that if G is a tree with a vertex of degree k, with $k \geq 2$, then G has at least k vertices with degree 1.

10.1.6B Prove that in a tree the number of vertices of degree 1 is

$$2 + \sum_{v \in W}(\delta(v) - 2),$$

where W is the set of vertices in the tree whose degree is at least 3.

10.1.7A Note that since a graph with n vertices is a tree if and only if it is a connected graph with $n - 1$ vertices, the algorithm of Exercise 9.4.5A that tests whether a graph is connected may also be used to decide whether a graph is a tree. Apply this algorithm to decide whether the following are trees.
  i. $G = (V,E)$, where $V = \{a,b,c,d,e,f\}$ and $E = \{af,bc,bf,df,ef\}$
  ii. $G = (V,E)$, where $V = \{a,b,c,d,e,f,g,h\}$ and $E = \{ad,be,de,df,dg,eh\}$

10.1.7B Determine whether the following graphs are trees.
  i. $G = (V,E)$, where $V = \{a,b,c,d,e,f,g,h,i,j\}$ and $E = \{aj,bf,bh,cd,ci,ei,fh,gj,ij\}$
  ii. $G = (V,E)$, where $V = \{a,b,c,d,e,f,g,h,i,j\}$ and $E = \{ab,ac,af,cg,ci,df,dg,ef,fg,gh\}$

10.2 LABELED TREES

Labeled trees are a particular case of *labeled graphs*. To explain what these are, we go back to the early days of graph theory, and to the problems from which the subject got its name. Early in the nineteenth century it was realized that chemical molecules are made up of elements. The properties of different substances could largely be explained by the different proportions of the elements that made them up. But in a few cases it was known that molecules that were made up of the same elements in the same proportions nevertheless had different properties. These related molecules are known as *isomers*. From these examples, it became apparent that knowing the constituents of substances was not enough. It was necessary also to know how the elements were arranged within the molecules.

Various notations were devised to represent the arrangement of atoms in molecules. The system that led to the method in use today seems to have originated with Alexander Crum Brown, a professor of chemistry at the University of Edinburgh. He described his system as providing a "graphic notation"* for molecules. His *graphic notation* became our *graphs*. We have said that this notation "seems" to have originated with Crum Brown because questions of priority are notoriously difficult to settle, and views on who first had a particular idea often vary a good deal and especially from one country to another. Our information is based on a very useful and interesting anthology of historic papers on graph theory.[†]

Anyway, Figure 10.4 gives an example of Crum Brown's graphic notation used to show two isomers with the same chemical formula C_3H_7OH. In his original notation, Crum Brown put circles around the letters denoting the atoms, but in line with modern notation, we have omitted them.

These diagrams are very much like those of trees. They consist of vertices joined by edges. The only real difference is that each vertex is associated with a letter that represents an element. In these diagrams the degree of each vertex has to be the same as the valency of the element associated with that vertex. Since hydrogen (H), oxygen (O), and carbon (C) have valencies one, two, and four, respectively, both diagrams in Figure 10.4 satisfy this condition.

Not every molecule can be represented by a tree. For example, in Friedrich Kekulé's famous model of the benzene atom, which he published in 1865, there is a "ring" of carbon

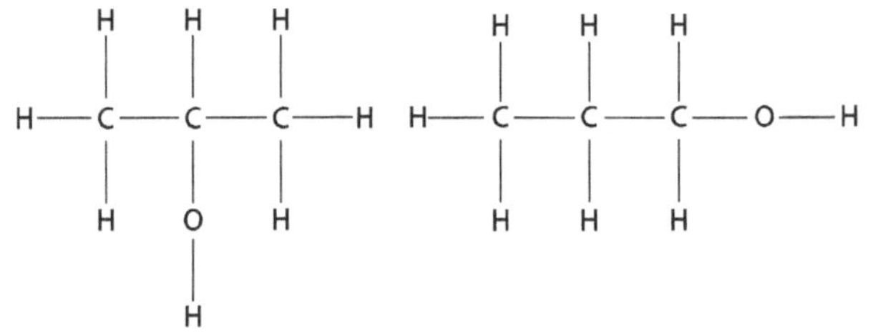

FIGURE 10.4

---

* A. Crum Brown, *On the theory of isomeric compounds*, Transactions of the Royal Society of Edinburgh, 23, 1864, pp. 707–719.
† Norman L. Biggs, E. Keith Lloyd, and Robin J. Wilson, *Graph Theory 1736-1936*, Clarendon Press, Oxford, 1976.

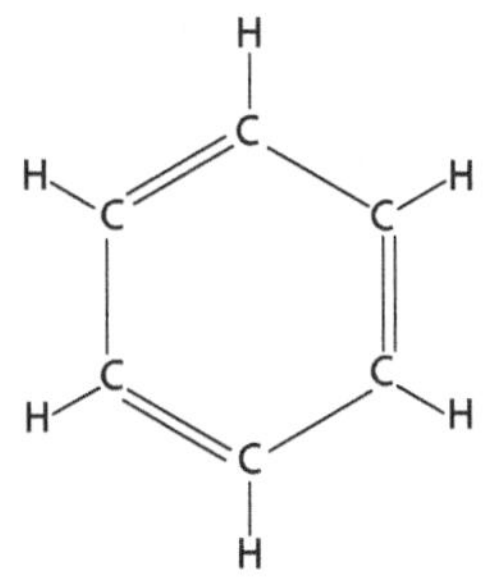

FIGURE 10.5

atoms and double bonds between some of the carbon atoms.* So the diagram representing the molecule shown in Figure 10.5 is that of a multigraph that has a closed path.

It is natural to ask how many different isomers there are with a particular chemical composition. In 1874 Cayley looked at this question in a paper "On the Mathematical Theory of Isomers."† He looked in particular at the paraffins, which have the chemical formula C_nH_{2n+2}. A graph corresponding to such a molecule has $3n + 2$ vertices, of which the n vertices corresponding to carbon must have degree 4, and the $2n + 2$ vertices corresponding to hydrogen must have degree 1. The handshaking lemma implies that such a graph has $\frac{1}{2}(4n+(2n+2)) = 3n+1$ edges, and hence must be a tree. Counting the trees corresponding to the paraffins turns out to be rather a hard problem. In 1889 Cayley first turned his attention to a somewhat easier problem, and we will follow in his footsteps. We begin by giving the formal definition.

DEFINITION 10.2

A *labeled graph* consists of a triple, say $G = (V,E,l)$, where (V,E) is a graph and $l: V \to L$ is a bijection that associates with each vertex an element of the set $L = \{1,2,\ldots,n\}$, where n is the number of vertices. A *labeled tree* is a labeled graph in which the graph is a tree.

The requirement that l should be a bijection means that, unlike chemical graphs, different vertices receive different labels. The following definition extends the notion of isomorphism for graphs (Definition 9.3) to labeled graphs. The idea is that labeled graphs are isomorphic if they "look the same" when we take into account the labeling of the vertices.

DEFINITION 10.3

Two labeled graphs (V_1,E_1,l_1) and (V_2,E_2,l_2) are said to be *isomorphic* if the mapping $\theta: V_1 \to V_2$, which associates with each vertex $x \in V_1$, the vertex $y \in V_2$ with the same label, is an isomorphism between the graphs (V_1,E_1) and (V_2,E_2). Thus θ is an isomorphism that satisfies the condition that for each $x \in V_1$, $l_1(x) = l_2(\theta(x))$.

---

* F. A. Kekulé, *Suela constitution des substances aromatiques*, Bulletin de la Société Chimique de Paris, 3, 1865, pp.98–110.
† *Philosophical Magazine* (4), 47, 1874, pp. 444–446.

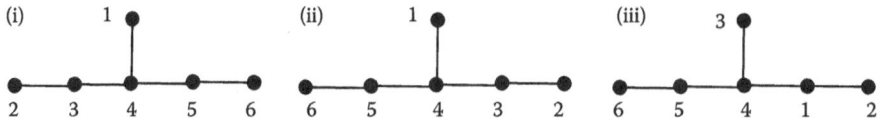

FIGURE 10.6

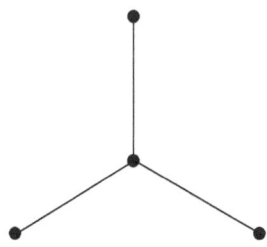

FIGURE 10.7

In Figure 10.6 the labeled graphs (i) and (ii) are isomorphic, but neither of them is isomorphic to (iii), even though the underlying graphs, without the labels, are isomorphic.

The definition means that two labelings of the same graph are regarded as the same if and only if the mapping that associates each vertex in the first labeling with the vertex that receives the same label in the second labeling is an isomorphism between the graph and itself. Thus the two labelings (i) and (ii) in Figure 10.6 are regarded as the same.

PROBLEM 10.1
How many different labelings are there of the tree shown in Figure 10.7?

Solution
The labels come from the set $\{1,2,3,4\}$. The label assigned to the vertex of degree 3 may be chosen in four ways, but then all the remaining ways of assigning labels to the vertices of degree 1 are the same. So there are four different ways to label the graph.

PROBLEM 10.2
How many different labelings are there of the tree shown in Figure 10.6?

Solution
The labels come from the set $\{1,2,3,4,5,6\}$. We may choose the label for the vertex of degree 3 in six ways, and then the label of the only vertex of degree 1 joined to it in five ways. There are then $C(4,2) = 6$ ways of choosing the labels for the vertices of degree 2, and then two further choices for assigning the remaining two labels to the two other vertices of degree 1. This gives $6 \times 5 \times 6 \times 2 = 360$ different labelings.

PROBLEM 10.3
Evaluate the number of different labeled trees with n vertices for $n = 1,2,3,4$.

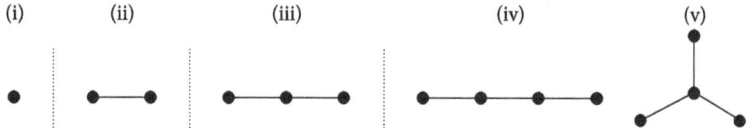

FIGURE 10.8

Solution

In Figure 10.8 we have shown all the trees with at most four vertices.

For $n = 1,2$ there is just one tree, which can be labeled in just one way. For $n = 3$, there is again just one tree, but it can be labeled in three different ways, corresponding to the three available choices for the label assigned to the vertex of degree 2.

For $n = 4$, there are two trees, shown as iv and v in Figure 10.8. In labeling the tree iv, there are $C(4,2) = 6$ ways to label the two vertices of degree 2, in each case leaving two choices for the labels assigned to the vertices of degree 1, making $6 \times 2 = 12$ different labelings altogether. We have already seen in Problem 10.1 that the tree v has four labelings. Hence there are $12 + 4 = 16$ different labeled trees with four vertices. Thus we obtain the following values:

Number of vertices	1	2	3	4
Number of labeled trees	1	1	3	16

Can you guess from this, admittedly rather scant, numerical evidence what the general formula is? Well, $3 = 3^1 = 3^{3-2}$ and $16 = 4^2 = 4^{4-2}$, so maybe there are n^{n-2} labeled trees with n vertices. This formula also fits the cases $n = 1$ and $n = 2$, but only because of our conventions for the meaning of x^{-1} and x^0. In fact, the formula n^{n-2} gives the correct value for all positive integers n. This result is due to Cayley in 1889.[*] We give an elegant proof due to the appropriately named Heinz Prüfer in 1918.[†]

Theorem 10.2 Cayley's Theorem on Labeled Trees

For each positive integer $n \geq 2$, there are n^{n-2} different labeled trees with n vertices.

Proof

There is just one labeled tree with two vertices, so the result holds for $n = 2$. Now suppose $n \geq 3$. We describe a one–one correspondence between labeled trees with n vertices and sequences of $n - 2$ positive integers from the range from 1 to n. We first describe the general method and then illustrate it by an example.

Given a labeled tree, T, with n vertices, we define the sequence $a_1, a_2, \ldots, a_n$ as follows. As T is a tree it has vertices of degree 1. Choose, from among the vertices of degree 1, the vertex with the smallest number as its label. Let v be this vertex, and let

---

[*] A. Cayley, A Theorem on Trees, *Quarterly Journal of Pure and Applied Mathematics*, 23, 1889, pp. 376–378; reprinted in his *Mathematical Papers*, Cambridge University Press, Cambridge, 13, 1897, pp. 26–28. Cayley did not give a complete proof but having dealt with the case $n = 5$ contented himself with saying, "It will be seen at once that the proof given for this particular case is applicable for any value whatever of n."

[†] Prüfer's proof was published in Neuer Beweiss eines Satzes über Permutationen, *Archiv der Mathematik und Physik*, (3) 27, 1918, pp. 142–144.

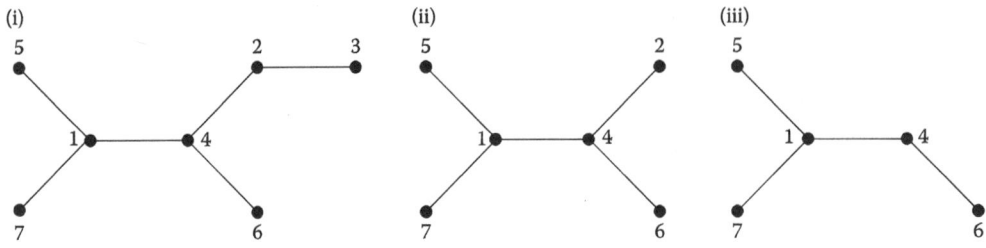

FIGURE 10.9

w be the one vertex to which it is joined. We let a_1 be the label that is assigned to the vertex w.

Now we delete the vertex v and the edge vw, and let T' be the resulting graph. T' will also be a tree. We apply the same process to T'. That is, we pick the vertex of degree 1 in T' with the lowest label, say v', and let a_2 be the label associated with the one vertex, say w', to which v' is joined by an edge. Then we obtain the tree T'' by deleting the vertex v' and the edge $v'w'$ from T', and so on, *until there are just two vertices left*. The resulting sequence $(a_1, a_2, \ldots, a_{n-2})$ is called the *Prüfer code* of the original tree T.

For example, we apply this process to the following labeled tree in Figure 10.9i.

The vertex of degree 1 with the lowest label is the vertex labeled 3. It is joined to the vertex labeled 2. Thus $a_1 = 2$. Deleting the vertex labeled 3 and its adjacent edge, we obtain the tree in Figure 10.9ii. In this tree the vertex of degree 1 with the lowest label is the vertex labeled 2. It is joined to the vertex labeled 4. So $a_2 = 4$. We delete the vertex labeled 2 and its adjacent edge to obtain the tree in Figure 10.9iii. Now the vertex of degree 1 with the lowest label is the vertex labeled 5, and we see that it is joined to the vertex labeled 1. So $a_3 = 1$. Proceeding this way we see that $a_4 = 4$ and $a_5 = 1$. So the Prüfer code of the original tree is $(2,4,1,4,1)$.

To complete the proof we need to show that there is a one–one correspondence between labeled trees and all possible Prüfer codes. We do this by showing how, given a Prüfer code, say $(a_1, a_2, \ldots, a_{n-2})$, we can construct a tree to which it corresponds.

The procedure is as follows. Throughout, "number" means one of the integers $1, 2, \ldots, n$. We join vertices labeled $1, 2, \ldots, n$ using the following process. First, find the least number, say b_1, that does not occur in the sequence $(a_1, a_2, \ldots, a_{n-2})$, and join the vertex labeled b_1 to the vertex labeled a_1. Next, find the least number, say b_2, other than b_1, that does not occur the sequence $(a_2, \ldots, a_{n-2})$, and join the vertex labeled b_2 to the vertex labeled a_2, and then find the least number, say b_3, other than b_1 and b_2, that does not occur in the sequence $(a_3, \ldots, a_{n-2})$, and so on. Then when there are no numbers left in the sequence we join the two vertices not in the list $b_1, \ldots, b_{n-2}$.

We illustrate this process by applying it to the Prüfer code $(2,4,1,4,1)$ as shown in Figure 10.10.

sequence = (2,4,1,4,1); $b_1 = 3, a_1 = 2$

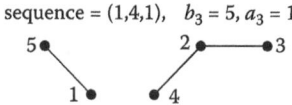

sequence = (4,1,4,1); $b_2 = 2, a_2 = 4$

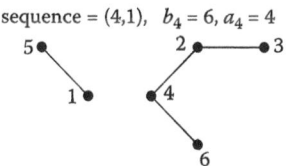

sequence = (1,4,1), $b_3 = 5, a_3 = 1$

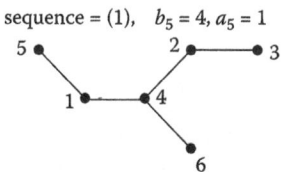

sequence = (4,1), $b_4 = 6, a_4 = 4$

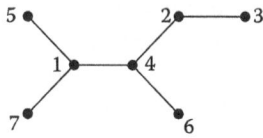

sequence = (1), $b_5 = 4, a_5 = 1$

This leaves the vertices labeled 1 and 7 to be joined.

FIGURE 10.10

We see that this process ends up with the original tree. It is not difficult to prove, by induction on n, that this always happens. We ask you do this in Exercise 10.2.3A.

It follows that there is a one–one correspondence between the labeled trees with n vertices and the Prüfer codes. A Prüfer code for a tree with n vertices is a sequence of $n-2$ positive integers taken from the set $\{1,2,\ldots,n\}$. So there are n^{n-2} different Prüfer codes, and hence there are n^{n-2} labeled trees with n vertices.

Exercises

10.2.1A Find the Prüfer codes for the labeled trees shown in Figure 10.11.
10.2.1B Find the Prüfer codes for the labeled trees shown in Figure 10.12.
10.2.2A Draw the trees whose Prüfer codes are (i) (1,2,1,2,1,2) and (ii) (1,2,3,1,2,3).
10.2.2B Draw the trees whose Prüfer codes are (i) (1,2,2,3,3,3) and (ii) (3,3,3,2,2,1).
10.2.3A Let T be a tree. Prove that the labeled tree constructed from the Prüfer tree, P, of T using the method described in the proof of Theorem 10.2 is isomorphic to T.

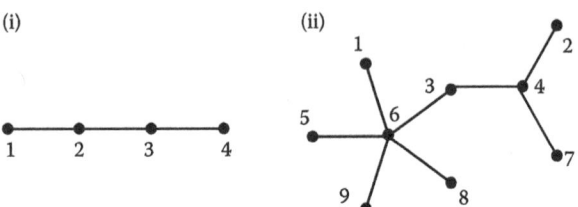

FIGURE 10.11

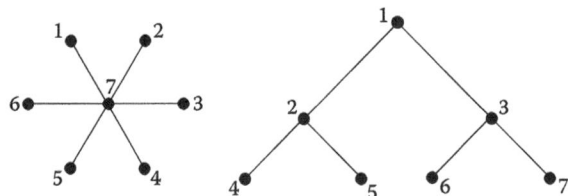

FIGURE 10.12

10.2.3B i. Two labeled trees have Prüfer codes (a) (1,2,3,4,5,6) and (b) (1,4,2,5,3,6), respectively. Are their underlying (unlabeled) graphs isomorphic?

ii. Prove that if, in a labeled graph, the vertex with label r has degree k, then the number r occurs $k-1$ times in its Prüfer code.

10.2.4A We define an *edge-labeled graph* to be a triple (V,E,l), where (V,E) is a graph and $l: E \to \{1,2,...,n\}$ is a bijection, where n is the number of edges. Two edge-labeled trees (V_1,E_1,l_1) and (V_2,E_2,l_2) are said to be *isomorphic* if there is bijection $\theta: V_1 \to V_2$ that is an isomorphism between the trees (V_1,E_1) and (V_2,E_2) and that matches edges with the same label. That is, for each edge, $uv \in E_1$, $l_2(\theta(u)\theta(v)) = l_1(uv)$. Show that there are m^{m-3} different (that is, nonisomorphic) edge-labeled trees with m vertices.

10.2.4B Use Cayley's Theorem to show that there are at least $n^{n-3}/(n-1)!$ (unlabeled) trees with n vertices.

10.3 SPANNING TREES AND MINIMAL CONNECTORS

In this section we discuss some practical problems to which graph theory contributes solutions. Suppose that you want to build bridges to connect four islands. If you want it to be possible to walk, or drive, directly from any one island to any other island, you would need to build six bridges altogether. Six bridges would be needed because this is the number of edges in the complete graph with four vertices. However, as bridges are expensive to build, it might be considered better just to build enough bridges to connect up all the islands. What is the smallest number of bridges that you would need to build?

It is easily seen that only three bridges are needed. They can be chosen in more than one way. One possibility is shown in Figure 10.13. In the language of graph theory we have constructed a tree, and we can regard this tree as a subgraph of the complete graph, K_4, with four vertices. We say that the tree *spans* K_4 since each vertex is connected to each other vertex using only the edges of this tree.

We are ready for a formal definition.

DEFINITION 10.4

Let G be a graph. We say that the subgraph G' *spans* G if it has the same vertices as G. A *spanning tree* of G is a subgraph that spans G and that is a tree.

Which graphs have spanning trees? If G is not connected, then no subgraph with the same set of vertices can be connected, and so G does not have a spanning tree. On the other hand it is not difficult to show that every connected graph has a spanning tree. In particular, for large n, the complete graph K_n has lots of spanning trees. If we label the vertices of the complete graph, K_n, with n vertices with the numbers $1,2,\ldots,n$, then the spanning trees for K_n correspond to the labeled trees with n vertices, and so by Cayley's theorem, K_n has n^{n-2} different spanning trees.

In many practical applications we seek more than just a spanning tree. We illustrate this with the following problem, before discussing the general case.

PROBLEM 10.4

It is desired to connect five islands by tunnels. The graph in Figure 10.14 shows the cost (in suitable units, for example, 1 unit = £10,000,000) of building a tunnel between pairs

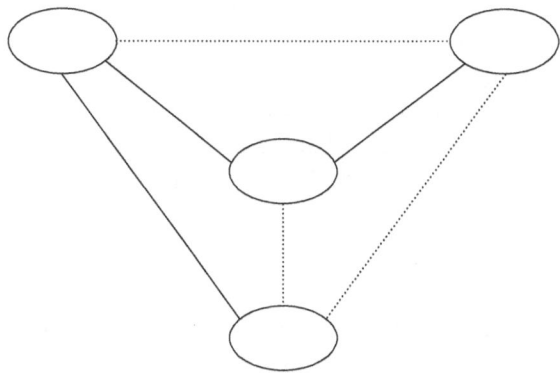

FIGURE 10.13

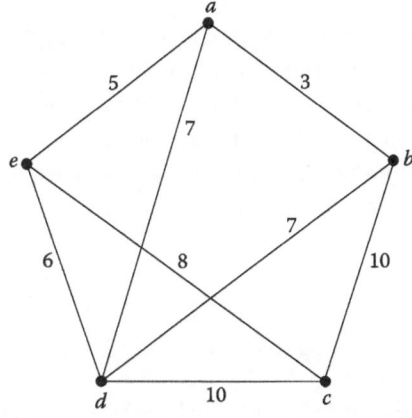

FIGURE 10.14

of islands for which this is feasible. Which tunnels should be built so as to connect the islands for the smallest total cost?

Solution

Tunnels should be built corresponding to the edges *ab*, *ae*, *de*, and *ce*. This incurs a total cost of $3 + 5 + 6 + 8 = 22$ units.

Comment

For the time being we just assert that this is the minimal cost of a spanning tree, say T, for the graph shown in Figure 10.14. We know that a tree with five vertices will have four edges, and so the solution must involve four tunnels. In a case as small as this, it is possible to find these tunnels just by trying possibilities, but even in this case, the solution is not completely obvious, as our solution does not simply consist of the four cheapest tunnels. In fact there is an algorithm, that is, a systematic procedure, for finding an optimal solution. (See Exercise 10.3.1A for the application of this algorithm to the solution to Problem 10.4.) Before we describe this, we need set up the theoretical framework for this and similar problems.

In the graph in Figure 10.14 each edge has associated with it a number. The magnitudes of these numbers are significant, unlike the case with labeled graphs where the numbers associated with the vertices serve merely as labels and so could be replaced by other symbols. In Figure 10.14 these numbers represent costs. In other cases they could represent distances or times. The general term for these numbers is *weights*, and a graph that has weights associated with its edges is called a *weighted graph*. In our example the weights are normally positive integers, but there is no need to impose this restriction. However, it is convenient to require that the weights are nonnegative numbers. Thus we phrase our definition as follows.

DEFINITION 10.5

A *weighted graph* is a pair (G,μ), where $G = (V,E)$ is a graph and $\mu : E \rightarrow \mathbb{R}$ is a mapping that associates with each edge, e, a nonnegative real number, $\mu(e)$. Given a weighted graph (G,μ), and a subgraph $G' = (V',E')$ of the graph $G = (V,E)$, the *weight* of G' is the sum of the weights of the edges in E'. We let $\mu(G')$ be the weight of G'. Therefore, $\mu(G') = \sum_{e \in E'} \mu(e)$.

Often, we represent a weighted graph by a picture of the graph with numbers alongside the edges to indicate their weights. For example, we can regard Figure 10.14 as a picture of a weighted graph. The spanning tree T given in our solution to Problem 10.4 has weight 22. Our claim in giving T as the solution to Problem 10.4 is that there is no other spanning tree with a lower weight. Since a spanning tree connects all the vertices of the original graph, we frame the following definition.

DEFINITION 10.6

Given a weighted graph (G,μ) and a spanning tree, T of G, we say that T is a *minimal connector* of G if for every spanning tree T' of G, $\mu(T) \leq \mu(T')$.

In general, a weighted connected graph will have more than one minimal connector. The extreme case is where all the edges have the same weight. Then all the spanning trees have the same weight and so they are all minimal connectors. In Exercise 10.3.3A you are asked to prove that when the different edges all have different weights, there is just one minimal connector.

We now describe an algorithm due to Joseph Kruskal* for finding a minimal connector in a graph.

We first show how this algorithm works in a particular example. Then we prove that it does construct a minimal connector. We end this section with some remarks about graph algorithms in general.

In our example the underlying graph will be K_5, the complete graph with five vertices. We suppose that these vertices are a, b, c, d, and e. It is convenient to give the weights of the edges in a table (Table 10.1), rather than indicate them on the graph itself. In this table we write xy rather than $\{x,y\}$ for the edge joining the vertices x and y.

The algorithm requires us to choose at each stage an edge of smallest possible weight satisfying conditions (i) and (ii) of the algorithm. It is therefore convenient first to reorder the edges in order of increasing weight, and then to look at each edge in this order

---

KRUSKAL'S ALGORITHM FOR MINIMAL CONNECTORS

Suppose we have a weighted connected graph with n vertices. Then we obtain a minimal connector by carrying out the following process:

For $k = 1,2,\ldots, n-1$ choose an edge e_k of the smallest possible weight satisfying the two conditions:

i. For $k \geq 2$, e_k is different from all of $e_1,\ldots,e_{k-1}$.
ii. For $k \geq 3$, there is no closed path using just some or all of the edges $e_1,\ldots,e_k$.

Then the edges $e_1,\ldots,e_{n-1}$ form a minimal connector.

TABLE 10.1

Edge	Weight
ab	1
ac	3
ad	5
ae	5
bc	2
bd	7
be	4
cd	8
ce	4
de	9

---

* Joseph B. Kruskal, Jr., On the Shortest Spanning Subtree of a Graph and the Traveling Salesman Problem, *Proceedings of the American Mathematical Society*, 7, 1956, pp. 48–50.

214 ■ How to Count: An Introduction to Combinatorics, Second Edition

TABLE 10.2

Edge	Weight	
ab	1	✓
bc	2	✓
ac	3	✗
be	4	✓
ce	4	✗
ad	5	✓
ae	5	
bd	7	
cd	8	
de	9	

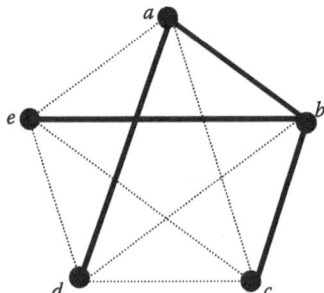

FIGURE 10.15

to see if it meets the conditions of the algorithm for selecting it. Thus in Table 10.2 we have put a tick (✓) if we can select that edge meeting conditions (i) and (ii), and a cross (✗) if we cannot. We stop after we have chosen four edges, as this is the number needed for a tree spanning a graph with five vertices. A good way to keep track as to whether choosing an edge violates condition (ii) by creating a closed path is to mark in the edges you have chosen by a heavy line. We have done this in Figure 10.15, which shows the underlying graph by dotted lines.

It will be seen that we have found a spanning tree made up of the edges *ab*, *bc*, *be*, and *ad* with weight $1 + 2 + 4 + 5 = 12$. In carrying out the algorithm we did not select the edge *ac* because we would then have chosen three edges that make up the closed path $a \to b \to c \to a$, and we did not select *ce* to avoid creating the closed path $b \to c \to e \to b$. Note also that we could have listed the edges in a different order with *ce* before *be* and that if we had done this we would have ended up with a different spanning tree with weight 12. For practice using this algorithm see Exercises 10.3.1A and 10.3.1B at the end of this section.

We now show that Kruskal's algorithm does indeed produce a minimal connector.

THEOREM 10.3

Kruskal's algorithm when applied to a weighted connected graph produces a minimal connector.

Proof

Let (G,μ) be a weighted connected graph with n vertices, and suppose that Kruskal's algorithm produces the subgraph T with edges $e_1, e_2, \ldots, e_{n-1}$ chosen in this order. T is a graph with $n-1$ edges and no closed paths, and therefore, by Theorem 10.1, T is a tree. Hence T is a spanning tree of G.

Now suppose T' is some other spanning tree of G. We aim to show that $\mu(T) \leq \mu(T')$, from which it will follow that T is a minimal connector. The tree T' also has $n-1$ edges, but since it is different from T it does not include all the same edges as T. Let r be the least integer, $1 \leq r \leq n-1$, such that e_r is not an edge of T'. Hence, by Theorem 10.1, the graph, say T^*, obtained from T' by adding the edge e_r has a closed path that must include the edge e_r. Since there are no closed paths in T, this closed path contains an edge, say e, that is not an edge of T, and so $e \neq e_r$. Then the graph, say T'', obtained from T^* by deleting the edge e is still a tree, and hence a spanning tree for G. Since $e_1, \ldots, e_{r-1}$ and e are all edges of T' they do not include a closed path. Hence, as at stage r in Kruskal's algorithm we chose the edge e_r and not e, it must be that $\mu(e_r) \leq \mu(e)$, and hence $\mu(T'') \leq \mu(T')$, where T'' includes one more edge of T than does T'. We can repeat this process, obtaining trees with more and more of the edges of T and with weights no larger than that of T'. Eventually we will have added all the edges of T, without increasing the total weight. So $\mu(T) \leq \mu(T')$. This completes the proof.

Note that we can use Kruskal's algorithm to find spanning trees for a connected graph, even where its edges are not assigned weights. All we need to do is assign each edge the same weight, and then carry out the algorithm. In other words, when choosing an edge e_k, the condition that it be of minimal weight can be ignored. We just make sure we choose an edge that satisfies conditions (i) and (ii) given in the algorithm. Thus Theorem 10.3 implies that every connected graph has a spanning tree.

We end this section with some remarks about algorithms. By an algorithm we mean a mechanical computational process that produces an output after a finite number of steps. Thus it is precisely the sort of process that can be implemented by a digital computer, subject to the idealization that there is no upper bound on the amount of storage space or time that may be used. To be more precise we would need to give a detailed specification of what we mean by a digital computer. This can be done, for example, in terms of Turing machines, but we do not go into the details here.

For a computer to find a minimal connector using Kruskal's algorithm, or to carry out some other graph algorithm, we need to have some way of inputting data that describes a graph. How this can be done will depend on what sort of programming language is being used. If we have a list-processing language we can input a graph by inputting a list of the edges. If we are restricted to a programming language that can handle only numbers, then we can input the description of graph G with n vertices as an $n \times n$ matrix, say $M = (a_{ij})$, where a_{ij} is 1 if the vertices v_i and v_j are joined by an edge, and 0 otherwise. The matrix M is called the *adjacency matrix* of the graph G.

Since we are dealing with graphs that are finite, in one sense it is immediate that given any definite question about graphs, there is an algorithm for answering it. A weighted graph has only a finite number of spanning trees, so to find a minimal connector, we need only list these, calculate their weights, and select a spanning tree whose weight is as small as possible. The practical snag with this approach is that, in general, the number of cases that need to be checked grows very rapidly as the number of vertices increases. For example, we have seen that the complete graph with n vertices has n^{n-2} spanning trees. For $n = 18$, this is more than 10^{20}, and to check this number of cases, even at the rate of 10^{12} a second, would take more than three years.

Thus, the point of Kruskal's algorithm, and similar algorithms, is that it is very efficient. As we have seen, for a graph with n vertices, the algorithm involves $n - 1$ steps. Of course, the efficiency of the algorithm depends on how many steps it takes to sort the vertices by weight, and on how many steps it takes to decide whether adding a particular edge creates a closed path. It is beyond the scope of this book to analyze the complexity of Kruskal's algorithm as measured by the number of steps needed to carry it out. Suffice it to say that there are algorithms for finding a minimal connector for a weighted connected graph with n vertices, where the number of steps needed grows no faster than An^2 for some constant A, and it is possible to do rather better.* Thus the algorithm can be implemented in a feasible time even for large graphs.

Exercises

10.3.1A a. Use Kruskal's algorithm to solve Problem 10.4.
b. Find a minimal connector for the graph of Figure 10.16.

10.3.1B Find a minimal spanning tree for the graph of Figure 10.17.

10.3.2A Show that if a weighted connected graph has one edge with a lower weight than every other edge, then this edge occurs in every minimal connector.

10.3.2B If a weighted connected graph has two edges that have a lower weight than all the other edges, must both these edges be included in every minimal connector?

10.3.3A Prove that if (G,μ) is a weighted connected graph in which each edge of G has a different weight, then G has just one minimal connector.†

10.3.3B Give an example of a weighted connected graph that has just one minimal connector, but in which there are two edges that have the same weight.

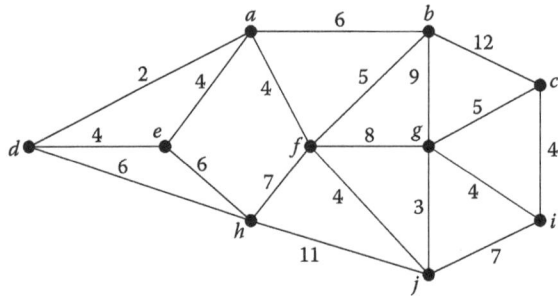

FIGURE 10.16

---

* These problems have generated a large technical literature because of their practical importance. See, for example, James A. McHugh, *Algorithmic Graph Theory*, Prentice Hall, Englewood Cliffs, New Jersey, 1990.
† In his paper cited in a footnote in Section 10.3, Kruskal attributes this result to Otakar Borůvka, On a Minimal Problem, *Práce Moravské Pridovedecké Spolecnosti*, 3, 1926.

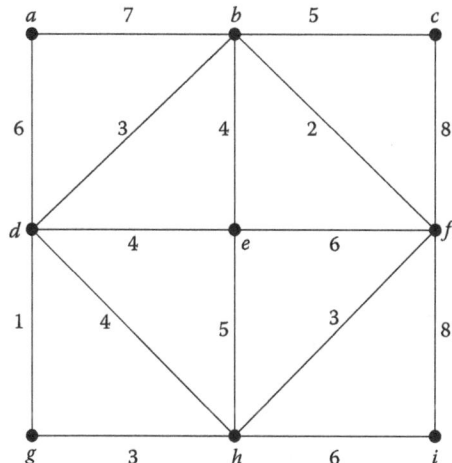

FIGURE 10.17

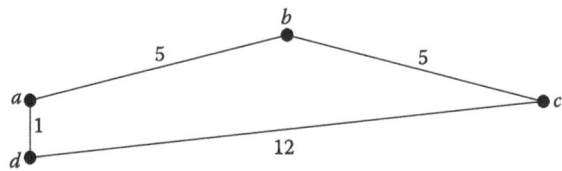

FIGURE 10.18

10.4 THE SHORTEST-PATH PROBLEM

A problem that is similar in spirit to that of finding a minimal connector for a graph is that of finding the path of least weight between two points in a graph. If we think of the weights as distance or times or costs, the corresponding problem is that of finding the shortest or quickest or cheapest route between two points. However, we will continue to talk in terms of distances or lengths. Thus in a weighted graph (G,μ), the *length* of the path

$$v_1 \to v_2 \to \ldots \to v_{n-1} \to v_n \qquad (P)$$

is the sum $\mu(v_1v_2) + \mu(v_2v_3) + \ldots + \mu(v_{n-1}v_n)$. We use $\mu(P)$ for this sum. The *distance* between two vertices, x and y, is the length of the shortest path from x to y if there is one, and ∞ otherwise. We let $d(x,y)$ be the distance from x to y.

The shortest-distance problem is rather more complicated than the minimal connector problem for the following reason. Notice that in carrying out Kruskal's algorithm we do not need to look ahead. Provided we are careful not to create any closed paths, we are bound to create a spanning tree of minimal weight. The shortest-distance problem is rather different. This can be seen from as simple a situation case as that given by the graph in Figure 10.18.

It will be seen that $a \to b \to c$ is the shortest path from a to c, and that although ad is the shortest edge, it is not included in this path. Setting off from a, it would be a mistake to choose the edge *ad*. Thus the algorithm we are about to describe to find the shortest path from a vertex *u* to a vertex *v* is more complicated than Kruskal's algorithm. It involves checking all possible paths from *u* to *v*. We follow the description of the algorithm with an example. You might find it helpful to consider the two together.

We first outline the method used. It involves both *temporary* and *permanent* labels assigned to the vertices. The temporary label associated with a vertex, y, has the form $(x, l(y))$, where x is either $*$ or a vertex other than y and $l(y)$ is a positive number or ∞. Starting with a temporary label attached to each vertex, at each stage one temporary label is made permanent, and this is indicated by replacing the round brackets by square brackets.

When the algorithm terminates, $l(v)$ is the length of the shortest path from u to v. Also, $u = y_1 \to y_2 \to \ldots \to y_n = v$ is a path of this length where, for $2 \le r \le n$, the permanent label attached to the vertex y_r is $[y_{r-1}, l(y_r)]$.

Before discussing why this algorithm works, we illustrate it by applying it to the weighted graph shown in Figure 10.19.

We set out the working of the algorithm in Table 10.3. The explanation is given below the table. The meaning of a *temporary* label, $(x, l(y))$, assigned to a vertex y is that the shortest path from u to y *found so far* has length $l(y)$ and x is the penultimate vertex in this path. The label becomes permanent when $l(y)$ has been established to be the length of the shortest path from u to y. By a *stage* of the algorithm we mean one single cycle of the steps 1 to 4.

DIJKSTRA'S ALGORITHM FOR SHORTEST PATHS*

The algorithm is assumed to operate on a weighted graph (G, μ) of which u and v are vertices.

Step 1. Initially label each vertex with the temporary label $(*, \infty)$, except that the starting vertex, u, is given the temporary label $(*, 0)$.

Step 2. Find a vertex, x, with a temporary label for which $l(x)$ is minimal, and make its label permanent.

Step 3. If $x = v$, stop. (If $l(x) = \infty$ there is no path from u to v.)

Step 4. Otherwise, for each vertex y with a temporary label, say $(w, l(y))$, that is joined to x by an edge, xy, if $l(x) + \mu(xy) < l(y)$, replace the temporary label attached to y by the *temporary* label $(x, l(x) + \mu(xy))$. Return to Step 2.

---

* E. Dijkstra, A Note on Two Problems in Connection with Graphs, *Numerische Mathematik*, 1, 1959, pp. 269–271.

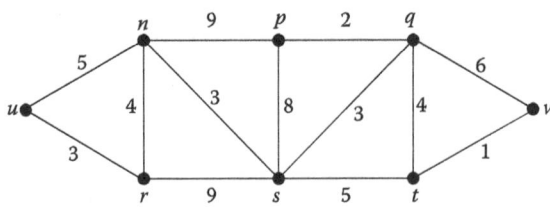

FIGURE 10.19

TABLE 10.3

Stage	(i)	(ii)	(iii)	(iv)	(v)	(vi)	(vii)	(viii)	(ix)
Vertex, x		u	r	n	s	q	p	t	v
u	(*,0)	[*,0]							
n	(*,∞)	(u,5)		[u,5]					
p	(*,∞)			(n,14)		(q,13)	[q,13]		
q	(*,∞)				(s,11)	[s,11]			
r	(*,∞)		(u,3)	[u,3]					
s	(*,∞)		(r,12)	(n,8)	[n,8]				
t	(*,∞)				(s,13)			[s,13]	
v	(*,∞)					(q,17)		(t,14)	[t,14]

We have used roman numerals for the stages to avoid confusion with the numbers used for the steps in the algorithm. At stage (i) we have assigned labels to the vertices in accordance with step 1 of the algorithm. Now at stage (ii), we have $y = u$ because in the label attached to u the distance is 0, whereas it is ∞ for all the other vertices with a temporary label. So the label attached to u is made permanent. Now u is joined by edges to the vertices n and r. We see that $l(u) + \mu(un) = 0 + 5 = 5$, and so n is assigned the temporary label $(u,5)$. And as $l(u) + \mu(ur) = 0 + 3 = 3$, r is assigned the temporary label $(u,3)$.

Of the current temporary labels, $(u,3)$ assigned to the vertex has the least second component. So at stage (iii) we assign the permanent label $[u,3]$ to the vertex r. The vertex r is joined to the vertices n and s, which do not have permanent labels. The vertex r has the permanent label $[u,3]$. The vertex n has the temporary label $(u,5)$. Since $l(r) + \mu(rn) = 3 + 4 = 7 > 5$, the temporary label attached to n is left unchanged. The vertex s has the temporary label $(*,\infty)$. Since $l(r) + \mu(rs) = 3 + 9 = 12 < \infty$, the temporary label attached to s is replaced by $(r,12)$.

Now consider stage (vi). At this stage the vertices with temporary labels are $(n,14)$ for p, $(s,11)$ for q, $(s,13)$ for t and $(*,\infty)$ for v. So the minimal value of l is associated with the vertex q. Therefore at step 2 in the algorithm we put $x = q$ and make the label attached to q permanent. The vertices with temporary labels that are joined to q by an edge are p, t, and v. The vertex p has a temporary label $(n,14)$. Since $l(q) + \mu(qp) = 11 + 2 = 13 < 14$, we replace the temporary label attached to p by $(q,13)$. The vertex t has a temporary label $(s,13)$. Since $l(q) + \mu(qt) = 11 + 4 = 15 > 13$, the temporary label attached to t is unchanged. The vertex v has a temporary label $(*,\infty)$. Since $l(q) + \mu(qv) = 11 + 6 = 17 < \infty$, the temporary label attached to v is replaced by $(q,17)$.

The algorithm terminates after stage (ix). From Table 10.3 we see that the shortest path from u to v has length 14. Also, we can trace the path back from v by following the vertex that forms the first part of the permanent label attached to v. In this way we obtain the path

$$v \leftarrow t \leftarrow s \leftarrow n \leftarrow u.$$

You should consolidate your understanding of how this algorithm works by doing the exercises at the end of this section. We now give a proof to show that Dijkstra's algorithm does produce a path from u to v, if there is one, of minimal length.

THEOREM 10.4

Dijkstra's algorithm applied to a weighted graph, with initial vertex u, determines whether there is a path from u to another vertex v and produces a path of minimal length from u to v when there is one.

Proof

Let (G,μ) be a weighted graph. We prove, by induction on the stage of Dijkstra's algorithm at which a vertex is assigned a permanent label, that if the algorithm ends with a vertex y assigned a permanent label $[x,l(y)]$, where $l(y)$ is finite, then there is a path from u to y in the graph and the shortest path has length $l(y)$.

This is certainly true at the first stage. Suppose that it is true at all stages before stage k and that at stage k a permanent label $[x,l(y)]$, is assigned to the vertex y, where $l(y)$ is finite. Then there is an edge from x to y and at an earlier stage, x was assigned a permanent label, say $[w,l(x)]$, where $l(x)$ is finite and $l(x) = d(u,x)$, that is, $l(x)$ is the length of the shortest path from u to x.

By the induction hypothesis there is a path from u to x. Hence, as xy is an edge, there is also a path from u to y, say

$$u \to \ldots \to x \to y \qquad (P),$$

where $u \to \ldots \to x$ is the shortest path from u to x. Since $l(y) = l(x) + \mu(xy)$, the path (P) has length $l(y)$. We need to show that $l(y) = d(u,y)$.

Let

$$u = y_1 \to y_2 \to \ldots \to y_n = y \qquad (P')$$

be some other path from u to y. Let r be the largest positive integer such that the vertex y_r is assigned a permanent label, $[y_{r-1}, l(y_r)]$, before stage k. If $y_{r+1} \neq y$, then at stage k, y_{r+1} has assigned to it a temporary label, say $(w,l(y_{r+1}))$, where $l(y) \leq l(y_{r+1})$, since otherwise we would not have assigned a permanent label to y at this stage before assigning a permanent label to y_{r+1}. By the way temporary labels are assigned, $l(y_{r+1}) \leq l(y_r) + \mu(y_r y_{r+1}) \leq \mu(P')$. Hence $l(y) \leq \mu(P')$. So $l(y)$ is the length of the shortest path from u to y, that is, $l(y) = d(u,y)$. The case where $y_{r+1} = y$ is similar. This completes the proof by induction.

Finally, we need to consider the case where v ends up with a permanent label $[*,\infty]$. Clearly there are no edges from vertices with permanent labels of the form $[x, l(y)]$ with $l(y)$ finite to those with permanent labels of the form $[*,\infty]$. Hence, if v ends up with a permanent label of this latter form, there is no path from u to v. This completes the proof of the theorem.

If you look at Table 10.3 you will see that in carrying out Dijkstra's algorithm for a graph with eight vertices, we need to calculate some of the entries in an 8 × 8 table. We hope that this makes it plausible that the number of steps needed to implement the algorithm grows proportionately to n^2 where n is the number of vertices in the table. For a detailed justification of this see the book *Algorithmic Graph Theory*.*

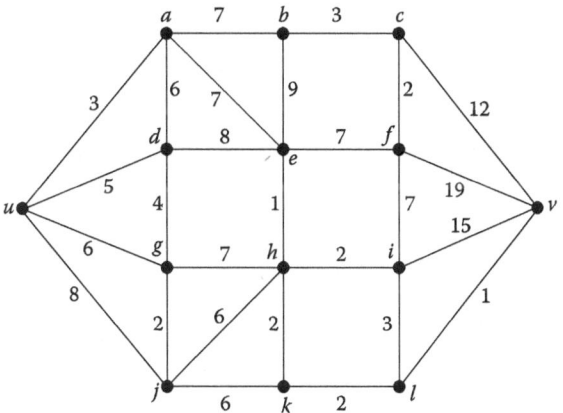

FIGURE 10.20

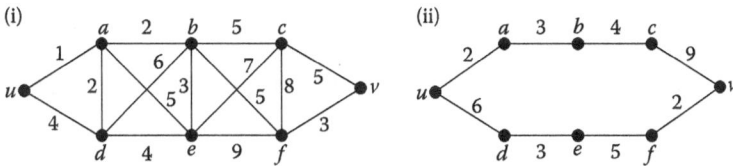

FIGURE 10.21

TABLE 10.4

Edge	Weight
ub	3
uc	4
ae	7
av	5
bd	7
bf	4
cd	2
cg	3
df	5
dg	7
ev	2
fg	6

---

* James A. McHugh, *Algorithmic Graph Theory*, Prentice Hall, Englewood Cliffs, New Jersey, 1990.

TABLE 10.5

Edge	Weight
ua	3
ud	4
ab	6
ac	2
ad	4
bc	3
be	2
bv	7
cd	5
ce	4
de	7
ev	3

We do not have space to discuss other graph algorithms in detail. We only mention briefly that for the traveling salesman problem, as described in Chapter 9, it is not known whether there is an efficient algorithm that gives the route with the least possible weight. Recall that this is the problem of finding a Hamiltonian path of minimal weight. However, a number of algorithms are known that run in feasible time and yield good approximations to the optimal solution.

Exercises

10.4.1A Use Dijkstra's algorithm to find the shortest path from u to v in the weighted graph shown in Figure 10.20.

10.4.1B Use Dijkstra's algorithm to find the shortest path from u to v in the weighted graphs shown in Figure 10.21.

10.4.2A In Table 10.4, a weighted graph is described by listing its edges and their weights. Use Dijkstra's algorithm to determine whether there is a path from u to v in this graph. If there is a path, find the shortest path from u to v.

10.4.2B In Table 10.5 a weighted graph is described by listing its edges and their weights. Use Dijkstra's algorithm to determine whether there is a path from u to v in this graph. If there is a path, find the shortest path from u to v.

CHAPTER 11

Groups of Permutations

11.1 PERMUTATIONS AS GROUPS

In Problem 11 of Chapter 1 we asked, "What is the most effective way to shuffle a pack of cards?" To answer this question we need to return to the ideas about permutations that we introduced in Section 2.6 of Chapter 2 where we discussed permutations and cycles. We will see that permutations provide an example of a mathematical structure called a *group*. Groups play an important role in many areas of mathematics. Our interest in them is because of their relation to problems involving counting patterns. So the work in this chapter is also used later. In particular, we see that it underlies our answer to Problem 13B about the number of ways to color a cube using three colors. We expect that many readers will have met some group theory before. However, we will not assume any previous knowledge. All the ideas about groups that we use will be introduced as we need them.

You will recall from Chapter 2 that you may think of a permutation of a set, X, as a bijection $\sigma: X \to X$. Usually X will be a set of the form $\{1, 2, \ldots, n\}$ for some positive integer n. We let S_n be the set of all permutations of the set $\{1, 2, \ldots, n\}$. In Section 2.6 we introduced two notations for these permutations. In the *bracket notation* we describe a permutation by writing the numbers 1 to n in one row, with the numbers that σ maps them to in a second row, enclosing both rows in one pair of brackets. So, for example, if we write

$$\sigma = \begin{pmatrix} 1 & 2 & 3 & 4 & 5 & 6 \\ 4 & 6 & 1 & 3 & 5 & 2 \end{pmatrix},$$

this means that σ is the permutation of the set $\{1,2,3,4,5,6\}$ such that $\sigma(1) = 4$, $\sigma(2) = 6$, $\sigma(3) = 1$, $\sigma(4) = 3$, $\sigma(5) = 5$, and $\sigma(6) = 2$. In the alternative, *cycle notation*, we write $\sigma = (1\ 4\ 3)(2\ 6)(5)$ or, leaving out the cycle of length 1, just

$$\sigma = (1\quad 4\quad 3)(2\quad 6).$$

TABLE 11.1

	(1 4 3)(2 6)		(1 5 2 4)(3 6)	
5	←	5	←	1
3	←	4	←	2
2	←	6	←	3
4	←	1	←	4
6	←	2	←	5
1	←	3	←	6

Since permutations are functions, we can compose them in the usual way. If σ and τ are permutations of the same set we can define the composite permutation, $\sigma \circ \tau$, by

$$\sigma \circ \tau(x) = \sigma(\tau(x)).$$

Permutations that are given to us in cycle notation can be composed by working out what happens to each element in turn. When doing these calculations it is important to remember that the composite permutation $\sigma \circ \tau$ means *first* τ, *then* σ. Here is an example of how it works out in practice.

Let $\sigma = (1\ 4\ 3)(2\ 6)$ and let $\tau = (1\ 5\ 2\ 4)(3\ 6)$ be two permutations from S_6. We can calculate the composite permutation $\sigma \circ \tau$ by the method illustrated in Table 11.1.

We work from right to left as $\sigma \circ \tau$ means carrying out the permutation τ first. For example, $\sigma \circ \tau(1) = \sigma(\tau(1)) = \sigma(5) = 5$ and $\sigma \circ \tau(4) = \sigma(\tau(4)) = \sigma(1) = 4$, as indicated in the table. Having worked out all the values of $\sigma \circ \tau$, we can read off from this table the bracket notation for $\sigma \circ \tau$, as

$$\sigma \circ \tau = \begin{pmatrix} 1 & 2 & 3 & 4 & 5 & 6 \\ 5 & 3 & 2 & 4 & 6 & 1 \end{pmatrix},$$

and we can then rewrite this in cycle notation as

$$\sigma \circ \tau = (1\ 5\ 6)(2\ 3).$$

After a bit of practice you will not find it necessary to write out the calculation of a composite permutation in full. The arrow diagram shown in Table 11.1 can be worked out mentally, and it should be possible to write down the composite permutation in cycle notation without the need to write down any intermediate steps. If you think you need some practice with these calculations, attempt Exercises 11.1.1A and 11.1.1B, or swap examples with a colleague, if possible.

The operation of *composition* of permutations has a number of important algebraic properties that we now describe. Although we are mainly interested in permutations of sets of the form $\{1, 2, \ldots, n\}$, these properties hold for arbitrary sets of permutations. So we give general proofs of these properties.

First, we introduce some notation. We use $S(X)$ for the set of all permutations of the set X. The *identity* map on X is the function $\iota_X : X \to X$, given by

$$\text{for all } x \in X, \iota_X(x) = x.$$

It should be clear that ι_X is a permutation of X; that is, it is a bijection. Since permutations are bijections, they have inverses. We use the standard notation σ^{-1} for the inverse of a permutation σ.

We usually omit the symbol $\circ$ for composition and write $\sigma\tau$ for the composite permutation $\sigma \circ \tau$.

THEOREM 11.1

For each set X, the operation of composition on the set, $S(X)$, of permutations of X has the following properties.

 a. For all $\sigma, \tau \in S(X)$, $\sigma\tau \in S(X)$.
 b. For all $\sigma \in S(X)$, $\sigma\iota_X = \sigma = \iota_X\sigma$.
 c. For all $\sigma \in S(X)$, $\sigma\sigma^{-1} = \iota_X = \sigma^{-1}\sigma$.
 d. For all $\sigma, \tau, \rho \in S(X)$, $(\sigma\tau)\rho = \sigma(\tau\rho)$.

Proof

 a. Suppose $\sigma, \tau \in S(X)$. Let $x, y \in X$ with $x \neq y$. Then, as τ is injective, $\tau(x) \neq \tau(y)$, and hence, as σ is injective, $\sigma(\tau(x)) \neq \sigma(\tau(y))$, that is, $\sigma\tau(x) \neq \sigma\tau(y)$. So $\sigma\tau$ is injective. Now suppose $z \in X$. Then, as σ is surjective, there is some $y \in X$ such that $\sigma(y) = z$, and as τ is surjective, there is some $x \in X$ such that $\tau(x) = y$. Thus there is some $x \in X$ such that $\sigma\tau(x) = \sigma(\tau(x)) = \sigma(y) = z$. So $\sigma\tau$ is surjective. We have therefore shown that $\sigma\tau$ is a bijection, and hence that $\sigma\tau \in S(X)$, as claimed.
 b. For $x \in X$ $\sigma\iota_X(x) = \sigma(\iota_X(x)) = \sigma(x)$. Consequently, $\sigma\iota_X = \sigma$. Similarly, $\iota_X\sigma = \sigma$.
 c. This follows immediately from the definition of σ^{-1}.
 d. Suppose $\sigma, \tau, \rho \in S(X)$ and $x \in X$. Then $((\sigma\tau)\rho)(x) = (\sigma\tau)(\rho(x)) = (\sigma(\tau(\rho(x))) = \sigma(\tau\rho(x)) = (\sigma(\tau\rho))(x)$. Since the composite permutations $(\sigma\tau)\rho$ and $\sigma(\tau\rho)$ have the same effect on each $x \in X$, we deduce that $= (\sigma\tau)\rho = \sigma(\tau\rho)$.

We need to mention two more notational points. Expressed in terms of cycles the identity permutation $\iota \in S_n$ consists of n cycles of length 1. If we adopt our standard convention of not bothering to write down cycles of length 1, ι would simply be represented by an empty space! This is not always convenient, so usually we write either ι for the identity permutation, or often e (from the German *einheit*).

Because we usually omit the symbol $\circ$ when we are writing composite permutations, there can be an ambiguity. For example, if we write

$$(1\ 2\ 4)(3\ 5)(1\ 4\ 3\ 5),$$

this could mean

$$(1\,2\,4)(3\,5) \circ (1\,4\,3\,5) \text{ or } (1\,2\,4) \circ (3\,5)(1\,4\,3\,5) \text{ or}$$

$$((1\,2\,4) \circ (3\,5)) \circ (1\,4\,3\,5) \text{ or } (1\,2\,4) \circ ((3\,5) \circ (1\,4\,3\,5)).$$

However, it follows from Theorem 11.1d, that these different expressions all represent the same permutation, and so this ambiguity of notation does not cause us any problems.

The properties of permutations given in Theorem 11.1 are so important that we give a special name to any collection of mathematical objects that can be combined in a way that satisfies them. This is embodied in the following definition.

DEFINITION 11.1

A *group* is a pair $(G, \bullet)$, where G is a set and $\bullet$ is an operation defined on G that satisfies the following four properties.

THE GROUP PROPERTIES

G1. Closure: For all $x, y \in G$, $x \bullet y \in G$.
G2. Identity: There is an element $e \in G$, such that for all $x \in G$, $x \bullet e \in x$ and $e \bullet x = x$.
G3. Inverses: For each $x \in G$, there is an element $x^{-1} \in G$, such that $x \bullet x^{-1} = e$ and $x^{-1} \bullet x = e$.
G4. Associativity: For all $x, y, z \in G$, $(x \bullet y) \bullet z = x \bullet (y \bullet z)$.

The notation e used in this definition carries with it an implication that there is just one element of G satisfying the *identity* property. It is not difficult to prove this. Indeed, suppose, to the contrary, that we have two elements e_1, e_2 in G with this property. Then for all $x \in G$, $x \bullet e_2 = x$ and so, in particular, $e_1 \bullet e_2 = e_1$. Also, for all $x \in G$, $e_1 \bullet x = x$ and so $e_1 \bullet e_2 = e_2$. Consequently $e_1 = e_1 \bullet e_2 = e_2$. The unique element of G satisfying the identity property is called the *identity element* of the group. As shown in the definition, we often use e for the identity element of a group. If we need to emphasize that it is the identity element of G, we write this element as e_G.

In a similar way the use of x^{-1} carries with it the implication that for each $x \in G$ there is just one element satisfying the *inverse* property. This is also easy to prove, as if x_1^{-1}, x_2^{-1} were both inverses of x, we have $x_1^{-1}, x_1^{-1} \bullet e = x_1^{-1} \bullet (x \bullet x_2^{-1}) = (x_1^{-1} \bullet x) \bullet x_2^{-1} = e \bullet x_2^{-1} = x_2^{-1}$.

In cases where it is clear from the context which operation is involved in a particular group, we write xy instead of $x \bullet y$. Similarly in such cases, we often talk about "the group G" rather than "the group $(G, \bullet)$."

It follows from Theorem 11.1 that for each set X $(S(X), \circ)$ forms a group. Permutation groups form a very important class of groups. Indeed, in a sense explained at the end of Section 11.2, all groups can be viewed as permutation groups. However, the richness of the group concept arises from the many other examples of groups that occur in mathematics. We list some of these examples, though they will not be of great interest to us in this book.

Examples of Groups

1. $(Z, +)$, the set of integers with the operation of addition, forms a group. Likewise, $(Q, +)$, the rational numbers; $(R, +)$, the real numbers; and $(C, +)$, the complex numbers, all with the operation of addition, form groups. In each case the identity element is the number 0, and the inverse of x is $-x$. It is easy to check that all these examples satisfy the definition of a group.

2. The sets Z, Q, R, and C do not form groups when the operation is multiplication. They are all closed under this operation, that is, *G1* holds. In each case the number 1 acts as an identity element. Also, multiplication satisfies the associativity property. However, there is a problem with the inverse property. When the operation is multiplication, the number 0 has no inverse, as, if there were any inverse, say z, it would have to satisfy $0 \times z = 1$, which is not possible. In the cases of Q, R, and C we can get around this difficulty by excluding 0. Thus, if we use Q^*, R^*, and C^* for the nonzero rational numbers, nonzero real numbers, and nonzero complex numbers, respectively, then $(Q^*, \times)$, $(R^*, \times)$, and $(C^*, \times)$ are all examples of groups, with 1 as the identity element and $1/x$ as the inverse of x. It is from these examples that we derive the general notation x^{-1} for the inverse of x. However, this does not work for the set, Z^*, of nonzero integers, since only for $x = 1$ and $x = -1$ is $1/x$ also an integer.

The examples in the next category are much more important.

3. For each positive integer n, and each *field** of numbers, F, let $M_n(F)$ be the set of $n \times n$ invertible (also called nonsingular) matrices, with entries from F, and let $\times$ be the usual operation of matrix multiplication. Then $(M_n(F), \times)$ is a group.

4. For each positive integer n, $(Z_n, +_n)$ is a group where Z_n is the set $\{0, 1, 2, \ldots, n-1\}$, and $+_n$ is the operation of addition modulo n.

In these examples, the group elements are familiar mathematical objects, and the operations are natural ones for those particular objects. Although, in a philosophical sense, mathematical entities are abstract objects, they seem very real to the mathematicians who work with them. Accordingly, groups of these kinds are sometimes referred to as *concrete groups*. In contrast, with *abstract groups* we do not specify what the group elements actually are, but only how they are combined. When the number of elements is small, this can be conveniently displayed by giving a *multiplication table*.

Here is an example of this type. The set G is $\{e, a, b, c, h, v, r, s\}$. The operation • is defined by Table 11.2. The value of $x • y$ is found by looking at the entry in the x-row and y-column. For example, we can see from the table that $v • c = s$.

---

* A *field* of numbers is a set of numbers closed under the operations of addition and multiplication, and in which these operations have the standard properties. The rational numbers, Q; the real numbers, R; and the complex numbers, C, are all examples of fields. The integers, Z, do not form a field. For the details see, for example, R. B. J. T. Allenby, *Rings, Fields and Groups*, Arnold, London, 1983.

TABLE 11.2

•	e	a	b	c	h	v	r	s
e	e	a	b	c	h	v	r	s
a	a	b	c	e	r	s	v	h
b	b	c	e	a	v	h	s	r
c	c	e	a	b	s	r	h	v
h	h	s	v	r	e	b	c	a
v	v	r	h	s	b	e	a	c
r	r	h	s	v	a	c	e	b
s	s	v	r	h	c	a	b	e

A table of this kind is called either a *group table* or sometimes a *Cayley table*.* It is not difficult to see from this table that the first three of the group properties are satisfied. To check *closure* we need only check that each entry in the table is one of the elements of the set G. The convention of using e for the identity element and putting it first in the table makes it easy to confirm that e acts as the identity element, but even if some differently named element had been the identity, and it had been placed somewhere else in the list, it would not have been difficult to spot from the table that it satisfies the $x • e = x = e • x$ property. To see that the inverse property $x • x^{-1} = e = x^{-1} • x$ holds, we need only check that the identity element, e, occurs once in each row and column and in positions that are symmetrical about the leading diagonal from the top left to the bottom right of the table.

We see, for example, that $a • c = e = c • a$, so that $a^{-1} = c$ and $c^{-1} = a$, and in fact all the other elements of this group are their own inverses.

To complete the check that Table 11.2 does indeed define a group, we need also to check that the operation • satisfies the associativity condition. Unfortunately, there is no very easy way to do this from the table, as this involves checking that $(x • y) • z = x • (y • z)$ for all choices of x, y, and z. In fact, it is not necessary to check the cases where at least one of x, y, and z is e, but this still leaves $7 \times 7 \times 7 = 343$ cases. If you are not willing to do all these calculations, we ask you take it on trust for the time being that the operation • is associative. We prove that it does have this property in the next section. Fortunately, as the proof of Theorem 11.1(d) shows, whenever the group elements are functions and the operation is composition, the associativity property always holds.

You may have noticed that in Table 11.2 each group element occurs exactly once in each row and each column. Tables of this kind are called *Latin squares*. In Exercise 11.1.2A you are asked to prove that a group table is always a Latin square. Latin squares are interesting and important combinatorial objects, but because of shortage of space we are not able to discuss them in this book.

Exercises

11.1.1A Evaluate the following compositions of permutations.

    i. (1 5 4)(3 6 7 2)∘(4 6 2)(1 5)(3 7)
    ii. (1 4 9 8 6)(2 3)∘(1 5 6)∘(7 1 3 2)

---

* After Arthur Cayley, whose biography is summarized in a footnote in Section 10.1.

11.1.1B Consider the permutations $\sigma = (1\ 5\ 8\ 2)(3\ 7)(4\ 6)$, $\tau = (1\ 7\ 3\ 5\ 2\ 4\ 8\ 6)$, and $\rho = (1\ 3\ 7)(4\ 6\ 8\ 2)$ from S_8. Find the permutations $\sigma \circ \tau$, $\tau \circ \rho$, $(\sigma \circ \tau) \circ \rho$ and $\sigma \circ (\tau \circ \rho)$ in cycle form, and hence verify that in this case $(\sigma \circ \tau) \circ \rho = \sigma \circ (\tau \circ \rho)$.

11.1.2A Prove that, if $(G, \bullet)$ is a group, then
  i. For all $x, y, z, \in G$,
     a. $x \bullet y = x \bullet z \Rightarrow y = z$ and
     b. $y \bullet x = z \bullet x \Rightarrow y = z$.
  ii. For all $x, y \in G$, there exist $w, z \in G$ such that $x \bullet w = y$ and $z \bullet x = y$.
  [Note that it follows from (i) that in a Cayley table there are no repetitions in any row or any column. Also it follows from (ii) that each group element occurs at least once in each row and in each column. Thus together (i) and (ii) imply that each row and each column form a permutation of the elements of the group. Each Cayley table is, therefore, a Latin square. However, as Exercise 11.1.4B shows, the converse is not, in general, true.]

11.1.2B A group, $(G, \bullet)$, is said to be *commutative* (or *Abelian*) if for all $x, y \in G$, $x \bullet y = y \bullet x$. For which positive integers n is the group $(S_n, \circ)$ commutative?

11.1.3A Find a permutation $\sigma \in S_5$ such that $(4\ 3\ 5\ 2\ 1) \circ \sigma = (1\ 4)(2\ 3)$.

11.1.3B Show that there is no permutation $\sigma \in S_3$ such that $(1\ 2\ 3) \circ \sigma = \sigma \circ (1\ 2)$.

11.1.4A Show that the following table can be completed in just one way so as to form a Latin square, and that when so completed, it is the Cayley table of a group.

	e	a	b
e	e	a	b
a	a		
b	b		

11.1.4B Give a multiplication table for the operation $\bullet$ on the set $X = \{e, a, b, c, d\}$ that forms a Latin square in such a way that $\bullet$ satisfies the *closure*, *identity*, and *inverse* properties, with e as the identity element, but $(X, \bullet)$ is not a group.

11.2 SYMMETRY GROUPS

Groups are very useful when it comes to studying the symmetries of geometric figures. As we see in the next chapter, we need to take symmetries into account when it comes to counting different patterns. We explain the idea of geometric symmetry with a simple example, the symmetries of a square, shown in Figure 11.1.

A square is a symmetrical figure, but what exactly do we mean by this? One way of explaining symmetry is to say that the square occupies the same space and looks the same if we transform it in certain ways. For example, if we give the square a quarter turn clockwise (that is, if we rotate the square through an angle $\frac{1}{2}\pi$ clockwise about the axis through the center of the square, and perpendicular to the plane of the square),* it looks exactly the same as it did originally. We now need to make this idea more mathematically precise.

---
* We use radian measure for angles. So $\frac{1}{2}\pi$ radians = 90°.

FIGURE 11.1

First, we need to be more precise about what transformations are allowed. If we want the figure to look the same after the transformation it must not distort either distances or angles. Since the angles in a triangle are determined once we know the lengths of its sides, a transformation that doesn't change distances also leaves angles unchanged. So all we need to specify in our definition is that the transformation leaves distances unchanged. As we want to consider both two- and three-dimensional figures, we frame our definition in terms of a space that could be either two-dimensional Euclidean space, R^2, or three-dimensional Euclidean space, R^3. In both these spaces we use $d(p,q)$ for the distance between two points p and q, measured in the standard way.* The *figures* that we mention in the following definition are subsets of either R^2 or R^3.

DEFINITION 11.2

Let S be either R^2 or R^3. A mapping $f: S \to S$ is said to be an *isometry* if for all $p,q \in S$, $d(f(p), f(q)) = d(p,q)$. A *symmetry* of a figure F in the space S is an isometry $f: S \to S$ such that $f(F) = F$.

It should be noted that an isometry is automatically injective, as we have that for any two points p, q, $p \ne q \Rightarrow d(p,q) > 0 \Rightarrow d(f(p), f(q)) = d(p,q) > 0 \Rightarrow f(p) \ne f(q)$. It can be shown that the isometries of R^2 are either rotations, reflections, translations, or glide reflections.[†] However, a bounded figure, such as a square, cannot be mapped to itself by a translation or glide reflection, and so we need only consider rotations and reflections of bounded figures in R^2. In R^3 we may also need to consider symmetries that are compositions of a reflection and a rotation. We should also not forget the identity mapping, e, which satisfies Definition 11.2 and so counts as a symmetry of every figure.

It is important to note that because we regard symmetries as functions, two transformations of a figure that are physically different but have the same effect on all the points are regarded as being the same symmetry. For example, rotating a figure through an angle $\frac{1}{2}\pi$ clockwise about an axis is physically different from rotating it through an angle $\frac{3}{2}\pi$ counterclockwise about the same axis. However, these different operations have the same effect on each point, so they will count as being the same symmetry.

---

* That is, in R^2 we measure distances by $d((x_1,y_1),(x_2,y_2)) = \sqrt{(x_1-x_2)^2 + (y_1-y_2)^2}$ and by the analogous formula in R^3. In fact, the definition that we give applies more generally, but we do not need to consider the more general context here.
† See, for example, David A. Brannan, Matthew F. Esplen, and Jeremy J. Gray, *Geometry*, Cambridge University Press, Cambridge, 1999.

Groups of Permutations ■ 231

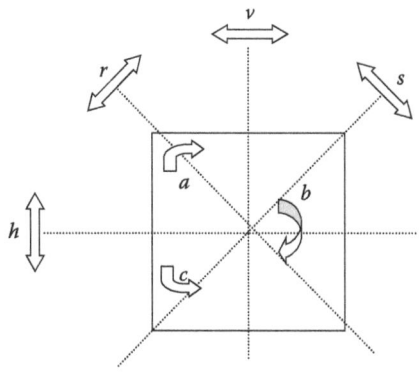

	The symmetries of a square
e	Identity
a	Rotation through ½π clockwise
b	Rotation through π clockwise
c	Rotation through ½π anticlockwise
h	Reflection in the horizontal axis
v	Reflection in the vertical axis shown
r	Reflection in the diagonal axis shown
s	Reflection in the diagonal axis shown

The rotations are about the axis through the centre of the square and perpendicular to it.

FIGURE 11.2

This is nothing more than our usual stipulation that if $f: D \to C$ and $g: D \to C$ are both functions with the same domain and codomain and for all $x \in D$, $f(x) = g(x)$ then we say that $f = g$.

Now that we have been careful to say what we mean by a symmetry of a figure we can see that a square has eight symmetries, as shown in Figure 11.2. The symbols used for these symmetries are somewhat arbitrary, but we will continue to use them.

We will use the notation $S(\square)$ for the group of symmetries of a square. It is now quite straightforward to work out the Cayley table for this group. We encourage you to do this. You might find it helpful to have an actual square to manipulate. Don't forget that in line with our usual convention for composing functions, when we compose two symmetries, say f and g, to obtain the composite fg this means *first* do g, *then* do f.

You should obtain exactly the same table as Table 11.2 in the previous section. Since the operation is composition of functions, we now know, without having to do lots of calculations, that the operation defined by Table 11.2 is associative. Hence it is the Cayley table of a group. In particular, we can also deduce that the symmetries of a square form a group. This is a special case of the more general result that the symmetries of any figure always form a group. This we now prove.

THEOREM 11.2

The set, G, of the symmetries of a figure, F, with the operation, denoted by $\circ$, of composition, forms a group.

Proof

We need to check that the four group properties are satisfied by $(G, \circ)$. Suppose that f and g are symmetries of F. Then as f and g are both isometries we have, for all points p and q (of the appropriate space),

$$d(fg(p), fg(q)) = d(f(g(p)), f(g(q))) = d(g(p), (g(q)) = d(p,q),$$

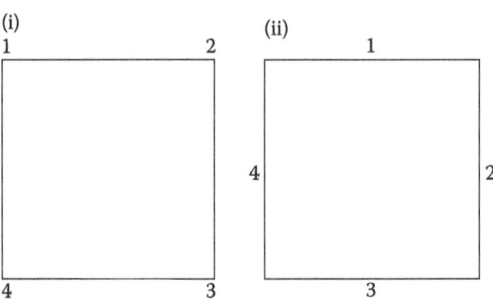

FIGURE 11.3

and hence fg is also an isometry. Since $f(F) = F$ and $g(F) = F$, $fg(F) = f(g(F)) = f(F) = F$. It follows that fg is also a symmetry of F. Thus the closure condition holds.

It is straightforward to check that the identity map is a symmetry of F, and hence that G has an identity element. Each symmetry is a bijection and so has an inverse. We leave as an exercise (Exercise 11.2.1A) to check that if f is an isometry of F then so also is f^{-1}, and hence that G satisfies the inverse condition. We already know that composition of symmetries is associative. Hence it follows that $(G, \circ)$ is a group.

Note that nothing in the above proof depended on the fact that we were dealing with figures in two- or three-dimensional space. The proof would work just as well in higher dimensional spaces. The only problem is that figures in four- and higher dimensional spaces are more difficult to picture!

We can relate groups of symmetries to permutation groups by adding numerical labels to the vertices or edges or, in three dimensions, the faces of a figure. For example, suppose we label the vertices of a square with 1, 2, 3, and 4 as shown in Figure 11.3i.

We can describe the symmetries of the square by specifying how the vertices move. For example, the symmetry h, which is the reflection in the horizontal axis of symmetry, moves the vertex in position 1 to position 4, the vertex in position 2 to position 3, and so on (notice that we regard the numbers as labeling positions that are fixed in space). Thus the permutation h corresponds to the permutation (1 4)(2 3) in S_4. You can readily see that each symmetry of the square corresponds to a permutation in S_4. In this way the eight symmetries of the square correspond to the eight permutations

$$e, (1\ 2\ 3\ 4), (1\ 3)(2\ 4), (1\ 4\ 3\ 2), (1\ 4)(2\ 3), (1\ 2)(3\ 4), (1\ 3), (2\ 4)$$

from S_4, which therefore, by themselves, form a group. This provides us with our first example of a *subgroup*, a concept that we describe in more detail in the next section. Notice that if we label the edges as shown in Figure 11.3ii, then we get a different correspondence between the symmetries of the square and permutations in S_4. For example, using the labeling of the edges, the symmetry h corresponds to the permutation (1 3).

In one sense the group of symmetries of a square is different from the group made up of the eight permutations listed above, since their elements are different, that is, symmetries in the first case and permutations in the second. However, in another sense they are different manifestations of the same group. We now make more precise what we mean by this.

In Figure 11.4 we show three Cayley tables; (i) is that of the rotational symmetries of a square, (ii) is the group Z_4 of the numbers 0, 1, 2, 3 with addition modulo 4,

(i)

o	e	a	b	c
e	e	a	b	c
a	a	b	c	e
b	b	c	e	a
c	c	e	a	b

(ii)

$+_4$	0	1	2	3
0	0	1	2	3
1	1	2	3	0
2	2	3	0	1
3	3	0	1	2

(iii)

o	e	(1 2 3 4)	(1 3)(2 4)	(1 4 3 2)
e	e	(1 2 3 4)	(1 3)(2 4)	(1 4 3 2)
(1 2 3 4)	(1 2 3 4)	(1 3)(2 4)	(1 4 3 2)	e
(1 3)(2 4)	(1 3)(2 4)	(1 4 3 2)	e	(1 2 3 4)
(1 4 3 2)	(1 4 3 2)	e	(1 2 3 4)	(1 3)(2 4)

FIGURE 11.4

and (iii) is the group of the corresponding permutations of the vertices of the square. Although the tables contain elements of different types, it is evident that the three tables display the same pattern. For example, the identity element occurs in the same positions in each table, and the element *a* occurs in the same position in the first Cayley table, as does the number 1 in the second table, and the permutation (1 2 3 4) in the third table.

We say that all three groups are *isomorphic* because they have the same structure in a sense that we now make precise. Note the similarities, and the differences, when this definition is compared with that of isomorphism of graphs (Definition 9.3).

DEFINITION 11.3 *Isomorphism of Groups*
Let $(G, \bullet)$ and $(H, *)$ be groups. We say that these groups are *isomorphic* if there is a bijection $\theta : G \to H$ such that for all $g_1, g_2 \in G$, we have

$$\theta(g_1 \bullet g_2) = \theta(g_1) * \theta(g_2). \tag{11.1}$$

Such a mapping θ is called an *isomorphism* between the two groups.

For example, the mapping θ from the permutations corresponding to the rotations of a square and the numbers modulo 4 [given by iii and ii in Figure 11.4], namely,

$$\theta(e) = 0, \; \theta((1\ 2\ 3\ 4)) = 1, \; \theta((1\ 3)(2\ 4)) = 2, \text{ and } \theta((1\ 4\ 3\ 2)) = 3,$$

is an isomorphism between the two groups. We check just one case of the Equation 11.1. We have

$$\theta((1\ 3)(2\ 4) \circ (1\ 4\ 3\ 2)) = \theta((1\ 3)(2\ 4)) +_4 \theta((1\ 4\ 3\ 2))$$

since the left-hand side is $\theta((1\ 2\ 3\ 4)) = 1$ and the right-hand side is $2 +_4 3 = 1$.

Note that in, Definition 11.3, the equation $\theta(g_1 \bullet g_2) = \theta(g_1) * \theta(g_2)$ expresses the fact that combining elements in the group $(G, \bullet)$ and then mapping to $(H, *)$ produces the same result as first mapping to $(H, *)$ and then combining in $(H, *)$. This condition implies that the Cayley tables of the two groups have the same pattern. As with isomorphisms between two graphs, an isomorphism between two groups is a *structure-preserving correspondence* of their elements. We regard two isomorphic groups as being essentially the same. So, for example, if we are asked, "How many different groups with four elements are there?" we are being asked to find the largest set, say X, of groups with four elements such that no two of the groups in X are isomorphic, but each group with 4 elements is isomorphic to one of the groups in X. Since the groups whose Cayley tables are shown in Figure 11.4 are isomorphic, any such set X can include at most one of them.

We have seen that each row of a Cayley table corresponds to a permutation of the group elements. For example, if we look at table (i) in Figure 11.4 we see from the a-row that a corresponds to the permutation that in bracket notation we can write as

$$\begin{pmatrix} e & a & b & c \\ a & b & c & e \end{pmatrix}$$

and in cycle notation as $(e\ a\ b\ c)$. A theorem due to Cayley says that this correspondence is always an isomorphism between a group $(G, \bullet)$ and the associated group of permutations of its elements. This is not difficult to prove once we notice that the permutation of G that corresponds to the element $g \in G$ is the permutation, θ_g, of G defined by $\theta_g(x) = g \bullet x$. Here is the proof.

THEOREM 11.3
Cayley's Theorem for Groups

Each group $(G, \bullet)$ is isomorphic to a group of permutations of its elements, with the operation of composition.

Proof

For each $g \in G$, we let $\theta_g : G \to G$ be the mapping defined by:

$$\text{for each } x \in G,\ \theta_g(x) = g \bullet x.$$

By Exercise 11.1.2A, each mapping θ_g is a permutation of the set G. We let G^* be the set of all these permutations, and we let $\Theta : G \to G^*$ be the mapping defined by $\Theta(g) = \theta_g$. Since $\Theta(g)(e) = \theta_g(e) = g \bullet e = g$, it follows that if $g \neq g'$, then $\Theta(g)(e) \neq \Theta(g')(e)$ and hence that $\Theta(g) \neq \Theta(g')$. Consequently, Θ is injective. By the definition of G^*, Θ is surjective, and hence Θ is a bijection.

We also have that, for all $g_1, g_2 \in G$, and all $x \in G$,

$$\Theta(g_1 \bullet g_2)(x) = \theta_{g_1 \bullet g_2}(x) = (g_1 \bullet g_2) \bullet x = g_1 \bullet (g_2 \bullet x) = \theta_{g_1}(\theta_{g_2}(x)) = \theta_{g_1} \circ \theta_{g_2}(x),$$

and therefore

$$\Theta(g_1 \bullet g_2) = \theta_{g_1} \circ \theta_{g_2} = \Theta(g_1) \circ \Theta(g_2).$$

Hence Θ is an isomorphism between the group $(G, \bullet)$ and the group $(G^*, \circ)$. This completes the proof.

In this sense every group may be viewed as a group of permutations. However, experience shows that this is not always a helpful way to think about groups. Note also that the group $(G^*, \circ)$ is generally forms only a small subset of the group $(S(G), \circ)$ of all permutations of the set G. For, if G is a group of n elements, there are $n!$ elements in $S(G)$.

Exercises

11.2.1A Prove that if f is a symmetry of the figure F, then the inverse, f^{-1}, of f is also a symmetry of f.

11.2.1B Prove that if $f: \mathbb{R}^2 \to \mathbb{R}^2$ is an isometry, then f is surjective.

11.2.2A i. How many symmetries does an equilateral triangle have?
  ii. Introduce some symbols for the symmetries of an equilateral triangle, and draw up the Cayley table for the group of these symmetries.
  iii. Express the symmetries of an equilateral triangle as elements of S_3 by using the numbers 1, 2, and 3 to label its vertices.

11.2.2B i. How many symmetries does a regular pentagon have?
  ii. How many symmetries does a regular n-gon (that is, a polygon with n sides, with all the sides equal and all the internal angles equal) have?

11.2.3A How many rotational symmetries does a cube have? Describe them.

11.2.3B Investigate the rotational symmetries of a regular tetrahedron.

11.2.4A Show that the group of rotational symmetries of an equilateral triangle is isomorphic to the group Z_3 of the integers 0, 1, 2 with addition modulo 3.

11.2.4B Show that every group with two elements is isomorphic to the group Z_2 and that every group with three elements is isomorphic to the group Z_3.

11.2.5A i. Show that each rectangle has four symmetries (note that we are taking "rectangle" to imply "not a square") and that the groups of symmetries of any two rectangles are isomorphic. It follows that we can talk about "*the* group of symmetries of a rectangle."
  ii. Show that the group of isometries of a rectangle is not isomorphic to the group Z_4.

11.2.5B Prove that every group with four elements is isomorphic to either the group Z_4 or to the group of symmetries of a rectangle.

11.3 SUBGROUPS AND LAGRANGE'S THEOREM

Group theory is a large subject with an enormous literature. We are going to confine ourselves to those aspects of the subject that are relevant to the combinatorial problems we are interested in. The following definition is important.

DEFINITION 11.4

Suppose that G is a group. A subset H of G is said to be *subgroup* of G, if H itself forms a group with respect to the same operation that makes G a group.

Note: Here it is very convenient to be able to suppress mention of the group operation. If we were being really pedantic, we would have distinguish between the group operation for G, say •, which, strictly speaking, is a mapping • : $G \times G \to G$, and the operation on H, which is the restriction of • to the set H.

Since H is a subset of G and we are using the same operation as for the group G, it automatically follows that H satisfies the associativity property. Thus, for H to be a subgroup of G, H must be a subset of G that satisfies the following properties, labeled "SG" for "subgroup."*

THE SUBGROUP PROPERTIES

SG1. H itself is closed under the operation $\bullet$; that is, for all $x, y \in H$, $x \bullet y \in H$.
SG2. H contains the identity element, e_G, of G.
SG3. H contains inverses of all its elements; that is, for each $x \in H$, also $x^{-1} \in H$.

We now present some examples of subgroups.

Examples of Subgroups

1. We have seen that Z, Q, R, and C are all examples of groups, with the operation of addition in each case. Since $Z \subseteq Q \subseteq R \subseteq C$, it follows that Z is a subgroup of Q, Q is a subgroup of R, and R is a subgroup of C.
2. Each group, G, with more than one element has two *trivial* subgroups. The set $\{e_G\}$ containing just the identity element of G can easily be seen to satisfy the subgroup conditions. At the other extreme, the whole group G also counts as being a subgroup of itself.
3. The eight permutations from S_4 corresponding to the symmetries of a square form a subgroup of S_4.
4. We consider the group, S_3, of all the six permutations of the set $\{1, 2, 3\}$. Rather than write out these permutations in cycle form, we introduce the following symbols for them. We put $p = (1\ 2\ 3)$, $q = (1\ 3\ 2)$, $r = (1\ 2)$, $s = (2\ 3)$, $t = (1\ 3)$, and e is the identity element, as usual. The Cayley table for this group is shown in Table 11.3.

It can be seen that the top left-hand corner of the table, taken by itself, is also the Cayley table of a group. In other words the subset $\{e, p, q\}$ forms a subgroup of S_3. In this case it is rather easy to spot this. In general, it is rather difficult to find subgroups just by examining Cayley tables. Can you find any more subgroups from this Cayley table? In fact, apart from the two trivial subgroups, and the subgroup $\{e, p, q\}$, there are just three other subgroups, each containing just the identity and one other element.

TABLE 11.3

	e	p	q	r	s	t
e	e	p	q	r	s	t
p	p	q	e	t	r	s
q	q	e	p	s	t	r
r	r	s	t	e	p	q
s	s	t	r	q	e	p
t	t	r	s	p	q	e

---

* See, for example, Allenby, *Rings, Fields and Groups*, Theorem 5.6.5, for a proof that H is a subgroup if and only if it satisfies these subgroup properties.

Some theory comes to our aid when it comes to finding subgroups. This has to do with the number of elements in a subgroup compared with the number of elements in the group. Group theorists have special terminology for the number of elements in a group or a subgroup.

DEFINITION 11.5

The *order* of a group, or a subgroup, is the number of elements in it.

For example, the order of S_3 is 6, and the group of symmetries of a square has order 8. Group theorists often use the notation $o(G)$ for the order of a group, but we have already introduced the notation $\#(G)$ for the number of elements in any set G, and so we will keep to this notation.

The key idea in what follows is that given a group G and a subgroup H we can use H to partition G into sets, called, in this context, *cosets*, all having the same number of elements as does H. They are defined as follows. To make the general idea more concrete, we use the group S_3 as our example.

DEFINITION 11.6

Let G be a group and let H be a subgroup of G. For each $g \in G$, the *coset gH* is defined to be the set $\{gh : h \in H\}$.

We use the notation gH for the coset, as it is obtained by combining the fixed element g of G with all the elements of H in turn, and so the notation gH is rather suggestive and a good aid to memory. We now look at the specific example of the group S_3 to make these ideas more concrete.

PROBLEM 11.1

Find the cosets of the subgroup $\{e, r\}$ of the group S_3. (Recall that the Cayley table of S_3 is given in Table 11.3.)

Solution

We calculate the cosets as follows:

$$eH = \{ee, er\} = \{e, r\}, \quad qH = \{qe, qr\} = \{q, s\}, \quad sH = \{se, sr\} = \{s, q\},$$

$$pH = \{pe, pr\} = \{p, t\}, \quad rH = \{re, rr\} = \{r, e\}, \quad tH = \{te, tr\} = \{t, p\}.$$

We see that the cosets of different elements can turn out to be the same set. For example, the cosets pH and tH both consist of the set $\{p, t\}$. Note that as we are dealing with sets, the order in which their elements are listed in our calculation does not matter. In other cases the cosets are completely different. For example, there is no element that is in both pH and rH. Thus the different cosets partition G into three disjoint sets, namely, $G = \{e, r\} \cup \{p, t\} \cup \{q, s\}$. Furthermore each coset contains the same number of elements as the subgroup H. Thus $\#(G) = 3 \times \#(H)$, and it follows that $\#(H)$ is a divisor of $\#(G)$.

Of course, we do not need group theory to work out that 2 is a divisor of 6. However, the point of this example is that we can prove that the facts we have noticed about the cosets of H are true for all subgroups of all groups. This we now prove.

It is convenient first to prove a lemma that gives a useful criterion for when two cosets, g_1H and g_2H, are identical.

LEMMA 11.4

If G is a group, H is a subgroup of G, and $g_1, g_2 \in G$, then

$$g_1H = g_2H \Leftrightarrow g_2^{-1}g_1 \in H. \tag{11.2}$$

Proof

First, suppose $g_1H = g_2H$. Since H is a subgroup, $e \in H$, and hence $g_1 = g_1 e \in g_1H$. Hence, as $g_1H = g_2H$, we deduce that $g_1 \in g_2H$, and so there is some $h' \in H$ such that $g_1 = g_2h'$. Therefore $g_2^{-1}g_1 = g_2^{-1}g_2h' = h'$ and hence $g_2^{-1}g_1 \in H$.

Second, suppose $g_2^{-1}g_1 \in H$. Let $h = g_2^{-1}g_1$. Then $g_1 = g_2h$. Now assume $x \in g_1H$. Then for some $h' \in H$, $x = g_1h' = g_2hh'$. As H is a subgroup of G, we have $hh' \in H$, and hence $x \in g_2H$. The argument to show that if $x \in g_2H$, then $x \in g_1H$ is similar. Therefore, $x \in g_1H \Leftrightarrow x \in g_2H$, and hence $g_1H = g_2H$.

We can now prove the main result of this section.

THEOREM 11.5

If G is a finite group and H is a subgroup of G, then the different cosets of H partition G into disjoint sets each containing the same number of elements as does H.

Proof

If $x \in G$, then as $e \in H$, $x = xe \in xH$. Thus each element of G is in at least one coset of H. Now suppose x is in two cosets, say $x \in g_1H$ and $x \in g_2H$. Then for some $h_1, h_2 \in H$, we have $x = g_1h_1 = g_2h_2$. It follows that $g_2^{-1}g_1 = h_2h_1^{-1}$. Now as $h_1, h_2 \in H$, and H is a subgroup of G, it follows that $h_2h_1^{-1} \in H$, and therefore $g_2^{-1}g_1 \in H$. It then follows from Lemma 11.4 that $g_1H = g_2H$. Therefore each element of G is in exactly one of the different cosets of H. Suppose $\#(H) = n$. Say that $H = \{h_1, \ldots, h_n\}$, where the elements h_i are all different. Then for each $g \in G$, $gH = \{gh_1, \ldots, gh_n\}$. Since $gh_i = gh_j$ implies $h_i = h_j$ (see Exercise 11.1.2A), the elements gh_i are all different. Therefore $\#(gH) = n = \#(H)$. Our key theorem is now an almost immediate consequence of Theorem 11.5.

THEOREM 11.6
Lagrange's Theorem*

If H is a subgroup of the finite group G, then the order of H is a divisor of the order of G.

---

* This theorem is named after Joseph Louis Lagrange (1736–1813), who was born in Turin in Italy. His mother was French and his father Italian. Eventually he moved to Paris, and in 1797 he became professor of mathematics at the École Polytechnique. Although Lagrange's theorem is now regarded as a fundamental result about finite groups, it was proved by Lagrange before the concept of a group had been isolated. Lagrange proved his theorem in a more special case dealing with the number of different polynomials that are obtained when its variables are permuted. For a good account of Lagrange's work, and the history of algebra more generally, see John Derbyshire, *Unknown Quantity, A Real and Imaginary History of Algebra*, John Henry Press, Washington DC, 2006, and Atlantic, London, 2007.

Proof

By Theorem 11.5, the different cosets of H completely partition G into disjoint sets each containing #(H) elements. So if there are k different cosets of H, we have that

$$k \times \#(H) = \#(G), \qquad (11.3)$$

and it follows immediately that #(H) is a divisor of #(G).

Although we have stated Lagrange's theorem for finite groups, it can be extended to the cases where the group G is infinite and H is either finite or infinite. The proof that the cosets of H partition G is just the same in this case. It is possible to interpret Equation 11.3 in these cases using Cantor's theory of infinite sets, but this does not lead to any very interesting conclusions.

Note also that in the finite case Lagrange's theorem tells us only about the *possible* orders of the subgroups of a given group, G. They must be divisors of the order of G. However, not all these possible orders need be realized. There are cases where n is a divisor of the order of a group G, but G has no subgroup of order n. An example of this kind may be found in Exercise 11.3.2B. There is an extensive theory originated by the Norwegian mathematician Ludwig Sylow (1832–1918), which specifies cases where a group does have subgroups of certain orders, but this is beyond the scope of thisbook.*

Exercises

11.3.1A Determine which of the following sets form subgroups of the group, $S(\square)$, of symmetries of a square (whose Cayley table is given in Table 11.2).

  i. $\{e,a,b,c\}$ ii. $\{e,a,b,c,h,v\}$ iii. $\{h,v,r,s\}$ iv. $\{e,r\}$

11.3.1B Which is the smallest subgroup of $S(\square)$ that contains both the symmetries b and h?

11.3.2A We have already noted that the set, Z, of integers forms a group with the operation of addition. Show that the subset, H, consisting of all integers that are multiples of 5, is a subgroup of Z. How many different cosets does H have?

11.3.2B Consider the following set, G, of 12 permutations from S_4:

$\{e,\ (1\ \ 2)(3\ \ 4),\ (1\ \ 3)(2\ \ 4),\ (1\ \ 4)(2\ \ 3),\ (1\ \ 2\ \ 3),\ (1\ \ 3\ \ 2),$
$(1\ \ 2\ \ 4),(1\ \ 4\ \ 2),(1\ \ 3\ \ 4),(1\ \ 4\ \ 3),(2\ \ 3\ \ 4),(2\ \ 4\ \ 3)\}.$

G forms a subgroup of S_4. This can be seen most easily by noticing that the permutations in G correspond to the rotational symmetries of a regular tetrahedron with its vertices labeled 1, 2, 3, and 4. Show that although G has order 12 and 6 is a divisor of 12, G does not have a subgroup of order 6. (You might find this easier after reading Section 11.4.)

---

* See, for example, Allenby, *Rings, Fields and Groups*, Chapter 6.

11.4 ORDERS OF GROUP ELEMENTS

Let G be a group, and let g be an element of G. Since G satisfies the closure condition, gg is also in G, and hence also $g(gg)$, and hence $g(g(gg))$, and so on. Because G satisfies the associativity condition, we can leave out the brackets, and write ggg and $gggg$ for these last two elements of G. At this point it is convenient to introduce the notation g^n for $\overline{gg...gg}^n$, that is, for the result of combining n g's together. Thus g^2 is gg, g^3 is ggg, and so on. We also use g^1 for g, and it will sometimes be convenient to write g^0 for the identity element e. It is immediately apparent that, when m and n are positive integers, $g^m g^n = g^{m+n}$, as both sides of this equation result from combining $m + n$ g's (but note that this index law depends on the associativity property). Also, if $g^m = g^n$ with $m > n$, then $g^{m-n} g^n = g^n$, and hence $g^{m-n} = e$. This observation will be useful in the sequel.

Now if G is a finite group, the elements g^n for $n = 0, 1, 2, \ldots$ cannot all be different. Suppose that $g^m = g^n$ with $m > n$. Then, as we have just noted, $g^{m-n} = e$, where $m - n$ is a positive integer. It follows that there is a smallest positive integer k, such that $g^k = e$. This integer k is given a special name.

DEFINITION 11.7

The least positive integer, k, if there is one, such that $g^k = e$, is called the *order* of the group element g, and is written $o(g)$. If there is no such k, we say that g is *an element of infinite order*.

PROBLEM 11.2

Calculate the orders of the elements of the group S_3, whose Cayley table is given in Table 11.3.

Solution

In each group $e^1 = e$, and so the identity element has order 1. It is, clearly, the only element of order 1. We see from the table that $r^2 = s^2 = t^2 = e$, and hence r, s, and t have order 2. We also have that $p^2 = q$, and hence $p^3 = p(p^2) = pq = e$. Thus $p \ne e$, $p^2 \ne e$, but $p^3 = e$, and so p has order 3. Similarly you can check that q has order 3.

We are now ready to explain the relationship between the meanings of *order* as used in Definitions 11.5 and 11.7. We first need a technical lemma.

LEMMA 11.7

Let G be a group and let g be an element of G of (finite) order k. Then for all integers m, n, we have

$$g^m = g^n \Leftrightarrow m \equiv n \pmod{k} \tag{11.4}$$

and, in particular,

$$g^n = e \Leftrightarrow n \equiv 0 \pmod{k}; \text{ that is, } k \text{ is a divisor of } n. \tag{11.5}$$

Proof

We first prove the equivalence 11.5. Then we show that we can deduce the equivalence in Equation 11.4.

Since g has order k, $g^k = e$, and hence, for each integer m, $g^{mk} = (g^k)^m = e^m = e$. So if n is a multiple of k, $g^n = e$. Conversely, suppose that $g^n = e$. Let r be the remainder when n is divided by k. Hence $0 \leq r < k$ and for some positive integer m, $n = mk + r$. Then $g^{mk+r} = e$ and hence $g^r = g^{mk} g^r = g^{mk+r} = g^n = e$. Since $0 \leq r < k$ and k is the least positive integer such that $g^k = e$, it follows that $r = 0$. So $n = mk$ and so k is a divisor of n. This proves the equivalence 11.5.

Now suppose m, n are positive integers with, say, $m \geq n$. Then using the equivalence 11.5, we have that $g^m = g^n \Leftrightarrow g^{m-n} = e \Leftrightarrow k$ is a divisor of $m - n \Leftrightarrow m \equiv n \pmod{k}$. This proves the equivalence 11.4.

It follows from this lemma that if g is an element of order k in a group G, there are only k different elements of the form g^n in G, namely, $e, g, g^2, \ldots, g^{k-1}$. It is not difficult to show that these elements form a subgroup of G (see Exercise 11.4.3A). This subgroup has order k.

In particular, in the case where G is finite, by Lagrange's theorem, k is a divisor of the order of G. We have thus proved the following useful consequence of Lagrange's theorem.

COROLLARY 11.8
Lagrange's Corollary

If G is a finite group, and $g \in G$, then the order of g is a divisor of the order of G.

As with Lagrange's theorem itself, this corollary only tells us about the possible orders of group elements. For example, you will see from the solution to Exercise 11.4.4B that in the group S_4, which has order 24, there are no elements of orders 6, 8, 12, or 24, even though these are all divisors of 24. We are now ready to give the first application of group theory to a combinatorial problem. We do this in the next section.

Exercises

11.4.1A Calculate the orders of the elements of
  i. The group of symmetries of a square,
  ii. The group of rotational symmetries of a cube.

11.4.1B Calculate the orders of the symmetries of a regular tetrahedron.

11.4.2A Calculate the orders of the elements of the group Z_{12} of the integers $\{0, 1, \ldots, 11\}$ with addition modulo 12.

11.4.2B Show that in the group of rotational symmetries of a circle, for each positive integer k, there is a symmetry of order k, and also that this group contains elements of infinite order.

11.4.3A Show that if g is an element of (finite) order k in a group G, then the subset $H = \{e, g, g^2, \ldots, g^{k-1}\}$ is a subgroup of G.

11.4.3B Suppose that g is an element of order 5 in a group G. Draw up the Cayley table for the subgroup $\{e, g, g^2, g^3, g^4\}$. Is this group isomorphic to another group you have already met? (It might help to write e as g^0 and g as g^1.)

11.4.4A Find all the subgroups of the group of symmetries of a square. (You may find it useful to use the Cayley table of this group as given in Table 11.2.)

11.4.4B Find as many subgroups of S_4 as you can.

11.5 THE ORDERS OF PERMUTATIONS

Shuffling a pack of cards amounts to permuting the cards. Generally the method is roughly to divide the pack into two piles and then more or less interleave the cards in one pile with those in the other pile. With very great skill, a standard pack of 52 cards can be shuffled by dividing the pack into two equal piles of 26 cards, and then alternating the cards from the two piles. This can be done in two ways depending on whether the card that is originally on top stays on top or ends up as the second card in the permuted pack. These two shuffles are called *a top riffle shuffle* and a *bottom riffle shuffle*, respectively, or, alternatively, an *out shuffle* and an *in shuffle*, respectively. These two shuffles are illustrated in Figure 11.5.

How many of these shuffles must take place before all the cards are restored to their original positions? It helps to answer this question if we write the relevant permutations in cycle notation. For example, we see that in the top riffle shuffle, the card originally in position 1 stays in position 1, the card originally in position 2 moves to position 3, ... the card in position 27 ends up in position 2, and so on. Thus this shuffle, which we denote by σ_T, corresponds to the permutation that is, in bracket notation,

$$\begin{pmatrix} 1 & 2 & 3 & . & . & 24 & 25 & 26 & 27 & 28 & 29 & . & . & 50 & 51 & 52 \\ 1 & 3 & 5 & . & . & 47 & 49 & 51 & 2 & 4 & 6 & . & . & 48 & 50 & 52 \end{pmatrix}.$$

In cycle notation this is

(1)(2 3 5 9 17 33 14 27)(4 7 13 25 49 46 40 28)(6 11 21 41 30 8 15 29)
(10 19 37 22 43 34 16 31) (12 23 45 38 24 47 42 32)(18 35)(20 39 26 51 50 48 44 36)(52).

We can now easily calculate the order of this permutation from its expression in cycle notation. We have already seen in Chapter 2, Section 2.6, that if we have a cycle of length n, then the numbers in the cycle are returned to their original positions after the permutation

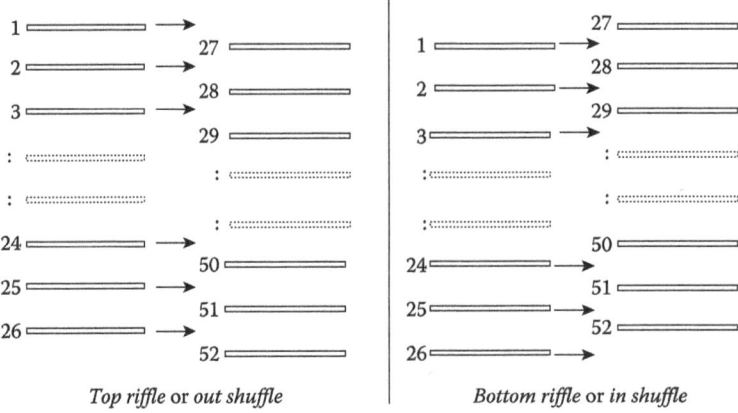

Top riffle or *out shuffle* | *Bottom riffle* or *in shuffle*

FIGURE 11.5

has been carried out n times. We see that σ_T is made up of six cycles each of length 8, one cycle of length 2, and two cycles of length 1.

Thus we need to carry out this permutation only eight times before each card is returned to its original position. In the language of group theory, we say that σ_T has order 8. Thus, *if* you could carry out a perfect riffle shuffle eight times in succession (a very big "if"!), you could give the appearance of having shuffled the pack very well, while returning all the cards to their original positions. A very useful skill to have. The bottom riffle shuffle is rather different. We ask you to calculate its order in Exercise 11.5.4A.

Here is another example. Consider the permutation $\sigma = (1\ 2)(3\ 4\ 5)(6\ 7\ 8\ 9)$ from S_9. Consider the effect of σ^k, that is, carrying out the permutation k times. The numbers 1 and 2 are returned to their original positions if k is a multiple of 2. To return 3, 4, and 5 to their original positions, k needs to be a multiple of 3, and to return 6, 7, 8, and 9 to their original positions, k must be a multiple of 4. So the least positive number k such that σ^k returns all of 1, 2, ..., 8, and 9 to their original positions is the least k that is a multiple of 2, 3, and 4. That is, the order of σ is the least common multiple of 2, 3, and 4, namely 12. It is easy to see how this generalizes. If a permutation in disjoint cycle form is made up of cycles of lengths $k_1, k_2, \ldots, k_s$, then the order of the permutation is the least common multiple of $k_1, k_2, \ldots, k_s$. We write this least common multiple as $lcm(k_1, \ldots, k_s)$.

Since the order of a permutation is determined by the structure of its disjoint cycle representation, it is useful to introduce some terminology and notation for this. We call this structure the *cycle type* of the permutation. We represent cycles of lengths 1, 2, 3, and so on, by the algebraic symbols x_1, x_2, x_3, and so on. We represent the number of cycles of a given length by writing these symbols with the appropriate exponents. So, for example, the cycle type of the permutation $\sigma = (1\ 2)(3\ 4)(5\ 6)(7\ 8\ 9\ 10\ 11)$ is $x_2^3 x_5^1$, indicating that σ is made up of three cycles of length 2 and one cycle of length 5. Often the exponent 1 is omitted, so that the cycle type of σ could be written as $x_2^3 x_5$.

In general, a permutation has cycle type $x_{k_1}^{r_1} x_{k_2}^{r_2} \ldots x_{k_s}^{r_s}$, where $k_1, k_2, \ldots, k_s$ is an increasing sequence of positive integers, and $r_1, \ldots, r_s$ are positive integers, if in disjoint cycle form it is made up of r_t cycles of length k_t, for $1 \leq t \leq s$. The usefulness of this notation will become apparent in Chapter 14.

Suppose σ is a permutation from S_n with cycle type $x_{k_1}^{r_1} \ldots x_{k_s}^{r_s}$. Then the total number of the positive integers in the cycles that make up σ is n. Hence we must have

$$r_1 k_1 + \ldots + r_s k_s = n. \tag{11.6}$$

If we rewrite Equation 11.6 as

$$\underbrace{k_1 + \ldots + k_1}_{r_1} + \underbrace{k_2 + \ldots + k_2}_{r_2} + \ldots + \underbrace{k_s + \ldots + k_s}_{r_s} = n, \tag{11.7}$$

we see that Equation 11.6 corresponds to a *partition* of n, as described in Chapter 6. Thus the different cycle types of permutations in S_n correspond to the partitions of n. Since two

TABLE 11.4

Partition	Order
4	4
3+1	3
2+2	2
2+1+1	2
1+1+1+1	1

permutations that have the same cycle structure have the same order, if we want to find the orders of the elements of S_n, all we need do is list the partitions of n. So, for example, the orders of the elements of S_4 are as given in Table 11.4.

Thus, as we remarked in the previous section, S_4 is a group of order 24 in which there are no elements of orders 6, 8, 12, and 24 even though these are divisors of 24.

Exercises

11.5.1A Find the orders of the following permutations.
   a. (1 2)(3 4 5)(6 7 8 9)(10 11 12 13 14)
   b. (1 2 3)(4 5 6 7)(8 9 10 11 12 13)

11.5.1B Find the order of the permutation $\sigma \in S_{20}$, which in bracket notation is

$$\begin{pmatrix} 1 & 2 & 3 & 4 & 5 & 6 & 7 & 8 & 9 & 10 & 11 & 12 & 13 & 14 & 15 & 16 & 17 & 18 & 19 & 20 \\ 7 & 8 & 14 & 1 & 2 & 6 & 11 & 16 & 10 & 17 & 4 & 15 & 20 & 9 & 12 & 18 & 13 & 5 & 3 & 19 \end{pmatrix}.$$

11.5.2A i. Find a permutation in S_{15} that has order 105.
   ii. Show that for $n < 15$ there is no permutation in S_n of order 105.

11.5.2B i. Find a permutation in S_{31} that has order 1001.
   ii. Show that for $n < 31$ there is no permutation in S_n of order 1001.

11.5.3A For $n = 4, 5, 6$, find the orders of the permutations in S_n.

11.5.3B For $1 \le n \le 10$ find the largest order of the permutations in S_n.

11.5.4A Express the permutation corresponding to a bottom riffle shuffle of 52 cards in disjoint cycle form, and use this to calculate its order.

11.5.4B Express the permutations corresponding to a bottom riffle shuffle and a top riffle shuffle of a pack of 40 cards, and calculate their orders.

11.5.5A Find the largest order of the permutations in S_{52}. (There are 281,589 partitions of 52, so that, without a computer, it is hardly practicable to answer this question by listing all the partitions and then calculating their least common multiples. Your search for a partition that corresponds to a permutation of the largest possible order should be guided by the fact that since for distinct primes p, q and positive integers k, l, $p^k + p^l < p^k p^l$, it is only necessary to consider partitions in which the parts are either powers of primes, or 1. We can regard the permutation of largest order in S_{52} as giving rise to the most effective way of shuffling a standard pack of 52 cards. Thus the answer to this exercise could be regarded as providing us with an answer to Problem 11 of Chapter 1.)

11.5.5B Find the largest order of the permutations in S_{50}.

CHAPTER 12

Group Actions

12.1 COLORINGS

In Chapter 1 we mentioned the following two problems.

PROBLEM 13A
Coloring a Chessboard
How many different ways are there to color the squares of a chessboard using two colors?

On a standard 8×8 chessboard, the squares are colored alternately black and white as shown in Figure 12.1i, but clearly they could be colored in many other ways. Alternative colorings are shown in Figure 12.1ii and iii. The problem is to decide exactly how many different colorings are possible.

PROBLEM 13B
Coloring a Cube
In how many different ways can you color a cube using three colors? One such coloring is shown in Figure 12.2.

These problems have a similar character. One difference is that the first problem is about a two-dimensional figure, and the second problem is about a three-dimensional figure. In this chapter we discuss only the first problem, as it is rather easier to work in two dimensions than in three. The solutions to both problems are given in the next chapter after we have described the ideas needed to solve them.

To make things as simple as possible for us, we begin by dealing with the case of a 2×2 chessboard; the case of a 1×1 chessboard is too simple for us to learn anything useful from it. A 2×2 chessboard has four squares, so with two choices of color for each square, there are altogether $2^4 = 16$ ways it can be colored. These 16 colorings are shown in Figure 12.3, where we have labeled them C1, C2, ..., C16, for future reference.

Are these colorings really all different? This depends on what we mean by "different." It seems reasonable to say that some of these colorings really are the same, and that they only look different because the chessboard has been rotated or reflected. For example, if we give the coloring C2 a quarter turn clockwise, it looks like C3. Also, the reflection in the vertical axis converts C2 into C3. From this point of view C2 and C3 are the same.

245

Indeed, in this light there are only six different patterns among these colorings. That is, we can put the colorings into the following six sets, with all the colorings in the same set having the same pattern.

{C1}, {C2, C3, C4, C5}, {C6, C7, C8, C9}, {C10, C11}, {C12, C13, C14, C15}, {C16}

Lurking in the background are the symmetries of the square. We have put two colorings in the same set if we can obtain one from the other by applying one of the symmetries of a square. As it happens, the case of a coloring of a 2×2 chessboard with two colors is so simple that it makes no difference whether we take into account reflections or not. In each case, whenever there is a reflectional symmetry taking one coloring to another coloring, there is also a rotational symmetry that does the same job. However, there is a difference as soon as we consider 3×3 and larger chessboards. Consider the two colorings in Figure 12.4. There is a reflectional symmetry that takes coloring (i) to coloring (ii), but no rotation of (i) produces coloring (ii).

We are free to choose whether we wish to regard these colorings as having the same pattern or not. This amounts to deciding which group of symmetries we are going to use, either the full group of all eight symmetries of the square, or just the subgroup consisting of only the identity and the rotational symmetries. The underlying theory turns out to be the same whichever group of symmetries we choose. The theory we are speaking about here deals with the interaction between a group and the members of some set. In the cases of interest to us the group will always be a group of symmetries of some figure, and the set will be a set of colorings of the same figure.

Exercises

12.1.1A How many colorings are there of
  i. A standard 8×8 chessboard, using two colors
  ii. The faces of a cube using three colors

12.1.1B How many colorings are there of an $n \times n$ chessboard using k colors?

12.1.2A Give an example of two colorings of a 2×2 chessboard, using *three* colors, such that one can be obtained from the other by a reflection but not by a rotation.

12.1.2B Are there two colorings of the faces of a cube using two colors such that one can be obtained from the other by a reflection but not by a rotation?

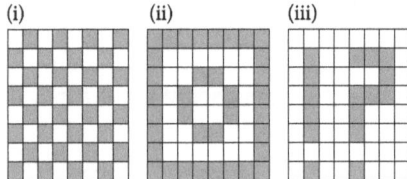

FIGURE 12.1

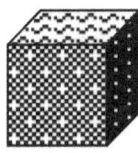

FIGURE 12.2

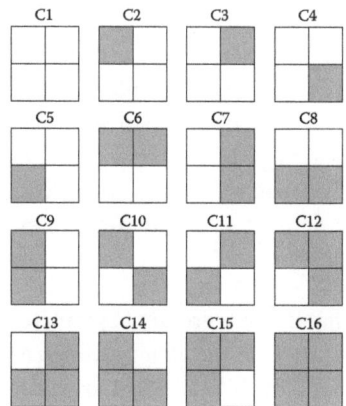

FIGURE 12.3

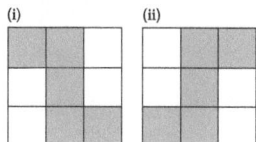

FIGURE 12.4

12.2 THE AXIOMS FOR GROUP ACTIONS

We wish to describe the general situation where we have an interaction between a group, say G, and a set, say X. In this abstract discussion you may find it helpful to keep in mind the particular example of the symmetries of a square and the colorings of a 2×2 chessboard, described in the previous section. We will often refer to this example.

The symmetries of a square in our example interact with each coloring to produce a (possibly) different coloring. For example, the quarter turn clockwise, for which we have used the symbol a, interacts with the coloring C2 to produce the coloring C3. We express this by writing $a \triangleright C2 = C3$, and in general, we use $g \triangleright x$ for the result of g acting on x. In order to develop a general theory we require that this interaction satisfies a couple of simple and natural properties, as given in the following definition.

DEFINITION 12.1

Let $(G, \bullet)$ be a group and let X be a set. We say that *G acts on X* if for each $g \in G$ and each $x \in X$, there is defined an element $g \triangleright x \in X$, in such a way that the following properties hold:

> **THE GROUP ACTION CONDITIONS**
>
> GA1. For each $x \in X$, $e_G \triangleright x = x$, where e_G is the identity element of G.
> GA2. For all $g_1, g_2 \in G$, and each $x \in X$, $g_1 \triangleright (g_2 \triangleright x) = (g_1 \bullet g_2) \triangleright x$.

Whenever a group acts on a set we say that we have a *group action*.

We stress that, in a group action, the elements of G are usually completely different from the elements of X. Thus the action of g on x to yield $g \triangleright x$ is very different from, for example, the combination of two group elements, which are objects of the same kind, to produce another group element. So, although we often omit the symbol for a group operation and write gh instead of $g \bullet h$, the symbol $\triangleright$ *should never be left out*. It helps to remind us of the disparity between g and x. [In those cases where the elements of the group G are functions with domain X, it may sometimes be appropriate to write $g(x)$ in place of $g \triangleright x$.]

It is not difficult to see that if G is a group of symmetries of a figure, and if X is the set of all colorings of the figure, then the action defined as in the previous section satisfies properties GA1 and GA2. In this case GA1 amounts to the fact that the identity element of G is the identity map, ι, and for each coloring x, $\iota \triangleright x = \iota(x) = x$, and GA2 to $g_1 \triangleright (g_2 \triangleright x) = g_1(g_2(x)) = g_1 \circ g_2(x) = (g_1 \circ g_2) \triangleright x$. Thus the examples of the previous section are group actions. For some examples of a different kind, see the exercises at the end of this section.

We conclude this section with a simple technical lemma about group actions that we need later on.

LEMMA 12.1

Let G be a group that acts on a set X. Then for each $g \in G$, and all $x, y \in X$,

$$g \triangleright x = y \Leftrightarrow g^{-1} \triangleright y = x.$$

Proof

Suppose that $g \triangleright x = y$. Then, $g^{-1} \triangleright y = g^{-1} \triangleright (g \triangleright x) = (g^{-1} \bullet g) \triangleright x$, by (GA2), $= e \triangleright x = x$, by GA1.

The converse implication, $g^{-1} \triangleright y = x \Rightarrow g \triangleright x = y$, is proved similarly.

Exercises

12.2.1A Let G be the group, $(\mathbb{R}, +)$, of real numbers with the operation of addition, and let X be the set, $\mathbb{R}^2$, of points in the plane. Suppose we define the action of $\mathbb{R}$ on $\mathbb{R}^2$ by specifying that for each $\theta \in \mathbb{R}$ and all $\boldsymbol{x} \in \mathbb{R}^2$, $\theta \triangleright \boldsymbol{x} =$ the point to which $\boldsymbol{x}$ is moved by a rotation through an angle θ counterclockwise about the origin. [Thus, $\theta \triangleright (x, y) = (x \cos\theta - y \sin\theta, x \sin\theta + y \cos\theta)$.] Show that this satisfies the group action conditions.

12.2.1B This question is about a more sophisticated example of a group action. Let G be any group. We define an action of the group G on G by: for $g \in G$ and $x \in G$, $g \triangleright x = gxg^{-1}$. Show that this satisfies the group action conditions. The action in this case is called *conjugation*, and gxg^{-1} is called the *conjugate of x by g*. This plays an important role in the theory of groups, but it is not greatly important from the point of view of this book.

12.2.2A Let n be a positive integer. We define the action of the group, S_n, of all permutations of the set $\{1, 2, \ldots, n\}$ on the set of edges of the complete graph, $K_n = (V_n, E_n)$, with n vertices, say $V_n = \{v_1, \ldots, v_n\}$, by: for $\sigma \in S_n$ and $\{v_i, v_j\} \in E_n$, $\sigma \triangleright \{v_i, v_j\} = \{v_{\sigma(i)}, v_{\sigma(j)}\}$. Show that this satisfies the group action conditions. This example of a group action plays an important role when it comes to counting the number of different simple graphs, as we do in Chapter 14.

12.2.2B Let G be a group, and let g_0 be a particular element of G. We define the action of the group, $(\mathbb{Z}, +)$, of the integers with addition, on G, as follows: For $n \in \mathbb{Z}$ and $g \in G$, $n \triangleright g = g_0^n g$, where we interpret g_0^0 as the identity element e_G of G, and for $n < 0$, we interpret g_0^n as $(g_0^{-n})^{-1}$. Show that this satisfies the group action conditions.

12.2.3A Suppose that the group G acts on the set X. Show that for all $g, h \in G$ and all $x \in X$, $g \triangleright x = h \triangleright x \Leftrightarrow g^{-1} h \triangleright x = x$.

12.2.3B Consider the group action of $\mathbb{R}$ on $\mathbb{R}^2$ as described in Exercise 12.2.1A. Determine the set

$$\{\theta \in \mathbb{R} : \theta \triangleright (1,0) = \frac{\pi}{4} \triangleright (1,0)\}.$$

12.3 ORBITS

We can now explain, in terms of group actions, what is meant by two colorings of some figure being regarded as "the same." We regard two colorings as being the same if some symmetry of the figure maps one to the other. Consequently, whether or not two colorings of a figure are regarded as being the same will depend heavily on which symmetries of the figure are taken into account. We can now explain this situation in general terms by using the language of group actions and giving a definition in this general context.

DEFINITION 12.2

Let G be a group that acts on a set X. We define the relation $\sim_G$ on X as follows:

For all $x, y \in X$, $x \sim_G y \Leftrightarrow$ there is some $g \in G$ such that $g \triangleright x = y$.

The notation suggests that $\sim_G$ is an equivalence relation. We now prove that this is indeed the case. Recall that this means showing that the relation $\sim_G$ is *reflexive*, *symmetric*, and *transitive*. The proof uses the group action conditions, and it is notable that the proof that $\sim_G$ has these properties uses the identity, inverses, and closure properties of a group, respectively, to do this.

LEMMA 12.2

The relation $\sim_G$ is an equivalence relation on X.

Proof

We check that $\sim_G$ has the three necessary properties to be an equivalence relation.

Reflexive: Suppose $x \in X$. Since G is a group, it has an identity element e, and by GA1 $e \triangleright x = x$. It follows that $x \sim_G x$. Therefore $\sim_G$ is reflexive.

Symmetric: Suppose $x, y \in G$ and $x \sim_G y$. Then there is some $g \in G$ such that $g \triangleright x = y$. As G is a group, g has an inverse, g^{-1}, which is also in G. By Lemma 12.1, $g^{-1} \triangleright y = x$. Hence $y \sim_G x$. Therefore $\sim_G$ is symmetric.

Transitive: Suppose $x,y,z \in X$ and both $x \sim_G y$ and $y \sim_G z$. Then there are $g,h \in G$ such that $g \triangleright x = y$ and $h \triangleright y = z$. As G is a group, $hg \in G$, and using GA2, we have that $hg \triangleright x = h \triangleright (g \triangleright x) = h \triangleright y = z$. Hence $x \sim_G z$. Therefore $\sim_G$ is transitive.

Because $\sim_G$ is an equivalence relation, it partitions X into disjoint equivalence classes. When G is a group of symmetries acting on a set of colorings, these equivalence classes are the colorings that we are regarding as being the same. Thus our question about how many *different* colorings there are may be restated as asking how many different equivalence classes there are. In this context the standard term for the equivalence classes is *orbits*. Here is the formal definition.

DEFINITION 12.3

Let the group G act on a set X. The equivalence classes of the relation $\sim_G$ are called *orbits*. For each $x \in X$, we let $Orb(x)$ be the orbit to which x belongs.

Since $y \in Orb(x) \Leftrightarrow x \sim_G y \Leftrightarrow$ for some $g \in G$, $g \triangleright x = y$, it follows that $Orb(x) = \{g \triangleright x : g \in X\}$. That is, the orbit of x consists of all the elements of X that we obtain from x by letting each element of G act on x.

So we are now interested in how to calculate the number of different orbits when a group acts on a set. Before we can give the theorem that answers this question we need to develop one more theoretical idea. We do this in the next section.

Exercises

12.3.1A Consider the group action described in Exercise 12.2.1A. Find the orbits of the points (1,0) and (0,0).

12.3.1B Find a group action on R^2 whose orbits are ellipses.

12.3.2A Let G be the group, $(R,+)$, of real numbers with addition, and let $X = R^2$. The action of G of X is defined by: for $t \in R$, and $(x,y) \in R^2$, $t \triangleright (x,y) = (x+t, y+2t)$. Show that this satisfies the group action conditions. Find the orbits of the points (0,0), (0,1), and (1,2).

12.3.2B Let G be the group, $S(\square)$, of symmetries of the square (recall that the Cayley table of this group is given in Table 11.2). Find the orbits of each element of G with respect to the group action of conjugation, as described in Exercise 12.2.1B.

12.4 STABILIZERS

Whenever a group G acts on a set X, each element of X is *fixed* by the identity element of G in the sense that $e \triangleright x = x$. This is built into the definition of a group action as condition GA1. Other elements of G may fix certain elements of X. For example, in our standard example of colorings of a 2×2 chessboard, the diagonal reflection r fixes each of the colorings C1, C2, C4, C10, C11, C13, C15, and C16, as these are symmetrical about the relevant diagonal. The type of symmetry of a particular coloring is determined by those symmetries that leave it unchanged. We give this set a special name.

DEFINITION 12.4

Let the group G act on a set X. For each $x \in X$, the set of elements of G that fix x is called the *stabilizer* of x. We let $Stab(x)$ be the stabilizer of x. Thus

$$Stab(x) = \{g \in G : g \triangleright x = x\}.$$

It can be seen that with our example of the group, $S(\square)$, of symmetries of a square acting on the colorings of a 2×2 chessboard, the stabilizer of C2 is $\{e,r\}$, and that of C10 is $\{e,b,r,s\}$. These are both subgroups of $S(\square)$. The next result tells us that this is not an accident.

LEMMA 12.3

If the group G acts on the set X, then for each $x \in X$, $Stab(x)$ is a subgroup of G.

Proof

We check that $Stab(x)$ satisfies the subgroup conditions.
  Closure: Suppose $g, h \in Stab(x)$, then $g \triangleright x = x$ and $h \triangleright x = x$. Hence, by GA2, $gh \triangleright x = g \triangleright (h \triangleright x) = g \triangleright x = x$, and hence $gh \in Stab(x)$. So the closure condition is satisfied.
  Identity: By GA1, $e \triangleright x = x$, and so $e \in Stab(x)$.

TABLE 12.1

Coloring	Orbit	Stabilizer
C1	{C1}	$\{e,a,b,c,h,v,r,s\}$
C2		$\{e,r\}$
C3		$\{e,s\}$
C4	{C2, C3, C4, C5}	$\{e,r\}$
C5		$\{e,s\}$
C6		$\{e,v\}$
C7		$\{e,h\}$
C8	{C6, C7, C8, C9}	$\{e,v\}$
C9		$\{e,h\}$
C10		
C11	{C10, C11}	$\{e,b,r,s\}$
C12		$\{e,s\}$
C13		$\{e,r\}$
C14	{C12, C13, C14, C15}	$\{e,s\}$
C15		$\{e,r\}$
C16	{C16}	$\{e,a,b,c,h,v,r,s\}$

Inverses: Suppose $g \in Stab(x)$. Then $g \triangleright x = x$, and so, by Lemma 12.1, $g^{-1} \triangleright x = x$. Hence $g^{-1} \in Stab(x)$, and the inverses condition is satisfied.

This completes the proof.

We are interested in stabilizers because, as we will soon see, they help us to count orbits. Group theorists are interested in stabilizers for other reasons. As Lemma 12.3 shows, they are useful in identifying subgroups. For example, the subgroup $\{e,b,r,s\}$ of $S(*)$ is not easy to find from the Cayley table (see Table 11.2), but can easily be identified as $Stab(C10)$.

Once again, it is instructive to return to our example of the group $S(\square)$ acting on the colorings of a 2×2 chessboard. In Table 12.1 we have listed the orbits and the stabilizers of the colorings.

We see that the larger the orbit then the smaller is the stabilizer. Indeed, in each case $\#(Orb(x)) \times \#(Stab(x)) = 8$, and 8 is the number of elements in the group $S(\square)$. This is not a coincidence, but an important result that is true in every case of a group action.

THEOREM 12.4
The Orbit-Stabilizer Theorem

If the group G acts on the set X, then for each $x \in X$,

$$\#(Orb(x)) \times \#(Stab(x)) = \#(G). \tag{12.1}$$

Proof

By Lemma 12.3, $Stab(x)$ is a subgroup of G. So Equation 12.1 is similar to the equation that occurs in our proof of Lagrange's theorem (Theorem 11.6), namely,

$$k \times \#(H) = \#(G), \tag{12.2}$$

where k is the number of different cosets of H.

Comparing Equations 12.1 and 12.2 we see that to prove Equation 12.1 it will be enough to prove that

$$\#(Orb(x)) = \text{the number of different cosets of } Stab(x).$$

Consequently, we need to show that there is a one–one correspondence between the elements of $Orb(x)$ and the cosets of $Stab(x)$. Now, $Orb(x) = \{g \triangleright x : g \in G\}$, and the set of cosets of $Stab(x)$ is $\{gStab(x) : g \in G\}$. We now establish the required one–one correspondence between these sets as follows:

For $g_1, g_2 \in G$, we have that

$$g_1 \triangleright x = g_2 \triangleright x \Leftrightarrow g_2^{-1} g_1 \triangleright x = x, \text{ by the result of Exercise 12.2.3A,}$$

$$\Leftrightarrow g_2^{-1} g_1 \in Stab(x), \text{ by the definition of } Stab(x),$$

$$\Leftrightarrow g_1 Stab(x) = g_2 Stab(x), \text{ by Lemma 11.4.}$$

This establishes the desired correspondence, and so completes the proof of Theorem 12.4. We can immediately deduce the following corollary.

COROLLARY 12.5

Let G be a finite group that acts on a set X. Then the number of elements in each orbit is a divisor of the order of G.

In particular, in the case of the group action of conjugation of a group on itself, the number of elements in each orbit, which in this context are also called *conjugacy classes*, is a divisor of the number of elements in the group. This is a very useful result when it comes to analyzing the structure of groups.

As we now see, we can easily deduce from the orbit-stabilizer theorem the following result, which gives a formula for the number of orbits.

THEOREM 12.6
The Orbit-Counting Theorem

Let G be a finite group that acts on the set X. Then the number of different orbits is

$$\frac{1}{\#(G)} \sum_{x \in X} \#(Stab(x)).$$

Proof

Suppose that there are k different orbits, $Orb(y_1), Orb(y_2), \ldots, Orb(y_k)$. For $1 \leq r \leq k$ and for each $x \in Orb(y_r)$, $Orb(x) = Orb(y_r)$. Hence, by the orbit-stabilizer theorem,

$$\#(Stab(x)) = \frac{\#(G)}{\#(Orb(x))} = \frac{\#(G)}{\#(Orb(y_r))}.$$

Hence

$$\sum_{x \in Orb(y_r)} \#(Stab(x)) = \sum_{x \in Orb(y_r)} \frac{\#(G)}{\#(Orb(y_r))}. \tag{12.3}$$

In Equation 12.3 each term in the sum on the right-hand side has the same value, and there are $\#(Orb(y_r))$ terms in this sum. We therefore deduce from Equation 12.3 that

$$\sum_{x \in Orb(y_r)} \#(Stab(x)) = \#(Orb(y_r)) \times \frac{\#(G)}{\#(Orb(y_r))} = \#(G).$$

Therefore, as $X = \bigcup_{r=1}^{k} Orb(y_r)$, where for $1 \leq r < s \leq k$ the sets $Orb(y_r)$ and $Orb(y_s)$ are disjoint,

$$\sum_{x \in X} \#(Stab(x)) = \sum_{r=1}^{k} \left(\sum_{x \in Orb(y_r)} \#(Stab(x)) \right) = \sum_{r=1}^{k} \#(G) = k \#(G).$$

It follows that

$$k = \frac{1}{\#(G)} \sum_{x \in X} \#(Stab(x)).$$

This completes the proof.

TABLE 12.2

x	C1	C2	C3	C4	C5	C6	C7	C8	C9	C10	C11	C12	C13	C14	C15	C16
#(Stab(x))	8	2	2	2	2	2	2	2	2	4	4	2	2	2	2	8

Is the formula given by this theorem useful? It is easy to use it in our standard example of the group $S(\square)$ acting of the set of colorings of a 2×2 chessboard. From Table 12.1, we obtain Table 12.2 of values of #(Stab(x)).

We thus see that in this case $\sum_{x \in X}$ #(Stab(x)) = 48, and hence, by the orbit-counting theorem, there are $\frac{1}{8} \times 48 = 6$ different orbits, which agrees with the value we have already found.

The calculation in this example is deceptively simple. We were able to use the orbit-counting theorem because we could easily list the 16 elements of X and work out the number of elements in their stabilizers. However, in the case we are really interested in, that of 8×8 chessboards, this method is completely impractical. There are 2^{64} colorings of an 8×8 chessboard using two colors, and there is no practical way we could list them all. If we could fit 50 colorings to a page, then we would need just over 10^{15} volumes with 360 pages each to list them all. However, although as we move from 2×2 chessboards to 8×8 chessboards the number of colorings becomes very large, the group of symmetries, $S(\square)$, remains the same and still has just eight elements. In the next chapter we show how this can be exploited.

Exercises

Note: In these exercises we show how the theory of group actions can be used to derive some combinatorial theorems about groups. They are not relevant for the rest of this book. We give only two applications. You will need to consult a book about groups for more.

12.4.1A   We say that two elements, x, y, of a group, $(G, \bullet)$, *commute* if $x \bullet y = y \bullet x$, and that a group is a *commutative* group if all pairs of its elements commute. The *center* of a group, written $Z(G)$, is defined to be the set of all those elements of the group that commute with every element of the group. That is, $Z(G) = \{g \in G:$ for all $x \in G, gx = xg\}$. Clearly, for every group G, we have $e_G \in Z(G)$. The example of the group of symmetries of an equilateral triangle (as given by Table 11.3) shows that it is possible to have $Z(G) = \{e\}$. The purpose of this exercise is to show that if #(G) is the power of a prime number, p, there are at least p elements in $Z(G)$.

a. Prove that $Z(G)$ is a subgroup of G.
b. Prove that $g \in Z(G) \Leftrightarrow$ the conjugacy class of g is $\{g\}$, that is, g is conjugate just to itself.
c. Prove that, if for some prime number p and some positive integer n, #(G) = p^n, then #($Z(G)$) ≥ p. (*Hint:* Make use of the remark after Corollary 12.5 that the number of elements in a conjugacy class divides the order of the group.)

12.4.1B   We have noted in exercise 11.3.2B that the converse of Lagrange's theorem is not, in general, true. That is, if k divides the order of a group, there need not be an element of the group of order k (nor even a subgroup of order k). However, this converse is true when k is a prime number. In this exercise you are asked to prove this result, which is due to the French mathematician Cauchy.*

We start with a group G and suppose that p is a prime number that is a divisor of $\#(G)$. We let X be the set of all ordered p-tuples of elements of G whose product is the identity element, e, of G. Thus

$$X = \{(g_1, g_2, \ldots, g_p): g_1, g_2, \ldots, g_p \in G \text{ and } g_1 g_2 \ldots g_p = e\}.$$

Recall that Z_p is the group of the integers $\{0, 1, \ldots, p-1\}$ with addition modulo p. We define an action of this group on X by

$$k \triangleright (g_1, g_2, \ldots, g_p) = (g_{k+1}, g_{k+2}, \ldots, g_{k+p}), \tag{12.4}$$

where the addition in the suffices is carried out modulo p.

i. Given that $\#(G) = n$, how many elements are there in X?
ii. Prove that Equation 12.4 defines an action that satisfies the group action conditions.
iii. Prove that

$$k \triangleright (g_1, \ldots, g_p) = (g_1, \ldots, g_p), \text{ for all } k \in Z_p \Leftrightarrow g_1 = g_2 = \ldots = g_p.$$

It follows from (iii) that the only orbits containing just one element comprise those elements of X of the form $(g, g, \ldots, g)$. Such a p-tuple is in X, if and only if $g^p = 1$, and hence either $g = e$, or g is an element of order p. Thus to prove that G contains at least one element of order p you need only show that:

iv. There is more than one orbit that consists of a single element of X.

---

* Augustine-Louis Cauchy was born in Paris on August 21, 1789, and died at Sceaux, near Paris, on May 23, 1857. He was an extremely distinguished and prolific mathematician who made important contributions to both pure and applied mathematics. He is best remembered as the originator, along with Gauss, of complex analysis where many of the standard results, for example, the Cauchy–Riemann equations and Cauchy's residue theorem, bear his name.

CHAPTER 13

Counting Patterns

13.1 FROBENIUS'S COUNTING THEOREM

In this chapter we are able to give the solutions to Problems 13A and 13B about counting the different patterns we get by coloring a chessboard and a cube. The key step is turning the orbit-counting theorem (Theorem 12.6) into a form in which it is much easier to use. Recall that this theorem tells us that when the group G acts on the set X, then the number of different orbits is

$$\frac{1}{\#(G)} \sum_{x \in X} \#(Stab(x)). \tag{13.1}$$

We noted that, in the case of coloring an $n \times n$ chessboard, this formula becomes impractical as n gets larger, as the number of colorings that make up the set X grows very rapidly. However, we also noticed that, in this case, the group G remains the same, as it is the group, $S(\square)$, of symmetries of the square, however large the chessboard. So the key idea is to replace the sum in expression 13.1 by a sum over the group of symmetries. We can see how this works in general by considering, once again, the action of the group $S(\square)$ on the set of colorings of a 2×2 chessboard.

In Table 13.1 we have put a tick in the row corresponding to a group element g and the column corresponding to a coloring x if and only if g fixes x; that is, if and only if $g \triangleright x = x$.

In Table 13.1 the ticks in each x column correspond to the group elements in $Stab(x)$, and so the number of ticks is $\#(Stab(x))$. Hence $\sum_{x \in X} \#(Stab(x))$ is the total number of ticks in the table. We can count these ticks in another way. Instead of adding up the column totals, 8, 2, 2, 8, ... , 2, 8, we can add up the row totals, 16, 2, 4, ... , 8, 8. As we have already noted, the number of columns in the corresponding table of this kind grows very rapidly as the chessboard gets larger, but the number of rows stays the same. So this seemingly simple idea of counting the total number of ticks by adding the row totals turns out to have lots of advantages. To state the result we need some terminology for the row totals.

TABLE 13.1

		Colorings																Total
		C1	C2	C3	C4	C5	C6	C7	C8	C9	C10	C11	C12	C13	C14	C15	C16	
Symmetries	e	✓	✓	✓	✓	✓	✓	✓	✓	✓	✓	✓	✓	✓	✓	✓	✓	16
	a	✓														✓		2
	b	✓							✓	✓						✓		4
	c	✓														✓		2
	h	✓						✓	✓							✓		4
	v	✓					✓	✓								✓		4
	r	✓	✓	✓							✓	✓		✓		✓	✓	8
	s	✓			✓	✓					✓	✓	✓		✓		✓	8
	Total	8	2	2	2	2	2	2	2	2	4	4	2	2	2	2	8	48

DEFINITION 13.1

Suppose that the group G acts on the set X. For each $g \in G$, the *fixed set* of g is the set of those elements of X that g fixes. Writing $Fix(g)$ for the fixed set of g, we have

$$Fix(g) = \{x \in X : g \triangleright x = x\}.$$

We usually refer to the fixed set of g as the *fix* of g.

It follows that the ticks in the g row correspond to the elements of $Fix(g)$, and so $\#(Fix(g))$ is the number of ticks in the g row. Since the sum of the row totals must be the same as the sum of the column totals, we have that

$$\sum_{g \in G} \#(Fix(g)) = \sum_{x \in X} \#(Stab(x)),$$

and we can use this to rewrite the orbit-counting theorem in the following much more useful form.

THEOREM 13.1
Frobenius's Counting Theorem*

If G is a finite group that acts on the set X, then the number of different orbits is

$$\frac{1}{\#(G)} \sum_{g \in G} \#(Fix(g)).$$

We see how useful this version of the theorem is in the next section.

---

* Because of a historical accident, this result is often called "Burnside's lemma." William Burnside attributed the result to Frobenius in the first edition of his book *Theory of Groups of Finite Order*, Cambridge University Press, Cambridge, 1897, but Frobenius's name was accidentally left out of the more widely read second edition of 1911. For the details of this history see P. M. Neumann, A Lemma That Is Not Burnside's, *Mathematics Scientist*, 4, 1979, pp. 133–141, and E. M. Wright, Burnside's Lemma: A Historical Note, *Journal of Combinatorial Theory B*, 25, 1981, pp. 89–90. Georg Ferdinand Frobenius was born in Berlin on October 26, 1849. He was the first to formulate the abstract concept of a group. He became professor of mathematics in Zurich in 1875, and he returned to Berlin as professor of mathematics at the university in 1892. He died in Charlottenberg on August 3, 1917.

FIGURE 13.1

Exercises

13.1.1A An equilateral triangle may be divided into four smaller equilateral triangles by the lines joining the midpoints of its edges, as shown in Figure 13.1.

Let X be the set of colorings of this figure in which each small triangle may be colored either black or white. Let the group, G, of symmetries of an equilateral triangle act on X in the standard way. For each $g \in G$ find the value of $\#(Fix(g))$ and hence work out the number of different orbits.

13.1.1B Let G be a group, and let G act on itself by conjugation. (Recall that this means that for all $g \in G$ and all $x \in G$, $g \triangleright x = gxg^{-1}$.) Show that for all $g \in G$, $Fix(g) = Stab(g)$.

13.2 APPLICATIONS OF FROBENIUS'S COUNTING THEOREM

We can now, at last, solve Problem 13A about the number of different colorings of an 8×8 chessboard that we introduced in Chapter 1. If we use Frobenius's counting theorem all we need to do is count the number of colorings that are fixed by each symmetry of the square.

SOLUTION TO PROBLEM 13A

In this problem, our underlying set, X, is the set of all 2^{64} colorings of an 8×8 chessboard using two colors. To apply Frobenius's counting theorem (Theorem 13.1), we need to calculate $\#(Fix(g))$ for each of the eight symmetries of a square (as given in Figure 11.2). Since the identity fixes every coloring, $\#(Fix(e)) = 2^{64}$.

We next consider $Fix(a)$, where a is the quarter-turn rotation clockwise. Which colorings are fixed by a; that is, which colorings look the same after the chessboard has been given a quarter turn?

The clockwise quarter turn moves the square labeled α to the square labeled β, β to γ, γ to δ, and δ to α (see Figure 13.2). So, if the coloring is to look the same after the rotation as it did before, all the four squares in this cycle must have the same color. More generally, once we have decided the colors of the 16 squares in the top left-hand quadrant of the board, then the colors of all the other squares are determined if the board is to look the same after the quarter-turn rotation. So $\#(Fix(a))$ is the number of different ways of coloring the 16 squares in the top left-hand corner of the board, that is, 2^{16}.

Another way to look at this, and one that will be very fruitful in the next chapter, is that corresponding to the rotation, there is a permutation, say π_a, of the 64 squares of the chessboard. This permutation is made up of 16 cycles each of length 4. For example, one of these cycles is $(\alpha\ \beta\ \gamma\ \delta)$. A coloring looks the same after the quarter-turn rotation if and only if, for each of these cycles, all the squares in the cycle are the same color. Therefore, in choosing a coloring that is fixed by the symmetry a, we are free to assign a color to each of these 16 cycles, and then all the squares in a given cycle must have the color assigned to that cycle. Since there are 16 cycles and two possible colors for each cycle, there are 2^{16} colorings that are fixed by the symmetry a. In other words, π_a

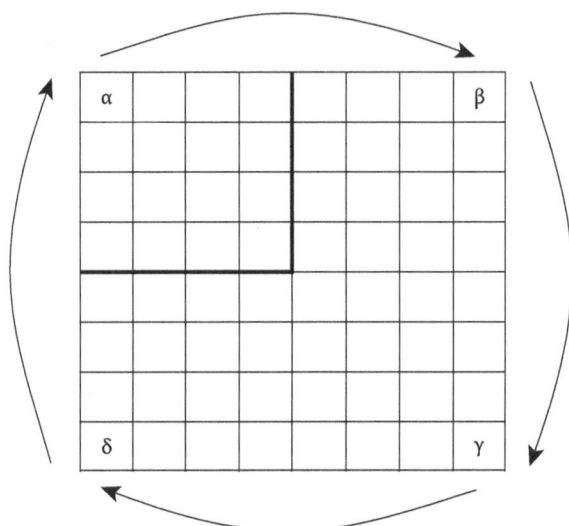

FIGURE 13.2

TABLE 13.2

g	e	a	b	c	h	v	r	s
#(Fix(g))	2^{64}	2^{16}	2^{32}	2^{16}	2^{32}	2^{32}	2^{36}	2^{36}

has cycle type x_4^{16}, showing that it is made up of 16 cycles, each of length 4, and hence #(Fix(a)) = 2^{16}.

We have stressed this point because it is crucial to using Frobenius's counting theorem. You are therefore encouraged to pause in your reading and to calculate #(Fix(g)) for each $g \in S(\square)$. Then check whether your answers agree with those in Table 13.2. You should tackle this problem by considering the permutation, π_g, of the 64 squares of the chessboard, corresponding to each symmetry, g. If the chessboard is to look the same after the symmetry has been carried out, all the squares in any given cycle of π_g must be assigned the same color. So if π_g has n cycles, #(Fix(g)) = 2^n.

Let us look more closely at Fix(r) and Fix(s). The permutation, π_r, of the 64 squares of the chessboard, corresponding to the diagonal reflection r, has cycle type $x_1^8 x_2^{28}$, as the 8 squares on the diagonal stay in their original positions and the other 56 squares are swapped round in pairs. This means that π_r is made up 8 cycles of length 1 and 28 cycles of length 2, making 36 cycles altogether. Hence #(Fix(r)) = 2^{36}. Similarly, #(Fix(s)) = 2^{36}.

We can now use Frobenius's counting theorem to calculate the total number of different patterns. All we have to do is to add up the values of #(Fix(g)) and then divide by the total number of symmetries. So the total number of different patterns is

$$\frac{1}{8}(2^{64} + 2(2^{36}) + 3(2^{32}) + 2(2^{16})) = 2,305,843,028,004,192,256.$$

Rather a large number!

It is easy to see that if we have more than two colors, all we need to do is to replace 2 in the above formula by the number of colors we are using. Thus the number of different patterns that can be obtained by coloring the squares of an 8 × 8 chessboard using c different colors is

$$\frac{1}{8}(c^{64} + 2c^{36} + 3c^{32} + 2c^{16}).$$

We now look at two more examples where we can use Frobenius's counting theorem. You are encouraged to try them for yourself before reading the solutions.

PROBLEM 13.1

In how many ways can the vertices of a regular hexagon be colored using c different colors?

Note that this is the same problem as that of counting the number of patterns that can be formed by arranging six beads in a ring, using beads of c colors.

Solution

The relevant set, X, here is the set of all ways of choosing colors for the six vertices. Since there are c colors to choose from for each vertex, $\#(X) = c^6$. The relevant group, G, is the group of symmetries of a regular hexagon. You will recall from Chapter 11 (see Exercise 11.2.2B) that there are 12 of these symmetries as follows:

e	the identity,
r_k, for $1 \leq k \leq 5$	rotation through an angle $2k\pi/6$ counterclockwise,
s_k, for $1 \leq k \leq 3$	reflection in the axis joining the vertices k and $k + 3$, and
t_k, for $1 \leq k \leq 3$	reflection in the axis, in the plane of the hexagon, perpendicular to the line joining the vertices k and $k + 3$.

We need to work out the number of colorings fixed by each symmetry. We can do this by considering the permutation of the vertices, say π_g, corresponding to each symmetry, g. If this permutation is made up of n disjoint cycles, then $\#(Fix(g)) = c^n$. We set this out in Table 13.3 using the labelings of the vertices shown in Figure 13.3.

Of course, it isn't really necessary to work these all out individually, as clearly s_1, s_2, and s_3 have the same cycle type, as also do t_1, t_2, and t_3.

It now follows immediately from Frobenius's counting theorem that the number of different patterns is $\frac{1}{12}(c^6 + 3c^4 + 4c^3 + 2c^2 + 2c)$.

PROBLEM 13.2

How many different ways are there to color the faces of a regular tetrahedron using c colors? (Two colorings should be regarded as the same if one can be obtained from the other by a rotational symmetry.)

TABLE 13.3

Symmetry g	Permutation π_g	Cycle type of π_g	#(Fix(g))
e	(1)(2)(3)(4)(5)(6)	x_1^6	c^6
r_1	(1 2 3 4 5 6)	x_6	c
r_2	(1 3 5)(2 4 6)	x_3^2	c^2
r_3	(1 4)(2 5)(3 5)	x_2^3	c^3
r_4	(1 5 3)(2 6 4)	x_3^2	c^2
r_5	(1 6 5 4 3 2)	x_6	c
s_1	(1)(4)(2 6)(3 5)	$x_1^2 x_2^2$	c^4
s_2	(2)(5)(1 3)(4 6)	$x_1^2 x_2^2$	c^4
s_3	(3)(6)(1 5)(2 4)	$x_1^2 x_2^2$	c^4
t_1	(1 4)(2 3)(5 6)	x_2^3	c^3
t_2	(1 6)(2 5)(3 4)	x_2^3	c^3
t_3	(1 2)(3 6)(4 5)	x_2^3	c^3

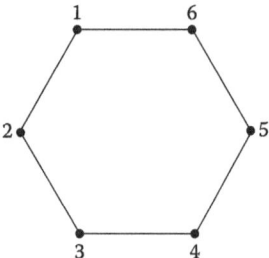

FIGURE 13.3

Solution

Here we think of the *faces* of the tetrahedron as the objects being permuted, as they are the objects to be colored. A regular tetrahedron has 12 rotational symmetries, as follows:

a. The identity, e. This fixes every coloring. As the tetrahedron has four faces, the corresponding permutation of the faces has cycle type x_1^4, and it follows that #(Fix(e)) = c^4.

b. Rotations through two-thirds of a turn (that is, through an angle $\frac{2}{3}\pi$), clockwise and counterclockwise about axes joining a vertex to the midpoint of the opposite face, as shown in Figure 13.4i. There are eight of these rotations. Each of the corresponding permutations of the faces has cycle type $x_1 x_3$, since it fixes one face and permutes the other three faces in a cycle of length 3. So each of them fixes c^2 colorings.

c. Rotations through a half turn ($\frac{1}{2}\pi$) through axes joining the midpoints of opposite edges, as shown in Figure 13.4ii. There are three of these rotations. Each of the corresponding permutations of the faces has cycle type x_2^2, and so each of them fixes c^2 colorings.

It therefore follows from Frobenius's counting theorem that there are $\frac{1}{12}(c^4 + 11c^2)$ different ways to color a regular tetrahedron using c colors.

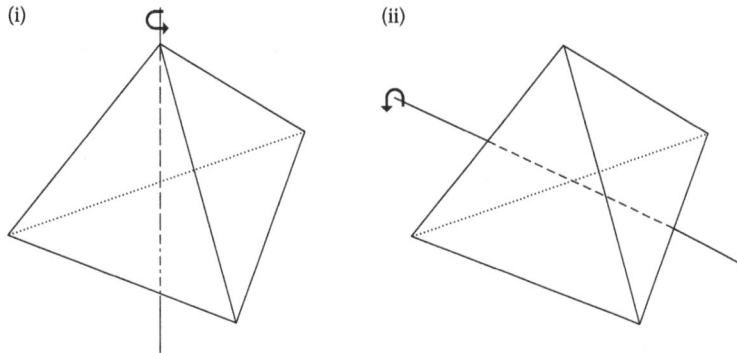

FIGURE 13.4

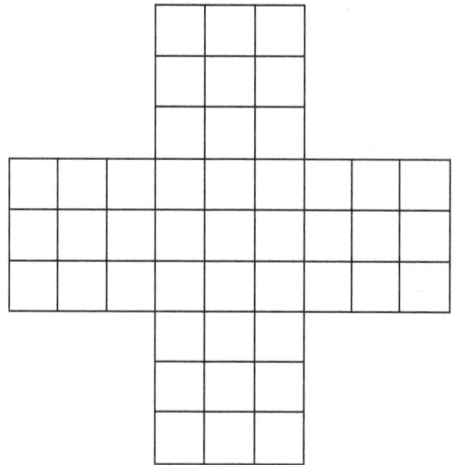

FIGURE 13.5

Frobenius's counting theorem enables us to count the total number of different patterns in the kind of examples given above and in the following exercises. If we wish to answer more specific questions such as finding the number of different ways to color a chessboard so that half the squares are black and half are white, we need to extend the theory. This we do in the next chapter.

Exercises

13.2.1A How many different ways are there to color the edges of an equilateral triangle using four colors?

13.2.1B How many different patterns may be formed by coloring the vertices of a regular octagon using c colors?

13.2.2A How many different patterns can be formed by coloring the squares of a 5×5 chessboard using three colors?

13.2.2B How many different patterns can be formed by coloring the small squares in Figure 13.5 using the colors red, white, and blue?

13.2.3A In how many different ways can you color the faces of a cube using the colors red, white, and blue? (Take into account just the rotational symmetries of a cube.)

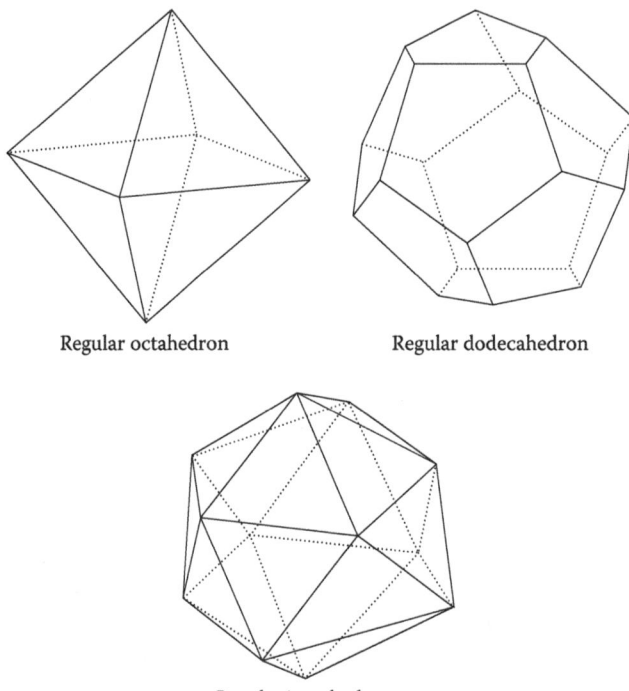

FIGURE 13.6

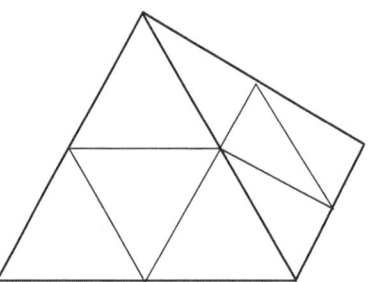

FIGURE 13.7

13.2.3B In how many different ways can the numbers 1 to 6 be placed on the faces of a cube? (Take into account just the rotational symmetries of a cube.)

13.2.4A There are five Platonic regular solids. In addition to the regular tetrahedron and the cube, which we have already considered, there are also the regular octahedron, dodecahedron, and icosahedron with 8, 12, and 20 faces, respectively. These are shown in Figure 13.6. For each of these three latter regular solids, calculate the number of different patterns that can be obtained by coloring their faces using c different colors. (Again, consider only the rotational symmetries of these figures.) It might help if you have models of Platonic regular solids at hand.

13.2.4B Each face of a regular tetrahedron is divided into four smaller equilateral triangles as shown in Figure 13.7. How many different patterns can be formed by coloring these small triangles using c colors? (Again, only take into account the rotational symmetries of a regular tetrahedron.)

CHAPTER 14

Pólya Counting

14.1 COLORINGS AND GROUP ACTIONS

We have now reached the stage where we can tackle Problem 14 of Chapter 1.

PROBLEM 14

In how many different ways can you color a cube using one red, two white, and three blue faces?

Frobenius's counting theorem, of Chapter 13, enables us to count the total number of different ways to color a cube, but it does not directly answer more delicate questions such as Problem 14.

A problem of a similar kind is that of counting the number of different positions that can arise in the game of noughts and crosses (also known as tic-tac-toe). This game is played on the grid in Figure 14.1.

This game involves two players who move alternately, the first putting a nought (O) and the second a cross (X) in one of the nine squares making up the grid. You win if you achieve a row (horizontally, vertically, or diagonally) of three noughts, or three crosses, before your opponent. In counting the different positions that can occur in this game we need to take account of the symmetries of the grid, which correspond to the symmetries of a square. In any given position each square can be occupied by a nought or a cross, or it can be empty. The numbers of noughts and crosses that can occur are constrained by the rules of the game. The number of noughts must at each stage be either equal to the number of crosses, or be one more than the number of crosses. Hence the problem of counting the number of positions that can occur in this game is rather more complicated than the problem of simply counting the number of different colorings of a chessboard.

We are now going to describe the theory that enables us to handle problems of this kind. First, we need to be a little more precise about what we mean by a coloring, and how the group of symmetries of the underlying figure acts on the set of colorings.

It is again helpful to look at a particular case. We therefore return to our favorite example of a 2×2 chessboard (Figure 14.2).

268 ■ How to Count: An Introduction to Combinatorics, Second Edition

A coloring of this chessboard, using black and white, may be regarded as a mapping from the set of squares that make up the chessboard to the set of colors. Thus if we let $\{\alpha,\beta,\gamma,\delta\}$ be the set of squares, then each coloring is a mapping

$$f : \{\alpha,\beta,\gamma,\delta\} \to \{black, white\}.$$

Hence the 16 different colorings correspond to the 16 different mappings of this kind. For example, the coloring that we have called C6 shown in Figure 14.3 corresponds to the mapping f_6 shown below it.

Corresponding to each symmetry, g, of the square there is a permutation, π_g, of the set of smaller squares. For example, the quarter-turn rotation of the square, a, corresponds to the permutation

$$\pi_a = (\alpha \quad \beta \quad \gamma \quad \delta).$$

How does it act on the set of colorings? We can see how the action is defined if we consider what happens when the permutation π_a acts on the coloring f_6 (Figure 14.4).

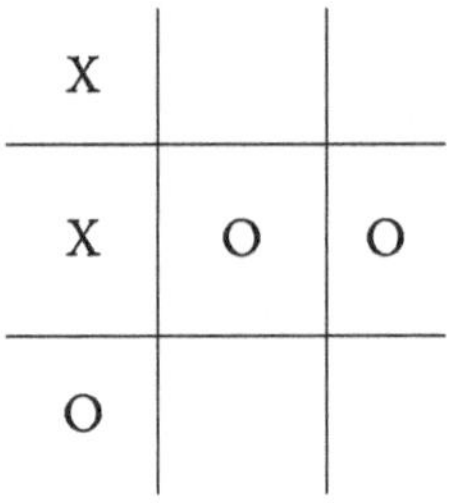

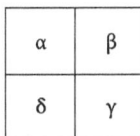

FIGURE 14.1

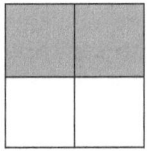

FIGURE 14.2

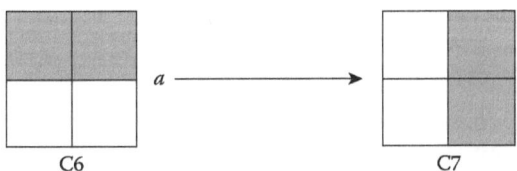

$f_6(\alpha) = black$, $f_6(\beta) = black$, $f_6(\gamma) = white$, $f_6(\delta) = white$

FIGURE 14.3

FIGURE 14.4

We see, for example, that the color assigned to a square, say in position χ, after the rotation is the same as the color that f_6 assigns to the square that the permutation that π_a moves into the position χ, that is, the square that was previously in position $\pi_a^{-1}(\chi)$. Thus the action of a on the coloring f_6 is given by

$$a \triangleright f_6 = f_6 \circ \pi_a^{-1}.$$

The analogous formula applies in the general case, which we now describe. Let D, C be two sets. We let X be the set of all mappings from D to C. Here "D" stands for "domain" and "C" for "codomain," but you can also think of C as being a set of colors. We call the mappings in X *colorings* of D.

Now suppose that G is a group of permutations of D. Thus G will be a subgroup of the group, $S(D)$, of all permutations of D. We define an action of G on the set of colorings X by

$$\text{for } \pi \in G, \text{ and } f \in X \text{ (that is, } f: D \to C\text{)}, \qquad \pi \triangleright f = f \circ \pi \qquad (14.1)$$

We need to check that this does define a group action. To do this, we first require a simple result about groups.

LEMMA 14.1

Let G be a group. Then for all $g_1, g_2 \in G$, $(g_1 g_2)^{-1} = g_2^{-1} g_1^{-1}$.

Proof

We have that $(g_1 g_2) g_2^{-1} g_1^{-1} = g_1(g_2 g_2^{-1}) g_1^{-1} = g_1 e g_1^{-1} = g_1 g_1^{-1} = e$. Similarly, it can be checked that $g_2^{-1} g_1^{-1}(g_1 g_2) = e$, and it follows that $g_2^{-1} g_1^{-1}$ is the inverse of $g_1 g_2$.

THEOREM 14.2

Equation 14.1 defines a group action of G on X.

Proof

The identity element of G is the identity map $\iota: D \to D$. So, for each $f \in X$, we have that $\iota \triangleright f = f \circ \iota^{-1} = f \circ \iota = f$. It follows that condition GA1 is satisfied.

Now, suppose $\pi_1, \pi_2 \in G$ and $f \in X$, then

$$\pi_1 \triangleright (\pi_2 \triangleright f)$$

$$= \pi_1 \triangleright (f \circ \pi_2^{-1}) = (f \circ \pi_2^{-1}) \circ \pi_1^{-1} = f \circ (\pi_2^{-1} \circ \pi_1^{-1}) = f \circ ((\pi_1 \circ \pi_2)^{-1}) = (\pi_1 \circ \pi_2) \triangleright f. \quad (14.2)$$

This shows that GA2 also holds. (Note that in Equation 14.2 we have used the associativity property for groups, Lemma 14.1, and the group action properties, as well as the definition in Equation 14.1 of the group action. In Exercise 14.1.2A you are asked to work out exactly which of these is used at each step.)

This completes the proof.

Exercises

14.1.1A Suppose $D = \{1,2,3,4,5\}$, $C = \{0,1,2\}$. Let $f: D \to C$ be the mapping defined by $f(n) =$ the remainder when n is divided by 3, and let $\pi \in S_5$ be the permutation given by $\pi = (1\ 4)(2\ 5\ 3)$. Find the mapping $f \circ \pi^{-1}$.

14.1.1B With D, C, and f as in Exercise 14.1.1A, find a permutation $\pi \in S_5$ such that $f \circ \pi^{-1}$ is the mapping defined by $f \circ \pi^{-1}(1) = 0$, $f \circ \pi^{-1}(2) = 1$, $f \circ \pi^{-1}(3) = 1$, $f \circ \pi^{-1}(4) = 2$, and $f \circ \pi^{-1}(5) = 2$.

14.1.2A In Equation 14.2 there are five steps. Write down the justification for each step.

14.1.2B Write down the justification for each step in the proof of Lemma 14.1.

14.2 PATTERN INVENTORIES

In this section we explain an algebraic method for describing colorings. This leads us to an algebraic expression telling us how many patterns of each kind there are.

Coloring a 2×2 chessboard black and white involves choosing, for each of the four squares, whether it is to be black or white. If we expand the algebraic expression

$$(b + w)^4 = (b + w)(b + w)(b + w)(b + w),$$

then we obtain each term by choosing from each bracket either b or w. Thus the choices are the same as when we are coloring the chessboard. So the terms obtained when we expand the above expression correspond to the different colorings of the chessboard. Since

$$(b + w)^4 = b^4 + 4b^3w + 6b^2w^2 + 4bw^3 + w^4, \tag{14.3}$$

we can deduce from the coefficients in the expression 14.3 that, for example, there are four of these colorings with one black and three white squares, and six colorings with two black and two white squares.

Thus we see that the algebraic expression in the expression 14.3 tells us how many colorings there are with a given number of black squares and a given number of white squares. However, the expression 14.3 does not tell us how many *different patterns* there are among these colorings. We are aiming toward a theorem that does provide a similar expression but where the coefficients count the different patterns rather than the colorings.

We first need to be more precise about what sort of expressions we are looking for. As usual, the abstract definition will be clearer if you keep in mind a particular example such as the one above. In that example the symbols b and w corresponded to the colors black and white. We will need to generalize this, and introduce some terminology for the resulting algebraic expressions.

DEFINITION 14.1

We suppose that X is the set of all mappings from a set D to a set C.

a. A *weight function* on C is a mapping w that assigns to each $c \in C$ an algebraic symbol $w(c)$.
b. The sum $\sum_{c \in C} w(c)$ is called the *store enumerator*.
c. For each coloring $f \in X$, the *weight* of f, written $W(f)$, is the algebraic expression

$$\prod_{d \in D} w(f(d));$$

that is, $W(f)$ is the product of algebraic symbols corresponding to the colors that f assigns to the elements of D.
d. If Y is a subset of X, the *inventory* of Y is the expression

$$\sum_{f \in Y} W(f).$$

e. If Y is a subset of X that includes exactly one coloring of each pattern, then the inventory of Y is called the *pattern inventory*.

We now illustrate these definitions in the context of our standard example of the colorings of a 2×2 chessboard using the colors black and white. Here $D = \{\alpha, \beta, \gamma, \delta\}$ and $C = \{black, white\}$.

a. The weight function w is defined by

$$w(black) = b \text{ and } w(white) = w,$$

and hence
b. The store enumerator is $w(black) + w(white) = b + w$. (Note that we are using "w" both for the weight function and for the weight assigned to the color white. The different contexts should, we hope, prevent any confusion.)
c. Let f_6 be the coloring given by

$$f_6(\alpha) = black, f_6(\beta) = black, f_6(\gamma) = white, f_6(\delta) = white.$$

Then

$$W(f_6) = \prod_{d \in D} w(f_6(d)) = w(f_6(\alpha))w(f_6(\beta))w(f_6(\gamma))w(f_6(\delta))$$

$$= w(black)w(black)w(white)w(white) = bbww = b^2w^2.$$

d. The inventory of the set X of all the mappings from D to C is given by Equation 14.3, and

e. The pattern inventory is

$$b^4 + b^3w + 2b^2w^2 + bw^3 + w^4,$$

as there are six different colorings, one with four black squares, one with three black and one white square, two with two black and two white squares, one with one black and three white squares, and one with four white squares.

One of the uses of inventories is illustrated by the following problems.

PROBLEM 14.1

An equilateral triangle is divided into three smaller congruent triangles labeled α, β, and γ in Figure 14.5. How many colorings are there using light red, dark red, and yellow paint (i) if light red and dark red are not both used, and (ii) if all three triangles are red?

Solution

Here D is the set of the three triangles labeled α, β, and γ, and C is the set of paints. Let w be the weight function that assigns the weights r_1, r_2, and y to the light red, dark red, and yellow paints, respectively. Thus the store enumerator is $r_1 + r_2 + y$, and the inventory of the set of all the colorings is

$$(r_1 + r_2 + y)^3 = r_1^3 + r_2^3 + y^3 + 3r_1r_2^2 + 3r_1^2r_2 + 3r_1y^2 + 3r_1^2y + 3r_2y^2 + 3r_2^2y + 6r_1r_2y.$$

For example, the coefficient 3 in the term $3r_1y^2$ tells us that there are three colorings with one light red and two yellow triangles. We can use the inventory to answer the above questions as follows.

i. The terms in the inventory that don't contain both r_1 and r_2 are

$$r_1^3 + r_2^3 + y^3 + 3r_1y^2 + 3r_1^2y + 3r_2y^2 + 3r_2^2y,$$

whose coefficients add up to 15. So there are 15 colorings in which light red and dark red are not both used.

ii. The terms in the inventory that do not include y are $r_1^3 + r_2^3 + 3r_1r_2^2 + 3r_1^2r_2$, whose coefficients add up to 8. So there are 8 colorings in which all three triangles

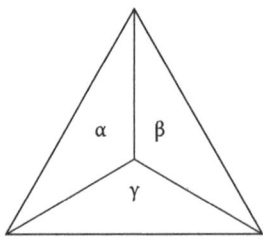

FIGURE 14.5

are red. An alternative approach to (ii) is to give both red paints the weight r. Then the store enumerator is $r + r + y$, that is, $2r + y$. Hence the inventory is

$$(2r + y)^3 = (2r)^3 + 3(2r)^2y + 3(2r)y^2 + y^3 = 8r^3 + 12r^2y + 6ry^2 + y^3,$$

from which we deduce that there are 8 colorings with three red triangles, 12 with two red and one yellow triangle, 6 with one red and two yellow triangles, and 1 with three yellow triangles.

PROBLEM 14.2

Each of 10 mathematics students has to do a project. The choice of project for each student is from six projects in pure mathematics, four projects in applied mathematics, two projects in statistics, and two projects in computing.

Two or more students may choose the same project, but because of the staff available to supervise the projects, they must be chosen so that four students do a pure mathematics project, three students do an applied mathematics project, two students do a statistics project, and one student does a computer project.

In how many different ways can the students choose their projects?

Solution

Here D is the set of 10 students, and C is the set of 14 projects. We assign the weights p, a, s, and c to the projects in pure mathematics, applied mathematics, statistics, and computing, respectively. Thus the store enumerator is $(6p + 4a + 2s + 2c)$, and the inventory of all the possible choices by the students, ignoring the constraints, is $(6p + 4a + 2s + 2c)^{10}$. The constraint on the numbers of projects implies that the relevant term in the expansion is the one involving $p^4 a^3 s^2 c$. By the multinomial theorem (Theorem 2.11), this term is

$$\frac{10!}{4!3!2!1!}(6p)^4(4a)^3(2s)^2(2c)^1 = 8{,}360{,}755{,}200 p^4 a^3 s^2 c,$$

and hence there are 8,360,755,200 ways in which the projects may be chosen.

Exercises

14.2.1A Write down the store enumerator and the inventory of the set of all mappings from D to C in the following case.
$D = \{\alpha, \beta, \gamma, \delta, \varepsilon\}$, $C = \{light\ blue,\ dark\ blue,\ navy\ blue,\ light\ red,\ crimson\}$, and the weight function $w : C \to \{b, r\}$ is defined by $w(light\ blue) = b$, $w(dark\ blue) = b$, $w(navy\ blue) = b$, $w(light\ red) = r$, and $w(crimson) = r$.

14.2.1B Six English-literature students have to choose a novel about which to write a dissertation. The novel must be by either Jane Austin, Charles Dickens, or Anthony Trollope. Two students may choose the same novel, but altogether they must be chosen so there are two by each author. Jane Austin published 6 novels, Charles Dickens 14, and Anthony Trollope 47. In how many ways can the students make their choices?

14.3 THE CYCLE INDEX OF A GROUP

We now begin our description of an algebraic method to calculate the pattern inventory in cases where a group acts on the set of colorings of a figure. Our starting point is the *cycle type* of a permutation, which we introduced in Chapter 11, Section 11.5. Recall that a permutation is said to have cycle type

$$x_{k_1}^{r_1}\ldots x_{k_s}^{r_s}$$

if, in its disjoint cycle form, it has r_t cycles of length k_t, for $1 \le t \le s$, including cycles of length 1. We use $ct(\pi)$ for the cycle type of a permutation π. Now comes a key definition.

DEFINITION 14.2

Let G be a group of permutations. The *cycle index* of G, written $CI(G)$, is defined to be the polynomial

$$\frac{1}{\#(G)}\sum_{\pi \in G} ct(\pi).$$

Notice that the cycle index of a group of permutations is a polynomial in the unknowns $x_1, x_2, x_3, \ldots$, and that the sum of the coefficients in this polynomial is 1, because there are $\#(G)$ terms in the sum, $\sum_{\pi \in G} ct(\pi)$, and we obtain the cycle index by dividing this sum by $\#(G)$.

We illustrate the above definition by calculating the cycle index of the groups S_n, for $n = 2, 3, 4$. (The case $n = 1$ is too simple to be of any interest.)

First consider the group S_2 of all permutations of the set $\{1,2\}$. There are two permutations in S_2: the identity, (1)(2), which has cycle type x_1^2; and the permutation (1 2), which has cycle type x_2. Hence the cycle index, $CI(S_2)$, of the group S_2, is the polynomial

$$\frac{1}{2}(x_1^2 + x_2).$$

The six permutations in the group S_3 and their cycle types are shown in Table 14.1. It follows that the cycle index, $CI(S_3)$, of the group S_3, is the polynomial

$$\frac{1}{6}(x_1^3 + 3x_1 x_2 + 2x_3).$$

TABLE 14.1

Permutation	Cycle Type
(1)(2)(3)	x_1^3
(1 2)	
(1 3)	$x_1 x_2$
(2 3)	
(1 2 3)	
(1 3 2)	x_3

TABLE 14.2

Partition	Cycle Type	Number of Permutations
4	x_4	6
3 + 1	$x_1 x_3$	8
2 + 2	x_2^2	3
2 + 1 + 1	$x_1^2 x_2$	6
1 + 1 + 1 + 1	x_1^4	1

As n becomes larger it becomes more and more cumbersome to calculate the cycle index of the group of permutations S_n by listing all the $n!$ permutations in the group. We have already seen in Chapter 11 that the different cycle types of the permutations in S_n correspond to the different partitions of n. So when it comes to the group S_4 we calculate its cycle index by listing the partitions of 4, the corresponding cycle type, and the number of permutations of this type (Table 14.2).

It follows that the cycle index, $CI(S_4)$, of the group S_4, is the polynomial

$$\frac{1}{24}(6x_4 + 8x_1 x_3 + 3x_2^2 + 6x_1^2 x_2 + x_1^4).$$

In order to use this method we need to be able to calculate the number of permutations of a given cycle type. We first explain how to do this in some specific cases, and then we derive the general formula. We consider first the number of permutations in S_4 of cycle type x_4. A permutation of this cycle type has the form $(a\ b\ c\ d)$, where a, b, c, and d are chosen from the set $\{1,2,3,4\}$. The numbers 1, 2, 3, and 4 can be arranged in order in 4! ways, but we need to remember that a permutation can be written in cycle form in more than one way. For example,

$$(1\ 2\ 3\ 4) = (2\ 3\ 4\ 1) = (3\ 4\ 1\ 2) = (4\ 1\ 2\ 3).$$

Each permutation of cycle type x_4 may be written in cycle form in four ways, and so there are altogether $4!/4 = 3! = 6$ permutations in S_4 of cycle type x_4. More generally, there are $k!/k = (k-1)!$ different permutations of k given objects that have cycle type x_k.

Next we consider the number of permutations in S_4 of cycle type x_2^2. Such a permutation has the form $(a\ b)(c\ d)$. The pair of numbers a, b may be chosen from the set $\{1, 2, 3, 4\}$ in $C(4,2) = 6$ ways. After a and b have been chosen, there is just one way they form a permutation, $(a\ b)$, of cycle length 2. Also, having chosen a and b, there is no further choice for the second cycle $(c\ d)$. However, before jumping to the conclusion that there are six permutations of cycle type x_2^2, we need to remember that the order in which we write the two cycles of length 2 does not matter, so that, for example, $(1\ 2)(3\ 4) = (3\ 4)(1\ 2)$. Since the two cycles of length 2 can be arranged in order in 2! ways, giving the same permutation in each case, there are only $6/2! = 3$ different permutations in S_4 of cycle type x_2^2.

We are now going to work out a general formula for the number of different permutations of a given cycle type. Before reading the following discussion, you might find it helpful to pause and work out for yourself the cycle index of the group S_5. You can check your answer in the solution to Exercise 14.3.3A, which is given at the end of the book.

We now calculate the number of permutations in S_n of cycle type

$$x_{k_1}^{r_1}...x_{k_t}^{r_t}, \qquad (14.4)$$

where

$$k_1 r_1 + ... + k_t r_t = n. \qquad (14.5)$$

We tackle this problem by counting the number of ways we can arrange the numbers from the set $\{1,2,...,n\}$ to form different permutations of the cycle type given by the expression 14.4. Suppose we first choose the numbers to make up the r_1 cycles of length k_1. The k_1 numbers that make up the first of these cycles can be chosen in $C(n,k_1)$ ways, and these k_1 numbers may be arranged to form $(k_1 - 1)!$ different cycles of length k_1. This leaves $n - k_1$ numbers from which to choose another k_1 numbers, and so these can be chosen in $C(n - k_1, k_1)$ ways and arranged to form $(k_1 - 1)!$ different cycles of length k_1. We continue in this way until we have chosen r_1 cycles of length k_1. These r_1 cycles can be arranged in order in $r_1!$ ways, each arrangement corresponding to the same permutation. Hence the number of ways we can choose the r_1 cycles of length k_1 so as to yield different permutations is:

$$\frac{1}{r_1!}\left((k_1-1)!C(n,k_1)\times(k_1-1)!C(n-k_1,k_1)\times...\times(k_1-1)!C(n-(r_1-1)k_1,k_1)\right)$$

$$= \frac{1}{r_1!}\left(\frac{(k_1-1)!n!}{k_1!(n-k_1)!}\times\frac{(k_1-1)!(n-k_1)!}{k_1!(n-2k_1)!}\times...\times\frac{(k_1-1)!(n-(r_1-1)k_1)!}{k_1!(n-r_1k_1)!}\right)$$

$$= \frac{n!}{r_1!k_1^{r_1}(n-r_1k_1)!}, \qquad (14.6)$$

after a lot of tidying up.

Having chosen r_1 cycles of length k_1, we are left with $n - r_1 k_1$ numbers from which to choose r_2 cycles of length k_2. The same reasoning that we used to obtain the formula in Equation 14.6 shows that we can do this in

$$\frac{(n-r_1 k_1)!}{r_2! k_2^{r_2}(n-r_1 k_1-r_2 k_2)!}$$

ways and so on.

It follows that the total number of permutations in S_n of the cycle type given by the expression 14.4 is

$$\frac{n!}{r_1!k_1^{r_1}(n-r_1 k_1)!}\times\frac{(n-r_1 k_1)!}{r_2!k_2^{r_2}(n-r_1 k_1-r_2 k_2)!}\times...\times\frac{(n-r_1 k_1-r_2 k_2-...-r_{t-1}k_{t-1})!}{r_t!k_t^{r_t}0!}.$$

All the terms in the numerator in this expression, other than $n!$, cancel with terms in the denominator, and so we are left with the much simpler expression

$$\frac{n!}{r_1!k_1^{r_1}r_2!k_2^{r_2}...r_t!k_t^{r_t}} \qquad (14.7)$$

as the number of permutations of the cycle type given by the expression 14.4 in S_n. Finally, the cycle index of S_n is obtained by multiplying the terms in Equation 14.7 by the corresponding cycle types, adding up all these products, and then dividing the total by $n!$.

PROBLEM 14.3
Calculate the number of permutations of the following cycle types in the group S_{24}.

i. $x_4^3 x_6^2$
ii. $x_1^3 x_2^4 x_4^2 x_5$

Solution
i. Using the notation of Equation 14.7, $t = 2$, $r_1 = 3$, $k_1 = 4$, $r_2 = 2$, and $k_2 = 6$, and so the number of different permutations is $24!/(3!4^3 2!6^2) = 24!/27{,}648 = 22{,}440{,}986{,}752{,}504{,}320{,}000$.
ii. Here $t = 4$, $r_1 = 3$, $k_1 = 1$, $r_2 = 4$, $k_2 = 2$, $r_3 = 2$, $k_3 = 4$, $r_4 = 1$, and $k_4 = 5$, and so the number of different permutations is $24!/(3!1^3 4!2^4 2!4^2 1!5^1) = 24!/368{,}640 = 1{,}683{,}074{,}006{,}437{,}824{,}000$.

Exercises

14.3.1A Find the cycle index of the subgroup of S_4 consisting of the permutations of the vertices of a square corresponding to the group of eight symmetries of the square.

14.3.1B Find the cycle index of the subgroup of S_6 consisting of the permutations of the vertices of a regular hexagon corresponding to the group of symmetries of the hexagon.

14.3.2A Find the cycle index of the subgroup of S_6 consisting of the permutations of the faces of a cube corresponding to the group of 24 rotational symmetries of the cube.

14.3.2B Find the cycle index of the subgroup of S_{12} consisting of the permutations of the edges of a cube corresponding to the group of 24 rotational symmetries of the cube.

14.3.3A Find the cycle indexes of the groups S_5 and S_6.

14.3.3B Find the cycle index of the group S_7.

14.3.4A How many different permutations are there in S_{12} with the following cycle types?
  i. $x_2^3 x_6$ ii. $x_1^4 x_3 x_5$ iii. $x_1 x_2^2 x_3 x_4$

14.3.4B How many different permutations are there in S_{16} with the following cycle types?
  i. x_4^4 ii. $x_2^3 x_5^2$ iii. $x_1^2 x_2^3 x_4^2$

14.4 PÓLYA'S COUNTING THEOREM: STATEMENT AND EXAMPLES

We are now ready to state the theorem that tells us how to calculate pattern inventories. We then illustrate the theorem with some examples. The proof of the theorem is postponed until the next section. As we shall see, applying the following rather complicated-looking theorem is not too difficult.

THEOREM 14.3
Pólya's Counting Theorem*

Let X be the set of all mappings from a set D to a set C, and let w be a weight function on C. Let G be a group of permutations of D whose action on X is given by:

$$\text{for } \pi \in G, \text{ and } f \in X, \pi \triangleright f = f \circ \pi^{-1}.$$

Then the pattern inventory is obtained from the cycle index of G, by substituting $\sum_{c \in C} w(c)^k$ for x_k, for $k = 1,2,3,\ldots$. That is, if $CI(G)(x_1,x_2,x_3,\ldots)$ is the cycle index of G, the pattern inventory is

$$CI(G)\left(\sum_{c \in C} w(c), \sum_{c \in C} w(c)^2, \sum_{c \in C} w(c)^3, \ldots\right). \tag{14.8}$$

The formula in the expression 14.8 means that we obtain the pattern inventory by taking the cycle index of G and replacing each occurrence of x_1 by the sum of the weights, each occurrence of x_2 by the sum of the squares of the weights, and so on. We now illustrate this with some examples.

We begin by returning again to our standard example of the colorings of a 2×2 chessboard. Of course, we already know how many patterns there are in this case, but it will be very helpful to start with a simple example where the calculation is very straightforward, and reassuring when we obtain the answer we expect!

PROBLEM 14.4

Use Pólya's counting theorem to calculate the pattern inventory for the number of different ways of coloring the squares of a 2×2 chessboard, using black and white.
Solution
Here $D = \{\alpha,\beta,\gamma,\delta\}$, $C = \{black, white\}$, and the weight function is given by $w(black) = b$ and $w(white) = w$. The group G is the group of permutations of D corresponding to the symmetries of a 2×2 chessboard with its squares labeled as shown in Figure 14.2. Thus

$$G = \{e, (\alpha\beta\gamma\delta), (\alpha\gamma)(\beta\delta), (\alpha\delta\gamma\beta), (\alpha\delta)(\beta\gamma), (\alpha\beta)(\gamma\delta), (\beta\delta), (\alpha\gamma)\},$$

and hence

$$CI(G) = \frac{1}{8}(x_1^4 + 2x_1^2 x_2 + 3x_2^2 + 2x_4). \tag{14.9}$$

---

* George Pólya, Kombinatorische Anzahlbestimmungen für Gruppen, Graphen und chemische Verbindungen, *Acta Mathematica*, 68, 1937, pp. 145–254. An English translation is given in G. Pólya and R. C. Read, *Combinatorial Enumeration of Groups, Graphs and Chemical Formulas*, Springer, Berlin, 1987. This book includes a survey by R. C. Read of work in this area. It turns out that Pólya's work was to some extent anticipated by a paper published by J. H. Redfield in 1927, which went unnoticed for many years. George Pólya was born in Budapest on December 13, 1887, and died in Palo Alto on September 7, 1985. For a biography of George Pólya see Gerald L. Alexanderson, *The Random Walks of George Pólya*, Mathematical Association of America, Washington, DC, 2000.

By Pólya's enumeration theorem, we now obtain the pattern inventory by substituting $b+w$ for x_1, b^2+w^2 for x_2, and b^4+w^4 for x_4 in Equation 14.9. Thus the pattern inventory is

$$\frac{1}{8}((b+w)^4 + 2(b+w)^2(b^2+w^2) + 3(b^2+w^2)^2 + 2(b^4+w^4))$$

$$= b^4 + b^3w + 2b^2w^2 + bw^3 + w^4.$$

From the terms in this pattern inventory we deduce that there is one pattern with four black squares, one pattern with three black and one white square, two different patterns with two black and two white squares, and so on.

If you do not have to hand a computer algebra package, you will need to calculate the pattern inventory by hand, using the binomial theorem and multiplying polynomials. A useful check is that the coefficients in the pattern inventory must be positive integers, as they count the numbers of different patterns. So if, after dividing by #(G), you are left with coefficients that are not positive integers, you know there must be slip somewhere. Unfortunately, the converse is not necessarily true!

The next problem is computationally more complicated.

PROBLEM 14.5

How many different patterns can be obtained by coloring the squares of a 3×3 chessboard with two red, three white, and four blue squares?

Solution

Here we take $D = \{1,2,3,\ldots,9\}$ and $C = \{red, white, blue\}$, with the elements of C assigned the weights r, w, and b, respectively. G is the group of permutations of D corresponding to the symmetries of a 3×3 board with its squares labeled as shown in Figure 14.6.

In Table 14.3, we have listed the symmetries of the square, using our standard notation introduced in Chapter 11, the corresponding permutations of D, and their cycle types.

We therefore see that the cycle index of G is

$$\frac{1}{8}(x_1^9 + x_1 x_2^4 + 4x_1^3 x_2^3 + 2x_1 x_4^2).$$

1	2	3
4	5	6
7	8	9

FIGURE 14.6

TABLE 14.3

Symmetry	Permutation	Cycle Type
e	e	x_1^9
a	(1 3 9 7)(2 6 8 4)(5)	$x_1 x_4^2$
b	(1 9)(2 8)(3 7)(4 6)(5)	$x_1 x_2^4$
c	(1 7 9 3)(2 4 8 6)(5)	$x_1 x_4^2$
h	(1 7)(2 8)(3 9)(4)(5)(6)	$x_1^3 x_2^3$
v	(1 3)(4 6)(7 9)(2)(5)(8)	$x_1^3 x_2^3$
r	(2 4)(3 7)(6 8)(1)(5)(9)	$x_1^3 x_2^3$
s	(1 9)(2 6)(4 8)(3)(5)(7)	$x_1^3 x_2^3$

Hence, by Pólya's counting theorem, the pattern inventory is

$$\frac{1}{8}((r+w+b)^9 + (r+w+b)(r^2+w^2+b^2)^4$$

$$+ 4(r+w+b)^3(r^2+w^2+b^2)^3 + 2(r+w+b)(r^4+w^4+b^4)^2). \quad (14.10)$$

In order to answer the question about the number of patterns with two red, three white, and four blue squares, we need to know the coefficient of the term $r^2 w^3 b^4$ in this pattern inventory. To calculate this coefficient by hand is feasible, but rather tedious, so we hope that you have available one of the many computer packages that do these calculations for us. Using one of these packages, we find that the relevant term in the above pattern inventory is $174 r^2 w^3 b^4$. So there are 174 different patterns with two red, three white, and four blue squares. If you have to do the calculation by hand, one or two shortcuts are sometimes available. For example, in Equation 14.10 we can see that the term $2(r + w + b)(r^4 + w^4 + b^4)^2$ cannot include terms involving $r^2 w^3 b^4$.

Exercises

14.4.1A How many different patterns can be formed by coloring the squares of a 5×5 chessboard black and white, so that there are 15 black squares and 10 white squares?

14.4.1B How many different patterns can be formed by coloring the squares of an 8×8 chessboard black and white, so that half the squares are black and half are white?

14.4.2A How many different patterns can be formed by coloring the triangles into which the square shown in Figure 14.7 is divided so that there are two red triangles, two white triangles, and four blue triangles?

14.4.2B How many different patterns can be formed by coloring the faces of a cube so that there are three red faces, two white faces, and one blue face? (Consider just the group of 24 rotational symmetries of the cube.)

14.4.3A How many different patterns can be formed by coloring the faces of a regular octahedron so that there are four red faces, two white faces, and two blue faces? (Consider just the group of rotational symmetries of the regular octahedron.)

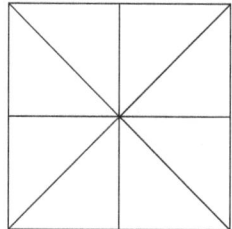

FIGURE 14.7

14.4.3B How many different positions can occur during a game of noughts and crosses? (Consider both the rotations and reflections of the noughts and crosses grid.)

14.5 PÓLYA'S COUNTING THEOREM: THE PROOF

We now set about giving a proof of Pólya's counting theorem. Throughout this section we suppose that D and C are two sets, that w is a weight function on C, and that G is a group of permutations of D that acts on the set X of all mappings from D to C in the standard way; that is,

$$\text{for } \pi \in G, \text{ and } f \in X, \quad \pi \triangleright f = f \circ \pi^{-1}.$$

The strategy of the proof is to partition X according to the weights of the mappings, to observe that G acts on each set of this partition separately, and to use Frobenius's counting theorem to count the number of different orbits in each case. Our proof proceeds by a series of lemmas.

In Chapter 13, when we calculated the values of $\#(Fix(g))$, we used the idea that a coloring of a chessboard is fixed by a particular permutation of its squares if and only if all the squares in any one cycle of the permutation have the same color. In general, a coloring f is fixed by a permutation π if and only if f takes a constant value on each cycle of π. That is, for each cycle $(d_1 \, d_2 \ldots d_k)$ of π, we must have $f(d_1) = f(d_2) = \ldots = f(d_k)$. Note also that $(d_1 \, d_2 \ldots d_k)$ is a cycle of π if and only if $d_2 = \pi(d_1)$, $d_3 = \pi(d_2)$, ..., $d_k = \pi(d_{k-1})$, and $d_1 = \pi(d_k)$. Thus each cycle of π has the form $(d \, \pi(d) \, \pi(\pi(d)) \ldots)$. It follows that f is constant on each cycle of π if and only if

$$\text{for all } d \in D, \quad f(d) = f(\pi(d)), \tag{14.11}$$

since Equation 14.11 is equivalent to $f(d) = f(\pi(d)) = f(\pi)(\pi(d))) = \ldots$.

We now give a general proof of this result.

LEMMA 14.4

Let π be a permutation of D. Then for each $f \in X$,

$$f \in Fix(\pi) \iff f \text{ is constant on each cycle of } \pi.$$

Proof

We have that

$$f \in \mathit{Fix}(\pi) \Leftrightarrow \pi \triangleright f = f$$

$$\Leftrightarrow f \circ \pi^{-1} = f$$

$$\Leftrightarrow \text{for all } x \in D, f(\pi^{-1}(x)) = f(x). \tag{14.12}$$

Now, as π is a permutation of D, $\{\pi^{-1}(x) : x \in D\} = D$, and hence, writing d for $\pi^{-1}(x)$, so that $x = \pi(d)$, we have

for all $x \in D, f(\pi^{-1}(x)) = f(x) \Leftrightarrow$ for all $d \in D, f(d) = f(\pi(d))$,

and so by Equation 14.12,

$$f \in \mathit{Fix}(\pi) \Leftrightarrow \text{for all } d \in D, f(d) = f(\pi(d))$$

$$\Leftrightarrow f \text{ is constant on each cycle of } \pi.$$

The next lemma leads us to a formula for the inventory of $\mathit{Fix}(\pi)$. By Lemma 14.4 this is the inventory of all functions that are constant on each cycle of π. The cycles of π partition D into disjoint sets. So we first consider the inventory of all functions that are constant on each set making up a partition of D.

LEMMA 14.5

Let $D_1 \cup D_2 \cup \ldots \cup D_k$ be a partition of D into disjoint sets. The inventory of those functions in X that are constant on each of the sets D_i is

$$\sum_{c \in C} w(c)^{\#(D_1)} \times \sum_{c \in C} w(c)^{\#(D_2)} \times \ldots \times \sum_{c \in C} w(c)^{\#(D_k)}. \tag{14.13}$$

Proof

Choosing a function, say f, that is constant on each set D_i is equivalent to choosing, for $1 \le i \le k$, an element, say c_i, from C to be the value of $f(d)$ for all $d \in D_i$. Since f then takes this value $\#(D_i)$ times, such a choice of c_i contributes $w(c_i)^{\#(D_i)}$ to the weight of f. So the weight of f is obtained by choosing, for $1 \le i \le k$, the appropriate term from the sum $\sum_{c \in C} w(c)^{\#(D_i)}$ and then multiplying all these terms together. So the inventory of all the functions in X that are constant on each set D_i is the sum of all the terms obtained in this way. Hence this inventory is given by the product in the expression 14.13.

The formula 14.13 is beginning to look like the formula in Pólya's theorem. The next lemma brings in the cycle types of the permutations. Recall that we use $ct(\pi)$ for the cycle type of a permutation π.

LEMMA 14.6

If $ct(\pi) = x_{k_1}^{r_1} x_{k_2}^{r_2} \ldots x_{k_t}^{r_t}$, then the inventory of $Fix(\pi)$ is

$$\left(\sum_{c \in C} w(c)^{k_1}\right)^{r_1} \times \left(\sum_{c \in C} w(c)^{k_2}\right)^{r_2} \times \ldots \times \left(\sum_{c \in C} w(c)^{k_t}\right)^{r_t} \quad (14.14)$$

or, equivalently,

$$ct(\pi)\left(\sum_{c \in C} w(c), \sum_{c \in C} w(c)^2, \sum_{c \in C} w(c)^3, \ldots\right). \quad (14.15)$$

Proof

The cycles of π partition D into r_1 sets of size k_1, r_2 sets of size k_2, and so on. By Lemma 14.4, $Fix(\pi)$ is the set of functions that are constant on all these sets, and hence, by Lemma 14.5, the inventory of $Fix(\pi)$ is as given by the formula 14.14. This formula is obtained from the cycle type of π by replacing each occurrence of x_k for $k = 1, 2, 3 \ldots$ by $\sum_{c \in C} w(c)^k$. Hence we can rewrite this formula as given in the formula 14.15. This completes the proof.

We now partition X into sets of functions with the same weight. Clearly, the relation $\sim$ defined on X by

$$f \sim g \Leftrightarrow W(f) = W(g)$$

is an equivalence relation. We let $X_1, X_2, \ldots, X_s$ be the equivalence classes of this relation. So they form a partition of X. (For example, in our example of the colorings of a 2×2 chessboard, the set X of all 16 colorings is partitioned according to the numbers of black and white squares in a particular coloring. So one of the equivalence classes is the set of all colorings with two black squares and two white squares, namely, $\{C6, C7, C8, C9, C10, C11\}$. Note that within this equivalence class there are two different patterns. You will also note that all the colorings of any one particular pattern lie in the same equivalence class. Our next result generalizes this observation.)

The next lemma implies that G acts on each set of this partition separately.

LEMMA 14.7

Suppose that $f, g \in X$ have the same pattern; that is, they occur in the same orbit of the group action. Then $W(f) = W(g)$.

Proof

Suppose f and g are in the same orbit of the group action. Then for some $\pi \in G$, $\pi \triangleright f = g$, and thus $f \circ \pi^{-1} = g$, and hence

$$W(g) = \prod_{d \in D} w(g(d)) = \prod_{d \in D} w(f(\pi^{-1}(d))) \quad (14.16)$$

Since π is a permutation of D, the set of values taken by $\pi^{-1}(d)$ as d runs through D is just D itself. Hence the product in the right-hand term of Equation 14.16 is the same as $\prod_{d \in D} w(f(d)) = W(f)$. Thus it follows that $W(g) = W(f)$.

We have now seen that for each set X_i of the partition of X determined by the weight of the functions, if $f \in X_i$, then $\pi \triangleright f \in X_i$. In this sense G acts separately on each set X_i. We use the notation π_i to indicate that we are restricting our attention to the action of the permutation π on the set X_i. We can then apply Frobenius's counting theorem to count the number of orbits of the action of G on each set X_i. This enables us to prove Pólya's counting theorem, which, for convenience, we restate.

THEOREM 14.3
Pólya's Counting Theorem

Let X be the set of all mappings from a set D to a set C, and let w be a weight function on C. Let G be a group of permutations of D whose action on X is given by:

$$\text{for } \pi \in G, \text{ and } f \in X, \quad \pi \triangleright f = f \circ \pi^{-1}.$$

Then the pattern inventory is obtained from the cycle index of G, by substituting $\sum_{c \in C} w(c)^k$ for x_k, for $k = 1,2,3,\ldots$. Thus if $CI(G)(x_1, x_2, x_3, \ldots)$ is the cycle index of G, the pattern inventory is

$$CI(G)\left(\sum_{c \in C} w(c), \sum_{c \in C} w(c)^2, \sum_{c \in C} w(c)^3, \ldots\right). \tag{14.8}$$

Proof

We let PI be the pattern inventory. Suppose that W_i is the common weight of all the functions in the set X_i of the partition, and that there are m_i different patterns represented by the functions in X_i. We have seen, from Lemma 14.7, that no pattern is represented by functions in more than one of the sets of the partition. It follows that

$$PI = \sum_{i=1}^{s} m_i W_i.$$

By Frobenius's counting theorem,

$$m_i = \frac{1}{\#(G)} \sum_{\pi \in G} \#(Fix(\pi_i)),$$

and hence

$$PI = \frac{1}{\#(G)} \sum_{i=1}^{s} \left(\sum_{\pi \in G} \#(Fix(\pi_i))\right) W_i. \tag{14.17}$$

Both the sums in Equation 14.17 are finite, and hence we can interchange the order of the summation to give

$$PI = \frac{1}{\#(G)} \sum_{\pi \in G} \left(\sum_{i=1}^{s} \#(Fix(\pi_i)) W_i \right). \qquad (14.18)$$

In Equation 14.18 the term inside the brackets is the inventory of $Fix(\pi)$, and hence, by Lemma 14.6, it follows from Equation 14.18 that

$$PI = \frac{1}{\#(G)} \sum_{\pi \in G} ct(\pi) \left(\sum_{c \in C} w(c), \sum_{c \in C} w(c)^2, \sum_{c \in C} w(c)^3, \ldots \right). \qquad (14.19)$$

Now adding up the expressions $ct(\pi)$ and then substituting $\sum_{c \in C} w(c)^k$ for x_k produces the same result as first making the substitutions and then adding up the resulting terms, as given by Equation 14.19. Thus Equation 14.19 may be rewritten as

$$PI = \left(\frac{1}{\#(G)} \sum_{\pi \in G} ct(\pi) \right) \left(\sum_{c \in C} w(c), \sum_{c \in C} w(c)^2, \sum_{c \in C} w(c)^3, \ldots \right). \qquad (14.20)$$

Now $(1/\#(G)) \sum_{\pi \in G} ct(\pi)$ is the cycle index, $CI(G)$, of the group G, and hence we can rewrite Equation 14.20 as

$$PI = CI(G) \left(\sum_{c \in C} w(c), \sum_{c \in C} w(c)^2, \sum_{c \in C} w(c)^3, \ldots \right), \qquad (14.21)$$

and this completes the proof of the theorem.

14.6 COUNTING SIMPLE GRAPHS

In this section we show how Pólya's counting theorem may be used to count the number of simple graphs with a given number of vertices and edges. The computation, though in principle straightforward, can be rather long. Since our attention will be restricted to simple graphs, we will use "graph" to mean "simple graph" throughout this discussion. (If multiple edges were allowed, then even with just two vertices, there are infinitely many different graphs, as the two vertices could be joined by any number of edges.)

To bring Pólya's counting theorem to bear we need to be able to view graphs in terms of colorings, that is, in terms of mappings from a set D to a set C. So we need to think about graphs slightly differently from when we viewed them as made up as a set of vertices together with a set of edges.

Consider, for example, the graph with five vertices in Figure 14.8.

It is convenient to suppose that the vertices are the first five positive integers. So $V = \{1,2,3,4,5\}$. The graph is determined by knowing for each two-element subset, say $\{i,j\}$, of V whether the vertices i and j are joined by an edge or not. So the graph is defined by a

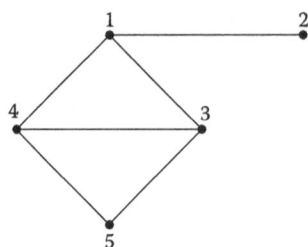

FIGURE 14.8

TABLE 14.4

$\{i,j\}$	$\{1,2\}$	$\{1,3\}$	$\{1,4\}$	$\{1,5\}$	$\{2,3\}$	$\{2,4\}$	$\{2,5\}$	$\{3,4\}$	$\{3,5\}$	$\{4,5\}$
$f\{i,j\}$	1	1	1	0	0	0	0	1	1	1

function whose domain is the set of all two-element subsets of $\{1,2,3,4,5\}$. We let D_5 be this set. Each graph with five vertices can then be described by a mapping

$$f: D_5 \to \{0,1\},$$

such that

$$f(\{i,j\}) = 1 \Leftrightarrow i \text{ and } j \text{ are connected by an edge.} \qquad (14.22)$$

For example, the graph shown in Figure 14.8 corresponds to the mapping f defined in Table 14.4.

In general, for each positive integer n, we let D_n be the set of all two-element subsets of $\{1,2,\ldots,n\}$. We will call D_n the set of *possible edges*. A graph with n vertices then corresponds to a mapping

$$f: D_n \to C, \quad \text{where } C = \{0,1\},$$

which specifies, through the condition in the equivalence 14.22, which pairs of vertices are joined by an edge. [In other words, we can think of the graph of Figure 14.8 as being derived from the complete graph K_5 by coloring some of the edges black (1) and some white (0). So simple graphs with five vertices correspond to edge colorings of K_5, using two colors.]

We will use the weight function w defined by

$$w(0) = 1 \text{ and } w(1) = c$$

so that a graph with k edges corresponds to a function with weight c^k.

Next we need to consider which is the relevant group of permutations of the set D_n of possible edges. Consider the straightforward example of a pair of isomorphic graphs shown in Figure 14.9.

These graphs correspond to the mappings $f_1, f_2: D_3 \to \{0,1\}$ as given in Table 14.5.

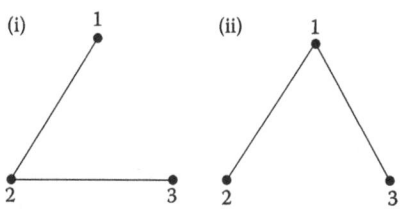

FIGURE 14.9

TABLE 14.5

$\{i,j\}$	$\{1,2\}$	$\{1,3\}$	$\{2,3\}$	$\{i,j\}$	$\{1,2\}$	$\{1,3\}$	$\{2,3\}$
$f_1(\{i,j\})$	1	0	1	$f_2(\{i,j\})$	1	1	0

The mapping $\pi: \{1,2,3\} \to \{1,2,3\}$ given by $\pi(1)=3, \pi(2)=1,$ and $\pi(3)=2$ is an isomorphism between these two graphs. Now π is a permutation of the set $\{1,2,3\}$, and corresponding to it there is a permutation, π^*, of the set, D_3, of possible edges given by $\pi^*(\{i,j\}) = \{\pi(i), \pi(j)\}$. For example, $\pi^*(\{1,2\}) = \{\pi(1), \pi(2)\} = \{1,3\}$. Because π is an isomorphism between the two graphs, for each possible edge $\{i,j\}$ we have that $\{i,j\}$ is an edge of the first graph if and only if $\pi^*(\{i,j\})$ is an edge of the second graph. That is, for each possible edge $\{i,j\}$,

$$f_2(\pi^*(\{i,j\})) = f_1(\{i,j\}).$$

It follows that $f_2 \circ \pi^* = f_1$, and hence $f_2 = f_1 \circ \pi^{*-1}$. So the action of the permutation π^* on the set of mappings $f: D_3 \to \{0,1\}$ is defined in exactly the same way as the group actions involved when we were considering the colorings of chessboards and other figures.

We are now ready to deal with the general case. For each permutation π of the set $\{1,2,\ldots,n\}$, that is, $\pi \in S_n$, there corresponds a permutation, π^*, of D_n defined by

$$\pi^*(\{i,j\}) = \{\pi(i), \pi(j)\}, \qquad \text{for } \{i,j\} \in D_n. \tag{14.23}$$

The collection of all these permutations, π^*, forms a subgroup of the group of all permutations of the set D_n. We let S_n^* be this subgroup. It acts on the set X of all the mappings $f: D_n \to \{0,1\}$ in the standard way. That is, for $\pi^* \in S_n^*$ and $f \in X$,

$$\pi^* \triangleright f = f \circ \pi^{*-1}.$$

We can now calculate the pattern inventory of the graphs with n vertices by working out the cycle index of the group S_n^* and then using Pólya's enumeration theorem. We need to be able to calculate for each permutation $\pi \in S_n$ the cycle type of the corresponding permutation $\pi^* \in S_n^*$. We do not need to do this for each permutation in S_n separately, because if two permutations $\pi_1, \pi_2 \in S_3$ have the same cycle type, then, clearly, so too do the corresponding permutations π_1^* and π_2^*. (Note the converse is *not*, in general, true. It is possible to have permutations π_1, π_2 that have different cycle types but where π_1^* and π_2^* have the same cycle type. See Exercises 14.6.4A and 14.6.4B.) So we only need to work out the cycle type of π^* in terms of the cycle type of π, a much smaller task!

The work in calculating the cycle type of π^* is simplified by the following observation, which we can illustrate by looking at a particular example. Consider the permutation $\pi = (1\ 2\ 3\ 4)(5\ 6\ 7)(8\ 9)$ from S_9. We need to work out the effect of the corresponding permutation π^* on each possible edge $\{i, j\}$. Our observation is that there are only two essentially different cases to consider. The first involves those possible edges, such as $\{1,4\}$ and $\{5,7\}$, whose two vertices occur in the same cycle of π. The second case is that of possible edges such as $\{1,5\}$ and $\{2,9\}$, whose vertices are in different cycles of π. We consider these cases in turn.

We first consider the case where the vertices of a possible edge occur in the same cycle. It helps to look at some simple particular example first. So we consider the case where π consists of a single cycle. As a specific example we let $\pi \in S_6$ be the permutation $(1\ 2\ 3\ 4\ 5\ 6)$, which has cycle type x_6.

It is helpful to think of the numbers $1, 2, \ldots, 6$ as labeling six points on the circumference of a circle as shown in Figure 14.10, and the permutation π as moving these points through one-sixth of a turn clockwise around the circle. If we now consider, for example, the effect of the permutation π^* on the possible edge $\{1,3\}$, we see that we need to apply the permutation π^* six times before this possible edge is mapped back to itself. That is, $\{1,3\}$ is in a cycle of length 6, corresponding to the fact that it takes a complete turn of the circle to move the black dots shown in Figure 14.10(i) back to their original position. This cycle is

$$\{1,3\} \to \{2,4\} \to \{3,5\} \to \{4,6\} \to \{1,5\} \to \{2,6\} \to \{1,3\}.$$

[Note that, for example, $\pi^*(\{4,6\}) = \{\pi(4), \pi(6)\} = \{5,1\} = \{1,5\}$.] More generally, each possible edge, with just three exceptions, is in a cycle of length 6. The possible edge $\{1,4\}$ is in a cycle of length 3,

$$\{1,4\} \to \{2,5\} \to \{3,6\} \to \{1,4\},$$

because it takes only three-sixths (that is, one-half) of a complete turn to move the black dots to a position where they are occupying the same places as they did to begin with.

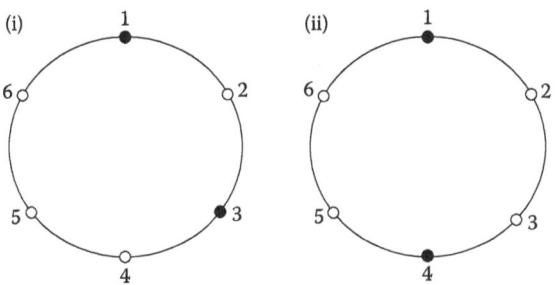

FIGURE 14.10

There are $C(6,2) = 15$ possible edges with vertices taken from the set $\{1,2,\ldots,6\}$. We see from the above calculation that the permutation π^* is made up of one cycle of length 3 of these possible edges and two cycles of length 6. So π^* has cycle type $x_3 x_6^2$. It is not difficult to generalize this. Suppose $\pi \in S_n$ has cycle type x_n. There are $C(n,2) = \frac{1}{2}n(n-1)$ possible edges with vertices taken from the set $\{1,2,\ldots,n\}$.

In the case where n is even, $\frac{1}{2}n$ of these edges are in a cycle of π^* length $\frac{1}{2}n$, and the remaining $\frac{1}{2}n(n-1) - \frac{1}{2}n = \frac{1}{2}n(n-2)$ possible edges form $\frac{1}{2}(n-2)$ cycles each of length n. So π^* has cycle type $x_{(1/2)n} x_n^{(1/2)(n-2)}$.

In the case where n is odd, with $n \geq 3$, there are no exceptions and all the possible edges are in cycles of length n. Since there are $J_s = Gtq$, possible edges, they make up $\frac{1}{2}(n-1)$ cycles of length n. So in this case π^* has cycle type $x_n^{(1/2)(n-1)}$.

We next need to consider possible edges whose vertices come from two different cycles. So we look at the cases of permutations that do not consist of a single cycle. As an example we take the permutation $\pi = (1\ 2\ 3\ 4)(5\ 6\ 7\ 8\ 9\ 10)$ from S_{10}. We see from what we have already done that the possible edges whose vertices come from the cycle $(1\ 2\ 3\ 4)$ contribute one cycle of length 2 and one cycle of length 4 to the cycle type of π^*. Likewise, the possible edges whose vertices come from the cycle $(5\ 6\ 7\ 8\ 9\ 10)$ contribute one cycle of length 3 and two cycles of length 6. We also need to consider possible edges where one vertex comes from the cycle $(1\ 2\ 3\ 4)$ and one from the cycle $(5\ 6\ 7\ 8\ 9\ 10)$. For such an edge, say $\{i, j\}$, with $i \in \{1,2,3,4\}$ and $j \in \{5,6,7,8,9,10\}$, $(\pi^*)^k (\{i, j\}) = \{\pi^k(i), \pi^k(j)\}$. Therefore $(\pi^*)^k (\{i, j\}) = \{i, j\} \Leftrightarrow \pi^k(i) = i$ and $\pi^k(j) = j$. This occurs if and only if k is a multiple of 4 and k is a multiple of 6. Hence, the least k such that $(\pi^*)^k (\{i, j\}) = \{i, j\}$ is the least common multiple of 4 and 6, that is, 12. So each edge $\{i, j\}$ is in a cycle of length 12. Since there are $4 \times 6 = 24$ possible edges of this type, they contribute two cycles of length 12 to π^*. Putting all this together we see that the cycle type of π^* is

$$x_2 x_4 \times x_3 x_6^2 \times x_{12}^2 = x_2 x_3 x_4 x_6^2 x_{12}^2.$$

This readily generalizes. If π is a product of two cycles, one of length r and the other of length s, then each of the rs possible edges, $\{i, j\}$, where i comes from the first cycle and j from the second, is in a cycle of π^* of length equal to the least common multiple of r and s, which we write as $\mathrm{lcm}(r,s)$. Hence, the total number of these cycles is $rs/\mathrm{lcm}(r,s)$, and this is equal to the greatest common divisor of r and s, which we write as $\gcd(r, s)$.

We can now describe the general rule for calculating the cycle type of π^* from the cycle type of π. As each possible edge has vertices that occur in just one or two cycles, all we need take into account when we calculate the cycle type of π^* are the contributions made by single cycles, and pairs of cycles of π. We have calculated these above, and they are set out in Table 14.6.

Using the rules given by Table 14.6 it is now straightforward, though a little tedious, to calculate the cycle type of π^* from that of π. Have a go at the examples in Problem 14.6 before reading the solution.

TABLE 14.6

Cycle Types in π	π^* is the Product of These Terms
x_r, where r is even	$x_{(1/2)r} x_r^{(1/2)(r-2)}$
x_r, where r is odd with $r \geq 3$	$x_r^{(1/2)(r-1)}$
the pair x_r, x_s	$x_{\text{lcm}(r,s)}^{\gcd(r,s)}$

PROBLEM 14.6

In each of the following cases calculate the cycle type of π^*.

a. $\pi \in S_6$, and π has cycle type $x_1 x_2 x_3$.
b. $\pi \in S_{10}$, and π has cycle type $x_2 x_4^2$.
c. $\pi \in S_{25}$, and π has cycle type $x_3^3 x_4 x_6^2$.

Solution

a. We first consider the single cycles of π. These are shown in Table 14.7.
   Next we consider the pairs of cycles of π. These are shown in Table 14.8.
   We now obtain the cycle type of π^* by multiplying all the terms in the second rows of these tables. This gives $x_1 \times x_3 \times x_2 \times x_3 \times x_6 = x_1 x_2 x_3^2 x_6$.
   Note from this calculation that a single cycle of length 1 in π by itself contributes nothing to the cycle type of π^*. This is because a possible edge has two vertices, and these cannot come from the same cycle of length 1. Also, a cycle of length 2 in π contributes just x_1 to the cycle type of π^*, since, although the cycle permutes two vertices, the possible edge with these vertices is mapped to itself by π^*.

b. Here we abbreviate the table by presenting the single cycles and pairs of cycles on one line. Because the cycle type of π is $x_2 x_4^2$, there are two cycles of type x_4, and two pairs x_2, x_4.
   We now obtain the cycle type of π^* by multiplying together all the terms in the second row of Table 14.9. This gives $x_1 x_2^2 x_4^{10}$ as the cycle type of π^*.

c. Here we abbreviate the table by just listing the different single cycles and the different pairs of cycles, with the number of each kind, instead of listing them all. For example, from the terms x_3^3 and x_6^2 we see that we have $3 \times 2 = 6$ pairs of cycle types x_3, x_6. See Table 14.10.

TABLE 14.7

Types of cycle in π	x_1	x_2	x_3
Contribution to cycle type of π^*	—	$x_1 x_2^0 = x_1$	$x_3^1 = x_3$

TABLE 14.8

Pairs of cycles of π	x_1, x_2	x_1, x_3	x_2, x_3
Contribution to cycle type of π^*	$x_2^1 = x_2$	$x_3^1 = x_3$	$x_6^1 = x_6$

TABLE 14.9

π	x_2	x_4	x_4	x_2, x_4	x_2, x_4	x_4, x_4
π^*	x_1	$x_2 x_4$	$x_2 x_4$	x_4^2	x_4^2	x_4^4

TABLE 14.10

π	x_3	x_4	x_6	x_3,x_3	x_3,x_4	x_3,x_6	x_4,x_6	x_6,x_6
Number	3	1	2	3	3	6	2	1
π^*	x_3	x_2x_4	$x_3x_6^2$	x_3^3	x_{12}	x_6^3	x_{12}^2	x_6^6

TABLE 14.11

Cycle type in S_5	x_1^5	$x_1^3x_2$	$x_1^2x_3$	$x_1x_2^2$	x_1x_4	x_2x_3	x_5
Number	1	10	20	15	30	20	24
Cycle type in S_5^*	x_1^{10}	$x_1^4x_2^3$	$x_1x_3^3$	$x_1^2x_2^4$	$x_2x_4^2$	$x_1x_3x_6$	x_5^2

We now obtain the cycle type of π^* by multiplying together the appropriate number of copies of the terms in the bottom line. That is, we multiply together these terms raised to the power given in the second line. This gives

$$(x_3)^3 \times (x_2x_4)^1 \times (x_3x_6^2)^2 \times (x_3^3)^3 \times (x_{12})^3 \times (x_6^3)^6 \times (x_{12}^2)^2 \times (x_6^6)^1 = x_2x_3^{14} x_4x_6^{28} x_{12}^7$$

for the cycle type of π^*.

Note that we have a check that may detect errors in these calculations. Since $\pi \in S_{25}$, we are dealing with graphs with 25 vertices, and so there are $C(25,2) = 300$ possible edges. This should equal the total length of the cycles of π^*. This check gives

$$(1 \times 2) + (14 \times 3) + (1 \times 4) + (28 \times 6) + (7 \times 12) = 300.$$

This agreement doesn't *prove* that our calculation is correct, but using this check will detect many blunders in calculations of this type.

We are now in a position to calculate the cycle indexes of the groups S_n^*. We list the cycle types of the permutations in S_n and their number, and then using the technique we have described above, we calculate the cycle types of the corresponding permutations in S_n^*. We illustrate this calculation in the case $n = 5$, leaving to you the easier case $n = 4$ and the harder case $n = 6$. The cycle index of S_5 is given in the solution to Exercise 14.3.3A, and we use this to give the first two rows of Table 14.11. The cycle types of the permutations, π^*, in the third row are calculated as described above, but we haven't given the details.

It follows from Table 14.11 that the cycle index of S_5^* is

$$\frac{1}{120}\left(x_1^{10} + 10x_1^4x_2^3 + 20x_1x_3^3 + 15x_1^2x_2^4 + 30x_2x_4^2 + 20x_1x_3x_6 + 24x_5^2\right).$$

Hence the pattern inventory for simple graphs with five vertices is

$$\frac{1}{120}[(1+c)^{10} + 10(1+c)^4(1+c^2)^3 + 20(1+c)(1+c^3)^3 + 15(1+c)^2(1+c^2)^4$$

$$+ 30(1+c^2)(1+c^4)^2 + 20(1+c)(1+c^3)(1+c^6) + 24(1+c^5)^2]$$

$$= 1 + c + 2c^2 + 4c^3 + 6c^4 + 6c^5 + 6c^6 + 4c^7 + 2c^8 + c^9 + c^{10}.$$

TABLE 14.12

Edges	0	1	2	3	4	5	6	7	8	9	10
Number of grapha	1	1	2	4	6	6	6	4	2	1	1

We can deduce from this how many different simple graphs there are with five vertices and a specified number of edges, as shown the Table 14.12.

Thus there are altogether 34 different graphs with five vertices. You are asked in Exercise 14.6.3A to draw all these graphs. We hope that this will convince you that Pólya's Counting Theorem, despite all the calculations involved, provides a better method for working out the number of different graphs than simply trying to list them. You will note the symmetry in Table 14.12. This is not an accident! (See Exercise 14.6.5A.)

It should be clear that, in principle, this calculation can be repeated for graphs with any finite number of vertices. The number of cycle types increases very rapidly as the number of vertices increases, so it becomes less and less feasible to do the calculations by hand. You might like to think about how to write a computer program to do all the hard work for you.

Exercises

14.6.1A In each of the following cases of the cycle type of a permutation $\pi \in S_8$, calculate the cycle type of the permutation $\pi^* \in S_8^*$.
  i. $x_1 x_2 x_5$ ii. x_2^4 iii. $x_2 x_6$ iv. $x_2^2 x_4$

14.6.1B In each of the following cases of the cycle type of a permutation $\pi \in S_{12}$, calculate the cycle type of the permutation $\pi^* \in S_{12}^*$.
  i. $x_2^2 x_4^2$ ii. $x_3^2 x_6$ iii. $x_4 x_8$ iv. $x_1^2 x_2^3 x_4$

14.6.2A Calculate the pattern inventory for simple graphs with four vertices.

14.6.2B Calculate the pattern inventory for simple graphs with six vertices.

14.6.3A Draw all the 34 different simple graphs with five vertices.

14.6.3B Draw all the different simple graphs with six vertices and seven edges.

14.6.4A Let π_1, π_2 be permutations in S_n that have cycles types $x_1^{n-2r} x_2^r$ and $x_1^{n-2s} x_2^s$, respectively, where $1 \leq r, s \leq \frac{1}{2}n$. Determine the relationship between r and s if the permutations π_1^* and π_2^* have the same cycle types.

14.6.4B Let π_1 and π_2 be permutations from S_n with cycle types $x_1^{n-2a-4b} x_2^a x_4^b$ and $x_1^{n-2c-4d} x_2^c x_4^d$, respectively, where $a, b, c,$ and d are nonnegative integers with $2a + 4b \leq n$ and $2c + 4d \leq n$. Find the condition on $a, b, c,$ and d for the permutations π_1^* and π_2^* from S_n^* to have the same cycle type.

14.6.5A i. In Exercise 9.3.2B of Chapter 9 we defined the *dual* of a graph $G = (V, E)$ to be the graph $G^* = (V, E^*)$, where for all $u, v \in V$ we have $\{u,v\} \in E^* \Leftrightarrow \{u,v\} \notin E$. (Thus G^* has the same vertices as G, and two vertices are joined by an edge in G^* if and only if they are not joined in G.) Prove that the graphs G_1, G_2 are isomorphic if and only if their duals G_1^*, G_2^* are isomorphic.
  ii. Deduce that for all positive integers n, e with $e \leq \frac{1}{2}n(n-1)$, the number of different (nonisomorphic) graphs with n vertices and e edges is the same as the number of different graphs with $\frac{1}{2}n(n-1) - e$ edges.

CHAPTER 15

Dirichlet's Pigeonhole Principle

15.1 THE ORIGIN OF THE PRINCIPLE

In this chapter we describe the counting technique known variously as the *pigeonhole principle*, *the box principle*, and the *drawer principle*, the last of these being a translation of the German term *Schubfachprinzip*. It was introduced by P. G. L. Dirichlet* in 1834 and used by him in his famous book on the theory of numbers in connection with rational approximations of irrational numbers. We mentioned this application in Chapter 1, where we stated the following problem.

PROBLEM 15A
Rational Approximations to Irrational Numbers

Show that, for each irrational number a, there exists a rational number p/q such that

$$\left|a - \frac{p}{q}\right| < \frac{1}{q^2}.$$

Even in its most complicated form, the principle seems so obvious that you might well suspect it could have no value. However, we shall see some really clever and pretty proofs, often with surprising outcomes, which show that such an assessment is hopelessly wrong!

---

* Peter Gustav Lejeune Dirichlet was born in Düren, Germany, on February 13, 1805. He was appointed professor of mathematics at the University of Berlin when he was 27 years old, and moved to Göttingen in 1855, where he died in 1859. Dirichlet made many contributions to mathematics. He is remembered for his work in analytic number theory, and especially for his theorem that when a and b are coprime integers, the sequence $\{an+b\}$ includes infinitely many prime numbers.

15.2 THE PIGEONHOLE PRINCIPLE

We begin by stating one version of the principle:

THEOREM 15.1
The Pigeonhole Principle (Version 1)

If $n + 1$ pigeons are placed in n pigeonholes then there must be at least one pigeonhole that contains at least two pigeons.

This seems so obvious as not really to require a proof. A trivial application is that among any 13 people there must be at least two who were born in the same month (though, of course, not necessarily in the same year). If this were challenged, we could argue as follows: "Take any 12 of the people (the pigeons!). If no 2 of these 12 were born in the same month of the year, then, between them, they account for ("occupy") all 12 months (the pigeonholes!) of the year. Since there are only 12 different months, the 13th person must have been born in the same month as one of the original 12." Clearly this argument could be generalized to give a proof of Theorem 15.1 if one were required.

As a first application of Theorem 15.1 we offer the following, possibly surprising theorem.

THEOREM 15.2

In every simple graph there are two vertices that have the same degree.*

Proof

Let G be a simple graph and suppose that G has n vertices. The degree of each vertex is an integer from the set $\{0, 1, 2,\ldots, n-1\}$. It looks at first as though there are n possible values for the degrees of the vertices, and so it would be possible for the n vertices to have different degrees. In fact, however, it is not possible to have both a vertex of degree 0 and a vertex of degree $n - 1$ in a graph with n vertices (why not?). Hence, as there are more vertices than possible degrees, by the pigeonhole principle, there must be at least two vertices with the same degree.

An amusing interpretation is that at a party, at which some pairs of guests shake hands, at any time there must be at least two people who have greeted the same number of the other guests with a handshake.

We have noticed that we need to have at least 13 people before we can be sure that there are least two people who were born in the same month. We can easily generalize this example. How many people must there be to ensure that at least four of them will share the same birth month? You might argue (this time concentrating more on the pigeonholes than on the pigeons) as follows.

If no month had *more* than three people born in it, there could be at most $12 \times 3 = 36$ people. Thus, we will ensure that there are at least four people with the same birth month, if we have 37 people.

---

* Recall from Chapter 9 that the degree of a vertex v is the number of vertices to which v is joined by an edge.

The principle behind this argument can be stated as follows (recall that if x is a real number, we use $\lfloor x \rfloor$ to denote the *integer part* of x, that is, the greatest integer not exceeding x).

THEOREM 15.3
Pigeonhole Principle (Version 2)

Suppose that $n + 1$ objects are placed in k boxes. Then there must be at least one box that contains at least $\lfloor n/k \rfloor + 1$ objects.

Proof

If no box contains more than $\lfloor n/k \rfloor$ objects, there can be at most $k \times \lfloor n/k \rfloor$ objects in total. Now, $\lfloor n/k \rfloor \leq (n/k)$ and hence $k \times \lfloor n/k \rfloor \leq n$, which contradicts the fact that there are $n + 1$ objects. So at least one box must contain at least $\lfloor n/k \rfloor + 1$ objects.

PROBLEM 15.1

One hundred cards, numbered 1 to 100, are distributed among six people. Show that there must be at least one person on whose cards there appear at least four 3s.

Solution

On the one hundred cards there are, in total, twenty 3s (namely, one on each of the cards 3, 13, 23, 30, 31, 32, 34, 35, 36, 37, 38, 39, 43, 53, 63, 73, 83, and 93 and two on 33). So here $n = 20$ and $k = 6$ and $\lfloor n/k \rfloor +1= \lfloor 20/6 \rfloor +1=4$. Hence it follows, from version 2 of the pigeonhole principle, that some person must hold at least four 3s.

It is surprising that the "obvious" pigeonhole principle can be used to help prove some statements that are by no means obvious. One famous result, due to Erdös and Szekeres,* is as follows.

THEOREM 15.4

Let n be a positive integer. Then each sequence of $n^2 + 1$ distinct numbers, when read from left to right, contains either an increasing subsequence of length at least $n + 1$ or a decreasing subsequence of length at least $n + 1$.

Proof

Let $a_1, a_2, ..., a_{n^2+1}$ be a sequence of $n^2 + 1$ distinct numbers. We shall assume that there is no increasing subsequence of length greater than n and show how to deduce that there must be a decreasing subsequence of length at least $n + 1$.

For each i, with $1 \leq i \leq n^2 + 1$, we let s_i be the length of the longest increasing subsequence that begins at a_i. By our assumption, for each i, $s_i \leq n$. So we can put the $n^2 + 1$ numbers, a_i, in n numbered boxes, where a_i is put in box k, if $s_i = k$. By the second version of the pigeonhole principle, there must be at least one box containing at least $n + 1$ terms. So, for some k, there are at least $n + 1$ numbers in the sequence, say $a_{i_1}, a_{i_2}, ..., a_{i_{n+1}}$,

---

* P. Erdös and G. Szekeres, A Combinatorial Problem in Geometry, *Compositio Mathematicae*, 2, 1935, pp. 463–470.

with $i_1 < i_2 < \ldots < i_{n+1}$ and $s_{i_1} = s_{i_2} = \ldots = s_{i_{n+1}} = k$. We show that the numbers $a_{i_1}, \ldots, a_{i_{n+1}}$ form a decreasing sequence.

Suppose that $1 \leq i < j \leq n^2 + 1$, and that $s_i = s_j = k$. Then there is an increasing subsequence of length k starting with the term a_j, say $a_j < \ldots < a_z$. If $a_i < a_j$, we would have an increasing sequence, $a_i < a_j < \ldots < a_z$ of length $k + 1$ starting with a_i. But this would contradict the fact that $s_i = k$. We therefore deduce that $a_i < a_j$. Thus we have that $a_{i_1} > a_{i_2} > \ldots > a_{i_{n+1}}$, and so there is a decreasing sequence of length $n + 1$.

Exercises

In Exercise 15.2.1A we have deliberately tried to make finding an increasing or decreasing subsequence of length 5 a little bit harder by our choice of the numbers in the given sequence. Since it is only the relative size of the numbers that matters, the 17 different numbers might just as well have been replaced by the integers 1 to 17. Indeed, if you describe the numbers in Exercise 15.2.1A as the fourth smallest, third smallest, etc., you will probably find the desired subsequence more rapidly. You may also get a good clue as to how to solve Exercise 15.2.5A(ii). This asks you to first think about a generalization of Theorem 15.4 in which $n^2 + 1$ is replaced by $mn + 1$ and then asks whether this could be replaced by mn. This exercise also answers the question as to whether we could replace $n^2 + 1$ by n^2 in Theorem 15.4, a question we hope that you have already asked yourself.

15.2.1A According to Theorem 15.4 the sequence 11, 8, 7, 3, 31, 27, 20, 15, 54, 43, 42, 35, 83, 78, 84, 64, 61 of 17 distinct integers must contain either an increasing or a decreasing subsequence of five integers. Can you find at least one such?

15.2.1B Find the longest increasing subsequence and the longest decreasing subsequence of the sequence 74, 59, 79, 26, 62, 95, 98, 83, 17, 64, 77, 32, 68, 4, 10, 38, 47, 23, 18, 35, 93, 29, 12, 21, 24, 92 of 26 integers. Hence verify that, in accordance with Theorem 15.4, there is either an increasing subsequence of length 6 or a decreasing subsequence of length 6.

15.2.2A In the large desert town of Sleed there are 777 people whose surname is Xerophyte. None of them has more than two forenames. Show that there must be at least two Xerophytes with the same initials.

15.2.2B Show that there is a positive integer k such that $k\pi$ differs from an integer by less than 1/1000. Then, just for fun, and with only pencil and paper, find an integer k such that $k\pi$ is within 1/1000 of some integer.

15.2.3A By a *lattice point* in two-dimensional space we mean a point whose coordinates (x, y) are both integers. Show that if you are given five lattice points, the midpoint of at least one of the line segments joining pairs of these points is also a lattice point.

15.2.3B We now consider lattice points in three-dimensional space, that is, points (x, y, z), where $x, y,$ and z are all integers. What is the least value of m that will guarantee that, given m lattice points in three-dimensional space, the midpoint of at least one of the $C(m, 2)$ lines joining these points in pairs is also a lattice point? Prove that your suggestion is correct.

15.2.4A An organizer of a party, restricted to those aged between 18 and 30 (inclusive), wanted to ensure that at least three were born in the same year. How many people must be invited to be sure this condition is fulfilled?

15.2.4B Your company's security officer wants you to invent a sequence of twenty 0s and 1s in which no subsequence of four successive symbols is repeated. Show that this cannot be done. Can you find a sequence of nineteen 0s and 1s in which no subsequence of four successive symbols is repeated?

15.2.5A i. Show that given any sequence of $mn + 1$ distinct integers there is either an increasing subsequence of length $m + 1$ or a decreasing subsequence of length $n + 1$.

  ii. Give an example of a sequence made up of the 15 integers 1,2,3,...,15 that has no decreasing subsequence of length 6 and no increasing subsequence of length 4.

15.2.5B Show that given a sequence, $a_1 < a_2 < ... < a_{mn+1}$, of $mn + 1$ distinct positive integers, there is either a subsequence consisting of $m + 1$ integers none of which divides any other or a subsequence consisting of $n + 1$ integers each of which divides the following one.

15.3 MORE APPLICATIONS OF THE PIGEONHOLE PRINCIPLE

We now give some more applications of the pigeonhole principle, many of which lead to rather surprising conclusions.

PROBLEM 15.2

Let $a_1, a_2,..., a_{100}$ be any 100 integers (positive or zero or negative, distinct or not). Then there exist integers r, s with $0 < r < s \leq 100$ such that the sum $a_{r+1} + a_{r+2} + ... + a_s$ is an exact multiple of 100. (Of course, there is nothing magical about the number 100.)

Solution

For each integer t, with $1 \leq t \leq 100$, let $u_t = a_1 + a_2 + ... + a_t$ and let k_t be the remainder when u_t is divided by 100. If any of the k_t is 0, then the sum $a_1 + a_2 + ... + a_t$ is a multiple of 100, and the desired result holds. Otherwise, for $1 \leq t \leq 100$ we have $1 \leq k_t \leq 99$, and hence by the pigeonhole principle, at least two of the terms k_t must be equal. That is, for some r, s with $1 \leq r < s \leq 100$, we have $k_r = k_s$. Thus $a_1 + a_2 + ... + a_r$ and $a_1 + a_2 + ... + a_s$ have the same remainder on division by 100. Hence $(a_1 + a_2 + ... + a_s) - (a_1 + a_2 + ... + a_r)$ is a multiple of 100; that is, $a_{r+1} + a_{r+2} + ... + a_s$ is a multiple of 100, and so the result holds also in this case. Hence the result holds for all sets of 100 integers.

Just as surprising but slightly more tricky to (see how to) prove is the following, also due originally to Erdös and Szekeres.

PROBLEM 15.3

Show that if we choose any 51 integers in the range from 1 to 100, there must be a pair of integers from those we have picked such that one of them is a multiple of the other.

Solution

Each integer, n, may be written uniquely in the form $2^k r$, where k is a nonnegative integer and r is a positive odd integer. If $1 \leq n \leq 100$, r must take one of the values in the set $\{1, 3,..., 99\}$ of the 50 odd integers in the range from 1 to 100. So given any set of

51 numbers in the range from 1 to 100, the pigeonhole principle shows that there must be at least two with the same odd factor, r, say $n_1 = 2^{k_1} r$ and $n_2 = 2^{k_2} r$, with $n_1 < n_2$. Then $k_1 < k_2$ and $n_2 = 2^{k_2 - k_1} n_1$. Hence n_2 is a multiple of n_1.

Now for a geometric example.

PROBLEM 15.4

Suppose that five points are chosen inside an equilateral triangle of side 2 cms. Show that there is (at least) one pair of points that lie within 1 cm of each other.

Solution

Split the given triangle into four equilateral triangles each of side 1 cm, as in Figure 15.1.

By the pigeonhole principle, at least one of the unit triangles must contain at least two points. For two points in the same small triangle to lie as much as one unit apart, each must be at a vertex of the triangle containing them. But that means that these two points lie on the perimeter of the large triangle, whereas we are told that all five points lie inside the large triangle.

Here is another example containing possibly helpful practical advice for any reader thinking of becoming an international chess grandmaster!

PROBLEM 15.5

A chess player plans to train for his next match by playing at least one game each day for 90 days, but at most five games over any period of three consecutive days. Show that there must be some period of consecutive days during which he plays 29 games.

Solution

Let c_i be the number of games completed by the end of the ith day of training. Since at least one game is played each day, and at most five games are played in any three consecutive days, we have that $0 < c_1 < c_2 < \ldots < c_{90} \leq 150$. Now consider the increasing sequences of positive integers $c_1 < c_2 < \ldots < c_{90}$ and $c_1 + 29 < c_2 + 29 < \ldots < c_{90} + 29$. Between them they contain 180 numbers, all in the range from 1 to 179. Hence, by the pigeonhole principle, there are two of them that are equal. Since the numbers in each of the two sequences are all different, this pair must consist of one number from the first sequence and one number from the second sequence. That is, for some integers r, s in the range from 1 to 90, we have $c_r = c_s + 29$. It follows that $c_r - c_s = 29$. Consequently that $s < r$, as $c_s < c_r$. Hence in the period of consecutive days from day $s + 1$ to day r the chess player completes exactly 29 games.

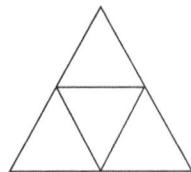

FIGURE 15.1

We now come to the problem due to Dirichlet, previously labeled Problem 15A, that we mentioned in Section 15.1.

PROBLEM 15.6

Show that, for each irrational number a, there exists a rational number p/q such that

$$\left| a - \frac{p}{q} \right| < \frac{1}{q^2}.$$

Solution

First we establish some notation. For each real number x we write $\{x\}$ for the number $x - \lfloor x \rfloor$. It follows that $0 \leq \{x\} < 1$ for each real number x. (For example, $\{\pi\} = 0.14159\ldots$ and $\{-8.7\} = 0.3$.)

For convenience we suppose that a is positive. The modification in the argument needed when a is negative is left to the reader. Now let t be any positive integer. Consider the $t + 1$ real numbers $\{a\}, \{2a\}, \ldots, \{ta\}, \{(t+1)a\}$, which are all in the (half-open, half-closed) interval $[0, 1)$, and also the t subintervals $[0, 1/t), [1/t, 2/t), \ldots, [(t-1)/t, 1)$ of $[0, 1)$. Since we have $t + 1$ numbers and only t intervals, by the pigeonhole principle, there are integers s, m, n with $0 \leq s \leq t$ and $1 \leq m < n \leq t + 1$ such that both $\{ma\}$ and $\{na\}$ lie in the interval $[s/t, (s+1)/t)$. It follows that $|\{ma\} - \{na\}| < 1/t$, that is $(na - \lfloor na \rfloor) - (ma - \lfloor ma \rfloor)| < 1/t$. Thus, if we put $q = n - m$ and $p = \lfloor na \rfloor - \lfloor ma \rfloor$, then p and q are integers and $|qa - p| < 1/t$. As $1 \leq m < n \leq t + 1$, it follows that $0 < q \leq t$ and hence, on dividing both sides of the inequality $|qa - p| < 1/t$ by q, we deduce that

$$\left| a - \frac{p}{q} \right| < \frac{1}{qt} \leq \frac{1}{q^2}.$$

Here is another example. Its conclusion seems almost unbelievable!

PROBLEM 15.7

There are nine people, aged from 18 to 58, at a family reunion. Show that it is possible to choose two groups of these people in such a way that the sums of the ages of the people in each group are equal.

Solution

Each nine-element set has $2^9 = 512$ subsets and hence 511 nonempty subsets. As everyone is aged between 18 and 58, in any nonempty subset of the nine people, the sum of the ages of the members ranges between 1×18 and 9×58, that is, between 18 and 522. So there are at most 505 possible different values for the age totals as we range over all the nonempty subsets of people present. As there are 511 nonempty subsets, it follows from the pigeonhole principle that there are two distinct subsets with the same total age. Should any person be included in both of these two subsets, we remove this person from each subset. Since these two new subsets will, clearly, also have the same age sum as each other, we have produced two subsets of the desired kind.

Finally we come to Problem 15B of Chapter 1, where we explained what we mean by "faultlessly."

PROBLEM 15.8

Can 18 dominoes be placed on a 6 × 6 board "faultlessly"?

Solution

Label the internal "struts," that is, the short lines that a domino might cover, by the numbers 1 to 60, as indicated in Figure 15.2. If 18 (nonoverlapping) dominos are placed on the board, then 18 of these struts will be covered, leaving 42 exposed. These lie on the 10 lines of struts and so, by the pigeonhole principle, at least one of the 10 lines will contain at least 5 exposed struts.

If one of these lines has exactly five exposed struts, then the sixth must be crossed by a domino as in Figure 15.3.

Apart from that one domino, each other domino occupies either two of the squares we have labeled with an A, or two labeled with a B. The squares labeled A make up a whole number of columns less one square, and hence there is an odd number of squares labeled A. So these squares cannot be entirely covered by nonoverlapping

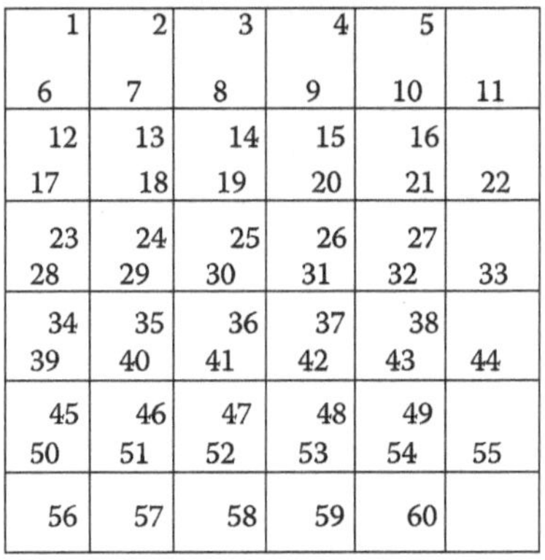

1	2	3	4	5	
6	7	8	9	10	11
12	13	14	15	16	
17	18	19	20	21	22
23	24	25	26	27	
28	29	30	31	32	33
34	35	36	37	38	
39	40	41	42	43	44
45	46	47	48	49	
50	51	52	53	54	55
56	57	58	59	60	

FIGURE 15.2

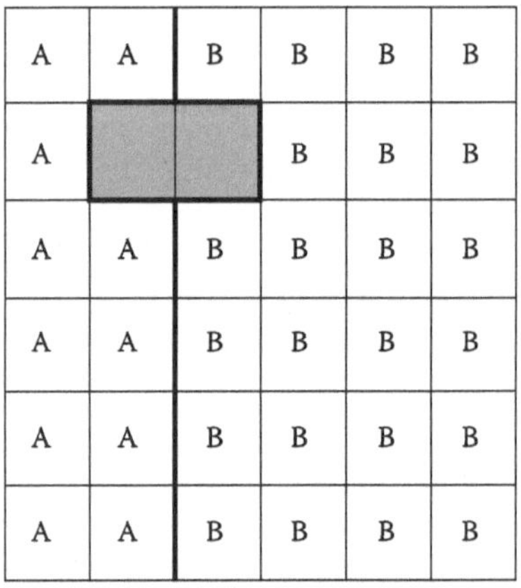

A	A	B	B	B	B
A			B	B	B
A	A	B	B	B	B
A	A	B	B	B	B
A	A	B	B	B	B
A	A	B	B	B	B

FIGURE 15.3

dominos. Hence, any row in which five struts are exposed must, in fact, have *all six* struts exposed. That is, the line is a fault line. This proves that a 6 × 6 board *cannot* be covered "faultlessly."

Exercises

15.3.1A Let A be a set of n numbers, let k be a fixed positive integer, with $k \leq n$, and let t be a fixed number. Suppose that the sum of the numbers in each k-element subset of A is at most t. Show that the sum of all the numbers in A is at most nt/k. Deduce that if the sum of the ages of the 85 residents of an old folks' home is more than 7000, then there is at least one group of 13 of the residents whose ages add up to more than 1066.

15.3.1B A student works 193 hours in 24 days in a supermarket – so just over eight hours per day on average. Show that there must be two *consecutive* days during which he works more than 16 hours, and likewise that there are three and four consecutive days during which he works more than 24 and 32 hours respectively. Show, however, that there may be *no* period of five *consecutive* days during which he works more than 40 hours.

15.3.2A Show that, given a set of $n + 1$ distinct integers, there are at least two integers in the set whose difference is divisible by n.

15.3.2B i. Show that given any 32 distinct integers, there exist two whose sum, or whose difference, is divisible by 60.

ii. Give an example of a set of 31 integers such that there are no two integers in the set whose sum or difference is divisible by 60.

15.3.3A Show that given 10 points inside an equilateral triangle of side 1 unit, there are two of these points whose distance apart is less than 1/3 of a unit.

15.3.3B A farmer has a field 100 m square. He wishes to plant 50 apple trees in the field, but bureaucratic regulations require that he plant no tree within 15 m of the edge of the field nor within 15 m of another tree. Show that there is no way he can plant all 50 trees without breaking the regulations.

15.3.4A Show that, if six points lie strictly inside a circle with radius 1 unit, then there is at least one pair of these points less than 1 unit apart. (*Hint:* Split the circle into six identical sectors. Then show that if no two points are within 1 unit of each other, then no point may lie at the center of C.)

15.3.4B Show that if there are nine points inside an equilateral triangle of side length 1 unit, then there are two of these points that lie within 1/3 of a unit of each other. (*Hint:* Split the triangle into 9 smaller equilateral triangles as in the solution to Problem 15.4. Consider the regular hexagon H resulting from cutting off the three corner equilateral triangles. Then use the result of Exercise 13.2.8A applied to the circle through all the vertices of H.)

15.3.5A A student has allocated 27 days to prepare for exams. He decides to study on each of these days for a whole number of hours, and for at least one hour on each day, but for not more than five hours in any period of three consecutive days. Show that there must be a period of successive days during which he will study for exactly eight hours.

15.3.5B A security guard watches his monitor for either 4, 5, 6, or 7 hours each day, but not more than 36 hours in any 7 consecutive days. He is to work 330 days in his first year.

Show that he may arrange his working hours so that on no set of consecutive working days does he work 111 hours. Show that he may also arrange his hours (perhaps differently each time) so that on no set of consecutive working days does he work 222, nor 333, nor 444, nor 555, nor 666, nor 777 hours.

Finally, show that no matter how his boss arranges his hours, there will be some period of consecutive days on which he works either 111 or 222 or ... or 777 hours.

15.3.6A (This generalizes the result of Problem 15.6.) Show that given any irrational number, a, any real number, x, and any positive real number, h, there exist integers u, v such that $|(va-u)-x|<h$.

15.3.6B Use the result of Exercise 15.3.6A to show that for some positive integer k, the integer 2^k begins with the digits 2009.

[*Hint:* You need to show that, for some integers k, t, we have $2009 \times 10^t \leq 2^k < 2010 \times 10^t$. These inequalities are equivalent to $\log_2(2009) + t\log_2 10 \leq k < \log_2(2010) + t\log_2 10$, that is, $10.97226185... + 3.321928095...t \leq k\ 10.97297979... + 3.321928095...t$.]

15.3.7A Show that if X is a set of nine distinct integers in the range from 1 to 60, there are two disjoint subsets, Y, Z, of X such that the sum of the integers in Y is equal to the sum of the integers in Z.

15.3.7B Let n be an even positive integer. You are given n boxes each containing between one and n chocolates, and between them a total of $2n$ chocolates. Show that it is always possible to divide the boxes between two piles so that the number of chocolates in each pile is n.

CHAPTER 16

Ramsey Theory

16.1 WHAT IS RAMSEY'S THEOREM?

In chapter 1 we raised the following problem.

PROBLEM 16A
Friends at a Party

There are six people at a party. The people in each pair are either friends or strangers. We claim that, among the six, there are (at least) three people who are all friends or all strangers. Are we correct?

Although this seems a fairly lighthearted problem it is related to other, much more significant questions. In 1930, Frank Ramsey,* in connection with a problem in mathematical logic, obtained the following very technical theorem. (Do not worry if it is a little bit tricky to grasp in full at the first—or even sixth!—reading. You won't need to know it. It just helps with setting the scene.)

RAMSEY'S THEOREM

Let positive integers $k, s, t_1, t_2, \ldots, t_k$ be given with $s \geq 2$. Then there exists an integer, F, such that if the set, C, of all subsets containing s numbers from the set $X = \{1, 2, \ldots, F\}$ of the first F positive integers is partitioned into k disjoint sets $C_1, C_2, \ldots, C_k$, in any manner you please, then either there are t_1 elements of X all of whose s-element subsets lie in C_1, or there are t_2 elements of X all of whose s-element subsets lie in C_2, or ... there are t_k elements of X all of whose s-element subsets lie in C_k.

---

* F. P. Ramsey, On a Problem in Formal Logic, *Proceedings of the London Mathematical Society*, 30, 1930, pp. 264–286. Frank Plumpton Ramsey was born in Cambridge on January 23, 1903, into an academic family. His father was Arthur Ramsey, an applied mathematician who published standard textbooks of their day on statics and dynamics, and who became president of Magdalene College, Cambridge. His younger brother, Michael, was Archbishop of Canterbury from 1961 to 1974. Frank Ramsey made distinguished contributions to economics, philosophy, and probability, as well as to mathematical logic. While an undergraduate he wrote a perceptive review of Wittgenstein's *Tractatus*. His combinatorial theorem was a preliminary result to enable him to solve a particular case of the decision problem for quantifier logic, but has proved more interesting and fruitful than the problem from logic that motivated it. He died tragically young in Guy's Hospital, London, on January 19, 1930, after an illness.

What has this to do with Problem 16A? Ramsey's theorem tells us that there is some number, say F, such that if we divided the two-element subsets of a set of F people into two classes (the pairs of *friends* and the pairs of *strangers*—let us call them *red* pairs and *blue* pairs!), then there will be either three people all of whose two-element subsets are red or three people all of whose two-element subsets are blue. Problem 16A asks whether we can take F to be 6.

The introduction of the colors *red* and *blue* into this discussion suggests a connection with edge colorings of graphs. We can represent a set of people by the vertices of a complete graph, coloring an edge *red* if it joins two people who are friends, and *blue* if it joins two people who are strangers. In these terms, Problem 16A asks whether, given an edge coloring of the complete graph K_6 using the two colors red and blue, there will always be either a red triangle or a blue triangle in the graph.

It is convenient to introduce some notation to help us to discuss problems of this kind.

DEFINITION 16.1

Given positive integers $k, s, t_1, t_2, \ldots, t_k$ with $s \geq 2$, we use the notation $F(s; t_1, t_2, \ldots, t_k)$ for the smallest integer F that has the property of Ramsey's theorem. (Note that it was not obvious, before Ramsey's proof, whether such a number F even existed!)

Using the notation introduced in this definition, we see that Problem 16A covers the case of Ramsey's theorem where $s = 2$ and $t_1 = t_2 = 3$, and claims that $F(2;3,3)$ may be taken to be 6. Whether or not $F(2;3,3)$ can be taken to be 5 or less is dealt with below.

As we have seen, in the case where $s = 2$, it is quite easy to interpret Ramsey's theorem geometrically, as the two-element subsets of X correspond to the edges of the complete graph whose vertices correspond to the elements of X, and the partition of X into the disjoint subsets $C_1, C_2, \ldots, C_k$ corresponds to a coloring of these edges. Any attempt to visualize Ramsey's theorem for $s = 3$ will be tricky (since we should have to imagine every plane determined by three points as being colored) and for $s > 3$ impossible for ordinary mortals. On the other hand, working with $k > 2$ (at least in the case of $s = 2$) is not as much of a problem since it is not difficult to picture the edges of a (complete) graph as being adorned by any number of different colors. Nevertheless we will mainly concentrate on the case $s = k = 2$ so that we can continue to use the appealing graphical interpretation in which the edges of a graph are painted in just two colors.

The Solution of Problem 16A

We shall solve Problem 16A easily by making a trivial application of the pigeonhole principle but will, simultaneously, stir up a bit of a hornets' nest of more difficult problems.

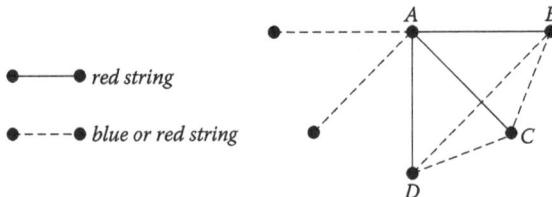

FIGURE 16.1

First imagine that each pair of the six people is holding the ends of a string, which is red if they are friends and blue if they are strangers as in Figure 16.1. Select any one of the six people present, say A (or any one of the six vertices of the graph, etc.). This person is holding five strings, each of which is red or blue. So A must be holding either at least three red strings or at least three blue strings. Let us suppose A holds at least three red strings. If not, then repeat the argument below but with the words *red* and *blue* interchanged.

Now, suppose B, C, D are three of the people who share a red string with A. If any two of B, C, D also share a red string, then this string, together with the red strings these two share with A, form a red triangle. If this is *not* the case, then each of the strings that B, C, D share must be blue, which, of course, provides a blue triangle.

With the number of people, six, apparently being plucked from nowhere, you may suspect that, if we could have "got away" with using only five people, we would have said so. And you would be right, as we will show below. Do notice that we are *not* saying that you will *never* get a set of three mutual friends or a set of three mutual strangers at your party if you invite five people. What we *are* saying is that it would be *possible* for you to invite five people so that, among them, there is no group of three mutual friends and no group of three mutual strangers.

To confirm this we only need produce one suitable *example*. That is most easily done with a graph, as in Figure 16.2, where the vertices represent the five people, the solid lines represent red strings, and the dotted lines represent blue strings. It is easily seen that, in this diagram, there is no solid-line triangle and no dotted-line triangle; hence there is no trio of mutual friends and no trio of mutual strangers.

In graph-theoretic terms, a triangle is nothing other than the complete graph, K_3, with three vertices. So we can summarize the conclusion of Problem 16A as follows: If the 15 edges of the complete graph, K_6, are each colored either red or blue, then there is either a subgraph isomorphic to K_3 all of whose edges are red, or a subgraph isomorphic to K_3 all of whose edges are blue. We then showed by an example that the number 6 could not be replaced by the number 5.

The significance of this way of formulating the result is that when we generalize it, as we are about to do, instead of going from triangles to quadrilaterals, pentagons, and so on, we generalize from K_3 to the other complete graphs K_4, K_5, and so on. Do note that K_4 does not have four edges but has six edges. Many a student has been caught out by

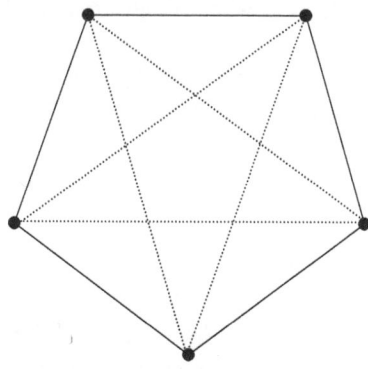

FIGURE 16.2

this after generalizing (wrongly) from "three vertices give a triangle" to "four vertices give an (ordinary) quadrilateral." Rather K_4 is what is called a "complete quadrilateral": a quadrilateral together with its diagonals.

It is rather a mouthful to have to keep talking about, for example, "a subgraph isomorphic to K_3 all of whose edges are red." So we introduce some terminology that enables us to abbreviate our statements.

DEFINITION 16.2

Suppose that the edges of the graph K_n are colored using red and blue and p is a positive integer. We say that K_n *has a red K_p* if there is a set, say P, of p vertices of K_n, such that all the edges joining pairs of vertices of P are red (that is, K_n has a subgraph isomorphic to K_p, all of whose edges are red). We say that K_n *has a blue K_p* if we can find a set P of p vertices such that all the edges joining them are blue.

In terms of this definition we can rephrase the result of Problem 16A by saying that if the edges of K_6 are colored using red or blue, then K_6 has either a red K_3 or a blue K_3.

We now introduce the following definition, which is the special case of the definition of $F(s : t_1, t_2, \ldots, t_k)$ with $s = k = 2$, $t_1 = p$, and $t_2 = q$.

DEFINITION 16.3

Let p,q be positive integers with $p,q \geq 2$ such that there exists an integer n with the property that, however the edges of the complete graph, K_n, are colored red and blue, K_n will always have either a red K_p or a blue K_q. Then there will be a smallest integer, n, with this property. This smallest integer n is called the (p,q) *Ramsey number* and written as $R(p,q)$.

For example, our discussion above shows that that $R(3,3) = 6$. Because of the symmetry between red and blue in Definition 16.3 the following result is obvious.

THEOREM 16.1

For all positive integers p,q, if $R(p,q)$ exists, then so also does $R(q,p)$, and we have that $R(p,q) = R(q,p)$.

We now face an infinity of questions! Does $R(p,q)$ exist for each pair of integers p,q? If so, what is its value? For example, what (if they exist) are $R(3,4)$ and $R(4,4)$?

Of course, Ramsey's theorem implies that $R(p,q)$ always exists, but we have not yet given a proof of this. We give a proof in the case $p, q \geq 3$ in Theorem 16.3, but first we determine the exact value of $R(2,q)$ for all $q \geq 2$.

THEOREM 16.2

For all $p,q \geq 2$, $R(p,2) = p$ and $R(2,q) = q$.

Proof

This is straightforward, since if each edge of the complete graph K_p is either red or blue, then either every edge is red, in which case we have p vertices joined by edges that are all red, or there is a blue edge, say $\{u, v\}$, in which case we have a set of two vertices joined by blue edges (actually only one edge). This shows that $R(p,2) \leq p$, but clearly it is not possible to have $R(p,2) < p$. So $R(p,2) = p$. Similarly, using Theorem 16.1, $R(2,q) = q$.

Before proceeding any further we show that $R(p,q)$ is worth searching for. In other words, it always exists! Having shown that $R(p,2)$ and $R(2,q)$ exist for all $p,q \geq 2$ we are only left with proving the next theorem.

THEOREM 16.3
(Erdös/Szekeres*)

For all integers $p,q \geq 3$, $R(p,q)$ exists and

$$R(p,q) \leq R(p-1,q) + R(p,q-1).$$

Comments

i. The main body of the proof is almost a repeat of the solution to Problem 16A.
ii. We need (at this point) to ask that $p,q \geq 3$ because of the quantities $p-1$ and $q-1$ appearing on the right-hand side of the inequality.

Proof

We prove by mathematical induction that for all integers $n \geq 6$ the result holds in the case that $p + q = n$.

First suppose that $p + q = 6$. Then since we are assuming that $p,q \geq 3$, we must have $p = q = 3$. We have seen that $R(3,3)$ exists, and $R(3,3) = 6$. Also, by Theorem 16.2, $R(3,2) = R(2,3) = 3$, and hence $R(3,3) \leq R(2,3) + R(3,2)$. Hence the result holds for $n = 6$.

We now assume, as our induction hypothesis, that the result holds whenever $p + q \leq k$, where p,q are integers with $p,q \geq 3$. Suppose $p + q = k + 1$. Using the induction hypothesis if $p,q > 3$, and Theorem 16.2 if either $p = 3$ or $q = 3$, we see that both $R(p-1,q)$ and $R(p,q-1)$ exist.

We now consider the complete graph, say G, with $R(p-1,q) + R(p,q-1)$ vertices, and suppose that each edge of G is colored either red or blue. Choose a vertex, v, of G, and let $R(v)$ be the set of vertices of G that are joined to v by a red edge, and $B(v)$ be the set of vertices joined to v by a blue edge. Then $\#(R(v)) + \#(B(v)) = R(p-1,q) + R(p,q-1) - 1$. Hence, by the pigeonhole principle, either $\#(R(v)) \geq R(p-1,q)$ or $\#(B(v)) \geq R(p,q-1)$.

We consider the case where $\#(B(v)) \geq R(p,q-1)$, the other case where $\#(R(v)) \geq R(p-1,q)$ being the same after interchanging "red" and "blue." By the definition of $R(p,q-1)$, the complete subgraph, say $G_{B(v)}$, with $B(v)$ as its set of vertices, has either a red K_p or a blue K_{q-1}.

If $G_{B(v)}$ has a red K_p, then so, too, does G. If not, then $G_{B(v)}$ has a blue K_{q-1}, and so there is a set, say X, of $q-1$ vertices of G joined by edges all of which are blue. As $v \notin B(v)$, $X \cup \{v\}$ is a set of q vertices of G joined by edges all of which are blue, and so forms a blue K_q in G.

---

* P. Erdös and G. Szekeres, A Combinatorial Problem in Geometry, *Compositio Mathematica*, 2, 1935, pp. 463–470.

This shows that $R(p,q)$ does indeed exist and that it is at most $R(p-1,q) + R(p,q-1)$. Thus we have completed the proof of the theorem.

So, what about the actual *values* of the (p,q) Ramsey numbers? In particular, what is the value of $R(3,4)$? That is, what is the smallest positive integer n so that, however the edges of the complete graph K_n are colored red or blue, there must be either a red K_3 or a blue K_4?

It turns out that $R(3,4) = 9$, as proved below. This requires us first to show that $R(3,4) > 8$. You are asked to do this by checking that the coloring of K_8 given in Figure 16.3 has neither a red K_3 nor a blue K_4.

THEOREM 16.4*

$$R(3,4) = 9.$$

Proof

We have already seen that the example given in Figure 16.3 shows that $R(3,4) > 8$. Now suppose that the edges of K_9 have been colored red and blue. We aim to show that K_9 has either a red K_3 or a blue K_4.

The first case we consider is where there is a vertex, say v, that is joined to four other vertices, say w, x, y, and z, by a red edge, as in Figure 16.4.

If any pair of the vertices w, x, y, and z is joined by a red edge, there would be a red K_3 made up of this pair of vertices together with v. Otherwise, the edges joining w, x, y, and z are all blue, and then these six edges form a blue K_4. So in this case K_9 has either a red K_3 or a blue K_4.

The second case is where there is a vertex, say v, that is joined to six vertices by blue edges. Because $R(3,3) = 6$ the subgraph made up of these six vertices and all the edges joining them has either a red K_3 or a blue K_3. If there is a blue K_3, its vertices together with v will form a blue K_4. So also in this case K_9 has either a red K_3 or a blue K_4.

Now each vertex is joined to eight other vertices. So, if there is no vertex joined to four other vertices by red edges, and no vertex is joined to six other vertices by

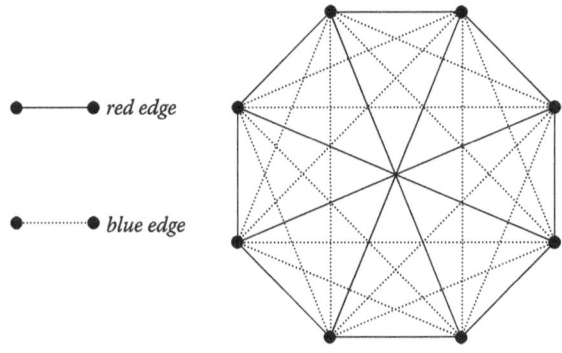

FIGURE 16.3

---

* R. E. Greenwood and A. M. Gleason, Combinatorial Relations and Chromatic Graphs, *Canadian Journal of Mathematics*, 7, 1955, pp. 1–7.

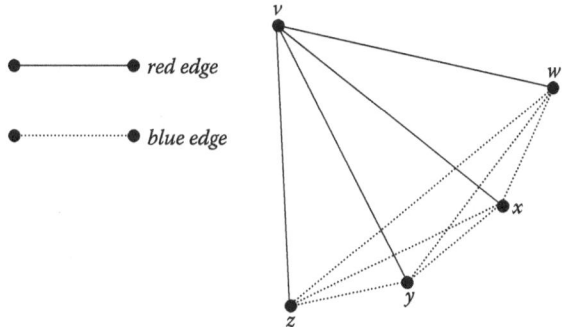

FIGURE 16.4

TABLE 16.1

$R(p, q)$	3	4	5	6	7	8	9
3	6	9	14	18	23	28	36
4		9	18	25	?	?	?	?

blue edges, it must be that each vertex is joined to three vertices by red edges and five vertices by blue edges. But then the subgraph consisting of just the red edges would have nine vertices each of degree 3, which is impossible, as, by the handshaking lemma (Theorem 9.4), the sum of the degrees must be an even number.

So either the first case or the second case must apply. It follows that, however the edges of K_9 are colored red and blue, there must be either a red K_3 or a blue K_4. Hence $R(3,4) \leq 9$. Since $R(3,4) > 8$, we deduce that $R(3,4) = 9$. This completes the proof.

Very few other exact values of $R(p,q)$ with $p,q \geq 3$ are known. The known values are given in Table 16.1.*

Exercises

16.1.1A Show that, however the edges of K_6 are colored red and blue, there must be at least two monochromatic triangles.

16.1.1B Show that, however the edges of K_7 are colored red and blue, there must be at least *three* monochromatic triangles. [*Hint:* Find monochromatic triangles using at most six of the seven vertices. Remove (a suitable) one of these vertices and find monochromatic triangles using the other six vertices.]

16.1.2A Mr. Friend invites six friends to tea. Show that, among all seven of them, there is either a group of four mutual friends or a group of three mutual strangers. (Assume that, given any two people, they are either friends or strangers.)

16.1.2B On another day Mr. Friend invites four friends to tea. Show that, among all five of them, there is either a group of three mutual friends or all four of Mr. Friend's friends are mutual strangers.

---

* Taken from a survey article by Stanislaw Radziszowski on the Web site http://www.combinatorics.org/Surveys. This article also gives the known lower and upper bound for other values of $R(p,q)$ where an exact value is not known, including the cases where there is a question mark in Table 16.1.

16.1.3A Show that, for every positive integer n, $R(n + 1,3) \geq 3n$.

16.1.3B Show that the inequality in Theorem 16.3 is a strict inequality if $R(p-1,q)$ and $R(p,q-1)$ are both even.

16.2 THREE LOVELY THEOREMS

We cannot show you here the methods that establish the most recent information, but three theorems that supply upper and lower bounds make for exciting reading and their proofs are well within our grasp. We present these now, starting with an upper bound that follows immediately from Theorem 16.3 and relates the Ramsey numbers to the binomial coefficients.

THEOREM 16.5

For all $p,q \geq 2$, we have that

a. $R(p,q) \leq C(p + q - 2, p - 1)$ or, equivalently, that
b. $R(p,q) \leq C(p + q - 2, q - 1)$.

Proof

For $q = 2$ the inequality amounts to $R(p,2) \leq C(p,1)$, and this follows from Theorem 16.2 as $C(p,1) = p$. Likewise, the result holds for $p = 2$. We may therefore assume that $p,q \geq 3$. Once again we use induction on the value of $p + q$. That is, we prove that, for all $n \geq 6$, the result holds whenever $p,q \geq 3$ and $p + q = n$.

Since, for $p = q = 3$, $R(p,q) = R(3,3) = 6$ and $C(p + q - 2, p - 1) = C(4,2) = 6$, the result holds for $p + q = 6$. Now assume that the desired result holds for all integers p,q such that $p,q \geq 3$ and $p + q = k$, where $k \geq 6$. Let $p,q \geq 3$ be integers such that $p + q = k + 1$. By Theorem 16.3, and the induction hypothesis,

$$R(p,q) \leq R(p - 1, q) + R(p, q - 1)$$
$$\leq C((p-1) + q - 2, (p-1) - 1) + C(p + (q-1) - 2, (q-1) - 1)$$
$$= C(p + q - 3, p - 2) + C(p + q - 3, p - 1)$$
$$= C(p + q - 2, p - 1),$$

by Theorem 2.7.

Hence the result holds also in this case and this completes the proof of (a). Then (b) follows immediately as, by Theorem 2.5,

$$C(p + q - 2, p - 1) = C(p + q - 2, q - 1).$$

It follows from Theorem 16.5 that $R(3,4) \leq C(5,2) = 10$. Quite close, but not exactly right! Also, by Theorem 16.5, $R(3,9) \leq C(10,2) = 45$, whereas $R(3,9) = 36$. Not quite so close!

What about lower bounds? The following result (and its proof) is very satisfying.

THEOREM 16.6

For all integers $p,q \geq 2$, we have $R(p,q) \geq (p-1)(q-1) + 1$.

Proof

In the case where $q = 2$ the inequality amounts to $R(p,2) \geq p$, and the truth of this follows from Theorem 16.2. The case $p = 2$ is similar.

We now suppose that $p,q \geq 3$. We let $t = (p - 1)(q - 1)$. We have to show that the edges of K_t can be colored red or blue so that K_t has neither a red K_p nor a blue K_q. Such a coloring is most easily visualized by taking the t vertices in a rectangular block comprising $p - 1$ rows and $q - 1$ columns. We join two vertices by a blue edge if they lie on the same row. Otherwise we join two vertices by a red edge. By the pigeonhole principle, if p of the vertices are chosen, then (at least) two must lie in the same row and so these must therefore be joined by a blue edge. Hence with this coloring K_t has no red K_p. Likewise, if q vertices are chosen, there must be a pair lying in the same column, so these must be joined by a red edge. So K_t has no blue K_q. It follows that $R(p,q) > t$; that is, $R(p,q) \geq (p - 1)(q - 1) + 1$.

The construction used to prove Theorem 16.6 is not complicated, and therefore it should not be a surprise that the lower bound it gives is not very good. For example, by Theorem 16.6, $R(3,4) \geq 7$, whereas $R(3,4) = 9$. Also, this theorem implies that $R(3,9) \geq 17$, compared with the actual value of 36.

The third theorem refers to *diagonal Ramsey numbers*, that is, numbers of the form $R(p,p)$. Our reason for including this theorem is solely that we wish to show you the delightful proof.

There is a nice little story credited to Paul Erdös.* Regarding the relative difficulty, as he saw it, of evaluating $R(5,5)$ and $R(6,6)$, he made the following assessment. "If aliens offered earthlings the choice of (i) determine $R(5,5)$ within one year or (ii) face intergalactic war, then we should make strenuous efforts to find $R(5,5)$. If the condition (i) were altered to that of finding $R(6,6)$ we should immediately prepare for war!"

One of Erdös's early contributions to intergalactic peace came via Theorem 16.8. In fact, Erdös proved this result by introducing a new probabilistic technique involving random colorings of the edges of complete graphs. We have recast the proof (so it is really the same proof!) without the probabilistic element. We need first a useful inequality for the binomial coefficients that somehow did not get included in Chapter 2.

LEMMA 16.7

Let t, p be positive integers with $t \geq p$, then $2C(t,p) < (t/p)^p e^p$.

Proof

We have

$$C(t,p) = \frac{t!}{(t-p)!\, p!} = \frac{t(t-1)\ldots(t-p+1)}{p!} < \frac{t.t\ldots t}{p!} = \frac{t^p}{p!} = \left(\frac{t}{p}\right)^p \frac{p^p}{p!}. \qquad (16.1)$$

---

* See Section 5.1 of Chapter 5 for a brief biography of Erdös.

Now

$$e^p = 1 + p + \frac{p^2}{2!} + \ldots + \frac{p^{p-1}}{(p-1)!} + \frac{p^p}{p!} + \frac{p^{p+1}}{(p+1)!} + \ldots$$

and hence, since all the terms on the right-hand side of the above equation are positive,

$$e^p > \frac{p^{p-1}}{(p-1)!} + \frac{p^p}{p!} = \frac{p \times p^{p-1}}{p \times (p-1)!} + \frac{p^p}{p!} = \frac{p^p}{p!} + \frac{p^p}{p!} = \frac{2p^p}{p!}.$$

Hence by Equation 16.1,

$$2C(t,p) < \left(\frac{t}{p}\right)^p e^p.$$

A natural reaction to seeing Lemma 16.7 is "I'd never thought of an inequality like that!" Of course, results of this kind do not come out of the blue. Generally they arise when they are suggested by other results, or are needed to complete the proof of another theorem. We are now ready for Erdös's theorem.

THEOREM 16.8
(Erdös 1947*)

If p is an integer with $p \geq 2$, then

$$R(p,p) \geq \frac{p(\sqrt{2})^{p-1}}{e}.$$

Proof

We let t be a positive integer, and we seek an inequality for t that rules out the possibility that K_t has neither a red K_p nor a blue K_p. Note first that as K_t has $C(t, 2)$ edges, it has $2^{C(t,2)}$ colorings of its edges using red and blue.

We first choose a specific set, say P, of p vertices from the t vertices of K_t. There are $C(t, p)$ different ways of making this choice. We color all the edges joining two vertices from P red. Let s be the number of edges of K_t that do not join two vertices in P. Then $s = C(t,2) - C(p,2)$. There are 2^s different ways in which these edges can be colored using red and blue. So there are 2^s ways of coloring the edges of K_t so that all the vertices in P are joined by red edges. Similarly, the number of ways of coloring the edges of K_t using red or blue so that all the edges with two vertices in P are blue is also 2^s. Since there are $C(t,p)$ ways to choose our set P, the total number of colorings of K_t so that there is either a red K_p or a blue K_p cannot exceed $2C(t,p)2^s$ (and may well be less since the chosen set

---

* P. Erdös, Some Remarks on the Theory of Graphs, *Bulletin of the American Mathematical Society*, 53, 1947, pp. 292–294.

P may not be the only set of p vertices such that all the edges joining them are red, or all these edges are blue, and hence this estimate may involve some double counting). Hence, if the inequality

$$2^{C(t,2)} > 2C(t,p)2^s \qquad (16.2)$$

holds, there must be at least one coloring of K_t without either a red K_p or a blue K_p.

Now,

$$2^{C(t,2)} > 2C(t,p)2^s \Leftrightarrow 2^{C(t,2)} > 2C(t,p)2^{C(t,2)-C(p,2)}, \text{by the definition of s,}$$

$$\Leftrightarrow 2^{C(p,2)} > 2C(t,p). \qquad (16.3)$$

It follows from Lemma 16.7 that,

$$\text{if } \left(\frac{t}{p}\right)^p e^p < 2^{C(p,2)}, \text{ then } 2C(t,p) < 2^{C(p,2)}. \qquad (16.4)$$

Now, as $C(p, 2) = (p(p-1))/2$,

$$\left(\frac{t}{p}\right)^p e^p < 2^{C(p,2)} \Leftrightarrow \left(\frac{t}{p}\right)^p e^p < \left(2^{\frac{p-1}{2}}\right)^p$$

$$\Leftrightarrow \frac{te}{p} < 2^{\frac{p-1}{2}} \Leftrightarrow t < \frac{p 2^{p/2}}{e\sqrt{2}} \qquad (16.5)$$

$$\Leftrightarrow t < \frac{p(\sqrt{2})^{p-1}}{e}.$$

So if t satisfies the inequality 16.5, it follows from the implication 16.4 that the inequality 16.3, and hence also the inequality 16.2 holds. That is, if t satisfies the inequality 16.5 there is a coloring of K_t with neither a red K_p nor a blue K_p. It follows that $R(p,p) \geq (p(\sqrt{2})^{p-1})/e$. This completes the proof.

As for upper bounds, we can easily derive an answer from Theorem 16.5.

THEOREM 16.9

For all integers $p \geq 2$, we have

$$R(p,p) \leq C(2p-2, p-1) \leq 4^{p-1}.$$

Proof

It follows immediately from Theorem 16.5 that $R(p, p) \leq C(2p - 2, p - 1)$. It is straightforward to prove, using mathematical induction, that for all positive integers n, $C(2n, n) < 4^n$ (you are asked to do this in Exercise 16.2.2A), and hence $C(2p - 2, p - 1) < 4^{p-1}$.

Exercises

16.2.1A i. Use Theorems 16.6 and 16.8 to determine lower bounds for $R(5,5)$ and $R(6,6)$.

ii. Explain how you can "see" that the lower bound for $R(p,p)$ given by Theorem 16.6 will be less than the lower bound given by Theorem 16.8, for all but finitely many values p (so that, for almost all integers p, Theorem 16.8 gives the better result).

16.2.1B Find the least value of p for which the lower bound for $R(p,p)$ given by Theorem 16.8 is larger than the lower bound given by Theorem 16.6.

16.2.2A Prove, using mathematical induction, that for every integer $n \geq 2$, $\frac{2}{3} 3^n \leq C(2n,n) \leq \frac{3}{8} 4^n$.

16.2.2B Prove that for every ε, with $0 < \varepsilon < 4$, and every positive constant K, there is a positive integer N such that for all integers $n > N$, $K(4-\varepsilon)^n < C(2n,n)$. [In fact, it can be shown that $C(2n,n)$ is asymptotic to $4^n/\sqrt{\pi n}$, using Stirling's formula* for $n!$.]

16.3 GRAPHS OF MANY COLORS

You will obviously be asking if similar results are available if we color complete graphs with more than two colors. Again Ramsey's theorem says that there are. The next theorem, which is essentially Ramsey's full theorem with $s = 2$, allows us to prove such results.

THEOREM 16.10

Let m and $n_1, n_2, \ldots, n_m$ be positive integers, and let $C_1, C_2, \ldots, C_m$ be m different colors. Then there is a positive integer M such that if the edges of the complete graph K_M are painted using these colors, then for some integer i with $1 \leq i \leq m$, K_M has a K_{n_i} of color C_i (that is, there are n_i vertices of K_M such that all the edges joining them are have been painted the color C_i).

Proof

We call the argument we use here the *tangerine argument*.[†] We give a proof using mathematical induction. The result is obviously true for $m = 1$. Suppose now that it holds for $m = k$, and consider the case where $m = k + 1$. We suppose that C_{k+1} is tangerine and, for the time being, describe all the other colors as *dull*.

There are k dull colors, $C_1, C_2, \ldots, C_k$. Hence, by the induction hypothesis, there is a positive integer L such that if the edges of K_L are painted using these colors, then for some integer i with $1 \leq i \leq k$, K_L has a K_{n_i} of colors C_i.

Now let $M = R(L, n_{k+1})$. If the edges of the complete graph K_M are painted either dull or tangerine, then K_M has either a K_L all of whose edges are dull or a tangerine $K_{n_{k+1}}$. Now, by the choice of L, in the first case for some i, with $1 \leq i \leq k$, K_M has a K_{n_i} of color

---

* Stirling's formula says that $n!$ is asymptotic to, that is, very closely approximated by, $\sqrt{2\pi n}(n/e)^n$. For a proof of this, see the first edition of this book. We have not managed to find space for it in this expanded edition.

† Tangerine is the color of Blackpool Football Club, of which one of the authors is a very keen follower. For some unfathomable reason, the other author cannot abide soccer (footnote added by the Blackpool supporter).

C_i. Thus for some i, with $1 \leq i \leq k+1$, K_M has a K_{n_i} of color C_i. Thus the result holds also for $m = k + 1$. This completes the proof.

The *least* such number M that satisfies the conditions of Theorem 16.10 is usually denoted by $R(n_1, n_2, \ldots, n_m)$ and is called the $(n_1, n_2, \ldots, n_m)$ *Ramsey number*. The only Ramsey number with $m \geq 3$ whose actual value is known is $R(3,3,3)$, which is 17 (see Exercise 16.3.1B).

As we have already noted, we could also generalize the above result by replacing the two-element subsets, that is, graph edges, by three-, four-, and more-element subsets of a given set. If we go in the other direction and consider one-element subsets of a set, that is, we color the elements of a set (rather than pairs, triples, quadruples, etc., of elements), we obtain Theorem 15.3, one of the generalizations of the pigeonhole principle. We therefore see that Ramsey theory is a vast generalization of Dirichlet's pigeonhole principle.

Exercises

16.3.1A a. Use the proof of Theorem 16.10 to show that for each positive integer k and all positive integers $n_1, \ldots, n_k, n_{k+1}$, $R(n_1, \ldots, n_k, n_{k+1}) \leq R(R(n_1, \ldots, n_k), n_{k+1})$, and deduce that for all positive integers p, q, and r,

$$R(p,q,r) \leq \min\{R(R(p,q),r), R(R(p,r),q), R(R(q,r),p)\}.$$

b. Prove that, for all positive integers $p, q, r \geq 2$,

$$R(p,q,r) \leq R(p-1,q,r) + R(p,q-1,r) + R(p,q,r-1) - 1.$$

16.3.1B Suppose that the edges of K_{17} are colored using red, blue, or green. Let v be a vertex of K_{17}. Show that
  i. At least six of the edges with v as endpoint must be of the same color (say red).
  ii. Consider the subgraph whose vertices are six vertices joined to v by red edges. Note that either this subgraph has at least one red edge, or all its edges are blue or green. Show that in the first case K_{17} has a red triangle, and in the second either a green or a blue triangle.
  iii. Deduce that $R(3,3,3) \leq 17$.

16.4 EUCLIDEAN RAMSEY THEORY

There are some fascinating Ramsey-type problems of a geometric nature. Here are some involving either coloring all points on the real line (or, if you prefer, the x-axis, or any other infinite straight line) or coloring all the points in the plane.

PROBLEM 16.1

Suppose that, with each point on the real line, we associate either the color red or the color blue (loosely, "we color each point of the line red or blue"). Show that we can always find points A, B, C, on the line, all of the same color and with B the midpoint of AC.

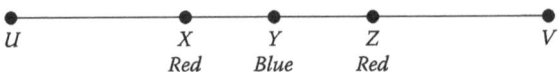

U X Y Z V
   Red Blue Red

FIGURE 16.5

Solution

We can certainly find points X, Z of the same color, say red. If the midpoint of XZ, say Y, is also red, then we can put $A = X$, $B = Y$, and $C = Z$, and there is nothing left to prove.

So, let us suppose that Y is blue. Now choose points U, V as in Figure 16.5 so that the lengths of the line segments UX, XZ, and ZV are all equal.

Now, if U is red, then $A = U$, $B = X$, $C = Z$ forms a red trio of the required kind. If V is red, then $A = X$, $B = Z$, $C = V$ is a red trio of the required kind. If neither U nor V is red, then each is blue and $A = U$, $B = Y$, $C = V$ is a blue trio of the required kind.

Another example of similar type is:

PROBLEM 16.2

Prove that if every point of the plane is colored red or blue, the plane contains a monochromatic rectangle, that is, a rectangle with all four vertices of the same color.

Solution

Draw any three (horizontal) parallel lines in the plane and, at right angles to those, seven vertical (and so mutually parallel) lines. (Do you wonder why seven? You'll see!) In our example in Figure 16.6 we have numbered the vertical lines to make it easy to refer to them.

On each of these seven vertical lines we write three letters, for example r, r, b, indicating the colors (in order from top to bottom) of the points of intersection of the line with the three horizontal lines. If two of these trios begin with r, r, then there is a rectangle with four red vertices. If two of these trios begin with b, b, there is a rectangle with four blue vertices. If neither of these cases occurs, at least five of these trios begin either r, b or b, r (on the vertical lines 2, 3, 4, 6, and 7 in our example). So there are either at least three lines where the trio begins b, r or at least three lines where the trio begins r, b. Without loss of generality we can assume there are three lines where the trio begins b, r (lines 2, 4, and 7 in our example). So the trios on these three lines are all either b, r, b or b, r, r, and hence at least two of these must be the same. If there are two lines with the trio b, r, b, there is a rectangle with all vertices blue (formed by the vertices shown on lines 2 and 7 in our example), and if there are two lines with the trio b, r, r, there is a rectangle with all vertices red. So in each case there is a monochromatic rectangle.

You will note that the hypothesis that *every* point of the plane is colored red or blue is much, much stronger than is needed to draw the conclusion that there is a monochromatic rectangle. Our solution shows that we can deduce this conclusion from the much weaker hypothesis that we have a 3×7 rectangular array of points of the plane each of which is colored red or blue. This suggests that from the assumption that every point of the plane is colored red and blue, more interesting conclusions can be drawn than just the existence of the monochromatic rectangle. We give a few examples.

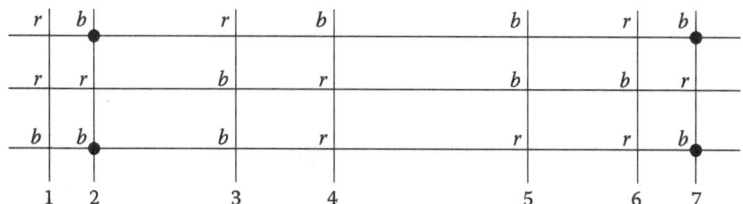

FIGURE 16.6

PROBLEM 16.3

Prove that if every point of the plane is colored red or blue, then there are lines of all possible lengths with monochromatic endpoints.

Solution

Suppose $x > 0$. Let ABC be an equilateral triangle in the plane with side length x. Two of the three points A, B, and C must be the same color, so there is a line of length x whose endpoints are the same color. In contrast, this result is not true if we restrict attention to points on a line.

PROBLEM 16.4

Show that it is possible to color the points of the real line red and blue, so that there are no points of the same color distant 1 unit from each other.

Solution

For each integer n, color the points in the interval $[n, n+\frac{1}{2})$ red, and those in the interval $[n+\frac{1}{2}, n+1)$ blue.

Exercise 16.4.3B asks you to prove that it is possible to color every point of the plane red or blue so that no equilateral triangle *with sides of unit length 1* has vertices all of the same color. In contrast, we can always find a monochromatic equilateral triangle of some size, as shown in the next problem.

PROBLEM 16.5

Prove that if every point of the plane is colored red or blue, then there is a monochromatic equilateral triangle.

Solution

Suppose every point of the plane is colored red or blue. Then by Problem 16.1, there are three points A, B, and C such that B is the midpoint of AC and such that the points A, B, and C are all the same color. Without loss of generality we can assume that they are all red.

Now let the point D be such that the triangle ACD is equilateral, let E be the midpoint of AD, and let F be the midpoint of CD. (See Figure 16.7.)

If D is red, the equilateral triangle ACD is monochromatic. If E is red, the equilateral triangle ABE is monochromatic. If F is red, the equilateral triangle BCF is

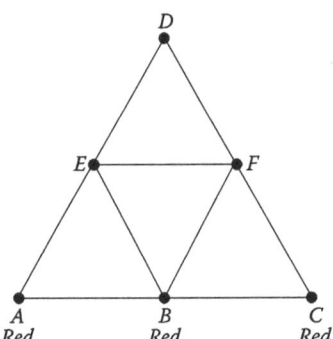

FIGURE 16.7

monochromatic. If none of D, E, and F is red, they are all blue, and so the equilateral triangle DEF is monochromatic. Thus, in every case there is a monochromatic equilateral triangle.

As a partial response to Problem 16B we have shown that, given a colouring of the plane using two colours, and an equilateral triangle, there is a similar (and hence, equilateral) triangle which is monochromatic. There may not be a *congruent* equilateral triangle as Problem 16.4.3B shows.

Given a non-equilateral triangle the same proof as for Problem 16.5 shows that there is a similar triangle which is monochromatic. That there is always a congruent monochromatic triangle is conjectured, but a proof is still awaited.

Exercises

16.4.1A Show that if all the points on a line are colored using red and blue, it will always be possible to find three points A, B, and C on the line that are given the same color, and such that the ratio of the lengths $AB:BC$ is 1:2.

16.4.1B Show that if all the points on a line are colored using red and blue, then for each pair of positive integers m, n, it is possible to find three points A, B, and C on the line that are given the same color and such that the ratio of the lengths $AB:BC$ is $m:n$.

16.4.2A Show that if all the points in the plane are colored using three colors, then there is a monochromatic rectangle (that is, a rectangle whose four vertices are all given the same color).

16.4.2B Show that if all the points in three-dimensional space are colored using two colors, then there is a rectangular box (that is, a cuboid) whose eight corners lie at points of the same color.

16.4.3A Describe a way to color the points of the plane using two colors so that there is no square with side length 1 unit all of whose vertices are given the same color.

16.4.3B Show that there is a coloring of the plane using two colors in which there is no monochromatic equilateral triangle of side 1 unit.

CHAPTER 17

Rook Polynomials and Matchings

17.1 HOW ROOK POLYNOMIALS ARE DEFINED

In Chapter 1 we raised the following problem, which we pointed out is relevant to many scheduling problems.

PROBLEM 17A
Nonattacking Rooks

Given the 5×5 board in Figure 17.1, in how many ways can 0, 1, 2, 3, 4, 5, or more nonattacking rooks (that is, no two rooks in the same row or column) be placed on the board so that *none of them lies on a black square*?

This is a generalization of Problem 2.3 of Chapter 2 where we determined the number of (different) ways that eight nonattacking rooks can be placed on an 8×8 chessboard. We will now look at the same problem for an $m \times n$ array of squares, but with the added restriction that some of the squares are "forbidden," in that we are not allowed to place rooks on them, as in Problem 17A, where we have colored the forbidden squares black. In fact, we could restrict our attention to $n \times n$ arrays, since a rectangular array can always be turned into a square array by adding extra rows or columns of black (forbidden) squares.

It is convenient to use the term *chessboard*, and usually just *board*, to mean an array of this kind. The standard chessboard is an 8×8 array in which alternate squares are black. Since throughout this chapter we are concerned with questions about nonattacking rooks, we shall sometimes omit the qualification "nonattacking" in discussing these questions.

A specific question of this type is: In how many ways can four rooks be placed on a standard chessboard so that (i) each rook is placed on a white square and (ii) no two rooks are attacking each other? We *could* answer the question merely by listing the possibilities. The trouble as usual with this sort of method is that in making the list we risk accidentally omitting some possible positions or, slightly less likely, counting some positions twice. In fact, the methods we shall develop will allow us to determine, all at once, exactly how many ways there are of placing k rooks on such a board, for any (nonnegative) integer k.

A more practical problem of this type is that of allocating, say, five jobs to five different employees, given that each employee can only do some of the jobs. As we saw in Chapter 1, we can represent this problem by a grid in which the rows correspond to the employees and the columns correspond to the jobs, and where we shade a square in a particular row and column to indicate that the corresponding employee cannot do the corresponding job. Figure 17.2 shows an example of this type.

Clearly, apart from the fact that we have now labeled the rows and columns, Figure 17.2 is essentially the same as Figure 17.1. The problem of placing nonattacking rooks in the white squares of Figure 17.1 is the same as that of putting ticks in the white squares of Figure 17.2 to indicate how the jobs are allocated, assuming that each employee is allocated one job. In a practical problem often the key question is *whether the jobs can be allocated at all*, rather than in how many different ways this can be done. This then becomes an existence question. We shall look at such questions in Section 17.2 of this chapter.

Given a board, B, we use the notation $r_k(B)$ for the number of ways we can place k nonattacking rooks on the white squares of B. Problem 2.3 in Chapter 2 shows that if B is an all white 8×8 board, $r_8(B) = 8!$. Clearly, for any board B, $r_1(B)$ is the number of white squares on the board. We also have that, for each board B, $r_0(B) = 1$, since there is precisely one way of doing nothing! This may seem a little strange, but you will soon see why it is, in any case, helpful to define $r_0(B)$ to be 1. Our remark that there is just one way of doing nothing is equivalent to the fact that given a board B, there is just one diagram showing the board with no rooks on it.

A general method, often employed in mathematics, is to "reduce" a given problem to more manageable "smaller" ones. We follow this method to show how, if we wish

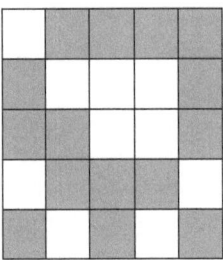

FIGURE 17.1

	Job 1	Job 2	Job 3	Job 4	Job 5
Employee A					
Employee B					
Employee C					
Employee D					
Employee E					

FIGURE 17.2

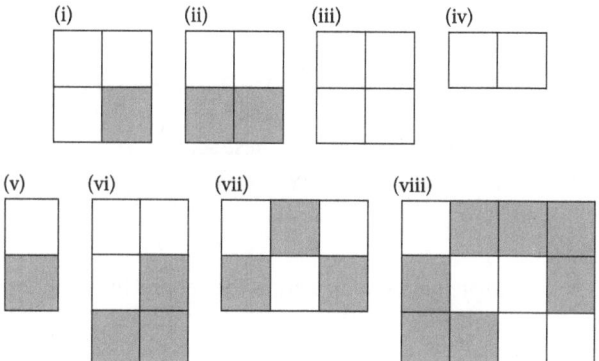

FIGURE 17.3

TABLE 17.1

		Boards							
		(i)	(ii)	(iii)	(iv)	(v)	(vi)	(vii)	(viii)
	0	1	1	1	1	1	1	1	1
k	1	3	2	4	2	1	3	3	5
	2	1	0	2	0	0	1	2	7
	3	0	0	0	0	0	0	0	3

to find obtain $r_k(B)$ for "large" boards, it is useful to have some information available concerning "smaller" boards. This we can gain merely by inspection, and it is with this that we begin.

PROBLEM 17.1

Consider the various boards in Figure 17.3.

Calculate the values of $r_k(B)$ for $0 \le k \le 3$, for each of these boards.

Solution

Most of the values in Table 17.1 are, we hope, obvious, so we confine ourselves to just a few words of explanation. The values in the first two rows arise from our observation above that for every board, B, $r_0(B) = 1$, and $r_1(B)$ is the number of white squares on the board. We can place k nonattacking rooks on a board only if we have available squares in at least k rows and at least k columns. This explains the zeros in rows 3 and 4 of Table 17.1. Since none of these boards has more than three rows, for each board $r_k(B) = 0$ for $k > 3$. Column (vi) is the same as column (i) because the third row of black squares in (vi) has no effect on the values of $r_k(B)$. The only case where some care is needed is board (viii), where we have obtained the answers by carefully enumerating all the possibilities. We use the notation S_{rs} for the square in row r and column s, with the rows and columns numbered so that the square in the bottom left-hand corner is S_{11}. Then the seven ways we can place two nonattacking rooks on board (viii) are given by the following seven pairs of squares:

$$(S_{13}, S_{22}), (S_{13}, S_{31}), (S_{14}, S_{22}), (S_{14}, S_{23}), (S_{14}, S_{31}), (S_{22}, S_{31}), \text{ and } (S_{23}, S_{31}).$$

Even in a comparatively simple case like that of board (viii), listing all the possibilities is a little tedious, and it requires care, as it is quite easy to overlook some possibilities. So we now introduce a more systematic method. We work with the generating functions for the sequences $\{r_k(B)\}$. Since $r_k(B) = 0$ if the board has fewer than k rows or fewer than k columns, all but finitely many of the terms in these sequences will be 0. Thus in this context all the generating functions are polynomials. We now give the formal definition.

DEFINITION 17.1
For each board B, the *rook polynomial*, $r(x, B)$, is the generating function of the sequence $\{r_k(B)\}$, that is,

$$r(x,B) = r_0(B) + r_1(B)x + r_2(B)x^2 + \ldots.$$

We note that, if B is an $m \times n$ board, then $r(x, B)$ is a polynomial of degree at most the minimum of m and n.

PROBLEM 17.2
Write down the rook polynomials of the boards in Problem 17.1.

Solution
From Table 17.1 we see that these polynomials are as follows.

a. $1 + 3x + x^2$ b. $1 + 2x$ c. $1 + 4x + 2x^2$ d. $1 + 2x$
e. $1 + x$ f. $1 + 3x + x^2$ g. $1 + 3x + 2x^2$ h. $1 + 5x + 7x^2 + 3x^3$

To prepare for our first reduction theorem, we need some definitions.

DEFINITION 17.2
  i. A *subboard* of a board B is a rectangular set of squares of B.
  ii. Two subboards, X and Y, of a board, B, are said to be *disjoint* if no square of X is in the same row or column as any square of Y. More generally, k subboards are *pairwise disjoint* if no square of one subboard is in the same row or column as any square of the other subboards.
  iii. We say that a board B can be *split up into the subboards* $B_1, B_2, \ldots, B_k$ if these subboards are pairwise disjoint and all the squares of B that are not in any of these subboards are black.

For example, in Figure 17.4 the board B can be split up into the subboards B_1 and B_2 as shown. Some more examples of split boards may be found in the solution to Problem 17.3.

THEOREM 17.1
The Disjoint Subboards Theorem
  a. Suppose that a given board B can be split up into the subboards B_1 and B_2. Then the rook polynomial of B is the product of the rook polynomials of B_1 and B_2; that is,

$$r(x, B) = r(x, B_1)r(x, B_2). \tag{17.1}$$

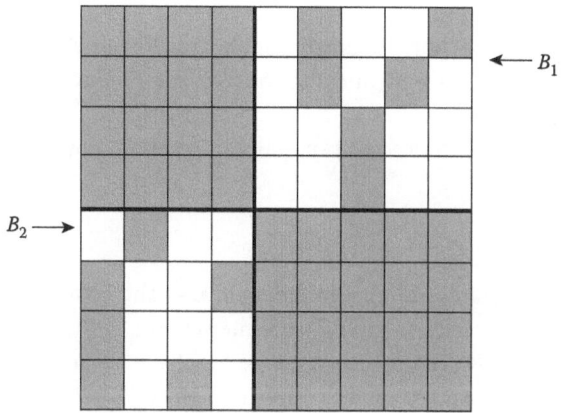

FIGURE 17.4

b. Suppose that a given board B can be split up into the k subboards $B_1, B_2, \ldots, B_k$, then

$$r(x, B) = r(x, B_1) r(x, B_2) \ldots r(x, B_k). \tag{17.2}$$

Comment

Before giving a general proof, we consider the particular case of the board in Figure 17.4. In how many ways can we place four nonattacking rooks on this board? Since the white squares of the subboard B_1 are in different rows and columns from those of B_2, the positions in which we put rooks on the board B_1 has no effect on the squares where we can put rooks on B_2 and vice versa. So, for example, if we choose to put three rooks on B_1 and one rook on B_2, the choices for B_1 are independent of the choices for B_2. You can check that there are 84 ways to place three rooks on B_1 and 10 ways to place one rook on B_2. Thus there are 84×10 ways to place four nonattacking rooks on the board B with three rooks on B_1 and one on B_2. It is more illuminating to write this product as $r_3(B_1) r_1(B_2)$.

In general, where the subboards B_1 and B_2 are disjoint, we can place i nonattacking rooks on B_1 and j on B_2 in $r_i(B_1) r_j(B_2)$ ways. We can place four rooks on the board B by putting i on B_1 and $4 - i$ on B_2, for $0 \leq i \leq 4$. Hence, for $0 \leq i \leq 4$, provided that we give $r_0(B_1)$ and $r_0(B_2)$ the value 1, we can put i rooks on B_1 *and* $4 - i$ on B_2 in $r_i(B_1) r_{4-i}(B_2)$ ways. Since different choices of i give different ways of placing the rooks, the total number of ways of placing four nonattacking rooks on B is given by the sum of these products. That is, it is

$$\sum_{i=0}^{4} r_i(B_1) r_{4-i}(B_2). \tag{17.3}$$

You will recognize the formula in Equation 17.3 as giving the coefficient of x^4 in the product of the polynomials $r_0(B_1) + r_1(B_1)x + r_2(B_1)x^2 + \ldots$ and $r_0(B_2) + r_1(B_2)x + r_2(B_2)x^2 + \ldots$, that is, in the product $r(x, B_1) r(x, B_2)$. (The formula corresponds to that in Equation 7.4 in Chapter 7.)

The same argument applies if we consider, in general, the number of ways of placing n rooks on a board B that is split into disjoint subboards B_1 and B_2. So the coefficients in the rook polynomial $r(x, B)$ are identical to those in the product $r(x, B_1)r(x, B_2)$. Hence $r(x, B) = r(x, B_1)r(x, B_2)$.

This discussion of a particular case generalizes. So we can now give a quick proof of the theorem.

Proof

a. Since the white squares of B_1 are in different rows and columns from those of B_2, nonattacking rooks may be placed on white squares of B_1 independently of how they are placed on B_2. Hence we may place i rooks on B_1 and j on B_2 in $r_i(B_1)r_j(B_2)$ ways. We may place n rooks on B by choosing an integer i with $0 \leq i \leq n$ and placing i rooks on B_1 and $n - i$ on B_2. Hence n nonattacking rooks may be placed on B in

$$\sum_{i=1}^{n} r_i(B_1)r_{n-i}(B_2) \tag{17.4}$$

ways.

The formula 17.4 gives the coefficient of x^n in the rook polynomial $r(x, B)$. As it is the formula for the coefficient of x^n in the product $r(x, B_1)r(x, B_2)$, it follows that Equation 17.1 in the statement of the theorem holds.

b. This is an obvious generalization of (a).

PROBLEM 17.3

Find the rook polynomials for the boards shown in Figure 17.5

Solution

We split the boards as shown in Figure 17.6.

i. The board has been split up into two subboards, which are the boards (a) and (b) of Figure 17.3. Hence by Theorem 17.1(a), and using the solution to Problem 17.2, we see that the rook polynomial of this board is $(1 + 3x + x^2)(1 + 2x)$.

ii. We see that the board can be split up into the boards (a), (e), and (g) of Figure 17.3. Hence, by Theorem 17.1(b), its rook polynomial is $(1 + 3x + x^2)(1 + x)(1 + 3x + 2x^2)$.

Now let us consider how we might find the rook polynomial for the board, say B, shown in Figure 17.1. This cannot be split up into subboards, so at first glance Theorem 17.1 doesn't appear to be any help at all. Theorem 17.2 will change that!

The key idea is that we can rearrange the rows and columns on the board shown in Figure 17.1 to get a board that can be split up into subboards. We illustrate this process in Figure 17.7.

We have numbered the rows and columns of the board to help you see the row and column permutations we have made. The middle diagram shows the effect of permuting the rows, as indicated, and the diagram on the right shows the final configuration after we have also permuted the columns.

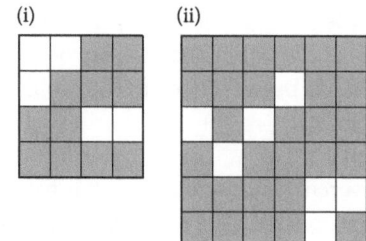

FIGURE 17.5

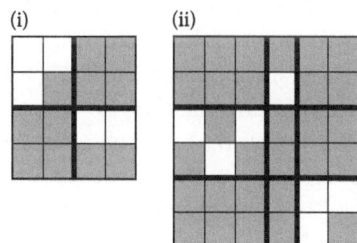

FIGURE 17.6

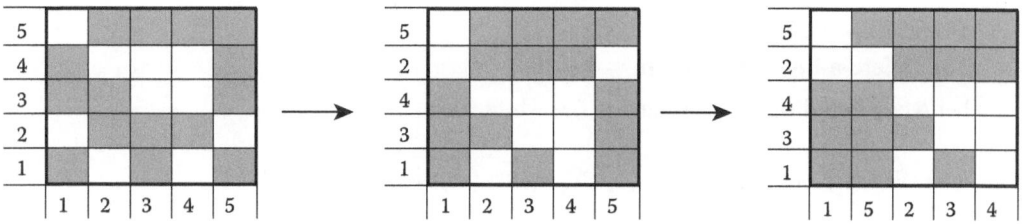

FIGURE 17.7

Since entire rows and columns are permuted, two squares that are initially in different rows and columns will end up in different rows and columns and vice versa. Thus there is a one–one correspondence between ways of placing a given number of nonattacking rooks on the initial board and the ways of placing them on the final board. Hence both boards have the same rook polynomials. We record this fact formally as follows.

THEOREM 17.2
Let B, B' be $m \times n$ boards, such that there is a permutation, say π, of the set $\{1, 2, \ldots, m\}$ and a permutation, say σ, of the set $\{1, 2, \ldots, n\}$ with the property that,
  for $0 \leq i \leq m$ and $0 \leq j \leq n$, the square in row i and column j of B is black
  if and only if the square in row $\pi(i)$ and column $\sigma(j)$ of B' is black.
Then the boards B and B' have the same rook polynomials.

As the board on the right-hand side of Figure 17.7 can be split up into the subboards B_1 in the top left-hand corner and B_2 in the bottom right-hand corner, it follows from

Theorems 17.1 and 17.2 that the rook polynomial of the original board B is the product of the rook polynomials of B_1 and B_2. Thus the problem of calculating the rook polynomial of the given 5×5 board has been reduced to the problem of calculating the rook polynomial of two smaller boards.

Our second reduction theorem is not so obvious. It is based on the observation that the different ways of placing n nonattacking rooks on a board can be divided into those that include putting a rook on a particular square, and those that do not. It is helpful to introduce some notation.

We use "square (r,s)" to mean the square in row r and column s. Let B be an $m \times n$ board with the square (r,s) white. We will use the notation $B-(r,s)$ for the board obtained from B by changing the color of this white square to black. The squares available to a rook on the board $B-(r,s)$ are the same as those for B except that the square (r,s) can no longer be used. This explains the choice of notation.

We will use the notation $B+(r,s)$ for the board obtained from B by coloring black *all* the squares that are in either column r or row s. This time the choice of notation may seem a little more mysterious. The squares available to a rook on the board $B+(r,s)$ are all those available on the board B after a rook has been put on the square (r,s).

An example to illustrate this notation is given in Figure 17.8. For clarity we have marked the square $(3,5)$ with a dot.

THEOREM 17.3
The Delete-a-Square Theorem
Let B be a board where the square (r,s) is white. Then

a. For each positive integer k, $r_k(B) = r_k(B - (r,s)) + r_{k-1}(B + (r,s))$.
b. $r(B, x) = r(B - (r,s), x) + xR(B + (r,s), x)$.

Proof
Let k be a positive integer. The set of all ways of placing k nonattacking rooks on the board B may be divided into two disjoint classes: (α) those in which a rook is not placed on the square (r,s), and (β) those in which a rook is placed on this square.

Clearly the number of placements in class (α) is the same as for the board obtained from B by coloring the square (r,s) black, which means that a rook cannot be placed on it. So there are $r_k(B - (r,s))$ placements in this class. If a rook is placed on the square

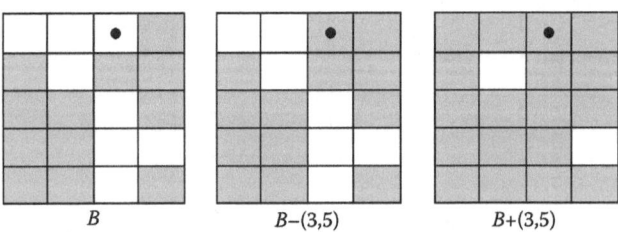

$\qquad\qquad B \qquad\qquad\qquad B-(3,5) \qquad\qquad B+(3,5)$

FIGURE 17.8

(r, s), then we are left with $k - 1$ rooks to be placed on the board, with none of them in column r or row s. The number of ways of placing these rooks is the same as the number of ways of placing $k - 1$ rooks on the board obtained from B by blacking out the squares in column r and those in row s, that is, $r_{k-1}(B + (r, s))$. This proves (a).

From (a) we can immediately deduce (b), as, for $k \geq 1$, the term involving x^k in the polynomial $r(B - (r, s), x) + xr(B + (r, s), x)$ is $r_k(B - (r, s))x^k + x(r_{k-1}(B + (r, s))x^{k-1})$, and hence the coefficient of x^k is $r_k(B - (r, s)) + r_{k-1}(B + (r, s))$, which by (a) is $r_k(B)$. It is also straightforward to check that the constant term in the polynomial $r(B - (r, s)) + xr(B + (r, s))$ is 1. Thus the polynomials on the left- and right-hand sides of the equation in (b) have the same coefficients. So they are the same polynomial.

PROBLEM 17.4

Calculate the rook polynomial for the board, B, shown in Figure 17.8.

Solution

By the disjoint subboards theorem the rook polynomial of the board $B - (3,5)$ of Figure 17.8 is the product of the rook polynomials of the subboards shown in Figure 17.9. These are quite easy to calculate, and are given under the boards.

So the rook polynomial of the board $B - (3,5)$ is $(1 + 3x + x^2)(1 + 4x + 2x^2)$. It is straightforward to see that the rook polynomial of the board $B + (3,5)$ is $(1 + x)^2$. Hence by the delete-a-square theorem, the rook polynomial of the board B is $(1 + 3x + x^2)(1 + 4x + 2x^2) + x(1 + x)^2$, that is, $1 + 8x + 17x^2 + 11x^3 + 2x^4$.

We now come to a third method of evaluating rook polynomials that is of most use in a case, such as that of the board B shown in Figure 17.10i, where there are more white squares than black squares. In such a case the disjoint subboards method will probably not be readily applicable and using the delete-a-square theorem may well require several steps. Fortunately, our third method may come to the rescue here. It is called the method of the *complementary board*.

DEFINITION 17.3

Given a board B, its *complementary board* is the board, B^C, obtained from B by changing all black squares to white and all white squares to black. Figure 17.10 shows the board B and its complementary board.

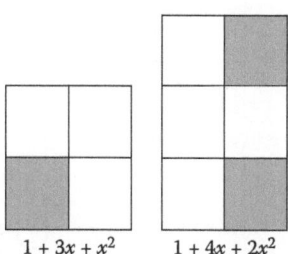

$1 + 3x + x^2$ $1 + 4x + 2x^2$

FIGURE 17.9

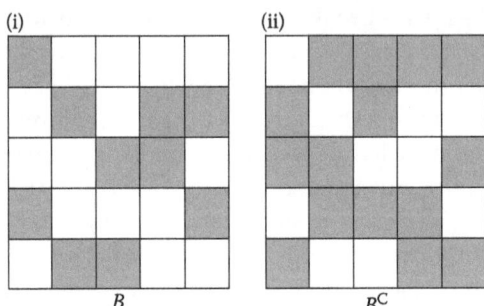

FIGURE 17.10

Here is the main theorem on complementary boards. We admit that its statement looks complicated, but the proof, if taken slowly, isn't *so* difficult. It is a good example of the use of the inclusion–exclusion theorem (Theorem 4.2 of Chapter 4).

THEOREM 17.4
The Complementary Board Theorem
Let B be an $m \times n$ board, with complementary board B^C. Then for each nonnegative integer k,

$$r_k(B) = \sum_{j=0}^{k} (-1)^j C(m-j, k-j) C(n-j, k-j)(k-j)! r_j(B^C). \tag{17.5}$$

Proof
We need first some notation. We let P be the set of all the different ways of placing k nonattacking rooks on an $m \times n$ board. Let $S_1, \ldots, S_r$ be different squares on the board, with $r \leq k$. We let $P(S_1, \ldots, S_r)$ be the set of ways of placing the k nonattacking rooks on the board with a rook on each of the squares $S_1, \ldots, S_r$. Note that $P(S_1, \ldots, S_r) = P(S_1) \cap \ldots \cap P(S_r)$ and that if any two of the squares $S_1, \ldots, S_r$ are in the same column or the same row, then $P(S_1, \ldots, S_r)$ is empty. We let $\beta(B)$ be the set of black squares on the board B.

By the result of Exercise 2.2.4A, there are $C(m,k)C(n,k)k!$ ways to place k nonattacking rooks on an $m \times n$ board. Those placings in which at least one rook is on a black square are those in the set

$$\bigcup_{S \in \beta(B)} P(S),$$

and hence

$$r_k(B) = C(m,k)C(n,k)k! - \#\left(\bigcup_{S \in \beta(B)} P(S)\right). \tag{17.6}$$

By the inclusion–exclusion theorem,

$$\#\left(\bigcup_{S\in\beta(B)} P(S)\right) = \sum_{j=1}^{k}(-1)^{j+1}\left(\sum_{S_1,\ldots,S_j\in\beta(B)} \#(P(S_1)\cap P(S_2)\cap\ldots\cap P(S_j))\right)$$

$$= \sum_{j=1}^{k}(-1)^{j+1} \sum_{S_1,\ldots,S_j\in\beta(B)} \#(P(S_1,\ldots,S_j)). \quad (17.7)$$

We now evaluate the sum

$$\sum_{S_1,\ldots,S_j\in\beta(B)} \#(P(S_1,\ldots,S_j)) \quad (17.8)$$

that occurs in Equation 17.7.

As we have already observed, if any of the squares $S_1,\ldots,S_r$ are in the same row or column, then $\#(P(S_1,\ldots,S_r))=0$. Thus the number of nonzero terms in the sum 17.8 is the number of ways of placing j rooks on the black squares of B so that no two of them are in the same row or column. But this is just the number of ways of placing j nonattacking rooks on the complementary board B^C, that is, $r_j(B^C)$.

Now suppose no two of the squares $S_1,\ldots,S_j$ are in the same row or column. So between them they lie in j rows and j columns. $P(S_1,\ldots,S_j)$ is the set of positions with j rooks on the squares $S_1,\ldots,S_j$ and hence $k-j$ nonattacking rooks in the remaining $m-j$ columns and $n-j$ rows in which none of these squares lie. Thus $\#(P(S_1,\ldots,S_j))$ is the same as the number of ways of placing with $k-j$ nonattacking rooks on an $(m-j)\times(n-j)$ board. So, by Exercise 2.2.4A, $\#(P(S_1,\ldots,S_j))=C(m-j,k-j)C(n-j,k-j)(k-j)!$

It follows that

$$\sum_{S_1,\ldots,S_j\in\beta(B)} \#(P(S_1,\ldots,S_j)) = C(m-j,k-j)C(n-j,k-j)(k-j)!\,r_j(B^C). \quad (17.9)$$

Now, by Equations 17.6, 17.7, and 17.9, we have that

$$r_k(B) = C(m,k)C(n,k)k! - \sum_{j=1}^{k}(-1)^{j+1}C(m-j,k-j)C(n-j,k-j)(k-j)!\,r_j(B^C). \quad (17.10)$$

and hence, as $r_0(B^C)=1$, we can rewrite Equation 17.10 as

$$r_k(B) = \sum_{j=0}^{k}(-1)^{j}C(m-j,k-j)C(n-j,k-j)(k-j)!\,r_j(B^C),$$

and this completes the proof.

PROBLEM 17.5

Calculate the rook polynomial for the board B of Figure 17.10.

Solution

Here, as we have said, it would appear that disjoint subboards theorem is not applicable while the delete-a-square theorem would take some while to implement. Do you agree? So let us try the complementary board method, and aim to calculate first the rook polynomial of the board B^C. Even this isn't completely straightforward. However, we can make progress by using the delete-a-square theorem and then permuting columns and rows, using Theorem 17.2 (Figure 17.11).

If we now take the board $B^C - (5,4)$ and apply the permutation (2 4) to the rows, and then the permutation (1 5 4 3 2) to the columns, we obtain the board in Figure 17.12.

We can apply the disjoint subboards theorem to the board in Figure 17.12. We have already seen in Problem 17.1(a) that the 2×2 board in the top right-hand corner has $1 + 3x + x^2$ as its rook polynomial. You can check by a direct calculation that the 3×3 board has rook polynomial $1 + 6x + 9x^2 + 2x^3$. It follows that the rook polynomial of the board in Figure 17.12, and hence of $B^C - (5,4)$, is $(1 + 3x + x^2)(1 + 6x + 9x^2 + 2x^3)$.

If we interchange rows 2 and 3 of the board $B^C + (5,4)$, we obtain the board in Figure 17.13, from which we can deduce that the rook polynomial of the board $B^C + (5,4)$ is $(1 + 2x)(1 + 4x + 3x^2)$.

We can now deduce from the delete-a-square theorem that the rook polynomial of the board B^C is

$$(1+3x+x^2)(1+6x+9x^2+2x^3)+x(1+2x)(1+4x+3x^2)$$

$$=1+10x+34x^2+46x^3+21x^4+2x^5. \quad (17.11)$$

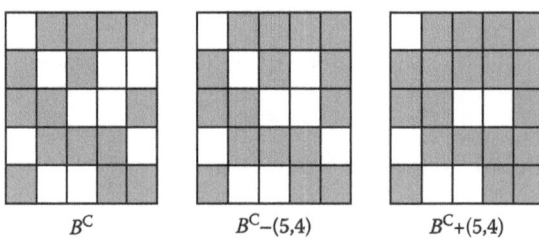

B^C $B^C-(5,4)$ $B^C+(5,4)$

FIGURE 17.11

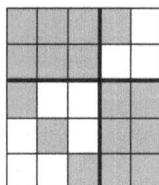

$B^C-(5,4)$, as above, with its rows and columns permuted

FIGURE 17.12

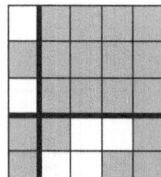

B^C+ (5,4) with its rows 2 and 3 permuted

FIGURE 17.13

We can now use the complementary board theorem to calculate the coefficients in the rook polynomial of the board B. We calculate the coefficient of x^4, that is, $r_4(B)$, and leave the other coefficients to you as an exercise. Since we are dealing here with a 5×5 board, we have $m = n = 5$. So, if we put $k = 4$ in Equation 17.5, we obtain

$$r_4(B) = \sum_{j=0}^{4} (-1)^j C(5-j, 4-j) C(5-j, 4-j)(4-j)! r_j(B^C)$$

$$= C(5,4)^2 4! r_0(B^C) - C(4,3)^2 3! r_1(B^C) + C(3,2)^2 2! r_2(B^C) - C(2,1)^2 1! r_3(B^C) + C(1,0)^2 0! r_4(B^C)$$

$$= 600 r_0(B^C) - 96 r_1(B^C) + 18 r_2(B^C) - 4 r_3(B^C) + r_4(B^C).$$

Now from Equation 17.11 we have that $r_0(B^C) = 1$, $r_1(B^C) = 10$, $r_2(B^C) = 34$, $r_3(B^C) = 46$, and $r_4(B^C) = 21$, and hence

$$r_4(B) = 600 \times 1 - 96 \times 10 + 18 \times 34 - 4 \times 46 + 21 = 89.$$

Although this has been rather a long calculation, you will see that it is much quicker done this way than by simply trying to list all the 89 different ways of placing four non-attacking rooks on the board B.

Exercises

17.1.1A Find, by a direct calculation, the rook polynomials of the boards shown in Figure 17.14.

17.1.1B Find the rook polynomial for the $n \times n$ board B_n on which all the squares are black other than those on the main diagonal. [The board B_3 is shown in Figure 17.14(i).]

17.1.2A Find the rook polynomial for the 4×4 board in which all the squares are white.

17.1.2B Find the rook polynomial for the 3×5 board in which all the squares are white.

17.1.3A Complete the solution to Problem 17.5 by calculating the values of $r_k(B)$ for $k = 0, 1, 2, 3$, and 5.

17.1.3B Find the rook polynomial of the boards whose complementary boards are those given in Problem 17.3.

17.1.4A Find the rook polynomials for the boards shown in Figure 17.15.

17.1.4B Find the rook polynomial for the board shown in Figure 17.16.

17.1.5A Your friendly combinatorics lecturer tells you that one of the 500 rook polynomials he set you to determine for homework is $1 + 4x + 7x^2 + 2x^3 + x^4$. Why must he must be wrong?

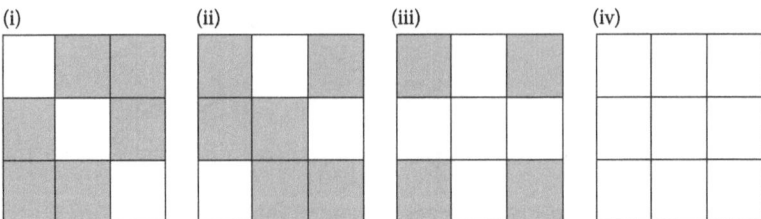

FIGURE 17.14

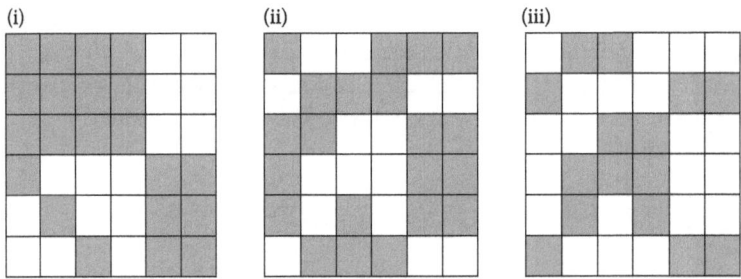

FIGURE 17.15

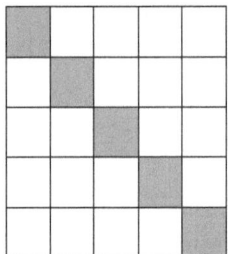

FIGURE 17.16

17.1.5B Show that $1+4x+6x^2+15x^3$ cannot be a rook polynomial.

17.1.6A Show that, if two 2×2 boards have the same rook polynomial, then they are (perhaps after a rotation) identical.

17.1.6B We know that two boards have the same rook polynomial if one can be obtained from the other by permuting rows and columns. Show that the converse isn't true. (*Hint:* There is a example with 2×3 boards.)

17.2 MATCHINGS AND MARRIAGES

While the manager seeking to allocate the jobs described by Figure 17.2 would no doubt be amused that there are three ways of allocating the jobs to the employees, two more pressing problems for most managers would probably be:

i. Can all the posts be suitably filled *at all*?
ii. Assuming that there is more than one way of filling the posts, how can these posts be filled in a way most advantageous to the firm?

Rook Polynomials and Matchings ■ 333

The first question is an example of an *existence problem:* Does a solution to the problem at hand actually exist? The second question is an example of an *optimization problem*. We confine ourselves to the first question.

Fortunately, there are theorems that can help the manager decide. This chapter describes one such. (See Theorem 17.5.) Although identically the same problem as that of matching people to jobs, the problem of pairing off one set of things with another (disjoint) set is more clearly and amusingly described in terms of marriage, hence it is often known as the *marriage problem*.

PROBLEM 17.6

Five men, Alan, Barry, Charlie, Dave, and Eric, and six women, Ursula, Vera, Wendy, Xena, Yolanda, and Zoe, (affectionately known, of course, as A, B, C, D, E, U, V, W, X, Y, and Z) enjoy the following friendships:

A is acquainted with U, Y, and Z; B is friendly with all the women; C with Y and Z; D with U, Y, and Z ; and E with Y and Z.

The question is whether it is possible—it is surely desirable(!)—that each of the men can be married to a woman with whom he is friendly. (We are assuming that a man can marry at most one woman, and a woman can marry at most one man.) Relationships such as friendships between men and women (between the elements of one set and those of a disjoint set) are sometimes conveniently (and clearly) expressed in terms of a bipartite graph. Figure 17.17 gives the bipartite graph corresponding to the friendships described above.

Solution

In fact, it is *not* possible for each man to marry a woman friend. This can be seen by observing that A, C, D, and E would need to have, between them, at least four friends, if each is to have a wife with whom he is friends. However, between them they know only three of the women, U, Y, and Z. It will follow from Theorem 17.5 that if only one of A, C, D, and E would get friendly with one of V, W, or X, then 5 happy(?) marriages (and one sad woman?) could result.

First, let us make a simple observation. That the four men require four wives emphasizes the fairly obvious remark that *for a pairing of all the men with friendly women to exist, it must necessarily be the case that, for each subset of k men, there must be (at least) k of the women with whom these k men are, between them, friendly.* This remark may

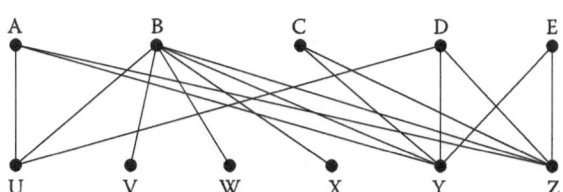

FIGURE 17.17

seem unremarkable until you learn that this necessary condition is also, remarkably, *sufficient*. That is, we have:

THEOREM 17.5
Hall's Marriage Theorem

Suppose that we have a set of m men and a set of at least m women such that for each integer k, with $1 \leq k \leq m$, each k men have between them at least k friends among the women. Then it is possible for each man to marry a woman with whom he is friendly.

Comment

We leave it to you, if you don't like mixing mathematics with everyday life, to restate this result more abstractly in terms of bipartite graphs.

Proof*

We prove the result by mathematical induction. It is clearly true in the case where $m = 1$. Suppose it holds for all $n < m$, and we have a set of m men who satisfy the condition of the theorem. We consider two cases:

a. For each $n < m$, each set of n men has, between them, at least $n + 1$ women they are friendly with.
b. For some $n < m$, there is a set, say S, of n men who have between them exactly n women they are friendly with.

In case (a), we can marry one of the men to one of his women friends. Let us call her $W(endy)$. We can then apply the induction hypothesis to the remaining set of $m - 1$ men. Given any k of them, they have between them at least $k + 1$ women friends, and hence at least k women friends other than W. So the induction hypothesis implies that each of these $m - 1$ men may marry one of their women friends.

In case (b), it follows from the induction hypothesis that each man in the set S of n men, with $n < m$, can marry a woman with whom he is friendly. Suppose there is a set, say T, of l of the remaining $m - n$, as yet unmarried, men with fewer than l friends between them among the remaining unmarried women. Then $S \cup T$ is a set of $n + l$ men with between them fewer than $n + l$ friends among all the women, contradicting our hypothesis that the men satisfy the condition of the theorem. It follows that, for $l \leq m - n$, each l of the remaining as yet unmarried men have between them at least l friends among the remaining women who are not yet married. Hence, we can apply the induction hypothesis to marry each of these remaining men to one of the remaining women with whom he is friendly. So each man can be married to one of his women friends.

Hence the result holds also for a set of m men that satisfy the given condition. This completes the proof of the theorem.

---

* P. Halmos and H. Vaughan, The Marriage Problem, *American Journal of Mathematics*, 72, 1950, pp. 214–215. The history of the theorem is given in this paper.

Let us go back to Problem 17.6 but now assume that A is also friendly with X. We can represent the position by the diagram in Figure 17.18, where a white square indicates that the man in the row is friendly with the woman in the column.

It is straightforward to check that the conditions of Hall's marriage theorem are satisfied, and hence this theorem implies that each man can marry one of his women friends. However, some care needs to be taken. If A marries either Y or Z, C and D cannot both marry a woman friend. However, the pairing of A with X, B with V, C with Y, D with U, and E with Z—leaving W unattached—works.

We can view this pairing in two other ways. First, as we have already noticed, it corresponds to placing five nonattacking rooks on the white squares of the board given in Figure 17.18. This is shown in Figure 17.19.

Another way to represent this situation is that we associate with each of the men with the set of women he is friendly with.

$$A \leftrightarrow \{U,X,Y,Z\}$$

$$B \leftrightarrow \{U,V,W,X,Y,Z\}$$

$$C \leftrightarrow \{Y,Z\}$$

$$D \leftrightarrow \{U,Y,Z\}$$

$$E \leftrightarrow \{Y,Z\}$$

The choice of U, V, X, Y, and Z, for the women, can be regarded as a selection of five letters, from the (union) of the above five sets, with one letter from each set and *with no letter being repeated*. Such a selection is called a *system of distinct representatives* or

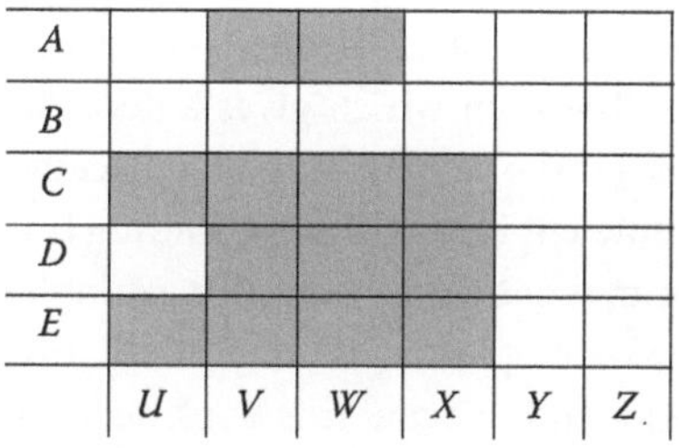

FIGURE 17.18

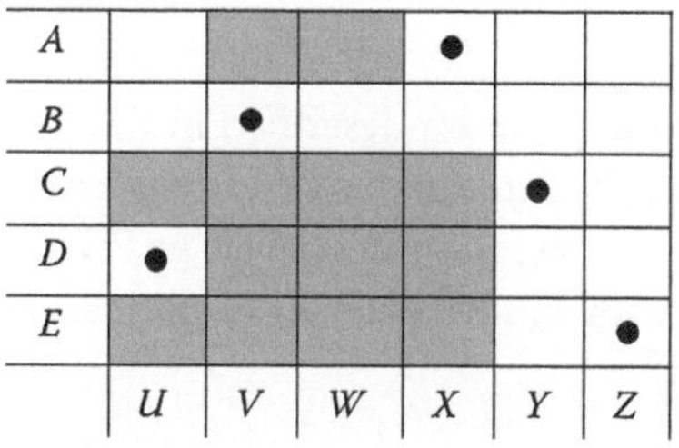

FIGURE 17.19

transversal for the five sets. Theorem 17.5 was originally stated in terms of this nomenclature, as follows.

THEOREM 17.6
Hall's Transversal Theorem*

The finite sets $S_1, \ldots, S_n$ have a transversal if and only if, for all positive integers $k \leq n$, for each choice of k integers, $n_1, n_2, \ldots, n_k$ with $1 \leq n_1 < n_2 < \ldots < n_k \leq n$,

$$\#(S_{n_1} \cup \ldots \cup S_{n_k}) \geq k.$$

We can now show that Problem 17B has a positive answer.

Let the 10 groups of four cards be $C_1, C_2, \ldots, C_{10}$. For $1 \leq i < 10$ we let S_i be the set of those distinct integers which appear on the cards in C_i. We now show that for $1 \leq k \leq 10$ and each choice of k distinct integers, $n_1, \ldots, n_k$ chosen from the range from 1 to 10 we have that $\#(S_{n_1} \cup \ldots \cup S_{n_k}) \geq k$.

For suppose that for some choice of k and $n_1, \ldots, n_k$, we have $\#(S_{n_1} \cup \ldots \cup S_{n_k}) < k$. This means that there are at most $k-1$ distinct integers represented among the $4k$ cards in the union $C_{n_1} \cup C_{n_2} \cup \ldots \cup C_{n_k}$. But this would imply, by the pigeon-hole principle, that at least one of these integers occurs on at least 5 of the cards, contrary to what we are given.

Hence, by Theorem 17.6, the sets $S_1, \ldots, S_{10}$ have a transversal, that is, we can pick one card from each of the groups $C_1, \ldots, C_{10}$ so that the ten cards that we pick carry the numbers $1, 2, \ldots, 10$.

There is a lot more that could be said about these topics. The theory of transversals has a wide range of applications. We have shown how to calculate rook polynomials. This enables us to work out how many ways we can place non-attacking rooks, or equivalently how many matchings or transversals there are in a given situation. We have also described Hall's Marriage Theorem which gives a necessary and sufficient condition for the existence of a transversal or a matching or a placement of non-attacking rooks. We haven't provided any efficient algorithms for finding transversals, nor have we considered how to find an optimal matching when, for example, there is a cost of assigning a particular job to a particular employee and we want to minimize the total cost. But all good things come to an end, and to avoid this book becoming too large, we end at this point. We have added some suggestions for further reading where you can read about topics we have not had space to cover.

Exercises

17.2.1A Show that in the situation described in Problem 17.6, if C is also friendly with X, then each man can marry a woman with whom he is friendly.

17.2.1B Show that in the situation described in Problem 17.6, if either D or E is also friendly with X, then each man can marry a women with whom he is friendly.

---

* P. Hall, On Representatives of Subsets, *Journal of the London Mathematical Society*, 10, 1935, pp. 26–30.

17.2.2A Suppose that the women A to G are friendly with the men a to h as follows:

A is friendly with $b, f,$ and h.
B is friendly with $a, b, c, d,$ and g.
C is friendly with $b, e,$ and f.
D is friendly with b and h.
E is friendly with $c, d, e, f,$ and g.
F is friendly with $e, f,$ and h.
G is friendly with b and h.

Draw the bipartite graph representing this situation. Show that it is not possible for each woman to marry a man she is friendly with.

17.2.2B The women in Exercise 17.2.2A wish to go on a cycle ride each initially choosing a partner who is a friend as follows:

$$A - b;\ B - d;\ C - e;\ D - h;\ E - g;\ F - f.$$

By Exercise 17.2.2A, this means that G cannot cycle with a friend. She would really like to cycle with b. Show that if C makes friends with either a or c, then by choosing a new partner if necessary, all seven women can cycle with a friend.

Solutions to the A Exercises

CHAPTER 2

2.2.1A The initial letter can be chosen in 26 ways. Each of the subsequent digits may be chosen in 10 ways (this assumes that a string of 11 zeros is possible). So the total number of different serial numbers for a €10 note is 26×10^{11}. Similarly, the number of different serial numbers for a £10 note is $26 \times 26 \times 10^8$. Since $26 < 10^3$, there are fewer of these than there are serial numbers for a €10 note.

2.2.2A There are 10 digits, and so they can be arranged in order in $10! = 3{,}628{,}800$ different ways. There are 3600 seconds in an hour and 24 hours in a day. So 10! seconds amount to $3{,}628{,}800/(3600 \times 24) = 42$ days, that is, six weeks exactly.

2.2.3A In the first race, with 10 horses, there are $10 \times 9 \times 8 = 720$ ways in which the first three positions can be filled. For the races with 8 and 6 horses, these positions can be filled in $8 \times 7 \times 6 = 336$ and $6 \times 5 \times 4 = 120$ ways, respectively. So the total number of predictions that can be made is $720 \times 336 \times 120 = 29{,}030{,}400$.

2.2.4A We call the square in row r and column s *square* (r, s). The square, say (r_1, s_1), for the first counter may be chosen in $m \times n$ ways. Then the second square, say (r_2, s_2), may be chosen in $(m-1) \times (n-1)$ ways, and so on. This gives $(mn)((m-1)(n-1))((m-2)(n-2))\ldots((m-k+1)(n-k+1)) = (m(m-1)\ldots(m-k+1))(n(n-1)\ldots(n-k+1)) = (m!n!)/((m-k)!(n-k)!)$ ways to choose the k squares for the counters. However, the order in which these squares are chosen does not matter. Since they can be chosen in order in $k!$ ways, there are $(1/k!)(m!n!/((m-k)!(n-k)!))$ different ways in which the counters may be placed on the grid. [Note that as

$$\frac{1}{k!}\left(\frac{m!n!}{(m-k)!(n-k)!}\right) = \frac{1}{k!}\left(\frac{m!}{(m-k)!} \times \frac{n!}{(n-k)!}\right) = k!\left(\frac{m!}{k!(m-k)!} \times \frac{n!}{k!(n-k)!}\right),$$

the answer could be given in either of the forms $(P(m, k)P(n, k))/k!$ or $k!C(m, k)C(n, k)$.]

2.3.1A A student can choose options from pure mathematics, applied mathematics, statistics, and computing in $C(12,3)$, $C(10,2)$, $C(6,2)$, and $C(4,1)$ ways, respectively. Hence, by the principle of multiplication of choices, the total number of ways in which students can choose their eight options is $C(12,3) \times C(10,2) \times C(6,2) \times C(4,1) = 220 \times 45 \times 15 \times 4 = 594{,}000$.

2.3.2A Choosing a subset, S, of a set, X, with n elements involves deciding for each element of X whether or not to include it in S. So for each element of X you have a choice of two options. Thus the total number of ways of choosing a subset is $2 \times 2 \times \ldots \times 2$ with n factors, that is, 2^n. On the other hand, the total number of subsets of a set of n elements is the total of the number of subsets containing zero elements, plus the number of subsets containing one element, plus the number of subsets containing two elements, and so on. As $C(n,r)$ is the number of subsets containing r elements, the total number of subsets of a set of n elements is $\sum_{r=0}^{n} C(n,r)$, and this sum must therefore be equal to 2^n.

2.3.3A Suppose we have a set X of n objects. The product $C(n,k)C(k,s)$ counts the number of ways of first choosing a subset, say Y, of X containing k objects, and then choosing a subset, say Z, of Y containing s of these elements. We can represent this by the following picture.

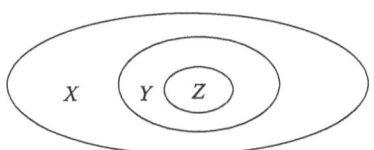

We can obtain exactly the same result by first choosing the subset Z of s objects from the n objects in X and then $k-s$ objects from the $n-s$ objects in $X \setminus Z$ to form the set $Y \setminus Z$. These choices can be made in $C(n,s)$ and $C(n-s, k-s)$ ways, respectively. So the total number of ways of making these choices is $C(n,s)C(n-s, k-s)$. Since these two expressions count the number of ways of arriving at the same situation, they must be equal. That is, $C(n,k)C(k,s) = C(n,s)C(n-s, k-s)$.

2.3.4A Let X be a set of n objects. For $0 \leq r \leq n$, there are $C(n,r)$ subsets of X containing r elements. Thus if Z is the set of all the subsets of X, the sum $\sum_{A \in Z} \#(A)$, which adds up the total of the numbers of elements in each subset of X, is equal to $\sum_{r=0}^{n} rC(n,r)$.

We now evaluate this sum in a second way. Each subset Y of X can be paired up with its *complement*, $X \setminus Y$, consisting of all the elements of X that are not in Y. Between them, Y and $X \setminus Y$ contain all the elements of X, and they have no elements in common. Thus $\#(Y) + \#(X \setminus Y) = \#(X) = n$. The 2^n subsets of X form 2^{n-1} pairs each consisting of a set and its complement. Hence the sum of the numbers of elements in all these sets is $n2^{n-1}$. Thus $\sum_{A \in X} \#(A) = n2^{n-1}$. Therefore $\sum_{r=0}^{n} rC(n,r) = n2^{n-1}$.

2.4.1A We need to count the number of ways six balls can be drawn with numbers that lead to your winning a fourth prize. For this to happen four of the balls must have numbers included among your six choices, and the remaining two must be among the 43 other numbers. So there are $C(6,4)C(43,2)$ ways in which you win a fourth prize. Hence the probability of winning a fourth prize is

$$\frac{C(6,4)C(43,2)}{C(49,6)} = \frac{15 \times 903}{13{,}983{,}816} = \frac{645}{665{,}896}.$$

To five decimal places this is 0.00097, which is approximately a probability of 1 in 1032.

2.4.2A With 100 balls in the bag there are $C(100,10)$ ways of selecting a sample of 10 balls. Five red and five blue balls can be chosen in $C(50,5) \times C(50,5)$ ways. So the required probability is

$$\frac{C(50,5)^2}{C(100,10)} = \frac{2,118,760}{8,170,019} = 0.259 \text{ (to three decimal places)}.$$

In this example there is only about a one in four chance that the sample reflects accurately the proportion of red and blue balls in the bag. In general, unless we look at every ball or they are all the same color, it is not likely that proportion of red balls in a sample will be the same as in the full bag of balls. If we think of the balls as people whose political opinions are being sampled, it follows that, even assuming that people are being asked sensible questions to which they are giving honest answers, it is unlikely that the results of an opinion poll reflect completely accurately the opinions of the electorate. Of course, providing that the sampling has been done carefully, the opinion polls are unlikely to be very far out. There is a whole branch of statistics that deals with the question of how accurate survey results are likely to be. The upshot of this theory is that for the sample sizes, usually around 1000, used for most national opinion polls, we can be 90% confident that the results are accurate to within 2% or 3% (this is on the assumption that the sampling has been done carefully). Thus an opinion poll that shows that support for a party has risen by 2% is compatible with the hypothesis that support for that party has actually fallen.

2.4.3A Ignoring leap years, there are 365 possible dates for each person's birthday. So there are 365^n possible birthdays for a set of n people. To count the number of these cases where at least two people share a birthday, it is easiest to count first the number of cases in which their birthdays are all different. Of course, for $n > 365$, there is no way in which n people can have different birthdays. For $n \leq 365$, there are 365 possibilities for the birthday of the first person. This leaves 364 possible dates for the birthday of the second person, and so on. So altogether there are $P(365,n) = 365!/(365 - n)!$ ways in which n people can have different birthdays. Hence there are $365^n - (365!/(365 - n)!)$ cases in which at least two people share a birthday. Hence the probability that at least two people share a birthday is

$$\left(365^n - \frac{365!}{(365-n)!}\right) \Big/ 365^n$$

$$= 1 - \frac{365 \times 364 \times \ldots (365-(n-1))}{365^n} = 1 - \frac{365}{365} \times \frac{364}{365} \times \ldots \times \frac{365-(n-1)}{365}.$$

The smallest integer, n, for which this probability is greater than 0.5 is $n = 23$. Thus with 23 people in a room, there is better than even chance that two of the people have the same birthday.

2.4.4A i. To select a hand with a 5-4-3-1 suit distribution from a pack of 52 cards, we first need to decide which suits have five, four, three, and one card respectively. The five-card suit may be chosen in four ways, then the four-card suit in three ways, and so on. So the suits may be chosen in 4! ways. Then the five cards from the given suit can be chosen in $C(13,5)$ ways, and so on. Thus there are $4! \times C(13,5) \times C(13,4) \times C(13,3) \times C(13,1)$ bridge hands with a 5-4-3-1 suit distribution. Hence the probability that a bridge hand has the suit distribution 5-4-3-1 is

$$\frac{4! \times C(13,5) \times C(13,4) \times C(13,3) \times C(13,1)}{C(52,13)} = \frac{82{,}111{,}732{,}560}{635{,}013{,}559{,}600}$$

$$= 0.129 \text{ (to three decimal places).}$$

ii. The suit with five cards can be chosen in four ways and then the suit with zero cards in three ways. The two suits with four cards in them are then automatically determined. Thus the required probability is

$$\frac{12 \times C(13,5) \times C(13,4)^2}{C(52,13)} = \frac{7{,}895{,}358{,}900}{635{,}013{,}559{,}600} = 0.012 \text{ (to three decimal places).}$$

iii. There are four ways to choose the four-card suit, and then the remaining suits must all have three cards in them. So the required probability is

$$\frac{4 \times C(13,4) \times C(13,3)^3}{C(52,13)} = \frac{66{,}905{,}856{,}160}{635{,}013{,}559{,}600} = 0.105 \text{ (to three decimal places).}$$

2.4.5A As can be seen from the answers below, because the different categories of hands are characterized in different ways, there is not a formula that covers all the cases. A good way to approach the question is to think about the choices involved in choosing a hand of the given type.

i. *Flush:* To select a hand that forms a flush, we need first to choose the suit and then five cards from this suit. The suit can be chosen in four ways. Then there are $C(13,5) = 1287$ ways to choose five cards from this suit. We need to subtract from 1287 the number of cases where the cards form a sequence (in which case the hand is a *straight flush*). Since an ace can count either high or low, there are 10 straight flushes in each suit with ranks A,2,3,4,5; 2,3,4,5,6; ...; 10,J,Q,K,A, respectively. So there are 1277 hands in each suit that are flushes but not straight flushes. So the total number of flushes is $4 \times 1277 = 5108$. Hence the probability that a hand dealt at random is a *flush* is

$$\frac{5108}{C(52,5)} = \frac{5108}{2{,}598{,}960} = 0.00197 \text{ (to five decimal places).}$$

ii. *Four of a kind:* There are 13 ranks to choose from. Once the rank has been chosen, 4 of the cards are determined, and there are 48 other cards to choose from. So there are $13 \times 48 = 624$ *four of a kind* hands. Hence the probability of getting such a hand is $624/2{,}598{,}960 = 0.00024$ (to five decimal places).

iii. *Full house*: The rank of the three cards can be chosen in 13 ways, and the three cards of this rank in $C(4,3) = 4$ ways, then the rank of the two cards in 12 ways, and the two cards of this rank in $C(4,2) = 6$ ways. So there are $13 \times 4 \times 12 \times 6 = 3744$ full houses, and the probability of getting such a hand is $3,744/2,598,960 = 0.00144$ (to five decimal places).

iv. *One pair*: The rank of the pair may be chosen in 13 ways, and we may then choose two cards of this rank in $C(4,2) = 6$ ways. The other cards must be of three other different ranks, which may be chosen in $C(12,3) = 220$ ways, and one card of each of these ranks may then be chosen in $4^3 = 64$ ways. Hence, the total number of *one-pair* hands is $13 \times 6 \times 220 \times 64 = 1,098,240$ and so the probability of being dealt such a hand is $1,098,240/2,598,960 = 0.42257$ (to five decimal places).

v. *Straight*: We saw in (a) that, as an ace can count either high or low, there 10 different sequences that the ranks of the cards in a straight may form. For any particular sequence, for example, 7,8,9,10,J, there are four choices for the suit of each card in the straight, giving $4^5 = 1024$ different choices altogether. However, from this we need to subtract the four cases in which all five cards are all of the same suit, as in this case we get a straight flush. Hence the total number of hands that a straight is $10 \times 1020 = 10,200$, and thus the probability that a hand dealt at random is a straight is $10,200/25,989,600 = 0.00392$ (to five decimal places).

2.4.6A The desired result is an immediate consequence of the binomial theorem. By this theorem, we have that $\sum_{r=0}^{n} C(n,r) a^r b^{n-r} = (a+b)^n$. Putting $a = p$ and $b = 1 - p$, it follows that $\sum_{r=0}^{n} C(n,r) p^r (1-p)^{n-r} = (p+(1-p))^n = 1^n = 1$, as required.

2.5.1A We can record each set of 12 faces that could come up as a sequence of numbers, with the first being the face that comes up on the first of the dice and so on. Since there are 6 possible outcomes for each of the dice, there are 6^{12} different possible sequences of outcomes. We need to determine the number of these sequences made up of two each of the numbers 1 to 6. By Theorem 2.10 there are $12!/2!^6$ such sequences. Hence the required probability is $(12!/2!^6)/6^{12}$. This is equal to $1,925/559,872$, which is 0.0034 to four decimal places.

2.5.2A ABRACADABRA has 11 letters, of which there are five As, two Bs, two Rs, one C, and one D. By Theorem 2.10, these letters may be arranged in $11!/(5!2!2!1!1!) = 83,160$ ways.

2.6.1A (1 3 8 2)(4 9 6)(5 7 10)

2.6.2A i. We can choose two numbers to make up the cycle of length 2 in $C(10,2)$ ways. Given two numbers, they can be used to make up just one cycle of length 2. We can then choose three of the remaining eight numbers to make up the cycle of length 3 in $C(8,3)$ ways. Given three numbers, they can be used to make up 2! different cycles of length 3. (If you are not sure why there are 2! different cycles, see the solution to Exercise 2.6.3A.) Once the cycles of lengths 2 and 3 have been chosen, the remaining five numbers can be used to make up 4! different cycles of length 5. So there are altogether $C(10,2) \times 1 \times C(8,3) \times 2! \times 4! = 120,960$ permutations of $\{1,2,3,4,5,6,7,8,9,10\}$ made up of disjoint cycles of lengths 2, 3, and 5.

ii. We can choose three numbers for the first cycle of length 3 in C(10,3) ways, and then arrange them to make 2! cycles of length 3. The second three numbers can then be chosen in C(7,3) ways, and they can be used to make 2! cycles of length 3. Since the order in which the two disjoint cycles of length 3 is chosen does not matter, these cycles can be chosen in $\frac{1}{2}$(C(10,3)×2×C(7,3)×2) = 8,400 ways. This leaves four numbers to make up the cycle of length 4. They can be used to make up 3! different cycles of length 4. Hence there are altogether 8400 × 3! = 50,400 permutations made of disjoint cycles of lengths 3, 3, and 4.

2.6.3A We can write a permutation of the set $\{1,2,3,\ldots,n\}$ consisting of a single cycle in cycle notation beginning with 1, that is, in the form (1 a_1 a_2 ... a_{n-1}) where a_1 a_2 ... a_{n-1} is a permutation of the numbers $2,3,\ldots,n$. Different permutations of these numbers give rise to different cycles of length n. So the number of permutations of $\{1,2,3,\ldots,n\}$ consisting of a single cycle of length n is the same as the number of permutations of the set $\{2,3,\ldots,n\}$, that is, $(n-1)!$ Since there are altogether $n!$ permutations of the set $\{1,2,\ldots,n\}$, the probability that such a permutation chosen at random consists of just one cycle is

$$\frac{(n-1)!}{n!} = \frac{1}{n}.$$

CHAPTER 3

3.1.1A The number of solutions in nonnegative integers of the equation $x + y + z + t \leq 20$ is the same as the number of such solutions of $x + y + z + t < 21$. So, by Theorem 3.3, there are C(21 − 4 + 1, 21 − 1) = C(24,20) = C(24,4) = 10,626 solutions.

3.1.2A We put $x' = x - 5$, $y' = y + 3$, $z' = z - 2$, $t' = t + 7$, and $w' = w - 4$. Then $x' + y' + z' + t' + w' = x + y + z + t + w - 1$. Thus $x + y + z + t + w = 14$ with $x \geq 5$, $y \geq -3$, $z \geq 2$, $t \geq -7$, and $w \geq 4$ if and only if $x' + y' + z' + t' + w' = 13$ with $x', y', z', t', w' \geq 0$. Therefore the number of solutions of $x + y + z + t + w = 14$ with $x \geq 5$, $y \geq -3$, $z \geq 2$, $t \geq -7$, and $w \geq 4$ is the same as the number of nonnegative integer solutions of $x' + y' + z' + t' + w' = 13$ By Theorem 3.1, this number is C(13 + (5 − 1),5 − 1) = C(17,4) = C(17,13) = 2380.

3.1.3A If we let the number of red, blue, green, and yellow marbles in a pack be r, b, g, and y, respectively, we see that we need to count the number of nonnegative integer solutions of $r + b + g + y = 50$. By Theorem 3.1, this number is C(50 + (4 − 1),4 − 1) = C(53,3) = C(53,50) = 23,426.

3.1.4A The number of ways of distributing 200 runs between the 11 batsmen is the same as the number of nonnegative integer solutions of the equation $r_1 + r_2 + \ldots + r_{11} = 200$. By Theorem 3.1, this number is C(210,10) = 36,976,937,738,226,486.

3.1.5A The number of ways of placing 20 identical balls in four distinct cups so that each cup contains an even number of balls is the same as the number of ways of placing 10 identical balls in four distinct cups (if this is not clear, imagine that the balls are stuck together

Solutions to the A Exercises ▪ 345

to form 10 pairs). This is the number of nonnegative solutions of $x + y + z + w = 10$, and, by Theorem 3.1, this is $C(13,3) = 283$.

3.1.6A By generalizing the solution of Problem 3.3 it can be seen that the number of ways of choosing k numbers from n so that no two numbers are consecutive is $C(n + 1 - k, k)$. So there are $C(91,10)$ ways of choosing 10 numbers from the set $\{1,2,...,100\}$ so that no two numbers are consecutive. Since there are altogether $C(100,10)$ ways of choosing 10 numbers from 100, the probability that no two numbers are consecutive is $C(91,10)/C(100,10)$. Therefore the probability that at least two numbers are consecutive is $1 - [C(91,10)/C(100,10)]$ = 0.629 to three decimal places.

3.2.1A We can think of the pupils (mouths) as being 30 distinct boxes into which the 30 identical chocolate bars are to be placed. It is possible for some pupils not to get a chocolate bar. So, from Table 3.2, we see that the number of ways to do this is $C(59,29)$. This is the rather large number 59,132,290,782,430,712.

3.2.2A The four identical black marbles can be placed in the five distinct boxes in $C(4+5-1, 5-1) = C(8,4)$ ways. The six distinct nonblack marbles may be placed in five distinct boxes in 5^6 ways. The answer is therefore $C(8,4) \cdot 5^6 = 70 \times 15{,}625 = 1{,}093{,}750$.

3.2.3A We can categorize the possible ways the manufacturer can make up a bag of marbles according to the number, k, of nontransparent marbles in the bag, where $0 \le k \le 20$. Since these have different colors they may be chosen in $C(20,k)$ ways. As the transparent marbles are identical these can be chosen in just one way. So the total number of ways of filling the bag of marbles is $\sum_{k=0}^{20} C(20,k) = 2^{20}$.

3.3.1A Using the triangle method we obtain the following values, beginning from line 7 of the table.

n \ k	1	2	3	4	5	6	7	8	9	10
7	1	63	301	350	140	21	1			
8	1	127	966	1701	1050	266	28	1		
9	1	255	3025	7770	6951	2646	462	36	1	
10	1	511	9330	34105	42525	22827	5880	750	45	1

3.3.2A Suppose we have partitioned the set $\{1,2,3,...,n\}$ into k nonempty disjoint sets $\{a_1, a_2, ..., a_{n_1}\}, \{b_1, b_2, ..., b_{n_2}\}, ...$, where the notation is chosen so that $a_1 < a_2 <..., b_1 < b_2 <...$, etc. We can associate with this partition a permutation, p, of the set $\{1,2,...,n\}$ given by $p(a_1) = a_2, p(a_2) = a_3, ..., p(a_{n_1}) = a_1, p(b_1) = b_2$, etc.

Clearly, different partitions correspond to different permutations, and, for $n > 2$, not all permutations of $\{1,2,...,n\}$ correspond to a partition in this way. Thus the number of partitions of $\{1,2,...,n\}$ is less than the number of permutations of $\{1,2,...,n\}$. The number of partitions of $\{1,2,...,n\}$ is the same as the number of ways of placing n distinct balls in any number of identical boxes, and this number is $S(n,1) + S(n,2) +...+ S(n,n)$. The number of permutations of n is $n!$ Therefore $S(n,1) + S(n,2) +...+ S(n,n) < n!$.

3.3.3A Given k distinct boxes, if we place n distinct balls in some of them, there will be r boxes that are not empty with $1 \leq r \leq k$. These r nonempty boxes can be chosen from the k distinct boxes in $C(k,r)$ ways. From Table 3.1 the n balls may be placed in these r distinct boxes with no empty boxes in $r!S(n,r)$ ways. So the n distinct balls may be placed in r of the k distinct boxes in $C(n,r) \times r!S(n,r)$ ways. Thus the total number of ways of placing n distinct balls in k distinct boxes is $\sum_{r=1}^{k} C(n,r)r!S(n,r)$ ways. Since, from Table 3.1, there are k^n ways to place n distinct balls in k distinct boxes, we deduce that $k^n = C(k,1)1!S(n,1) + C(k,2)2!S(n,2)+\ldots+C(k,k)k!S(n,k)$.

3.3.4A A function $f: X \to Y$ is determined by specifying for each element x of $X = \{1,2,3,4,5,6,7,8,9,10\}$ one of the five numbers in Y for the value of $f(x)$. This corresponds to deciding which of five boxes to put x in. So the number of surjective functions, $f: X \to Y$, is the same as the number of ways of placing 10 distinct balls in five distinct boxes, with no box empty. From Table 3.1, this number is $5!S(10,5)$. The value of $S(10,5)$ is given in the solution to Exercise 3.3.1A as 42,525. This gives $5!S(10,5) = 5,103,000$.

3.3.5A $S(n + 1,n)$ is the number of ways of placing $n + 1$ distinct balls in n identical boxes, with no box empty. The only way this can be done is if one box contains two balls, and the other $n - 1$ boxes contain one ball each. Since the boxes are identical, all that distinguishes one placing from another is which two balls are together in the same box. We can choose two balls from $n + 1$ in $C(n + 1,2)$ ways. So this is the number of ways to place the $n + 1$ distinct balls in the n identical boxes.

An alternative solution, using mathematical induction, runs as follows:

Base. For $n = 1$, $S(n + 1,n) = S(2,1) = 1$ and $[n(n + 1)]/2 = 1$ so the result holds in this case.

Induction step. Suppose the result holds for $n = k$, that is, assume that $S(k+1,k) = [k(k+1)]/2$. By Theorem 3.4, $S(k + 2,k +1) = S(k + 1,k) + (k + 1) S(k + 1,k +1) = \{[k(k + 1)]/2\} + (k + 1)$, by the induction hypothesis, and Theorem 3.5(i). Therefore $S(k + 2,k +1) = [(k + 1)(k + 2)]/2$, and hence the result holds also for $n = k + 1$.

Therefore, by mathematical induction, the result holds for all integers $n \geq 1$.

3.3.6A Here, as we are given the formula for $S(n,3)$, a proof by mathematical induction, using Theorem 3.3, is straightforward. We indicate how the formula may be derived at the end of this answer.

Base. When $n = 3$, $S(n,3)= S(3,3) = 1$ and $\frac{1}{2}(3^{n-1} - 2^n +1) = \frac{1}{2}(9-8+1) = 1$, and hence the result is true in this case.

Induction step. Suppose the result holds for $n = k$ with $k \geq 3$, that is, $S(k,3) = \frac{1}{2}(3^{k-1} - 2^k +1)$.

By Theorem 3.4, $S(k+1,3) = S(k,2) + 3S(k,3) = (2^{k-1} - 1) + 3(\frac{1}{2}(3^{k-1} - 2^k +1)) = \frac{1}{2}(3^k - 2^{k+1} +1)$, and hence the result holds also for $n = k + 1$.

Therefore, by mathematical induction, the result holds for all integers $n \geq 3$.

Finding the formula. If we put, say $a_k = S(k,3)$, then we can rewrite the equation $S(k+1,3) = S(k,2) + 3S(k,3)$ as $a_{k+1} - 3a_k = S(k,2)$ and thus, $a_{k+1} - 3a_k = 2^{k-1} - 1$. This is a *linear recurrence relation* of a type whose method of solution we describe in Chapter 7.

3.3.7A Arranging m 0s and n 1s in line so that no two 1s are adjacent, amounts to inserting 1s in the gaps between the 0s, so that no two 1s are placed in the same gap, for example, with $m = 8, n = 3$:

$$1\,0\quad 0\,1\,0\quad 0\quad 0\,1\,0\quad 0\quad 0.$$

With m 0s there are $m + 1$ gaps, including the spaces before the first 0 and after the last 0. Therefore the number of arrangements is the same as the number of ways of choosing n of these gaps in which to put the 1s. So the number of arrangements is $C(m + 1, n)$.

3.3.8A Since the order in which the balls are chosen is not taken into consideration, all that matters is how many times each ball is chosen. Thus, if we use b_i for the number of times ball i is chosen, we are seeking the number of nonnegative integer solutions of the equation $b_1 + \ldots + b_n = r$. By Theorem 3.1, this number is $C(r + n - 1, n - 1)$.

CHAPTER 4

4.1.1A We let P_k be the set of integers in the range from 1 to 1,000,000 that are perfect kth powers. We seek the value of $\#(P_2 \cup P_3)$. By the inclusion–exclusion theorem, $\#(P_2 \cup P_3) = \#(P_2) + \#(P_3) - \#(P_2 \cap P_3) = \#(P_2) + \#(P_3) - \#(P_6)$, since a number is both a perfect square and a perfect cube if and only if it is a perfect sixth power. The squares in the range 1–1,000,000 are $1^2, 2^2, \ldots, 1000^2$ and so $\#(P_2) = 1000$. Similarly $\#(P_3) = 100$ and $\#(P_6) = 10$. Hence

$$\#(P_2 \cup P_3) = 1000 + 100 - 10 = 1090.$$

4.1.2A We let D_k be the set of numbers in the range from 1 to 1,000,000 that are divisible by k. Then $\#(D_k) = \lfloor 1,000,000/k \rfloor$. We seek that value of $1,000,000 - \#(D_2 \cup D_3 \cup D_5 \cup D_7)$. By the inclusion–exclusion theorem,

$$\#(D_2 \cup D_3 \cup D_5 \cup D_7) = \big(\#(D_2) + \#(D_3) + \#(D_5) + \#(D_7)\big)$$

$$-\big(\#(D_2 \cap D_3) + \#(D_2 \cap D_5) + \#(D_2 \cap D_7) + \#(D_3 \cap D_5) + \#(D_3 \cap D_7) + \#(D_5 \cap D_7)\big)$$

$$+\big(\#(D_2 \cap D_3 \cap D_5) + \#(D_2 \cap D_3 \cap D_7) + \#(D_2 \cap D_5 \cap D_7) + \#(D_3 \cap D_5 \cap D_7)\big)$$

$$-\#(D_2 \cap D_3 \cap D_5 \cap D_7)$$

$$= \big(\#(D_2) + \#(D_3) + \#(D_5) + \#(D_7)\big) - \big(\#(D_6) + \#(D_{10}) + \#(D_{14}) + \#(D_{15}) + \#(D_{21}) + \#(D_{35})\big)$$

$$+\big(\#(D_{30}) + \#(D_{42}) + \#(D_{70}) + \#(D_{105})\big) - \#(D_{210})$$

$$= (500{,}000 + 333{,}333 + 200{,}000 + 142{,}857) - (166{,}666 + 100{,}000 + 71{,}428$$

$$+ 66{,}666 + 47{,}619 + 28{,}571) + (33{,}333 + 23{,}809 + 14{,}285 + 9{,}523) - 4761 = 771{,}429.$$

Hence the number of integers in the range from 1 to 1,000,000 not divisible by any of 2, 3, 5, and 7 is 1,000,000 − 771,429 = 228,571.

4.1.3A We let V_1, V_2, V_3, and V_4 be the sets of bridge hands with no spades, no hearts, no diamonds, and no clubs, respectively. We seek the value of $\#(V_1 \cup V_2 \cup V_3 \cup V_4)$. By the inclusion–exclusion theorem, this is equal to

$$\sum_{1 \le i \le 4} \#(V_i) - \sum_{1 \le i < j \le 4} \#(V_i \cap V_j) + \sum_{1 \le i < j < k \le 4} \#(V_i \cap V_j \cap V_k) - \#(V_1 \cap V_2 \cap V_3 \cap V_4). \tag{1}$$

For $1 \le i \le 4$, a hand in V_i consists of 13 cards drawn from the 39 cards in the three suits that are not excluded. Therefore $\#(V_i) = C(39,13)$. For $1 \le i < j \le 4$, a hand in $V_i \cap V_j$ consists of 13 cards drawn from the 26 cards in the two suits that are not excluded. Therefore, $\#(V_i \cap V_j) = C(26,13)$. Similarly, for $1 \le i < j < k \le 4$, $\#(V_i \cap V_j \cap V_k) = C(13,13)$. There are no bridge hands with voids in all four suits, so $\#(V_1 \cap V_2 \cap V_3 \cap V_4) = 0$. Therefore, by Equation 1,

$$\#(V_1 \cup V_2 \cup V_3 \cup V_4) = 4 \times C(39,13) - 6 \times C(26,13) + 4 \times C(13,13) - 0 = 32{,}427{,}298{,}180.$$

4.1.4A A positive integer less than or equal to n has a prime factor in common with n if and only if it is divisible by one of $p_1, p_2, \ldots, p_k$. We let E_i, for $1 \le i \le k$, be the set of integers in the range from 1 to n that are divisible by p_i. Thus $\phi(n) = n - \#(\bigcup_{i=1}^{k} E_i)$. By the inclusion–exclusion theorem,

$$\#\left(\bigcup_{i=1}^{k} E_i\right) = \sum_{1 \le i \le k} \#(E_i) - \sum_{1 \le i < j \le k} \#(E_i \cap E_j) + \ldots = \sum_{1 \le i \le k} \frac{n}{p_i} - \sum_{1 \le i < j \le k} \frac{n}{p_i p_j} + \ldots$$

and hence

$$\phi(n) = n - \sum_{1 \le i \le k} \frac{n}{p_i} + \sum_{1 \le i < j \le k} \frac{n}{p_i p_j} - \ldots = n \left(1 - \sum_{1 \le i \le k} \frac{1}{p_i} + \sum_{1 \le i < j \le k} \frac{1}{p_i p_j} - \ldots \right).$$

Therefore, using the algebraic identity $(1-x_1)(1-x_2) \ldots (1-x_k) = 1 - \Sigma_{1 \le i \le k} x_i + \Sigma_{1 \le i < j \le k} x_i x_j - \ldots$, it follows that

$$\phi(n) = n\left(1 - \frac{1}{p_1}\right)\left(1 - \frac{1}{p_2}\right) \ldots \left(1 - \frac{1}{p_k}\right).$$

4.1.5A Throwing a die amounts to sampling with replacement the six numbers on the face of the die. Hence, we can use the formula given in the solution to Problem 4.3 for the probability, $\theta(n,s)$, that, after s samples have been drawn with replacement from n objects, each object has been sampled at least once. Putting $n = 6$ in this formula, we see that the probability that each of the numbers 1–6 occurs at least once in s throws is

$$\theta(6,s) = \frac{1}{6^s}\sum_{k=0}^{6}(-1)^k C(6,k)(6-k)^s = 1 - 6\left(\frac{5}{6}\right)^s + 15\left(\frac{4}{6}\right)^s - 20\left(\frac{3}{6}\right)^s + 15\left(\frac{2}{6}\right)^s - 6\left(\frac{1}{6}\right)^s.$$

We seek the least value of s such that $\theta(6,s) > 0.5$. By trying $s = 1,2,3,\ldots$ we eventually find that $\theta(6,12) = 1{,}654{,}565/3{,}779{,}136$ and $\theta(6,13) = 485{,}485/944{,}784$, so $\theta(6,12) < 0.5 < \theta(6,13)$ and hence the required value of s is 13.

4.1.6A We let A_i be the set of those sequences of s samples that does not include any elements from X_i. There are np elements in the set $\bigcup_{i=1}^{n} X_i$, and there are altogether $(np)^s$ sequences of s elements. Hence the probability, say P, that a sequence of s samples includes at least one element from each set X_i is

$$\frac{1}{(np)^s}\left((np)^s - \#\left(\bigcup_{i=1}^{n} A_i\right)\right).$$

By the inclusion–exclusion theorem,

$$P = \frac{1}{(np)^s}\left((np)^s - \sum_{k=1}^{n}(-1)^{k+1}\sum_{1 \le i_1 < \ldots < i_k \le n}\#(A_{i_1} \cap A_{i_2} \cap \ldots \cap A_{i_k})\right).$$

For $1 \le i_1 < i_2 < \ldots < i_k \le n$, the set $A_{i_1} \cap A_{i_2} \cap \ldots \cap A_{i_k}$ consists of those sequences of s samples whose elements are drawn from the sets, X_j, other than $X_{i_1},\ldots,X_{i_k}$, that is, from $n - k$ of these sets. Hence in which there are $(n - k)p$ choices for each element. Thus $\#(A_{i_1} \cap A_{i_2} \cap \ldots \cap A_{i_k}) = ((n-k)p)^s$. Hence

$$P = \frac{1}{(np)^s}\left((np)^s - \sum_{k=1}^{n}(-1)^{k+1}C(n,k)((n-k)p)^s\right) = \frac{1}{n^s p^s}\sum_{k=0}^{n}(-1)^k C(n,k)(n-k)^s p^s$$

$$= \frac{1}{n^s}\sum_{k=0}^{n}(-1)^k C(n,k)(n-k)^s = \theta(n,s).$$

We therefore see that the probability is, indeed, independent of the value of p.

4.1.7A By the result of Exercise 4.1.6A the probability that after s cards have been sampled, at least one card of each of the 13 ranks has been drawn is $\theta(13,s)$, which is given by $\theta(13,s) = \frac{1}{13^s}\sum_{k=0}^{13}(-1)^k C(13,k)(13-k)^s$. We seek the least value of s such that $\theta(13,s) > 0.99$. A computer calculation shows that the required value is $s = 90$.

4.2.1A The number of derangements of 10 objects is

$$10!\sum_{k=0}^{10}\frac{(-1)^k}{k!},$$

350 ■ Solutions to the A Exercises

and hence the probability that all 10 letters are put in the wrong envelopes is

$$\sum_{k=0}^{10} \frac{(-1)^k}{k!} = \frac{16,481}{44,800} \sim 0.368.$$

4.2.2A i. The two Os in *ROBOT* must replace two of the other three letters, and this pair may be chosen in three ways. Consider, for example, the case where they replace *R* and *B*, so that we have

$$\begin{pmatrix} R & O & B & O & T \\ O & ? & O & ? & ? \end{pmatrix}.$$

Then there are two choices for the position of *T*, and then the remaining letters *R* and *B* can be placed in two ways. Thus the permutation above may be completed in $2 \times 2 = 4$ ways. Hence there are altogether $3 \times 4 = 12$ anagrams of *ROBOT*.

ii. The three *A*s must occupy the positions of three of the four letters *G, M, N,* and *R*. Thus these positions can be chosen in $C(4,3) = 4$ ways. The position for the letter in the set $\{G,M,N,R\}$ that is not replaced by an *A* can then be chosen in three ways, since it can occupy the position of any of the *A*s. The remaining three letters from this set can then be placed in 3! ways. So the total number of anagrams of *ANAGRAM* is $4 \times 3 \times 3! = 72$.

iii. We consider the anagrams of *TENNESSEE* according to which letter replaces *T*.

  a. *T* is replaced by one of the *N*s. Then the four *E*s must replace the *N*s and the *S*s, and the remaining *N*, the *S*s, and the *T* must replace the *E*s. There are four possible positions for the *N*, and then three choices for the *T*, so there are $4 \times 3 = 12$ anagrams of this type.

  b. *T* is replaced by one of the *S*s. Exactly as in case (a), there are 12 anagrams of this type.

  c. *T* replaced by one of the *E*s. We divide this case into subcases according to which letter *T* now replaces.

   α. *T* replaces one of the *N*s. This can happen in two ways. Then the other *N* and the two *S*s must be replaced by the remaining *E*s. There are then $C(4,2) = 6$ ways in which the 2 *N*s and 2 *S*s can replace the *E*s. So there are $2 \times 6 = 12$ anagrams of this type.

   β. *T* replaces one of the *S*s. There are similarly 12 anagrams of this type.

   γ. *T* replaces one of the *E*s. This can happen in four ways. Then there are four ways in which the remaining *E*s can replace *N*s and *S*s. If one *S* is not replaced by an *E*, it must then be replaced by an *N*, and then there are three ways in which the remaining N and the two Ss can replace the remaining Es. Likewise, if one N is not replaced by an E, there are three ways in which the remaining letters can be replaced. So there are 4 × 12 = 48 anagrams of this type.

Thus there are 12 + 12 + (12 + 12 + 48) = 96 different anagrams altogether.

Unfortunately, there does not appear to be a quicker method—but please let us know if you find one.

4.3.1A The formula in Equation 4.7 of Theorem 4.5 gives

$$S(4,2) = \frac{1}{2!}\sum_{s=0}^{1}(-1)^s C(2,s)(2-s)^4 = \frac{1}{2!}\left(C(2,0)2^4 - C(2,1)1^4\right) = \frac{1}{2!}(16-2) = 7,$$

and also

$$S(8,4) = \frac{1}{4!}\sum_{s=0}^{3}(-1)^s C(4,s)(4-s)^8 = \frac{1}{4!}\left(C(4,0)4^8 - C(4,1)3^8 + C(4,2)2^8 - C(4,3)1^8\right)$$

$$= \frac{1}{4!}(65{,}536 - 26{,}244 + 1{,}536 - 4) = \frac{40{,}824}{24} = 1{,}701,$$

both of which agree with the values given in Table 3.2.

4.3.2A As $S(1,1) = 1$, the result is true for $n = 1$. Now assume, as our induction hypothesis, that it holds for $n = m$. Suppose $1 \leq k \leq m + 1$. As $S(m + 1,1) = 1$, the formula holds when $n = m + 1$ and $k = 1$. So we need only consider the cases where $2 \leq k \leq m + 1$. Then, using Theorem 3.4 and the induction hypothesis, we have that

$$S(m+1,k) = S(m,k-1) + kS(m,k) =$$

$$\frac{1}{(k-1)!}\sum_{s=0}^{k-2}(-1)^s C(k-1,s)(k-1-s)^m + \frac{k}{k!}\sum_{s=0}^{k-1}(-1)^s C(k,s)(k-s)^m. \tag{1}$$

Now

$$\sum_{s=0}^{k-2}(-1)^s C(k-1,s)(k-1-s)^m = \sum_{s=1}^{k-1}(-1)^{s-1}C(k-1,s-1)(k-s)^m. \tag{2}$$

[Do not let this change in the range of the summation variable, s, throw you. To see that Equation 2 is true, just check that both the left- and right-hand sides of Equation 1 are different ways of writing the expression

$$C(k-1,0)(k-1)^m - C(k-1,1)(k-2)^m + \ldots + (-1)^{k-2}C(k-1,k-2)1^m.]$$

Substituting from Equation 2 into Equation 1, we obtain

$$S(m+1,k) = \frac{1}{(k-1)!} \sum_{s=1}^{k-1} (-1)^{s-1} C(k-1,s-1)(k-s)^m + \frac{1}{(k-1)!} \sum_{s=0}^{k-1} (-1)^s C(k,s)(k-s)^m$$

$$= \frac{1}{(k-1)!} \sum_{s=1}^{k-1} (-1)^{s-1} C(k-1,s-1)(k-s)^m + \frac{1}{(k-1)!} \left(C(k,0)k^m + \sum_{s=1}^{k-1} (-1)^s C(k,s)(k-s)^m \right)$$

$$= \frac{1}{(k-1)!} \sum_{s=1}^{k-1} (-1)^{s-1} (k-s)^m [C(k-1,s-1) - C(k,s)] + \frac{1}{(k-1)!} k^m. \qquad (3)$$

Now by Theorems 2.7 and 2.4,

$$C(k-1,s-1) - C(k,s) = -C(k-1,s) = \frac{k-s}{k} C(k,s). \qquad (4)$$

By Equations 3 and 4,

$$S(m+1,k) = \frac{1}{(k-1)!} \sum_{s=1}^{k-1} (-1)^{s-1} (k-s)^m \left(-\frac{k-s}{k} \right) C(k,s) + \frac{1}{(k-1)!} k^m$$

$$= \frac{1}{k!} \sum_{s=1}^{k-1} (-1)^s C(k,s)(k-s)^{m+1} + \frac{1}{(k-1)!} k^m. \qquad (5)$$

Finally, $(1/(k-1)!)k^m = (1/k!)k^{m+1}$, which is the same as $(1/k!)(-1)^s C(k,0)(k-s)^{m+1}$ when $s = 0$. So Equation 5 is equivalent to

$$S(m+1,k) = \frac{1}{k!} \sum_{s=0}^{k-1} (-1)^s C(k,s)(k-s)^{m+1},$$

and hence the result holds also for $n = m + 1$.

This completes the proof, by mathematical induction, that the result is true for all integers $n \geq 1$.

CHAPTER 5

5.1.1A $[x]_6 = x(x-1)(x-2)(x-3)(x-4)(x-5) = x^6 - 15x^5 + 85x^4 - 225x^3 + 274x^2 - 120x$ (either by direct expansion, or by reading off the coefficients from Table 5.3).

5.1.2A All we need do is multiply the relevant matrices, and check that we get the appropriate identity matrix in each case. Although matrix multiplication is not in general commutative, the theory tells us that if P, Q are square matrices with $PQ = I$, then $QP = I$, so we need only check one product in each case. Here we have

$$P_2Q_2 = \begin{pmatrix} 1 & 0 \\ -1 & 1 \end{pmatrix}\begin{pmatrix} 1 & 0 \\ 1 & 1 \end{pmatrix} = \begin{pmatrix} 1 & 0 \\ 0 & 1 \end{pmatrix} \text{ and } P_3Q_3 = \begin{pmatrix} 1 & 0 & 0 \\ -1 & 1 & 0 \\ 2 & -3 & 1 \end{pmatrix}\begin{pmatrix} 1 & 0 & 0 \\ 1 & 1 & 0 \\ 1 & 3 & 1 \end{pmatrix} = \begin{pmatrix} 1 & 0 & 0 \\ 0 & 1 & 0 \\ 0 & 0 & 1 \end{pmatrix}.$$

5.1.3A $s(n,1)$ is the modulus of the coefficient of x in the polynomial $x(x-1)(x-2)...(x-(n-1))$. Thus $s(n,1) = |(-1)(-2)...(-(n-1))| = (n-1)!$

5.1.4A $x(x-1)(x-2)...(x-(n-1)) = s(n,n)x^n - s(n,n-1)x^{n-1} + ... + (-1)^{n-1}s(n,1)x$. Putting $x = -1$, this gives $(-1)(-2)...(-n) = (-1)^n s(n,n) + (-1)^n s(n,n-1) + ... + (-1)^n s(n,1)$. That is, $(-1)^n n! = (-1)^n (s(n,n) + s(n,n-1) + ... + s(n,1))$ and therefore $s(n,1) + ... + s(n,n) = n!$

5.1.5A Again, we begin with the equation $x(x-1)(x-2)...(x-(n-1)) = s(n,n)x^n - s(n,n-1)x^{n-1} + ... + (-1)^{n-1}s(n,1)x$, but this time the answer we are seeking suggests that we put $x = -k$. This gives

$$-k(-(k+1))(-(k+2))...(-(k+n-1)) = (-1)^n(s(n,n)k^n + s(n,n-1)k^{n-1} + ... + s(n,1)k).$$

That is, $(-1)^n k(k+1)...(k+n-1) = (-1)^n \sum_{r=1}^n s(n,r)k^r$, and hence

$$\sum_{r=1}^n s(n,r)k^r = \frac{(k+n-1)!}{(k-1)!} = n!\frac{(k+n-1)!}{n!(k-1)!} = n!C(k+n-1,n).$$

5.1.6A $s(n,k)$ is the modulus of the coefficient of x^k in $x(x-1)(x-2)...(x-(n-1))$. The terms involving x^k in this product are obtained by choosing x from k of the brackets, and $n-k$ of the constants $\{0,1,2,...,n-1\}$ from the remaining brackets. If 0 is one of the chosen constants, the product is 0. Thus $s(n,k)$ is the sum of all the products of $n-k$ different integers chosen from $\{1,2,...,n-1\}$.

5.1.7A $\theta^5 = \theta(\theta^4) = xD(xD + 7x^2D^2 + 6x^3D^3 + x^4D^4)$ (using the solution to Problem 5.3) $= x(D + xD^2 + 14xD^2 + 7x^2D^3 + 18x^2D^3 + 6x^3D^3 + 4x^3D^4 + x^4D^5) = xD + 15x^2D^2 + 25x^3D^3 + 10x^4D^4 + x^5D^5$. Thus we see from Table 5.2 that the coefficients are the required Stirling numbers.

5.2.1A A permutation of the set $\{1,2,...,n\}$, with $n \geq 4$, made up of $n - 2$ cycles, must consist of either (a) one cycle of length 3 and $n - 3$ cycles of length 1, or (b) two cycles of length 2 and $n - 4$ cycles of length 1. To get a permutation of type (a), we first choose the three numbers to make up a cycle of length 3. This can be done in $C(n,3)$ ways. Each three numbers can be arranged to form two different cycles of length 3, and there is no further choice involved, as the remaining $n - 3$ numbers must all be in separate cycles of length 1. So there are $2C(n,3) = (n(n-1)(n-2))/3$ permutations of this type. To get a permutation of type (b), we first choose two numbers to make up the first cycle of length 2, and then two further numbers to make up the second cycle of length 2. The order in which we make these choices does not matter. So there are

$$\frac{1}{2}C(n,2)C(n-2,2) = \frac{1}{2} \times \frac{n!}{(n-2)!2!} \times \frac{(n-2)!}{(n-4)!2!} = \frac{n(n-1)(n-2)(n-3)}{8}$$

permutations of this type. Hence

$$p(n,n-2) = \frac{n(n-1)(n-2)}{3} + \frac{n(n-1)(n-2)(n-3)}{8} = \frac{n(n-1)(n-2)(3n-1)}{24}.$$

5.2.2A We start from the identity of Theorem 5.1, $x^n = \sum_{k=1}^{n} S(n,k)[x]_k$, that is, $x^n = \sum_{k=1}^{n} S(n,k)x(x-1)...(x-(k-1))$. If we divide both sides by x, we obtain $x^{n-1} = \sum_{k=1}^{n} S(n,k)(x-1)...(x-(k-1))$, and now, putting $x = 0$, we deduce that $0 = \sum_{k=1}^{n} (-1)^{k-1} S(n,k)(k-1)!$.

5.2.3A $S(n+1, k+1)$ is the number of ways of placing $n+1$ distinct balls in $k+1$ identical boxes. Consider one particular ball, say b. We first choose r of the other n balls to go in different boxes from that in which we put b. Since there are k other boxes that all must have at least one ball in them, $k \le r \le n$. These r balls can be chosen in $C(n,r)$ ways, and they can be placed in the remaining k boxes in $S(r,k)$ ways. Thus, for $k \le r \le n$, we can place the $n+1$ balls in $k+1$ boxes with r balls not in the same box as b, in $C(n,r)S(r,k)$ ways. Hence

$$S(n+1,k+1) = \sum_{r=k}^{n} C(n,r)S(r,k).$$

5.3.1A i. $C(2n-1, n-1) - C(2n-1, n+1) = \dfrac{(2n-1)!}{(n-1)!n!} - \dfrac{(2n-1)!}{(n+1)!(n-2)!}$

$$= \frac{(2n-1)!(n(n+1)-n(n-1))}{n!(n+1)!}$$

$$= \frac{(2n-1)!(2n)}{n!(n+1)!} = \frac{1}{n+1}\left(\frac{(2n)!}{n!n!}\right) = \frac{1}{n+1}C(2n,n) = C_n.$$

ii. As $C(2n-1, n-1)$ and $C(2n-1, n+1)$ are both integers, it follows from (i) that C_n is an integer.

5.3.2A We prove this result by induction. Since $C_1 = 1$, the result holds for $n = 1$. Now suppose that the result holds for all $m \le n$. By Theorem 5.8, $C_{n+1} = \sum_{k=0}^{n} C_k C_{n-k}$. We first consider the case where $n + 1$ is even, say $n+1 = 2s$, so that $n = 2s-1$, and thus $C_{n+1} = \sum_{k=0}^{2s-1} C_k C_{2s-k-1} = \sum_{k=0}^{s-1} C_k C_{2s-k-1} + \sum_{k=s}^{2s-1} C_k C_{2s-k-1}$. Now putting $l = 2s - k - 1$, we see that

$\sum_{k=s}^{2s-1} C_k C_{2s-k-1} = \sum_{l=0}^{s-1} C_{2s-l-1} C_l = \sum_{k=0}^{s-1} C_k C_{2s-k-1}$, on replacing the "dummy variable" l by k. Hence $C_{n+1} = 2\sum_{k=0}^{s-1} C_k C_{2s-k-1}$, and consequently is even. Next, when $n + 1$ is odd, say $n = 2s$, we have similarly that $C_{n+1} = 2\sum_{k=0}^{s-1} C_k C_{2s-k} + C_s C_s$. It follows that C_{n+1} is odd, if and only if n is even and $C_{(1/2)n}$ is odd. Thus, by the induction hypothesis, C_{n+1} is odd if and only if, for some integer t, $\frac{1}{2}n = 2^t - 1$. Now, this holds if and only if $n+1 = 2(2^t-1) + 1 = 2^{t+1}-1$, for some integer t. Hence the result holds also for $n+1$. This completes the proof.

5.3.3A We consider a path, joining the corners of a $n \times n$ grid, that apart from its endpoints lies below the diagonal. Such a path must never cross the "subdiagonal" marked by a dotted line in the diagram below.

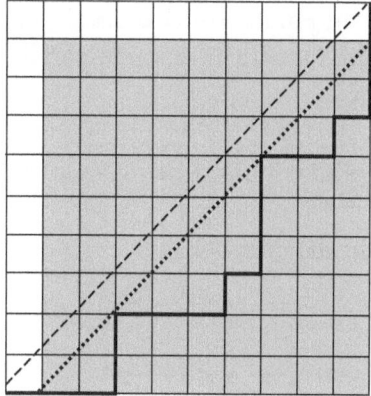

Thus the number of such paths is the same as the number of paths joining the corners of the shaded $(n-1)\times(n-1)$ grid that do not cross the diagonal; that is,

$$C_{n-1} = \frac{1}{n} C(2n-2, n-1).$$

5.3.4A We can prove these results using mathematical induction on the number of pairs of brackets in an expression.

i. The result is clearly true for expressions with no brackets and hence consisting of a single X. Suppose the result holds for expressions with fewer than k pairs of brackets. An expression with k pairs of brackets will have the form $(\Gamma\Delta)$ where Γ and Δ are expressions with fewer than k pairs of brackets. Hence, by the induction hypothesis, there is an X in Δ after the last left-hand bracket. This X also occurs after the last left-hand bracket in the expression $(\Gamma\Delta)$. So the result holds also for expressions with k pairs of brackets.

ii. The sequence of 1s and -1s associated with a single X is an empty sequence that counts as a Catalan sequence, so the result is true for expressions with no brackets. Suppose the result holds for expressions with fewer than k pairs of brackets. We show that the result holds also for an expression (ΓΔ) with k pairs of brackets, and hence in which Γ and Δ are expressions with fewer than k pairs of brackets. Let S_Γ and S_Δ be the sequences of 1s and −1s, associated with Γ and Δ, respectively. Since, by (i), the final X in Γ occurs after the final left-hand bracket, the sequences of 1s and −1s associated with the expression (ΓΔ) is $1S_\Gamma - 1S_\Delta$, and it should be clear that as S_Γ and S_Δ satisfy the "$s_k \geq 0$" condition, so also does the sequence $1S_\Gamma - 1S_\Delta$. Hence, (ΓΔ) is also a Catalan sequence.

5.3.5A Given a sequence $a_1, \ldots, a_n$ of n nonnegative integers we define a path of points on an $n \times n$ grid, by considering the elements of the sequence in turn as follows. We start at the point $(0,0)$. Suppose the last point we have defined is (x,y) and the next term we consider is a_k. If $a_k > 0$, we add to our path the points $(x+1,y)$ and $(x+1, y+a_k)$, while if $a_k = 0$, we just add the point $(x+1, y)$. For example, in the case $n = 6$, corresponding to the sequence 1,0,0,3,0,2, we obtain the sequence of points

$$(0,0), \ (1,0), \ (1,1), \ (2,1), \ (3,1), \ (4,1), \ (4,4), \ (5,4), \ (6,4), \ (6,6),$$

and hence the path shown in the diagram below.

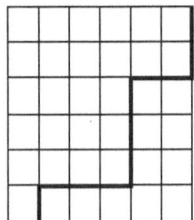

The condition that $a_1 + a_2 + \ldots + a_n = n$ means that the path ends at the point (n,n), and the condition that $a_1 + a_2 + \ldots + a_k = k$ ensures that the path does not go above the bottom-left to top-right diagonal. Hence, by the result of Problem 5.8, there are C_n sequences with the specified properties.

5.3.6A If each up-sloping arrow is replaced by 1 and each down-sloping arrow by −1 we see that there is a one–one correspondence between "mountain paths" made up of $2n$ arrows, and Catalan sequences of length $2n$ So, by Theorem 5.6, there are C_n such mountain paths.

5.3.7A With H corresponding to 1 and T corresponding to −1, each sequence of five heads and five tails, with at least as many heads as tails at each stage, corresponds to a Catalan sequence of length 10 and vice versa. Thus, there are $C_5 = \frac{1}{6}C(10,5) = 42$ such sequences.

5.3.8A The votes, as they are counted, correspond to a sequence of a As and a Bs. The total number of these sequences is the number of ways of choosing the positions for the

As, that is, $C(2a,a)$. We now consider the number of these sequences with at least as many As as Bs at each stage. These correspond to Catalan sequences of length $2a$. Hence there are $C_a = (1/(a+1)C(2a,a))$ such sequences. Hence the required probability is $C_a/C(2a,a) = 1/(a+1)$.

CHAPTER 6

6.2.1A The partitions of n into at most three parts, for $1 \le n \le 8$, are given in the following table. The values of $p_3(n)$ are given by the number of partitions in each column. It will be seen these are the values given in Table 6.1.

n	1	2	3	4	5	6	7	8
p	1	2	3	4	5	6	7	8
a		1+1	2+1	3+1	4+1	5+1	6+1	7+1
r			1+1+1	2+2	3+2	4+2	5+2	6+2
t				2+1+1	3+1+1	4+1+1	5+1+1	6+1+1
i					2+2+1	3+3	4+3	5+3
t						3+2+1	4+2+1	5+2+1
i						2+2+2	3+3+1	4+4
o							3+2+2	4+3+1
n								4+2+2
s								3+3+2
$p_3(n)$	1	2	3	4	5	7	8	10

6.2.2A The partitions of 7 are as follows:

At most Four Parts not all of Size < 4	At most Four Parts and Size at Most 4	At Least Five Parts and Size at Most 4
7	4+3	3+1+1+1+1
6+1	4+2+1	2+2+1+1+1
5+2	4+1+1+1	2+1+1+1+1+1
5+1+1	3+3+1	1+1+1+1+1+1+1
	3+2+2	
	3+2+1+1	
	2+2+2+1	

From this table we see that $p_4(7) = q_4(7) = 11$.

6.2.3A Partitioning n into k parts where the order matters corresponds to inserting $k-1$ plus symbols (+) into a string of n 1s. For example, the string of symbols

$$1\ 1\ 1\ 1\ 1\ +\ 1\ +\ 1\ 1\ +\ 1\ 1$$

corresponds to the partition 5+1+2+2 of 10. When we have n 1s, there are $n - 1$ gaps in which the $k - 1$ plus symbols may be inserted. So there are $C(n - 1, k - 1)$ ways to do this.

6.2.4A The partitions of n into unequal parts of size at most k may be divided into two disjoint classes.

i. Those partitions not containing a part of size k. Each of these partitions is therefore a partition of n into unequal parts of size at most $k - 1$. So there are $u_{k-1}(n)$ partitions of this type.

ii. Those partitions containing a part of size k. Each of these partitions consists of a part of size k and a partition of $n - k$ into unequal parts of size at most $k - 1$. So there are $u_{k-1}(n - k)$ partitions in this class.

It follows that $u_k(n) = u_{k-1}(n) + u_{k-1}(n - k)$.

6.2.5A We use $[x]$ for the integer nearest to x. Then, by Theorem 6.4(iii), we need to show that for each positive integer n,

$$\left\lfloor \frac{n^2}{12} + \frac{n}{2} + 1 \right\rfloor = \left[\frac{(n+3)^2}{12} \right].$$

We need to consider separately the cases $n = 6s + r$, where r, s are nonnegative integers, and $r \le 5$. We have that

$$\left\lfloor \frac{(6s+r)^2}{12} + \frac{6s+r}{2} + 1 \right\rfloor = \left\lfloor 3s^2 + (r+3)s + \frac{r^2}{12} + \frac{r}{2} + 1 \right\rfloor = 3s^2 + (r+3)s + \left\lfloor \frac{r^2}{12} + \frac{r}{2} + 1 \right\rfloor,$$

and

$$\left[\frac{(6s+r+3)^2}{12} \right] = \left[3s^2 + (r+3)s + \frac{r^2 + 6r + 9}{12} \right] = 3s^2 + (r+3)s + \left[\frac{r^2}{12} + \frac{r}{2} + \frac{3}{4} \right].$$

Thus we need only show that for each integer r with $0 \le r \le 5$,

$$\left\lfloor \frac{r^2}{12} + \frac{r}{2} + 1 \right\rfloor = \left[\frac{r^2}{12} + \frac{r}{2} + \frac{3}{4} \right].$$

This is a matter of routine. For example, with $r = 5$,

$$\left\lfloor \frac{r^2}{12} + \frac{r}{2} + 1 \right\rfloor = \left\lfloor \frac{67}{12} \right\rfloor = 5, \text{ and } \left[\frac{r^2}{12} + \frac{r}{2} + \frac{3}{4} \right] = \left[\frac{16}{3} \right] = 5.$$

The other cases are left to the reader.

6.2.6A When n is even, for some integer k, $n = 2k$. Hence, in this case, $\lfloor \frac{1}{2}n+1 \rfloor + \lfloor \frac{1}{2}(n-3)+1 \rfloor = \lfloor k+1 \rfloor + \lfloor k-\frac{1}{2} \rfloor = (k+1)+(k-1) = 2k = n$. When n is odd, for some integer k, $n = 2k + 1$. So $\lfloor \frac{1}{2}n+1 \rfloor + \lfloor \frac{1}{2}(n-3)+1 \rfloor = \lfloor k+\frac{3}{2} \rfloor + \lfloor k \rfloor = (k+1)+k = 2k+1 = n$.

6.4.1A Let X be the set of those partitions of n with no parts of size 1. By Theorem 6.7 $\#(X) = p(n) - p(n-1)$. Let X_1 be the set of those partitions in X that have at most k parts, and let X_2 be the set of those partitions in X with more than k parts. Since $p_k(n)$ is the number of partitions of n into at most k parts, $\#(X_1) \leq p_k(n)$. For each partition P in X_2, we let P' be the partition obtained from P by deleting 1 from the k smallest parts of P. Since each part of P is of size at least 2, the number of parts in P' is the same as the number of parts in P. So P' is a partition of $n - k$ into more than k parts. Clearly, if $P_1 \neq P_2$, then $P_1' \neq P_2'$. So $\#(X_2) \leq p(n-k)$. It follows that $p(n) - p(n-1) \leq p_k(n) + p(n-k)$.

6.4.2A The proof is a variant of that of Theorem 6.1. If D is a dot diagram for a partition of n into exactly k parts, then its dual is a partition of n whose maximum part has size exactly k and vice versa. So there are equal numbers of partitions of each type.

6.4.3A Consider a dot diagram, D, for a partition of n into l parts with $l \leq k$. Add k dots, by adding one dot to each of the rows and, when $l < k$, $k - l$ further rows each containing one dot, and let D' be the resulting partition. The diagram illustrates this in the case $n = 13$, $k = 6$, and $l = 4$.

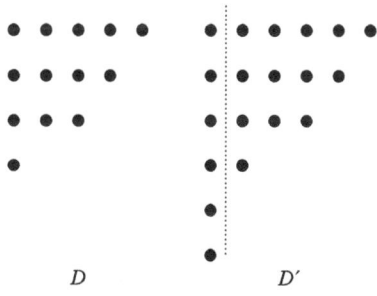

$$D \qquad\qquad D'$$

In the case illustrated, D is a partition of 13 into at most six parts, and D' is a partition of 19 $= 13 + 6$ into exactly six parts. More generally, the mapping $D \mapsto D'$ is a bijection between the partitions of n into at most k parts, and the partitions of $n + k$ into exactly k parts. So there are equal numbers of partitions of each type.

6.4.4A Given a partition,

$$a_1 + a_2 + \ldots + a_k = n, \qquad (1)$$

we can construct the following two different partitions of $2n$:

$$a_1 + a_1 + a_2 + a_2 + \ldots + a_k + a_k = 2n \qquad (2)$$

and

$$2a_1 + 2a_2 + \ldots + 2a_k = 2n. \qquad (3)$$

Since, for $n \geq 2$, we can easily find other partitions of $2n$, we have $p(2n) > 2p(n)$.

6.4.5A Given a dot diagram, D, for a partition of n in which all parts have size at least 2, let D' be the dot diagram obtained by adding 1 to the longest row in D. Thus D' corresponds to a partition of $n + 1$ in which all parts have size at least 2. Clearly, if $D_1 \neq D_2$, then $D_1' \neq D_2'$. So the number of partitions of $n + 1$ with no parts of size 1 is at least as many as the number of partitions of n with no parts of size 1. Hence, by Theorem 6.10, $p(n) - p(n-1) \leq p(n+1) - p(n)$. This inequality can be rearranged to give $p(n) \leq \frac{1}{2}(p(n+1) + p(n-1))$.

6.4.6A By Theorem 6.10, $p(n) - p(n-1)$ is the number of partitions of n in which there are no parts of size 1. If D is a dot diagram for such a partition, then each part has size at least 2, and so its dual D^* is the dot diagram of a partition where the two largest parts are equal, and vice versa.

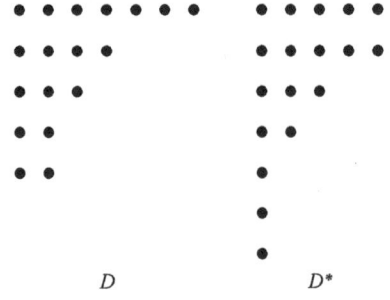

$$D \qquad\qquad D^*$$

It follows that there are equal numbers of partitions of each type, and hence there are $p(n) - p(n-1)$ partitions of n in which the two largest parts are equal.

CHAPTER 7

7.1.1A The terms involving x^n in the product $(1 + x + x^2 + x^3 + \ldots)^2$ are the $n + 1$ terms $1.x^n$, $x.x^{n-1}$, $x^2 x^{n-2}, \ldots, x^n.1$, and hence the coefficient of x^n is $n + 1$.

7.1.2A Consider the product

$$(a_0 + a_1 x + a_2 x^2 + a_3 x^3 + \ldots)(1 + x^3 + x^6 + x^9 + \ldots) = (1 + x + x^2 + x^3 + \ldots).$$

Equating the constant terms gives $a_0 = 1$. Equating the coefficients of x gives $a_1 = 1$, and equating the coefficients of x^2 gives $a_2 = 1$. For $n \geq 3$, equating the coefficients of x^n gives

$$a_n + a_{n-3} + a_{n-6} + \ldots + a_{n-3t} = 1, \qquad (1)$$

where $t = \lfloor n/3 \rfloor$. We can now prove, using mathematical induction, that for all $n \geq 3$, $a_n = 0$. By Equation 1, $a_3 + a_0 = 1$, and hence, as $a_0 = 1$, it follows that $a_3 = 0$. So the result is true for

$n = 3$. Now suppose that $n > 3$ and the result holds for all m, with $3 \le m < n$. It follows from Equation 1 that $a_n + a_r = 1$, where r is the remainder when n is divided by 3. Since $a_0 = a_1 = a_2 = 1$, it follows that $a_n = 0$. Therefore, by mathematical induction, for all $n \ge 3$, $a_n = 0$.

7.2.1A i. a. We have that $\sum_{n=0}^{\infty} x^n = 1/(1-x)$. Hence, differentiating both sides we obtain $\sum_{n=0}^{\infty} nx^{n-1} = 1/(1-x)^2$ and, differentiating again, $\sum_{n=0}^{\infty} n(n-1)x^{n-2} = 2/(1-x)^3$ consequently $\sum_{n=0}^{\infty} n(n-1)x^n = 2x^2/(1-x)^3$. Thus $x \mapsto 2x^2/(1-x)^3$ is the generating function for the sequence $\{n(n-1)\}$.

b. Since $\sum_{n=0}^{\infty} nx^{n-1} = 1/(1-x)^2$, it follows that $\sum_{n=0}^{\infty} nx^n = x/(1-x)^2$, and differentiating both sides gives $\sum_{n=0}^{\infty} n^2 x^{n-1} = (1+x)/(1-x)^3$, and hence $\sum_{n=0}^{\infty} n^2 x^n = [x(1+x)]/(1-x)^3$, and so the generating function of the sequence $\{n^2\}$ is $x \mapsto [x(1+x)]/(1-x)^3$.

c. From (b), $\sum_{n=0}^{\infty} n^2 x^n = [x(1+x)]/(1-x)^3$. Differentiating both sides gives $\sum_{n=0}^{\infty} n^3 x^{n-1} = (1+4x+x^2)/(1-x)^4$, and hence the generating function of the sequence $\{n^3\}$ is $x \mapsto [x(1+4x+x^2)]/(1-x)^4$.

ii. Since the generating functions, say f and g, for the sequences $\{n\}$ and $\{n^2\}$, respectively, are given by $f(x) = x/(1-x)^2$ and $g(x) = [x(1+x)]/(1-x)^3$, the generating function, say h, for the sequence $\{\frac{1}{2}n(n+1)\}$, that is, the sequence $\{\frac{1}{2}(n^2+n)\}$ is given by

$$h(x) = \frac{1}{2}(f(x)+g(x)) = \frac{1}{2}\left(\frac{x(1+x)}{(1-x)^3} + \frac{x}{(1-x)^2}\right) = \frac{1}{2}\left(\frac{x(1+x)+x(1-x)}{(1-x)^3}\right) = \frac{x}{(1-x)^3},$$

which agrees with Equation 7.12.

7.2.2A Let f be the generating function for the sequence $\{a_n\}$ that is given by $a_0 = 0$ and $a_{n+1} = a_n + (n+1)^2$. Thus $a_{n+1} = a_n + n^2 + 2n + 1$. Hence

$$\sum_{n=0}^{\infty} a_{n+1} x^n = \sum_{n=0}^{\infty} a_n x^n + \sum_{n=0}^{\infty} n^2 x^n + 2\sum_{n=0}^{\infty} nx^n + \sum_{n=0}^{\infty} x^n. \qquad (1)$$

Because $a_0 = 0$,

$$\sum_{n=0}^{\infty} a_{n+1} x^n = \frac{1}{x}\sum_{n=0}^{\infty} a_{n+1} x^{n+1} = \frac{1}{x} f(x).$$

Thus using our knowledge of the generating functions for the sequences $\{n^2\}$, $\{n\}$, and $\{1\}$, it follows from Equation 1 that

$$\frac{1}{x} f(x) = f(x) + \frac{x(1+x)}{(1-x)^3} + \frac{2x}{(1-x)^2} + \frac{1}{(1-x)} = f(x) + \frac{1+x}{(1-x)^3},$$

and hence $f(x) = [x(1 + x)]/(1 − x)^4$. Note that, using this method, we have found the generating function without needing to use the formula for the sum of the first n squares.

7.3.1A We let X_n be the set of those sequences of n digits in which there are not consecutive even digits. Thus $\#(X_n) = a_n$. A sequence in X_n has either the form $x\$$, where $x \in \{1,3,5,7,9\}$ and $\$ \in X_{n-1}$, or the form $yx\$$, where $y \in \{0,2,4,6,8\}$, $x \in \{1,3,5,7,9\}$, and $\$ \in X_{n-2}$. There are $5a_{n-1}$ sequences of the first type and $25a_{n-2}$ of the second type. Therefore the sequence $\{a_n\}$, satisfies the recurrence relation $a_n = 5a_{n-1} + 25a_{n-2}$. The initial conditions are as follows. We have $a_0 = 1$, because there is one sequence with zero digits in it, namely, the empty sequence, and this sequence does not include consecutive even digits. Also, $a_1 = 10$, because each sequence consisting of a single digit does not include consecutive even digits.

7.4.1A Let A be the generating function for the sequence $\{a_n\}$. Since $a_n = 8a_{n-1} + 15a_{n-2}$, with $a_1 = 2$ and $a_2 = 16$, we have that $\sum_{n=3}^{\infty} a_n x^n = 8x \sum_{n=3}^{\infty} a_{n-1} x^{n-1} - 15x^2 \sum_{n=3}^{\infty} a_{n-2} x^{n-2}$. Therefore $A(x) - 2x - 16x^2 = 8x(A(x) - 2x) - 15x^2 A(x)$. Hence $(1 - 8x + 15x^2)A(x) = 2x$. Therefore

$$A(x) = \frac{2x}{(1-8x+15x^2)} = \frac{2x}{(1-5x)(1-3x)} = \frac{1}{1-5x} - \frac{1}{1-3x} = \sum_{n=0}^{\infty} (5x)^n - \sum_{n=0}^{\infty} (3x)^n.$$

Hence, equating coefficients, we deduce that $a_n = 5^n - 3^n$.

7.4.2A We let $A(x) = \sum_{n=1}^{\infty} a_n x^n$. From the recurrence relation $a_n = a_{n-2} + a_{n-1}$, which holds for $n \geq 3$, we have that $\sum_{n=3}^{\infty} a_n x^n = \sum_{n=3}^{\infty} a_{n-2} x^n + \sum_{n=3}^{\infty} a_{n-1} x^n$. Therefore

$$\left(\sum_{n=1}^{\infty} a_n x^n\right) - a_1 x - a_2 x^2 = x^2 \sum_{n=3}^{\infty} a_{n-2} x^{n-2} + x\left(\left(\sum_{n=2}^{\infty} a_{n-1} x^{n-1}\right) - a_1 x\right).$$

That is, $A(x) - 2x - 3x^2 = x^2 A(x) + x(A(x) - 2x)$. Hence $A(x) = (x^2 + 2x)/(1 − x − x^2)$.

7.4.3A We prove the result using mathematical induction. From the definition of g_n, and the fact that $f_1 = f_2 = 1$, we have $g_3 = g_1 + g_2 = \alpha + \beta = \alpha f_1 + \beta f_2$. So the result holds when $n = 3$. Now, suppose $n > 3$, and the result holds for all $k < n$. Then $g_k = g_{k-2} + g_{k-1} = (\alpha f_{k-4} + \beta f_{k-3}) + (\alpha f_{k-3} + \beta f_{k-2})$, by the induction hypothesis. Hence $g_k = \alpha(f_{k-4} + f_{k-3}) + \beta(f_{k-3} + f_{k-2}) = \alpha f_{k-2} + \beta f_{k-1}$, using the definition of the Fibonacci numbers. Hence the result holds also for $n = k$. Therefore, by mathematical induction, the result holds for all $n \geq 3$.

7.5.1A The general solution has the form $A\alpha^n + B\beta^n$, where α, β are the roots of the quadratic equation $x^2 - 3x - 4 = 0$, that is, $(x + 1)(x - 4) = 0$, giving $\alpha, \beta = -1, 4$. So the general solution is $a_n = A(-1)^n + B4^n$. Since $a_1 = 1$ and $a_3 = 3$, $-A + 4B = 1$ and $A + 16B = 3$, giving $A = -\frac{1}{5}$ and $B = \frac{1}{5}$. Therefore $a_n = \frac{1}{5}((-1)^{n+1} + 4^n)$.

7.5.2A Let X_n be the set of sequences of 0s, 1s, and 2s of length n in which a 0 can only be followed by a 1. Such a sequence will have either the form $x\$$, with $x \in \{1,2\}$ and $\$ \in X_{n-1}$, or the form $01\$$, with $\$ \in X_{n-2}$. It follows that $a_n = 2a_{n-1} + a_{n-2}$. Also, it can be seen that $a_1 = 3$ and $a_2 = 7$. The general solution of the recurrence relation has the form $A\alpha^n + B\beta^n$, where

α, β are the roots of the quadratic equation $x^2 - 2x - 1 = 0$, from which it follows that α, β are $1 \pm \sqrt{2}$. Thus, the general solution of the recurrence relation is $a_n = A(1+\sqrt{2})^n + B(1-\sqrt{2})^n$. Since $a_1 = 3$, we have

$$(1+\sqrt{2})A + (1-\sqrt{2})B = 3. \tag{1}$$

Since $a_2 = 7$ and $(1 \pm \sqrt{2})^2 = 1 \pm 2\sqrt{2} + 2 = 3 \pm 2\sqrt{2}$,

$$(3+2\sqrt{2})A + (3-2\sqrt{2})B = 7. \tag{2}$$

Solving Equations 1 and 2 gives $A = \frac{1}{2}(1+\sqrt{2})$ and $B = \frac{1}{2}(1-\sqrt{2})$. Hence $a_n = \frac{1}{2}((1+\sqrt{2})^{n+1} + (1-\sqrt{2})^{n+1})$.

7.5.3A By expanding the determinant down the first column, we see that

$$\det(A_n) = \begin{vmatrix} 1 & 1 & 0 & 0 & \cdots & 0 & 0 & 0 \\ 1 & 1 & 1 & 0 & \cdots & 0 & 0 & 0 \\ 0 & 1 & 1 & 1 & \cdots & 0 & 0 & 0 \\ \vdots & \vdots & \vdots & \vdots & & \vdots & \vdots & \vdots \\ \vdots & \vdots & \vdots & \vdots & & \vdots & \vdots & \vdots \\ 0 & 0 & 0 & 0 & \cdots & 1 & 1 & 0 \\ 0 & 0 & 0 & 0 & \cdots & 1 & 1 & 1 \\ 0 & 0 & 0 & 0 & \cdots & 0 & 1 & 1 \end{vmatrix} - \begin{vmatrix} 1 & 0 & 0 & 0 & \cdots & 0 & 0 & 0 \\ 1 & 1 & 1 & 0 & \cdots & 0 & 0 & 0 \\ 0 & 1 & 1 & 1 & \cdots & 0 & 0 & 0 \\ \vdots & \vdots & \vdots & \vdots & & \vdots & \vdots & \vdots \\ \vdots & \vdots & \vdots & \vdots & & \vdots & \vdots & \vdots \\ 0 & 0 & 0 & 0 & \cdots & 1 & 1 & 0 \\ 0 & 0 & 0 & 0 & \cdots & 1 & 1 & 1 \\ 0 & 0 & 0 & 0 & \cdots & 0 & 1 & 1 \end{vmatrix} \tag{1}$$

$$= \det(A_{n-1}) - \begin{vmatrix} 1 & 1 & 0 & \cdots & 0 & 0 & 0 \\ 1 & 1 & 1 & \cdots & 0 & 0 & 0 \\ \vdots & \vdots & \vdots & & \vdots & \vdots & \vdots \\ \vdots & \vdots & \vdots & & \vdots & \vdots & \vdots \\ 0 & 0 & 0 & \cdots & 1 & 1 & 0 \\ 0 & 0 & 0 & \cdots & 1 & 1 & 1 \\ 0 & 0 & 0 & \cdots & 0 & 1 & 1 \end{vmatrix} = \det(A_{n-1}) - \det(A_{n-2}),$$

on expanding the second determinant in Equation 1 along the top row. Hence, the required recurrence relation is

$$d_n = d_{n-1} - d_{n-2}, \text{ with } a_1 = |1| = 1 \text{ and } a_2 = \begin{vmatrix} 1 & 1 \\ 1 & 1 \end{vmatrix} = 0. \tag{2}$$

The auxiliary equation associated with Equation 2 is $x^2 - x + 1 = 0$. This has the complex roots $\alpha, \beta = \frac{1}{2}(1 \pm \sqrt{3}i)$. Hence the general solution is $a_n = A\left(\frac{1}{2}(1+\sqrt{3}i)\right)^n + B\left(\frac{1}{2}(1-\sqrt{3}i)\right)^n$ for some constants A, B. The initial conditions $a_1 = 1$, $a_2 = 0$ give $\frac{1}{2}(1+\sqrt{3}i)A + \frac{1}{2}(1-\sqrt{3}i)B = 1$ and $\frac{1}{2}(-1+\sqrt{3}i)A + \frac{1}{2}(-1-\sqrt{3}i)B = 0$, from which it follows that $A = \frac{1}{2}\left[1-(1/\sqrt{3})i\right]$ and $B = \frac{1}{2}\left[1+(1/\sqrt{3})i\right]$. Using the fact that $\frac{1}{2}(1+\sqrt{3}i) = e^{i\pi/3}$ and $\frac{1}{2}(1-\sqrt{3}i) = e^{-i\pi/3}$, it can be shown that $a_n = \cos(n\pi/3) + (1/\sqrt{3})\sin(n\pi/3)$, or, equivalently, $a_n = (2/\sqrt{3})\sin(\frac{1}{3}(n+1)\pi)$.

7.6.1A We first need to solve the homogeneous recurrence relation

$$a_n - 2a_{n-1} - 8a_{n-2} = 0. \tag{1}$$

The quadratic equation $x^2 - 2x - 8 = 0$ is equivalent to $(x - 4)(x + 2) = 0$ with solutions $x = 4$, -2. Hence the general solution of Equation 1 has the form $A4^n + B(-2)^n$, where A and B are constants. We next seek a particular solution of the recurrence relation

$$a_n - 2a_{n-1} - 8a_{n-2} = 18 - 9n \tag{2}$$

of the form $a_n = Cn + D$, where C and D are constants. Substituting in Equation 2 gives

$$(Cn + D) - 2(C(n-1) + D) - 8(C(n - 2) + D) = 18 - 9n. \tag{3}$$

Equating the coefficients of n in Equation 3 and equating the constants in Equation 3 gives $-9C = -9$ and $-9D + 18C = 18$. Hence $C = 1$ and $D = 0$. It follows that the general solution of Equation 2 is given by $a_n = A4^n + B(-2)^n + n$. Since $a_1 = 1$ and $a_2 = 3$, we have $4A - 2B + 1 = 1$ and $16A + 4B + 2 = 3$, which gives $A = \frac{1}{24}$ and $B = \frac{1}{12}$. Hence the required solution is $a_n = \frac{1}{24}4^n + \frac{1}{12}(-2)^n + n$.

7.6.2A i. A sequence of length 1 contains an even number of vowels if and only if it contains zero vowels, and so just consists of a single consonant. So $a_1 = 21$. Let E_n, O_n be the sets of sequences of n letters of the English alphabet containing an even and an odd number of vowels, respectively. Then $\#(E_n) = a_n$, and as there are altogether 26^n sequences of n letters, $\#(O_n) = 26^n - a_n$. A sequence in E_n is either of the form $c\$$, where c is one of the 21 consonants and $\$ \in E_{n-1}$, or of the form $v\$$, where v is one of the five vowels and $\$ \in O_{n-1}$. It follows that $\#(E_n) = 21\#(E_{n-1}) + 5\#(O_n)$, and therefore $a_n = 21a_{n-1} + 5(26^{n-1} - a_{n-1}) = 16a_{n-1} + 5(26^{n-1})$, as required.

ii. We need to solve the recurrence relation $a_n - 16a_{n-1} = 5(26^{n-1})$ with $a_1 = 21$. The general solution of the homogeneous recurrence relation $a_n - 16a_{n-1} = 0$ is given by $A(16^n)$, where A is some constant. We try to find a particular solution of the form $a_n = \alpha(26^n)$. Substituting this into our recurrence relation gives $\alpha(26^n) - 16\alpha(26^{n-1}) = 5(26^{n-1})$. Canceling the common factor of 26^{n-1}, we deduce that $26\alpha - 16\alpha = 5$, and so $\alpha = \frac{1}{2}$. Therefore the general solution is $a_n = A(16^n) + \frac{1}{2}(26^n)$. Since $a_1 = 21$, we have $16A + 13 = 21$, giving $A = \frac{1}{2}$. Hence the solution is $a_n = \frac{1}{2}(16^n) + \frac{1}{2}(26^n)$.

7.6.3A We let $R(m,n)$ be the number of arithmetic operations needed to row-reduce an $m \times n$ matrix to echelon form, where $m < n$. Consider such a matrix:

$$\begin{bmatrix} a_{11} & a_{12} & \cdots & a_{1n} \\ a_{21} & a_{22} & \cdots & a_{2n} \\ \vdots & \vdots & & \vdots \\ a_{m1} & a_{m2} & \cdots & a_{mn} \end{bmatrix}.$$

At the first stage we ensure, by interchanging rows if necessary, that $a_{11} \neq 0$. (If all the entries in the first column are 0, we are really dealing with an $m \times (n-1)$ matrix, but we are aiming to calculate $R(m,n)$ in the worst possible case.) This does not require any arithmetic. We then divide the top row by a_{11} to get 1 in the first row and column. This takes $n-1$ divisions, as we do not need to do any arithmetic to work out that a_{11} divided by a_{11} is 1. Then, for $2 \leq s \leq m$, we multiply the top row by a_{s1} and subtract the result from the sth row. This makes the first entry in each of these rows 0. For each of the remaining $n-1$ entries in each row, this involves doing one multiplication and one subtraction, and so, altogether $2(n-1)(m-1)$ arithmetic operations. So we have now carried out $(n-1) + 2(n-1)(m-1) = (n-1)(2m-1)$ arithmetic operations and we have obtained a matrix in the form

$$\begin{bmatrix} 1 & a'_{12} & \cdots & a'_{1n} \\ 0 & a'_{22} & \cdots & a'_{2n} \\ \vdots & \vdots & & \vdots \\ 0 & a'_{m2} & \cdots & a'_{mn} \end{bmatrix}.$$

At the second stage we ensure that $a'_{22} \neq 0$ by a row interchange if necessary. We then divide the second row by a'_{22} and then subtract suitable multiples of the resulting second row from the remaining rows, so that all the entries in the second column other than a'_{22} are 0. This takes the same amount of arithmetic as the first stage except that we can now ignore the first column, so that we are essentially dealing with an $m \times (n-1)$ matrix. Hence this requires $(n-2)(2m-1)$ arithmetic operations. Assuming the worst case where the matrix has rank m, there are m stages in this process and so

$$R(m,n) = (n-1)(2m-1) + (n-2)(2m-1) + \ldots + (n-m)(2m-1) = \frac{1}{2}m(2n-m-1)(2m-1).$$

7.6.4A *The Tower of Hanoi*
When there is just one disk, only one move is needed to transfer it to the third peg, so $a_1 = 1$. To move n disks to the third peg, we need first to transfer the top $n-1$ disks to the second peg, which takes a_{n-1} steps. Then in one move we transfer the largest disk to the third peg. Finally we transfer the remaining $n-1$ disks from the second peg to the third peg, which takes a further a_{n-1} moves. Hence $a_n = a_{n-1} + 1 + a_{n-1}$, that is,

$$a_n = 2a_{n-1} + 1. \tag{1}$$

The general solution of the associated homogeneous recurrence relation $a_n - 2a_{n-1} = 0$ is $A(2^n)$, where A is a constant. If we try a particular solution of Equation 1 of the form $a_n = \alpha$, we see that the constant α has to satisfy the equation $\alpha = 2\alpha + 1$ and hence $\alpha = -1$. Thus the general solution of Equation 1 is $a_n = A(2^n) -1$. Since $a_1 = 1$, $2A -1 = 1$, and hence $A = 1$. Therefore $a_n = 2^n -1$. It follows that with 64 disks the process takes $2^{64} -1$ moves. You can check that $2^{64} -1$ seconds comes to about 5.8×10^{11} years.

7.6.5A i. Suppose the recurrence relation $r_n - r_{n-1} = 2n^2 - 3n + 2$ has a solution of the form $r_n = En^2 + Dn + C$, where C, D, and E are constants. Substituting this into the recurrence relations gives that, for each positive integer n,

$$(En^2 + Dn + C) - (E(n-1)^2 + D(n-1) + C) = 2n^2 - 3n + 2.$$

Therefore, for each positive integer n,

$$2En + D - E = 2n^2 - 3n + 2.$$

This is clearly impossible as a linear polynomial cannot be identically equal to a quadratic polynomial. This shows that the recurrence relation has no solution of the given form.

ii. If we substitute $r_n = En^2 + Dn + C$ in the recurrence relation $r_n - 2r_{n-1} = 2n^2 - 3n + 2$, we obtain

$$(En^2 + Dn + C) - 2(E(n-1)^2 + D(n-1) + C) = 2n^2 - 3n + 2.$$

This gives

$$-En^2 + (4E - D)n + (-2E + 2D - C) = 2n^2 - 3n + 2.$$

Equating the coefficients of n^2 and n and the constant terms gives $-E = 2$, $4E - D = -3$, and $-2E + 2D - C = 2$, giving $E = -2$, $D = -5$, and $C = -8$. It follows that $r_n = -2n^2 - 5n - 8$ is a solution of the recurrence relation.

7.6.6A i. We prove the result by mathematical induction. We have $r_2 - r_1 = f(2)$. Hence $r_2 = r_1 + f(2)$, and so the result is true for $n = 2$. Suppose that the result holds for $n = m$, so that $r_m = r_1 + \sum_{k=2}^{m} f(k)$. Then, as $r_{m+1} - r_m = f(m+1)$, we have

$$r_{m+1} = r_m + f(m+1) = r_1 + \sum_{k=2}^{m} f(k) + f(m+1) = r_1 + \sum_{k=2}^{m+1} f(k).$$

Hence the result holds also for $n = m + 1$. Therefore, by mathematical induction, it holds for all $n \geq 2$.

ii. It follows that the solution of Equation 7.49 is given by

$$r_n = 0 + \sum_{k=2}^{n} 2k^2 - 3k + 2 = 2\left(\frac{n(n+1)(2n+1)}{6} - 1\right) - 3\left(\frac{n(n+1)}{2} - 1\right) + 2(n-1),$$

using the standard formulas for $\sum_{k=1}^{n} k^2$ and $\sum_{k=1}^{n} k$, adapted to the case where the sums begin with $k = 2$. It follows, after doing the algebra, that $r_n = \frac{2}{3}n^3 - \frac{1}{2}n^2 + \frac{5}{6}n - 1$. This agrees with the solution to Problem 7.8 that we have already given.

7.7.1A i. Suppose that $\beta_1, \beta_2, \ldots, \beta_k$ are constants such that

$$\beta_1\{r_1^n\} + \beta_2\{r_2^n\} + \ldots + \beta_k\{r_k^n\} = \{0\}. \tag{1}$$

It follows that all the terms of the sequence on the left-hand side of Equation 1 are equal to 0. In particular, we have that

$$\beta_1 r_1 + \beta_2 r_2 + \ldots + \beta_k r_k = 0$$
$$\beta_1 r_1^2 + \beta_2 r_2^2 + \ldots + \beta_k r_k^2 = 0$$
$$\vdots$$
$$\beta_1 r_1^k + \beta_2 r_2^k + \ldots + \beta_k r_k^k = 0$$

It follows that

$$\mathbf{A}_k \mathbf{v}_k = \mathbf{0}, \tag{2}$$

where

$$\mathbf{A}_k = \begin{pmatrix} r_1 & r_2 & \ldots & r_k \\ r_1^2 & r_2^2 & \ldots & r_k^2 \\ \vdots & \vdots & & \vdots \\ r_1^k & r_2^k & \ldots & r_k^k \end{pmatrix} \text{ and } \mathbf{v}_k = \begin{pmatrix} \beta_1 \\ \beta_2 \\ \vdots \\ \beta_k \end{pmatrix}.$$

It is straightforward to show (or a standard algebraic fact that you already know) that $\det(\mathbf{A}_k) = r_1 r_2 \ldots r_k \prod_{1 \le i < j \le k}(r_j - r_i)$, and hence, if the numbers $r_1, r_2, \ldots, r_k$ are all different, $\det(\mathbf{A}_k) \ne 0$. It follows that the matrix $\mathbf{A}_k$ is invertible. Hence by Equation 2, $\mathbf{v}_k = \mathbf{A}_k^{-1} \mathbf{0} = \mathbf{0}$. Therefore $\beta_1 = \beta_2 = \ldots = \beta_k = 0$. It follows that the vectors $\{r_1^n\}, \{r_2^n\}, \ldots, \{r_k^n\}$ are linearly independent.

ii. Let $\{z_n\}$ be a solution of the recurrence relation $\alpha_0 a_n + \alpha_1 a_{n-1} + \ldots + \alpha_k a_{n-k} = 0$. Since the sequences $\{y(1)_n\}, \{y(2)_n\}, \ldots, \{y(k)_n\}$ are solutions of this recurrence relation, so too is the sequence $\{w_n\}$, where, for each positive integer n, $w_n = z_1 y(1)_n + z_2 y(2)_n + \ldots + z_k y(k)_n$. Then, for $1 \le i \le k$, $w_i = z_1 y(1)_i + z_2 y(2)_i + \ldots + z_i y(i)_i + \ldots + z_k y(k)_i = z_i$. Since subsequent terms of the sequences $\{w_n\}$ and $\{z_n\}$ are given by the same recurrence relation, the sequences $\{w_n\}$ and $\{z_n\}$ are identical. Therefore $\{z_n\}$ is a linear combination of the sequences $\{y(1)_n\},\{y(2)_n\},\ldots,\{y(k)_n\}$. This shows that these sequences span S.

iii. From (i) and (ii) it follows that the dimension of S is k. From Theorem 7.1 and (a), it follows that the sequences $\{x_1{}^n\},\{x_2{}^n\},\ldots,\{x_k{}^n\}$ form a linearly independent set of k sequences from S, and hence it is a basis for S. So each solution of the recurrence relation is a linear combination of these sequences, as claimed.

7.7.2A i. The auxiliary equation is $x^2 - 10x + 25 = 0$, that is $(x - 5)^2 = 0$, with the repeated root $x = 5$. It follows that the general solution of the recurrence relation is $a_n = A5^n + Bn5^n$. The initial conditions $a_1 = 15$, $a_2 = 325$ give $5A + 5B = 15$ and $25A + 50B = 325$. These equations have the solution $A = -7$, $B = 10$. Hence the solution of the recurrence relation is $a_n = -7(5^n) + 10n5^n = (10n - 7)5^n$.

ii. The auxiliary equation is $x^2 + 4x + 4 = 0$, that is, $(x + 2)^2 = 0$, with the repeated root $x = -2$. Hence the general solution is $a_n = A(-2)^n + Bn(-2)^n$. The initial conditions give $-2A - 2B = 4$, $4A + 8B = 4$, with the solution $A = -5$, $B = 3$. So the solution of the recurrence relation is $a_n = -5(-2)^n + 3n(-2)^n = (3n - 5)(-2)^n$.

iii. The auxiliary equation is $x^3 - 7x^2 + 16x - 12 = 0$, that is, $(x - 3)(x - 2)^2 = 0$, with the root $x = 3$ and the repeated root $x = 2$. Hence the general solution is $a_n = A3^n + B2^n + Cn2^n$. The initial conditions give $3A + 2B + 2C = 8$, $9A + 4B + 8C = 42$, and $27A + 8B + 24C = 142$. These equations may be solved to give $A = 2$, $B = -4$, and $C = 5$. Hence the general solution is $a_n = 2(3^n) - 4(2^n) + 5n(2^n) = 2(3^n) + (5n - 4)2^n$.

iv. The auxiliary equation is $x^3 - 6x^2 + 12x - 8 = 0$, that is, $(x - 2)^3 = 0$, with the triple root $x = 2$. Hence the general solution is $a_n = A2^n + Bn2^n + Cn^22^n$. The initial conditions give $2A + 2B + 2C = 0$, $4A + 8B + 16C = 8$, and $8A + 24B + 72C = 16$. These equations have the solution $A = -4$, $B = 5$, $C = -1$. Hence the solution of the recurrence relation is $a_n = -4(2^n) + 5n(2^n) - n^2(2^n) = (-4 + 5n - n^2)2^n$.

7.8.1A Using the binomial theorem for $(1 + x)^\alpha$ with x replaced by $-4x$ and α by $\tfrac{1}{2}$ gives

$$(1-4x)^{1/2} = 1 + \left(\frac{1}{2}\right)(-4x) + \frac{(1/2)(-1/2)}{2!}(-4x)^2 + \frac{(1/2)(-1/2)(-3/2)}{3!}(-4x)^3 + \ldots.$$

The term involving x^n for $n \geq 3$ in this expansion is

$$\frac{(1/2)(-1/2)(-3/2)(-5/2)\ldots(-(2n-3)/2)}{n!}(-4x)^n.$$

Hence the coefficient of x^n is

$$-\frac{(1)(3)(5)\ldots(2n-3)2^{2n}}{2^n n!} = -\frac{(2n-2)!2^n}{(2)(4)\ldots(2n-2)n!} = -\frac{(2n-2)!2^n}{2^{n-1}(n-1)!n!} = -\frac{2(2n-2)!}{(n-1)!n!}.$$

Thus to obtain positive coefficients in the generating function for the sequence $\{C_n\}$ of Catalan numbers as given by the formula $(1\pm\sqrt{1-4x})/2x$ given in Equation 7.67 we need to take the minus sign. This gives the series

$$\frac{1}{2x}(1-(1-4x)^{1/2}) = \frac{1}{2x}\left(1-(1-2x-2x^2-\sum_{n=3}^{\infty}\frac{2(2n-2)!}{(n-1)!n!}x^n)\right) = 1+x+\sum_{n=3}^{\infty}\frac{(2n-2)!}{(n-1)!n!}x^{n-1}$$

$$= 1+x+\sum_{n=2}^{\infty}\frac{(2n)!}{n!(n+1)!}x^n.$$

So the coefficient of x is C_1 and for $n \geq 2$, the coefficient of x^n is

$$\frac{(2n)!}{n!(n+1)!} = \frac{1}{n+1}\left(\frac{(2n)!}{n!n!}\right) = \frac{1}{n+1}C(2n,n) = C_n.$$

CHAPTER 8

8.1.1A i. The coefficient of x^{100} in the product $(1+x^7+x^{14}+\ldots)(1+x^{11}+x^{22}+\ldots)$ is the number of solutions of the equation $7a + 11b = 100$, where a and b are nonnegative integers. It can be checked that there is just the one solution $a = 8, b = 44$. So the coefficient of x^{100} is 1.

ii. In this case we seek the number of solutions of $3a + 14b = 100$, where a and b are nonnegative integers. There are just two solutions: $a = 10, b = 5$ and $a = 24, b = 2$. So the coefficient of x^{100} is 2.

8.1.2A i. The generating function, say Q, is given by

$$Q(x) = (1+x+x^2+x^3+\ldots)(1+x^2)(1+x^3+x^6+x^9+\ldots)(1+x^4)\ldots$$

$$= \frac{1}{1-x}\times(1+x^2)\times\frac{1}{1-x^3}\times(1+x^4)\times\ldots = \frac{(1+x^2)(1+x^4)(1+x^6)\ldots}{(1-x)(1-x^3)(1-x^5)\ldots}.$$

ii. Similarly, the generating function, say R, is given by

$$R(x) = \frac{(1+x)(1+x^3)(1+x^5)\ldots}{(1-x^2)(1-x^4)(1-x^6)\ldots}.$$

8.1.3A i. The relevant partitions and the values of $p_{d,o}(n)$ for $1 \leq n \leq 10$ are given in the following table.

n	Partitions of n into distinct odd parts	$p_{d,o}(n)$
1	1	1
2	–	0
3	3	1
4	3 + 1	1
5	5	1
6	5 + 1	1
7	7	1
8	7 + 1, 5 + 3	2
9	9, 5 + 3 + 1	2
10	9 + 1, 7 + 3	2

b. The generating function is $x \mapsto (1+x)(1+x^3)(1+x^5)\ldots$, that is, $x \mapsto \prod_{k=0}^{\infty} 1+x^{2k+1}$.

8.1.4A We have that $P(x) = \sum_{n=1}^{\infty} p(n)x^n$, and hence $P(x)/(1-x) = P(x) \times [1/(1-x)] = (p(1)x + p(2)x^2 + p(3)x^3 + p(4)x^4 + \ldots)(1 + x + x^2 + x^3 + \ldots)$. The term involving x^n in this product is $p(1)x \cdot x^{n-1} + p(2)x^2 \cdot x^{n-2} + \ldots + p(n)x^n \cdot 1$, and hence the coefficient of x^n is $p(1) + p(2) + \ldots + p(n)$, that is, $t(n)$. Hence $x \mapsto P(x)/(1-x)$ is the generating function for the sequence $\{t(n)\}$.

8.1.5A We have seen from the solution to 8.1.2A(ii) that the generating function for the sequence $\{a(n)\}$, where $a(n)$ is the number of partitions in which only the even parts can be repeated is given by

$$R(x) = \frac{(1+x)(1+x^3)(1+x^5)\ldots}{(1-x^2)(1-x^4)(1-x^6)\ldots} = \frac{1+x}{1-x^2} \times \frac{1}{1-x^4} \times \frac{1+x^3}{1-x^6} \times \frac{1}{1-x^8} \times \ldots$$

$$= \frac{1}{1-x} \times \frac{1}{1-x^4} \times \frac{1}{1-x^3} \times \frac{1}{1-x^8} \times \ldots = \frac{1}{(1-x)(1-x^3)(1-x^4)(1-x^5)(1-x^7)(1-x^8)\ldots},$$

which is the generating function for the sequence $\{b(n)\}$, where $b(n)$ is the number of partitions in which the even parts must be multiples of 4. It follows that for all n, $a(n) = b(n)$.

8.1.6A i. When the product $(1+x^{2^0})(1+x^{2^1})\ldots(1+x^{2^n})$ is expanded, each term is obtained by multiplying 1s chosen from some of the brackets and terms x^{2^a}, with $0 \leq a \leq n$, chosen from the remaining brackets. Thus the terms have the form x^k, where $k = 2^{a_1} + 2^{a_2} + \ldots + 2^{a_r}$, with $0 \leq a_1 < a_2 < \ldots < a_r \leq n$. Now every integer in the range from 1 to $2^{n+1} - 1$ has a unique representation of this form, and so the product is equal to the sum $1 + x + x^2 + x^3 + \ldots + x^{2^{n+1}-1}$.

ii. By the obvious extension of (i), $(1+x^{2^0})(1+x^{2^1})(1+x^{2^2})\ldots = 1 + x + x^2 + \ldots = 1/(1-x)$ and therefore

$$1-x = \frac{1}{(1+x^{2^0})(1+x^{2^1})(1+x^{2^2})\ldots}.$$

c. Since $1/(1+x^{2^k}) = 1 - x^{2^k} + x^{2 \cdot 2^k} - x^{3 \cdot 2^k} + x^{4 \cdot 2^k} - \ldots$, we see that the positive terms correspond to choosing an even number of parts of size 2^k and the negative terms correspond to choosing an odd number of parts of size 2^k. Hence the coefficient of x^n in the product $1/\left[(1+x^{2^0})(1+x^{2^1})(1+x^{2^2})\ldots\right]$ is the difference between the number of ways of writing n as the sum of an *even* number of powers of 2, and the number of ways of writing n as the sum of an *odd* number of powers of 2. By (b), for $n \geq 2$, the coefficient of x^n in this product is 0. So for $n \geq 2$ the number of ways of writing n as the sum of an even number of powers of 2 is equal to the number of ways of writing it as an odd number of powers of 2.

8.2.1A Using Theorem 8.6, and the values of $p(n)$ that we have already calculated (see Problem 8.3), we have

$p(16) = p(15) + p(14) - p(11) - p(9) + p(4) + p(1) = 176 + 135 - 56 - 30 + 5 + 1 = 231$,

$p(17) = p(16) + p(15) - p(12) - p(10) + p(5) + p(2) = 231 + 176 - 77 - 42 + 7 + 2 = 297$,

$p(18) = p(17) + p(16) - p(13) - p(11) + p(6) + p(3) = 297 + 231 - 101 - 56 + 11 + 3 = 385$,

$p(19) = p(18) = p(17) - p(14) - p(12) + p(7) + p(4) = 385 + 297 - 135 - 77 + 15 + 5 = 490$,

and

$p(20) = p(19) + p(18) - p(15) - p(13) + p(8) + p(5) = 490 + 385 - 176 - 101 + 22 + 7 = 627$.

8.2.2A The relevant partitions and the corresponding values of $u_{e,d}(n)$ and $u_{o,d}(n)$ for $1 \leq n \leq 10$ are given in the following table.

n	Partitions into an Even Number of Distinct Parts	$u_{e,d}(n)$	Partitions into an Odd Number of Distinct Parts	$u_{o,d}(n)$
1	—	0	1	1
2	—	0	2	1
3	$2+1$	1	3	1
4	$3+1$	1	4	1
5	$4+1, 3+2$	2	5	1
6	$5+1, 4+2$	2	$6, 3+2+1$	2
7	$6+1, 5+2, 4+3$	3	$7, 4+2+1$	2
8	$7+1, 6+2, 5+3$	3	$8, 5+2+1, 4+3+1$	3
9	$8+1, 7+2, 6+3, 5+4$	4	$9, 6+2+1, 5+3+1, 4+3+2$	4
10	$9+1, 8+2, 7+3, 6+4,$ $4+3+2+1$	5	$10, 7+2+1, 6+3+1,$ $5+4+1, 5+3+2$	5

We see that $u_{d,o}(n) = u_{d,e}(n)$ for $n = 3,4,6,8,9,10$, $u_{d,o}(1) - u_{d,e}(1) = u_{d,o}(2) - u_{d,e}(2) = 1 = (-1)^2$, and $u_{d,o}(5) - u_{d,e}(5) = u_{d,o}(7) - u_{d,e}(7) = -1 = (-1)^3$. Since $1, 2 = \frac{1}{2}(3(1^2) \pm 1)$, and $5, 7 = \frac{1}{2}(3(2^2) \pm 2)$, these values correspond to those given by Theorem 8.5.

CHAPTER 9

9.1.1A The pictures tell us which pairs of vertices are joined by an edge. Our precise definition of when two sets of vertices joined by edges are really the same is given by Definition 9.3 of *isomorphic* graphs.

9.3.1A The graphs (i) through (v) all have five vertices, and graphs (vi) through (x) have six vertices. So none of the first five graphs is isomorphic to any of the last five graphs. In the following table we show the number of vertices of each degree in these graphs.

degree	(i)	(ii)	(iii)	(iv)	(v)	(vi)	(vii)	(viii)	(ix)	(x)
0	1	0	0	0	0	0	0	0	0	0
1	1	2	3	3	2	0	0	0	0	0
2	2	1	1	1	1	0	0	0	0	0
3	1	2	1	1	2	6	6	6	6	6

We see from this table that among the graphs with five vertices there are just two possible pairs of isomorphic graphs, namely, the pair (ii) and (v), and the pair (iii) and (iv).

(ii) and (v). The mapping θ given by $\theta(a) = b$, $\theta(b) = c$, $\theta(c) = e$, $\theta(d) = d$, and $\theta(e) = a$ is an isomorphism from graph (ii) to graph (v).

(iii) and (iv). The mapping ϕ given by $\phi(a) = c$, $\phi(b) = b$, $\phi(c) = a$, $\phi(d) = e$, and $\phi(e) = d$ is an isomorphism from graph (iii) to graph (iv).

The case of the graphs (vi) to (x) with six vertices is more complicated, as in each of these graphs all the vertices have degree 3. However, they are not all isomorphic. In the graphs (vi), (vii), and (x), we can separate the vertices into two sets of three vertices with all the edges joining the vertices in one of these sets to all the vertices in the other. For graph (vi) these vertex sets are $\{a,b,c\}$ and $\{d,e,f\}$; for graph (vii) they are $\{a,c,e\}$ and $\{b,d,f\}$; and for graph (x) they are $\{a,d,f\}$ and $\{b,c,e\}$. Any mapping that matches up the two sets of one graph with those of the second graph will be an isomorphism between the relevant pair of graphs. For example, the mapping θ defined by $\theta(a) = a$, $\theta(b) = c$, $\theta(c) = e$, $\theta(d) = b$, $\theta(e) = d$, and $\theta(f) = f$ is an isomorphism between graphs (vi) and (vii). Similarly, the mapping ϕ defined by $\phi(a) = a$, $\phi(b) = b$, $\phi(c) = d$, $\phi(d) = c$, $\phi(e) = f$, and $\phi(f) = e$ is an isomorphism between graphs (vii) and (x), and the mapping Ψ defined by $\Psi(a) = a$, $\Psi(b) = d$, $\Psi(c) = f$, $\Psi(d) = b$, $\Psi(e) = c$, and $\Psi(f) = e$ is an isomorphism between graphs (vi) and (x). The graphs (viii) and (ix) each includes a triangle, that is, three vertices each joined by an edge to the other two. In graph (viii) one such triangle is $\{a,c,e\}$ and in graph (ix) one triangle is $\{a,b,e\}$, whereas the other graphs with six vertices do not have triangles. So graphs (viii) and (ix) are not isomorphic to any of the graphs (vi), (vii), and (x). The graphs (viii) and (ix) are isomorphic. For example, the mapping σ defined by σ(a) = a, σ(b) = d, σ(c) = b, σ(d) = c, σ(e) = e, and σ(f) = f is an isomorphism between these two graphs.

9.3.2A There are no examples with fewer than five vertices. The two graphs shown below have five vertices, of which two have degree 1 and three have degree 2, but they are (clearly) not isomorphic.

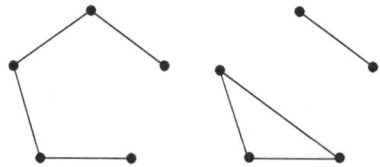

9.3.3A   i. 2,2,2,3,3,4 ⇒ 2,1,1,2,2 ~ 1,1,2,2,2 ⇒ 1,1,1,1 ⇒ 1,1,0 ~ 0,1,1 ⇒ 0,0. So 2,2,2,3,3,4 is the degree sequence of a graph. One such graph is shown below, on the left.

   ii. 1,1,3,3,5,5 ⇒ 0,0,2,2,4. Clearly 0,0,2,2,4 is not the degree sequence of a graph, and hence 1,1,3,3,5,5 is also not the degree sequence of a graph.

   iii. 3,3,3,3,3,3,3,3 ⇒ 3,3,3,3,2,2,2 ~ 2,2,2,3,3,3,3 ⇒ 2,2,2,2,2,2 ⇒ 2,2,2,1,1 ~ 1,1,2,2,2 ⇒ 1,1,1,1 ⇒ 1,1,0 ~ 0,1,1 ⇒ 0,0, and hence 3,3,3,3,3,3,3,3 is the degree sequence of a graph. One such graph is shown below, on the right.

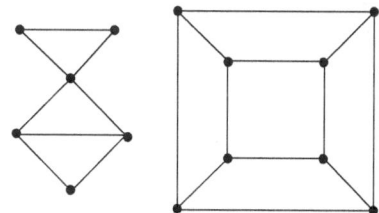

9.3.4A In a simple graph with n vertices and $\frac{1}{2}n(n-1)$ edges, each pair of vertices must be joined by an edge. So any bijection between the vertices of two such graphs is an isomorphism, and hence the graphs are isomorphic.

9.3.5A   i. The sequence 1, 2, 3, 4, 5 is not a multigraph degree sequence as 1+2+3+4+5 is odd, which, is not possible for the sum of the degrees of a graph. (Why? see the handshaking lemma in section 9.4.)

   ii. The sequence 1, 2, 2, 3, 5, 5 is a multigraph degree sequence, as shown by the graph below. (You can check that it is not a simple graph degree sequence.)

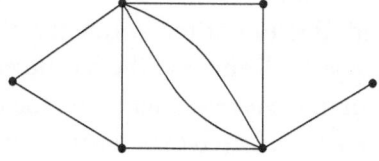

iii. The sequence 2, 2, 2, 8 is not a multigraph degree sequence, as the vertex of degree 8 needs to have eight edges joining it to three vertices each of degree 2, which is not possible.

9.4.1A We prove, using mathematical induction, that for every positive integer n, a connected graph with n vertices and no closed paths has $n - 1$ edges. Clearly this holds when $n = 1$. Now suppose the result holds for $n = k$ and that G is a connected graph with $k + 1$ vertices that has no closed paths. It follows from Theorem 9.5 that G has a vertex, say a, which has degree 1. The graph, say G', that results from G by deleting the vertex a and the one edge adjacent to it is still connected and has k vertices. Hence, by the induction hypothesis, it has $k - 1$ edges. Hence the graph G has k edges. So the result holds also for $v = k + 1$. Therefore, by mathematical induction, the result is true for all positive integers n.

9.4.2A Suppose first that the removal of the edge $\{u, v\}$ results in a graph that is still connected. It follows that in G there is a path from u to v that does not use the edge $\{u, v\}$. Let this path be $u = v_1 \to v_2 \to \ldots \to v_{k-1} \to v_k = v$. Then $u \to v_2 \to \ldots \to v_{k-1} \to v \to u$ is a closed path in G in which the edge $\{u, v\}$ occurs. Conversely, suppose that G is connected and that there is a closed path in G, say $u \to v_2 \to \ldots \to v_{k-1} \to v \to u$, which includes the edge $\{u, v\}$. Let G' be the graph that results from G by deleting the edge $\{u, v\}$. Because G is connected, each pair of vertices a, b of G is connected by a path in G. If there is such a path that does not include the edge $\{u, v\}$, then there is a path in G' that connects a and b. This must be the case. For suppose $a \to u_2 \to \ldots \to u \to v \to \ldots \to u_{l-1} \to b$ is a path connecting a and b which includes the edge $\{u,v\}$. Then $a \to u_2 \to \ldots \to u \to v_2 \to \ldots \to v_{k-1} \to v \to \ldots \to u_{l-1} \to b$ is a walk in G' from a to b. It follows that there is a path in G' from a to b. Hence G' is connected.

9.4.3A It can be seen, for example, that $a \to b \to c \to h \to j \to e \to a$ is a closed path in the Petersen graph of length 6.

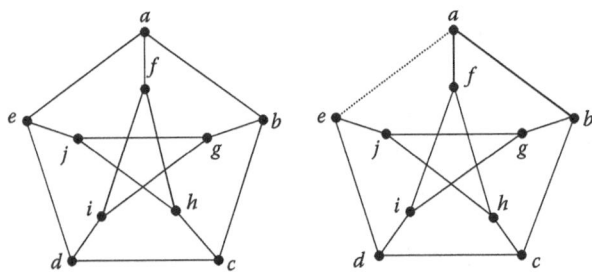

A closed path of length 7 must include at least two vertices from each of the two sets of vertices $\{a,b,c,d,e\}$ and $\{f,g,h,i,j\}$, and hence must include at least two of the edges leading from the first set to the second. Because of the symmetry of the graph, we can assume that such a path includes the edge $\{a,f\}$, and that the second edge adjacent to the vertex a that it includes is $\{a,b\}$, as shown in the above diagram on the right. If the closed path also includes the edge $\{b,g\}$, then there would need to be a path of length 4 from g to f. It is easy to check that no such path of length 4 exists. Similarly, we can rule out the use of the edges {c,h}, {d,i}, and {e,j}. Therefore there is no closed path of length 7.

9.4.4A Let $G = (V,E)$ be a simple graph with n vertices that is not connected. Select a vertex v_0 of G. Let V_1 be the set of vertices consisting of v_0 together with all the vertices connected to v_0 by an edge, and let $V_2 = V \backslash V_1$. Let $\#(V_1) = r$. Then $\#(V_2) = n - r$. Since G is not connected, $1 \leq r < n$. There are no edges joining a vertex in V_1 to a vertex in V_2. So the maximum number edges G can have is the number of edges joining each pair of vertices in V_1 plus the number of edges joining each pair of vertices in V_2. So G has at most e_r edges, where $e_r = \frac{1}{2}r(r-1) + \frac{1}{2}(n-r)(n-r-1)$ edges. Now $e_r = \frac{1}{4}n^2 - \frac{1}{2}n + (r - \frac{1}{2}n)^2$. It follows that for $1 \leq r < n$, e_r achieves a maximum value when $|r - \frac{1}{2}n|$ is as large as possible, that is, for $r = 1$ and $r = n-1$. It follows that $e_r \leq \frac{1}{2}(n-1)(n-2)$. Therefore, if G has $\frac{1}{2}(n-1)(n-2)+1$ edges, then G must be connected.

9.4.5A a. The following algorithm applied to a graph $G = (V,E)$ determines whether the graph is connected.

> **ALGORITHM FOR DECIDING WHETHER A GRAPH $G = (V,E)$ IS CONNECTED**
>
> 1. Choose a vertex $v \in V$ and put $i = 0$ and $V_0 = \{v\}$
> 2. Let V_{i+1} be the set of those vertices in $V \backslash \bigcup_{s=0}^{i} V_s$ that are joined to a vertex in V_i.
> 3. If $V_{i+1} = \emptyset$, stop. Otherwise replace i by $i + 1$, and return to step 1.
>
> If, when the algorithm terminates, $i = k$ and $V = \bigcup_{s=0}^{k} V_s$, the graph is connected; otherwise, the graph is not connected.

It is easy to see that when the algorithm terminates, $\bigcup_{s=0}^{k} V_s$ is the set of vertices connected to v by a path, and hence the graph is connected if and only if $\bigcup_{s=0}^{k} V_s$ is the set of all the vertices of G.

b. We apply the algorithm of (a) to each graph in turn.

i.
i	V_i
0	$\{a\}$
1	$\{d,e,f\}$
2	$\emptyset$

$V_1 \cup V_2 \cup V_3 \neq V$, and so the graph is not connected.

ii.
i	V_i
0	$\{a\}$
1	$\{b,c,d,e,f\}$
2	$\{g,h\}$
3	$\emptyset$

$V_0 \cup V_1 \cup V_2 = V_3 = V$, and so the graph is connected.

iii.

i	V_i
0	$\{a\}$
1	$\{e,f\}$
2	$\{c,d\}$
3	$\{b,h\}$
4	$\{g,j\}$
5	$\varnothing$

$\bigcup_{i=0}^{5} V_i = V$, and so the graph is connected.

9.5.1A In Figure 9.23, graph (i), we have $v = 8$, $e = 15$, and $f = 9$. So $v-e+f = 8-15+9 = 2$; in graph (ii) we have $v = 7$, $e = 9$, and $f = 4$. So $v-e+f = 7-9+4 = 2$.

9.5.2A i. Suppose that G is drawn in the plane with f faces. Since each closed path of G contains at least four edges, each face is bounded by at least four edges and, as each edge is a boundary of at most two faces, hence $4f \leq 2e$ and so $2f \leq e$. By Euler's formula $f = e-v+2$, and hence $2(e-v+2) \leq e$; from it follows that $e \leq 2v-4$.

ii. The graph $K_{3,3}$ is bipartite, and hence each closed path must contain at least four edges. However for this graph $v = 6$ and $e = 9$, and so $e > 2v-4$. So, by (a), $K_{3,3}$ is not planar.

iii. By the symmetry of the graph, it does not matter which edge is deleted. On the left below is $K_{3,3}$ with the edge $\{b,e\}$ deleted. In the middle this graph has been drawn to show that it is planar, and on the right it is redrawn with all the edges being straight line segments.

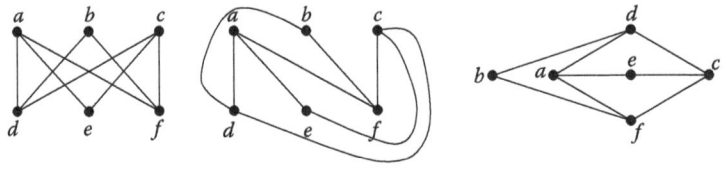

9.5.3A Consider the subgraph of the Petersen graph obtained by deleting the edges $\{c,d\}$ and $\{g,j\}$, as shown below, on the left. If we now drop the vertices c, d, g, and j of degree 2, we obtain the graph on the right. It can be seen that this is isomorphic to $K_{3,3}$.

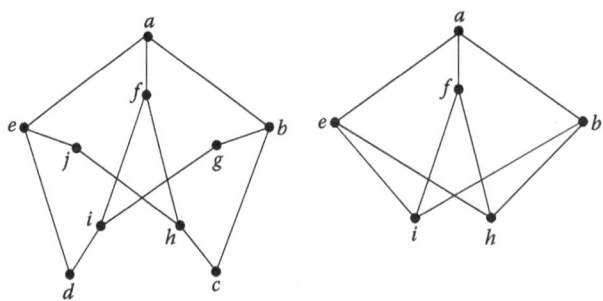

9.5.4A Suppose first that G is a connected bipartite graph whose vertex set can be partitioned into disjoint sets V_1 and V_2 in such a way that every edge joins a vertex in V_1 to a vertex in V_2. Then in any path the vertices must come alternately from V_1 and V_2, and so any closed path must have even length. Conversely, suppose that G is a connected graph in which each closed path has even length. Let v_0 be one vertex of G. We let V_1 be the set of vertices, v, such that there is a path from v_0 to v of odd length, and let V_2 be the set of vertices, v, such that there is a path from v_0 to v of even length. Since G is connected, for each vertex v of G there is a path from v_0 to v. Thus every vertex is in $V_1 \cup V_2$. (We can regard the vertex v_0 by itself as forming a path of length 0 from v_0 to v_0, so that $v_0 \in V_2$.) We show that no vertex is in both V_1 and V_2. Let $v_0 \to v_1 \to \ldots \to v_k = v$ and $v_0 \to u_1 \to \ldots \to u_l = v$ be two paths from v_0 to v of lengths k and l, respectively. These paths may have some edges in common, and, if they diverge, they must come back together again. Suppose that for some r, s we have $v_r = u_s$, the paths diverge at this vertex, and rejoin at the vertex $v_{r+t} = u_{s+u}$, for some t, u. We can picture this as follows:

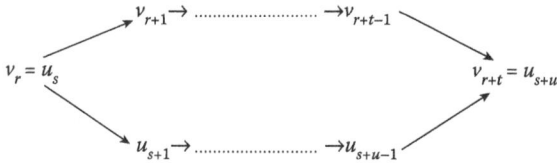

We thus have a closed path

$$v_r = u_s \to u_{s+1} \to \ldots \to u_{s+u-1} \to u_{s+u} = v_{r+t} \to v_{r+t-1} \to \ldots \to v_{r+1} \to v_r = u_s$$

in G of length $t+u$. Therefore $t+u$ is even, and hence $t \equiv u \pmod{2}$. Since this holds whenever the paths $v_0 \to v_1 \to \ldots \to v_k = v$ and $v_0 \to u_1 \to \ldots \to u_l = v$ diverge and then rejoin, it follows that $k \equiv l \pmod{2}$, so no vertex can be in both V_1 and V_2. Now suppose uv is an edge of G. Either there is a path from v_0 to u that does not include v, or a path from v_0 to u that goes through v, in which case there is a path from v_0 to v that does not include u. Suppose $v_0 \to \ldots \to u$ is a path of length k that does not include v. Then $v_0 \to \ldots \to u \to v$ is a path of length $k+1$. Now since one of $k, k+1$ must be odd and the other even, the edge uv joins a vertex in V_1 to a vertex in V_2. The same conclusion follows if there is a path from v_0 to v that does not go through u. We deduce that G is a bipartite graph.

9.5.5A This graph is planar. It can be redrawn as shown below with the edges meeting only at vertices.

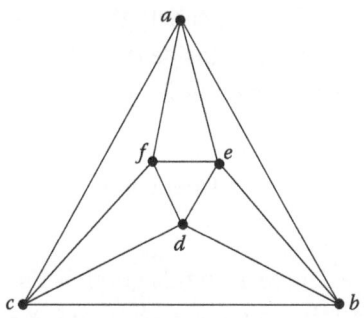

9.5.6A

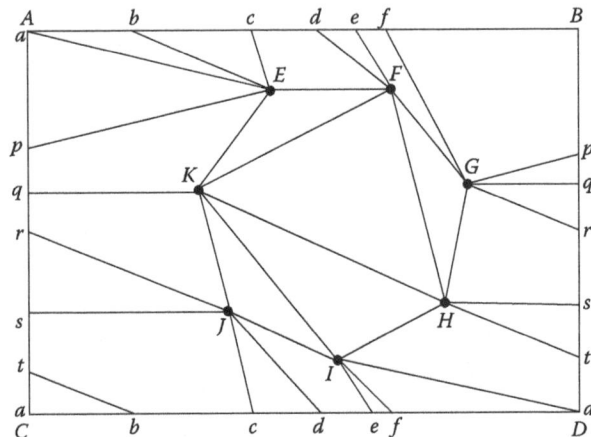

This becomes a torus if the edge *AB* is glued to *CD*, and then *AC* to *BD*. In this way the points on *AB* labeled *a*, *b*, *c*, *d*, *e*, and *f* are identified with the points with the same label on *CD*, and the points on *AC* with labels *p*, *q*, *r*, *s*, and *t* are identified with the points with the same label on *BD*. Then the seven vertices *E*, *F*, *G*, *H*, *I*, *J*, and *K* are all joined to each other by edges represented on the torus by lines that meet only at these vertices.

9.5.7A In this case we still have six points on the circumference of the circle each with degree 7, but instead of 15 points each of degree 4 inside the circle, we now have 12 of degree 4 and 1 of degree 6. So the sum of the degrees is 6×7+12×4+1×6 = 96. So, by the handshaking lemma there are 48 edges. Also, there are now 6+12+1 = 19 vertices. Hence, using the formula $r = e-v+1$, where r is the number of regions inside the circle, as in Problem 9.10, we have that $r = 48-19+1 = 30$. So in this case there are 30 regions.

9.5.8A We prove the result by mathematical induction. We can let P_1, P_2, P_3 be any three distinct points on the sphere. So the result is true for $n \leq 3$. Now suppose the result is true for $n = s$ and the sequence of points $P_1, P_2,\ldots, P_s$ satisfies the stated conditions. The planes through three of these points make at most $C(s,3)$ different angles with the *xy*-plane. Hence there is a plane, say Π, distinct from all these planes and that meets the sphere. Let P_{s+1} be any point where this plane meets the sphere. Then the points $P_1, P_2,\ldots, P_s, P_{s+1}$ also satisfy the condition. So the result holds also for $n = s+1$. Therefore, by mathematical induction, it holds for all *n*. If the points $P_1, P_2,\ldots, P_n$ satisfy this condition, the straight line segments joining pairs of these points do not meet inside the sphere. So the complete graph, K_n, and hence every subgraph of it, may be drawn in R^3 with the edges represented by straight line segments.

9.6.1A i. This graph has four vertices of degree 3, and so is neither Eulerian nor semi-Eulerian.

ii. All the vertices have degree 4, so this graph is Eulerian. One example of a closed trail that includes all the edges is $a \to b \to c \to d \to e \to a \to c \to e \to b \to d \to a$.

iii. This graph has two vertices of degree 3 (*c* and *f*) with all the other vertices having degree 4. So it is semi-Eulerian. One example of a trail that includes all the edges is $c \to d \to e \to f \to a \to b \to c \to e \to a \to d \to b \to f$.

9.6.2A Let *G* be a connected graph with just two vertices, say *a* and *b*, whose degrees are odd numbers. Let *G'* be the graph obtained from *G* by adding an edge joining *a* and *b*. This increases the degrees of *a* and *b* by 1 so every vertex of *G'* has even degree. Therefore, by Theorem 9.9(a), *G'* is Eulerian and so has a closed trail that includes every edge of *G'*. If we delete the added edge joining *a* and *b* from this closed trail, we obtain a trail that includes every edge of *G*. Hence *G* is semi-Eulerian.

9.6.3A The graph has four vertices whose degrees are odd. We need to increase each of their degrees by 1 to make the graph Eulerian. So two more bridges are needed. These could correspond to the edges *ab* and *cd*, or *ac* and *bd*, or *ad* and *bc*.

9.7.1A i. In the diagram below we have labeled the vertices so we can refer to them. This graph is Hamiltonian. For example, $a \to f \to e \to d \to c \to b \to h \to i \to j \to k \to l \to m \to g \to a$ is a Hamiltonian path in the graph (there are lots of others).

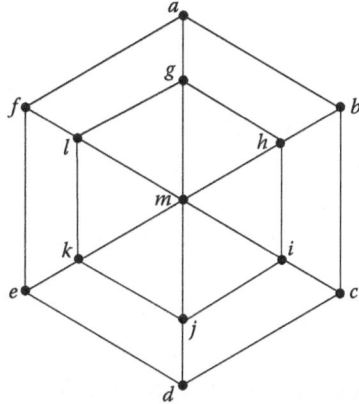

ii. This graph is not Hamiltonian. An argument to show this is given below the diagram where we have labeled the vertices so that we can refer to them.

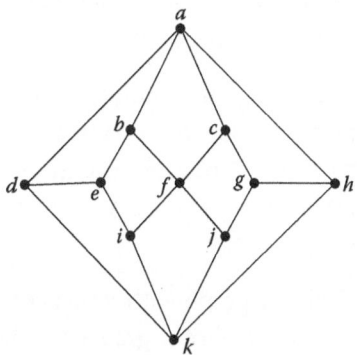

The graph is bipartite with each edge joining a vertex in the set $V_1 = \{a,e,f,g,k\}$ to a vertex in the set $V_2 = \{b,c,d,h,i,j\}$. A Hamiltonian path would have to alternate between a vertex in V_1 and a vertex in V_2. This is impossible as there are more vertices in V_2 than in V_1. Hence there is no Hamiltonian path in this graph.

9.7.2A The Petersen graph is not Hamiltonian. In the following argument to prove this we use the labeling of the vertices given in Figure 9.14. Please draw your own diagram.

We note first that since a Hamiltonian path must visit each vertex, and each vertex of the Petersen graph has degree three, a Hamiltonian path must use exactly two of the edges adjacent to each vertex. We note next that a Hamiltonian path must include at least one edge between a vertex in the set $\{a,b,c,d,e\}$ and a vertex in the set $\{f,g,h,i,j\}$. So, because of the symmetry of the graph, we may assume that any Hamiltonian path includes the edges *af* and *ae*. It follows that the edge *ab* is not included, and hence the edges *bc* and *bg* are used. We now consider which of the edges adjacent to *d* are used.

If *de* is used, then *ej* cannot be used, and hence the edge *gj* must be used. Since *bg* and *gj* are used, *gi* is not used, and so *di* and *fi* are used. This means that we already have a closed path $a \to f \to i \to d \to e \to a$ that does not include all the vertices. Thus a Hamiltonian path cannot include the edge *de*. Similarly, the edge *cd* cannot be used. But this is impossible as two edges adjacent to *d* must be used. We deduce that there is no Hamiltonian path.

9.7.3A Let G be a graph with n vertices and $\frac{1}{2}(n-1)(n-2)+2$ edges, where $n \geq 3$. Suppose u and v are two vertices of G not joined by an edge, and that $\delta(u) + \delta(v) = d$. Then there are at most $C(n-2,2) = \frac{1}{2}(n-2)(n-3)$ edges joining the $n-2$ vertices of G other than u and v, and there are d edges joining these $n-2$ vertices to u or v or both. So the total number of edges is at most $\frac{1}{2}(n-2)(n-3)+d$. Therefore $\frac{1}{2}(n-1)(n-2)+2 \leq \frac{1}{2}(n-2)(n-3)+d$ and hence $n \leq d$. It therefore follows from Theorem 9.10 that G is Hamiltonian.

For each n let G_n be the graph obtained from the complete graph K_{n-1} with $n-1$ vertices by adding one additional vertex v that is joined to just one vertex of K_{n-1}. Then G_n has $\frac{1}{2}(n-1)(n-2)+1$ edges but is not Hamiltonian as v has degree 1.

9.7.4A The graph is connected, and each vertex has an even degree. So, by Theorem 9.9, the graph is Eulerian. Any Hamiltonian path would have to include, for each vertex of degree 2, both edges adjacent to it. However, the six edges that are adjacent to the vertices of degree 2 form a closed path that does not go through all the other vertices. So there is not a Hamiltonian path. Hence the graph is not Hamiltonian.

9.8.1A The graph can be colored using three colors. The vertices at the corners of the square can be colored using the first color, the vertex at the center with a second color, and the other four vertices with a third color. As the graph includes three vertices each joined to the other two, it cannot be colored with fewer than three colors. So its chromatic number is 3.

9.8.2A i. Let G be a connected planar graph with v vertices and no vertices of degree less than 5. Suppose G has k vertices of degree 5, and hence $v-k$ vertices of degree at least 6, and that G has e edges. By Theorem 9.7, $e \leq 3v-6$, and by the handshaking lemma, $2e \geq 5k + 6(v-k) = 6v-k$. Therefore $6v-k \leq 2e \leq 6v-12$ and hence $12 \leq k$.

ii. A regular icosahedron has 12 vertices each of degree 5, and hence when projected onto the plane provides a graph with the required properties. It can be drawn as shown in the diagram below.

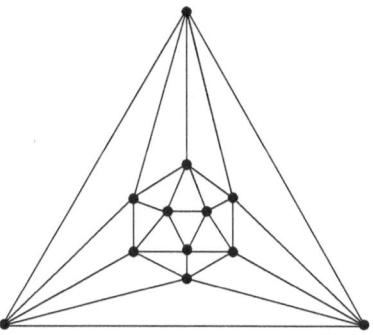

9.8.3A In the graph shown below the vertices represent the subjects. They are joined by an edge if at least one student takes both subjects.

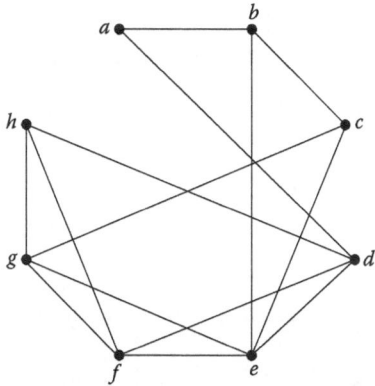

Because the graph includes a triangle, at least three time slots are needed. If we assign a, c, and f to one hour; b, d, and g to a second hour; and e and h to a third hour, then vertices joined by an edge are assigned different hours. So three different hours are needed.

9.8.4A

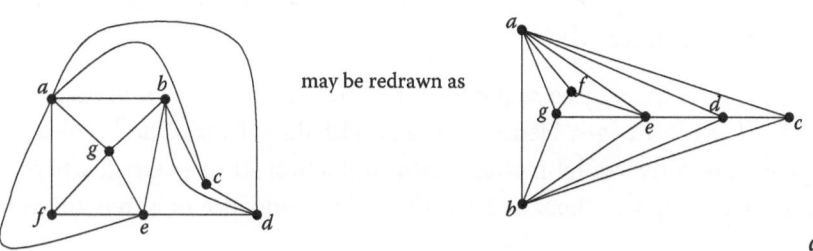

may be redrawn as

CHAPTER 10

10.1.1A There is just one tree with two vertices, one with three vertices, two with four vertices, and three with five vertices, as shown below.

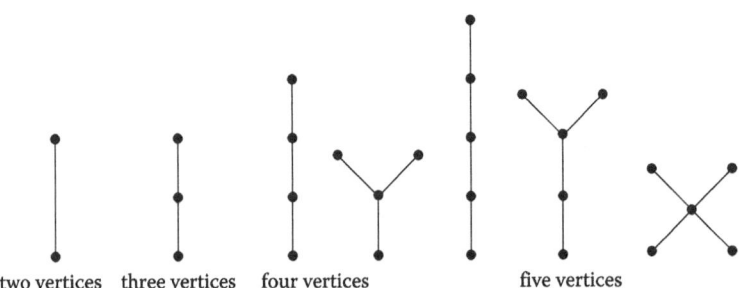

two vertices   three vertices   four vertices   five vertices

10.1.2A We prove this result by mathematical induction. It is clearly true for a tree with two vertices. Suppose it is true for trees with k vertices where $k \geq 2$. Now let T be a tree with $k + 1$ vertices. T is therefore a connected graph. A connected graph in which each vertex has degree at least 2 must have a closed path, by Theorem 9.5. So T must contain at least one vertex, say v_0, of degree 1. Let T' be the graph obtained from T by deleting v_0 and the edge adjacent to it. T' will still be connected. Hence it is a tree with k vertices, so by the induction hypothesis, it has at least two vertices of degree 1. At least one of these must also have degree 1 in the graph T. So T has at least two vertices with degree 2. So the result holds also for trees with $k + 1$ vertices. So, by mathematical induction the result is true for all trees.

10.1.3A The two trees below each have degree sequences 1, 1, 1, 2, 2, 3 but are clearly not isomorphic. (From the solution to Exercise 10.1.1A, we see that there are no examples with fewer than six vertices.)

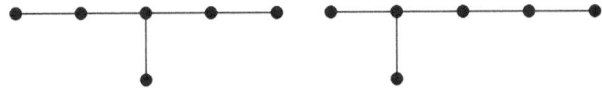

10.1.4A Let G be a graph with n vertices, $n - 1$ edges, and no closed paths, where $n \geq 3$. Suppose that G has k components with $n_1, n_2, \ldots, n_k$ vertices, respectively, with $k \geq 1$. So $n_1 + n_2 + \ldots + n_k = n$. The connected components have no closed paths, and hence are themselves trees. So by Theorem 10.1(b), the components have $n_1 - 1, n_2 - 1, \ldots, n_k - 1$ edges, respectively. So G has $(n_1 - 1) + (n_2 - 1) + \ldots + (n_k - 1) = n - k$ edges. Since G has $n - 1$ edges, $k = 1$, and hence G is connected.

10.1.5A If $a_1, \ldots, a_n$ is the degree sequence for a tree, this tree has n vertices, and so, by Theorem 10.1, it has $n - 1$ edges. Hence, by the handshaking lemma, $a_1 + \ldots + a_n = 2(n-1) = 2n-2$. We prove the converse by mathematical induction. If $a_1 + a_2 = 2$, where a_1, a_2 are positive integers, $a_1 = a_2 = 1$. Then, as 1,1 is the degree sequence of a tree, the result holds for $n = 2$. Suppose the result holds for all $n \leq k$, and that $a_1,\ldots, a_k, a_{k+1}$ is a sequence of positive integers where $a_1 \leq \ldots \leq a_k \leq a_{k+1}$ and $a_1 + \ldots + a_k + a_{k+1} = 2k$. It follows that $a_1 = 1$, and for some r, with $2 \leq r \leq k + 1$, $a_r \geq 2$. Let l be the least positive integer such that $a_l \geq 2$. Now consider the sequence of k positive integers

$$a_2,\ldots,a_{l-1},a_l-1,a_{l+1},\ldots,a_k,a_{k+1}, \text{ (that is, } 1,\ldots,1,a_l-1,a_{l+1},\ldots,a_k,a_{k+1}).$$

Then $a_2 \leq \ldots \leq a_{l-1} \leq a_l-1 \leq a_{l+1} \leq \ldots \leq a_{k+1}$ and $a_2 + \ldots + a_{l-1} + (a_l-1) + a_{l+1} + \ldots + a_{k+1} = 2(k-1)$. Therefore, by the induction hypothesis, $a_2,\ldots, a_{l-1}, a_l-1, a_{l+1},\ldots, a_{k+1}$ is a degree sequence of a tree. Thus there is a tree, say T, with k vertices, $v_2,\ldots, v_l,\ldots, v_{k+1}$ with degrees $a_2,\ldots, a_{l-1}, a_l-1, a_{l+1},\ldots, a_{k+1}$, respectively. Let T' be the tree obtained from T by adding one vertex, v_1, which is joined by an edge just to the vertex v_l. Then T' is a tree with $k + 1$ vertices with degrees $a_1, a_2,\ldots, a_{l-1}, a_l, a_{l+1},\ldots,a_{k+1}$. So this is the degree sequence of a tree. Hence the result is also true for $n = k + 1$. This completes the proof by mathematical induction.

Let G be a tree with a vertex of degree k, and n vertices altogether. Suppose G has m vertices of degree 1. Then the other $n - (m + 1)$ vertices have degrees ≥ 2. G has $n - 1$ edges, so by the handshaking lemma $2(n - 1) \geq m + k + 2(n - (m + 1))$. Hence $m \geq k$, as requried.

10.1.7A i. This graph has six vertices and five edges and so could be a tree. Applying the algorithm for connectedness, we have:

i	V_i
0	$\{a\}$
1	$\{f\}$
2	$\{b,d,e\}$
3	$\{f\}$
4	$\emptyset$

$\bigcup_{s=0}^{4} V_s = \{a,b,c,d,e,f\}$, so the graph is connected and hence it is a tree.

ii. This graph has eight vertices and six edges and so cannot be a tree.

10.2.1A i. (2,3)   ii. (6,4,6,4,3,6,6)

10.2.2A We use the method described in the proof of Theorem 10.2.

i. Here $n - 2 = 6$. So $n = 8$, and we begin with eight isolated vertices, say $v_1,\ldots, v_8$ with labels 1, 2, ..., 8, respectively. By "number" we will mean one of the numbers 1, 2, ..., 8. The least number not in the sequence (1,2,1,2,1,2) is 3. So we begin by joining the vertex v_3 to the vertex v_1, and we delete 1 from the beginning of the sequence. The least number, other than 3, not in the remaining sequence (2,1,2,1,2) is 4, and so next we join v_4 to v_2 and delete 2 from the sequence. The least number, other than 3 and 4, not in the sequence (1,2,1,2) is 5, so we join

v_5 to v_1 and delete 1 from the sequence. The remaining sequence is now (2,1,2). Proceeding in this way we join v_6 to v_2 and v_7 to v_1. This leaves the sequence (2). The least number, other than 3,4,5,6,7, not in this sequence is 1, so now we join v_1 to v_2. The two numbers not in the list 3,4,5,6,7,1 are 2 and 8, so we end by joining v_2 to v_8. The resulting labeled tree is shown below.

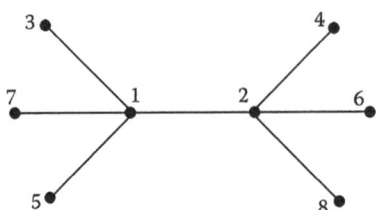

ii. We set out the solution in the form of a table. At each stage the vertex labeled a_k is joined to the vertex labeled b_k.

k	Sequence	a_k	b_k
1	(1,2,3,1,2,3)	1	4
2	(2,3,1,2,3)	2	5
3	(3,1,2,3)	3	6
4	(1,2,3)	1	7
5	(2,3)	2	1
6	(3)	3	2

The two numbers not in the sequence $b_1, b_2, \ldots, b_6$ are 3 and 8, so we end by joining the vertices with these labels. This results in the following tree.

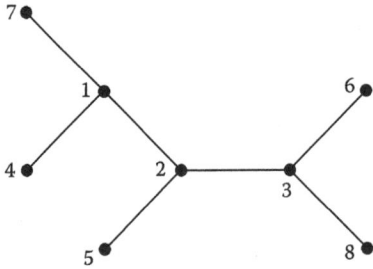

10.2.3A We prove, using mathematical induction, that for all $n \geq 3$, the result holds for all trees with n vertices. It is straightforward to check that the result holds for $n = 3$. Now suppose that the result holds for $n = k$, and let T be a labeled tree with $k + 1$ vertices. Let v_0 be the vertex of degree 1 in T with the lowest number, say r, as its label, and let v_0 be joined to the vertex with label s. Let T' be the tree that results from T by deleting v_0 and the one edge adjacent to it. T' does not quite meet our requirement for a labeled tree, because the numbers assigned to its vertices are $1, 2, \ldots, r-1, r+1, \ldots, k, k+1$, with r omitted. We can rectify this situation by replacing the labels $r + 1, \ldots, k + 1$ by the labels $r, \ldots, k$, respectively. That is, we let

$$t' = \begin{cases} t, & \text{if } 1 \leq t < r, \\ t-1, & \text{if } r < t \leq k+1. \end{cases}$$

and we replace each label t by t'. Then T' is a labeled tree with k vertices. The Prüfer code for T is thus $P = (s, t_1, t_2, \ldots, t_{k-2})$, where T' has Prüfer code $(t_1', \ldots, t_{k-2}')$.

Now consider the tree that we construct from the Prüfer code P. This consists of the vertex labeled s joined to the vertex whose label is the least number not included in the sequence P together with the tree constructed from the sequence $(t_1, \ldots, t_{k-2})$. By the induction hypothesis, this is the tree T' together with the edge joining v_0 to the vertex with label s, that is, the tree T.

10.2.4A Consider a tree, T, with m vertices. To each labeling of the vertices of T with the numbers $0, 1, \ldots, m - 1$, there corresponds a labeling of the edges of T with the numbers $1, 2, \ldots, m - 1$ as follows. For each vertex of T with label $r \neq 0$, by Theorem 10.1d, there is a unique path from the vertex labeled 0 to the vertex labeled r. We assign the label r to the final edge in this path. This process is illustrated below with a tree with 11 vertices labeled $0, 1, \ldots, 10$ on the left, with the corresponding edge labeling on the right.

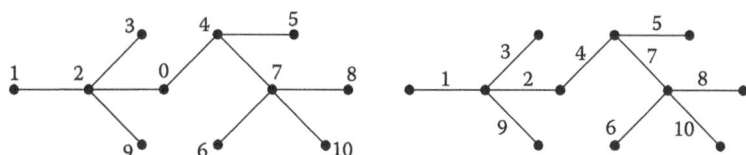

It is easy to check that if the label 0 is switched with a label, say r, associated with a vertex joined by an edge to the vertex labeled 0, the corresponding labeling of the edges is unchanged, but otherwise different vertex labelings corresponding to different edge labelings. By switching the labels 0 and r in the way indicated, the label 0 can be associated with any of the m vertices without changing the edge labeling. It follows that the number of edge labelings is $1/m \times$ the number of vertex labelings. Hence, by Cayley's theorem (Theorem 10.2), there are $(1/m)(m^{m-2}) = m^{m-3}$ different edge-labeled trees with m vertices.

10.3.1A

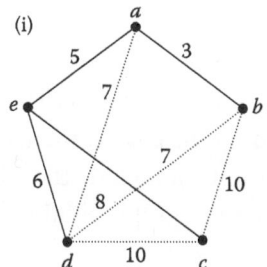

(i)

Edge	Weight	
ab	3	✓
ae	5	✓
de	6	✓
ad	7	✗
bd	7	✗
ce	8	✓
bc	10	
cd	10	

(ii)

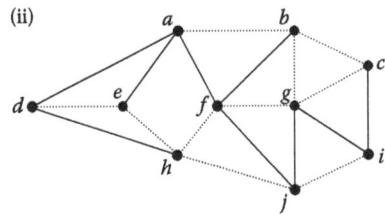

Edge	Weight	
ad	2	✓
gj	3	✓
ae	4	✓
af	4	✓
ci	4	✓
de	4	✗
fj	4	✓
gi	4	✓
bf	5	✓
cg	5	✗
ab	6	✗
dh	6	✓
eh	6	
fh	7	
ij	7	
fg	8	
bg	9	
hj	11	
bc	12	

10.3.2A Let (G, μ) be a weighted connected graph with an edge e with lower weight than any other edge. Suppose T is a spanning tree for G that does not include the edge e. If T' is the graph obtained from T by adding the edge e, then, by Theorem 10.1(e), T' has a closed path. As T is a tree, this closed path must include the edge e. We let e' be some other edge in this closed path, and let T'' be the graph obtained from T' by deleting the edge e'.

T'' will also be a spanning tree for G. Since $\mu(e) < \mu(e')$, and the edges of T'' are the same as those of T except that e' has been replaced by e, $\mu(T'') < \mu(T)$. Hence T is not a minimal connector. So every minimal connector must include the edge e.

10.3.3A Let (G, μ) be a weighted connected graph, with n vertices, in which each edge has a different weight. Let $e_1, e_2, \ldots, e_{n-1}$ be the edges picked out by Kruskal's algorithm in this order. So e_1 is the edge of least weight and, for $2 \leq k \leq n-1$, e_k is the edge of least weight such that there is no closed path made up of edges from the set $\{e_1, e_2, \ldots, e_k\}$. We show that every minimal connector must include all the edges $e_1, \ldots, e_{n-1}$. By the result of Exercise 10.3.2A, the edge e_1 is included in every minimal connector. Suppose now that $k < n-1$ and that all the edges $e_1, \ldots, e_k$ must be included in every minimal connector. We show that the same is also true for the edge e_{k+1}. If not, there is a minimal connector T that does not include the edge e_{k+1}. If we add this edge to T to form the graph T', T' will have a closed path. By the choice of e_{k+1} this closed path must include an edge e not in the set $\{e_1, \ldots, e_k, e_{k+1}\}$. By deleting the edge, say e, in this closed path we obtain a spanning tree, say T''. So T'' is obtained from T by removing the edge e and replacing it with e_{k+1}. Also, by the choice of e_{k+1}, $\mu(e_{k+1}) < \mu(e)$. Thus $\mu(T'') < \mu(T)$, which contradicts the fact that T is a minimal connector. So every minimal connector also includes the edge e_{k+1}. We can therefore deduce that there is just one minimal connector, namely, the one made up of the edges $e_1, \ldots, e_{n-1}$.

10.4.1A The stages of Dijkstra's algorithm are set out in the table below. To save space the parentheses have been omitted from the labels. Instead, we have used bold type for permanent labels.

	i	ii	iii	iv	v	vi	vii	viii	ix	x	xi	xii	xiii	xiv	xv
x		u	a	d	g	j	b	e	h	c	i	k	f	l	v
u	*,0	*,0													
a	*,∞	u,3	**u,3**												
b	*,∞		a,10				**a,10**								
c	*,∞						b,13			**b,13**					
d	*,∞	u,5		**u,5**											
e	*,∞		a,10					**a,10**							
f	*,∞							e,17		c,15			**c,15**		
g	*,∞	u,6			**u,6**										
h	*,∞				g,13			e,11	**e,11**						
i	*,∞								h,13		**h,13**				
j	*,∞	u,8			u,8	**u,8**									
k	*,∞					j,14			h,13			**h,13**			
l	*,∞										i,16	k,15		**k,15**	
v	*,∞									c,25				l,16	**l,16**

From this table we see that the shortest path from u to v has length 16. Tracing the path back from v we see that $u \to a \to e \to h \to k \to l \to v$ is a path of length 16 from u to v.

10.4.2A As we see from the table below Dijkstra's algorithm terminates after eight stages with $x = a$ and $l(a) = \infty$. It follows that there is no path from u to v.

	(i)	(ii)	(iii)	(iv)	(v)	(vi)	(vii)	(viii)
x		u	b	c	d	f	g	a
u	(*,0)	[*,0]						
a	(*,∞)							[*,∞]
b	(*,∞)	(u,3)	[u,3]					
c	(*,∞)	(u,4)		[u,4]				
d	(*,∞)		(b,10)	(c,6)	[c,6]			
e	(*,∞)							
f	(*,∞)		(b,7)			[b,7]		
g	(*,∞)			(c,7)			[c,7]	
v	(*,∞)							

CHAPTER 11

11.1.1A i. (1 4 7 6 3 2)(5) or, simply, (1 4 7 6 3 2). [Of course, this permutation can also be written as (4 7 6 3 2 1), or (7 6 3 2 1 4), etc. The permutations in subsequent solutions can also be written in different forms, so your solution may look different from those given here, yet still be correct.]

ii. (1 2 7 5)(3)(4 9 8 6) or (1 2 7 5)(4 9 8 6).

11.1.2A i. a. Suppose that

$$x \bullet y = x \bullet z. \tag{1}$$

Since G is a group, x has an inverse x^{-1} and it follows from Equation 1 that

$$x^{-1} \bullet (x \bullet y) = x^{-1} \bullet (x \bullet z). \tag{2}$$

Using the associativity and identity properties, $x^{-1} \bullet (x \bullet y) = (x^{-1} \bullet x) \bullet y = e \bullet y = y$. Similarly, $x^{-1} \bullet (x \bullet z) = z$. Therefore from Equation 2, $y = z$.

b. Suppose that $y \bullet x = z \bullet x$. Then $y = y \bullet e = y \bullet (x \bullet x^{-1}) = (y \bullet x) \bullet x^{-1} = (z \bullet x) \bullet x^{-1} = z \bullet (x \bullet x^{-1}) = z \bullet e = z$.

ii. Since G is a group, x has an inverse x^{-1} in G. Since y is also in G, it follows from the closure property that the element $w = x^{-1} \bullet y$ is in G. Then, $x \bullet w = x \bullet (x^{-1} \bullet y) = (x \bullet x^{-1}) \bullet y = e \bullet y = y$. Similarly, $z = y \bullet x^{-1}$ is in G and $z \bullet x = y$.

11.1.3A We see from the table below that $\sigma = (1)(2\ 4)(3\ 5)$ or $(2\ 4)(3\ 5)$.

	(4	3	5	2	1)			σ	
4		←				1		←	1
3		←				4		←	2
2		←				5		←	3
1		←				2		←	4
5		←				3		←	5
		←				(1	4)(2 3)	←	

11.1.4A In the table on the left, x cannot be a or b, and so the only choice is that x is e. It follows that w has to be b, and then y must be e and z has to be a. So the table can be completed to form a Latin square in just one way, as shown on the right. The table below is the Cayley table of the group of permutations $\{e, (1\ 2\ 3), (1\ 3\ 2)\}$. It has the same pattern as that of the second table, which we therefore deduce is the Cayley table of a group.

	e	a	b
e	e	a	b
a	a	w	x
b	b	y	z

	e	a	b
e	e	a	b
a	a	b	e
b	b	e	a

∘	e	$(1\ 2\ 3)$	$(1\ 3\ 2)$
e	e	$(1\ 2\ 3)$	$(1\ 3\ 2)$
$(1\ 2\ 3)$	$(1\ 2\ 3)$	$(1\ 3\ 2)$	e
$(1\ 3\ 2)$	$(1\ 3\ 2)$	e	$(1\ 2\ 3)$

11.2.1A If f is a symmetry of the figure F, $f: F \to F$ is a bijection, and hence $f^{-1}: F \to F$ is also a bijection. Also, f is an isometry. Hence for all points $p, q \in F$, $d(f(f^{-1}(p)), f(f^{-1}(q))) = d(f^{-1}(p), f^{-1}(q))$; that is, $d(p, q) = d(f^{-1}(p), f^{-1}(q))$, and so f^{-1} is an isometry. It follows that f^{-1} is also a symmetry of F.

11.2.2A

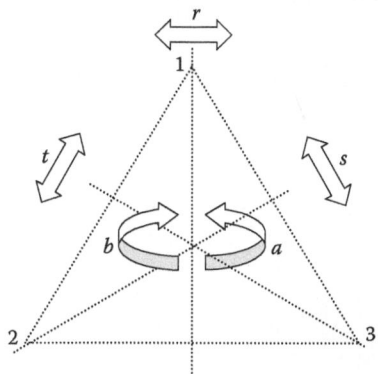

i. An equilateral triangle has six symmetries. These are the identity, rotations through one-third of a turn (that is, through $\frac{2}{3}\pi$ radians, or 60°) counterclockwise and clockwise, and three reflections in axes through a vertex and the midpoint of the opposite side.

ii. We let e be the identity, a be the counterclockwise rotation through $\frac{2}{3}\pi$, b the clockwise rotation through $\frac{2}{3}\pi$, and r, s, and t be the reflections shown in the diagram above. Using this notation, the Cayley table for this group of symmetries is as shown below.

	e	a	b	r	s	t
e	e	a	b	r	s	t
a	a	b	e	t	r	s
b	b	e	a	s	t	r
r	r	s	t	e	a	b
s	s	t	r	b	e	a
t	t	r	s	a	b	e

iii. With the vertices labeled as shown, a corresponds to the permutation (1 2 3), b to (1 3 2), r to (2 3), s to (1 3), and t to (1 2).

11.2.3A A cube has three different types of axes of rotational symmetry.

i. Axes joining the midpoint of opposite faces. There are three of these axes, and there are three rotational symmetries about each of them, namely, through $\frac{1}{4}$, $\frac{1}{2}$, and $\frac{3}{4}$ of a full turn. So there are altogether nine rotational symmetries of this type.

ii. Axes joining the midpoints of opposite edges. There are six of these axes, and there is just one rotational symmetry about each of them, namely, through $\frac{1}{2}$ of a full turn. So there are six rotational symmetries of this type.

iii. Axes joining opposite vertices. There are four of these axes and two rotational symmetries about each of them, namely, through $\frac{1}{3}$ and $\frac{2}{3}$ of a complete turn. So there are eight rotational symmetries of this type.

These rotations are illustrated by the following pictures.

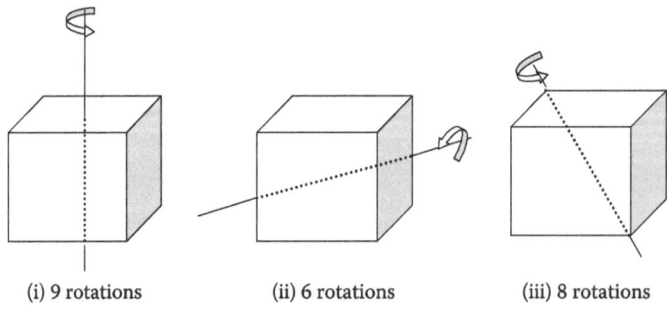

(i) 9 rotations   (ii) 6 rotations   (iii) 8 rotations

Hence, if we also include the identity, we see that a cube has 24 rotational symmetries.

11.2.4A From the solution to Exercise 11.2.2A, we see that the Cayley table for the group of rotational symmetries of an equilateral triangle is as shown on the left below. The Cayley table for the group, Z_3, of integers 0, 1, 2 with addition modulo 3 is as shown on the right.

∘	e	a	b
e	e	a	b
a	a	b	e
b	b	e	a

$+_3$	0	1	2
0	0	1	2
1	1	2	0
2	2	0	1

These Cayley tables have the same pattern, so the groups are isomorphic. More formally, the mapping θ, defined by θ(e)=0, θ(a)=1, and θ(b)=2, is an isomorphism between the two groups.

11.2.5A i. A rectangle has four symmetries: the identity, e; a half-turn rotation, a; a reflection, h, in the horizontal axis; and a reflection, v, in the vertical axis of symmetry. The Cayley table for this group is shown below on the left. This is the same for all rectangles, and so all the groups arising in this way are isomorphic. So we can talk about "*the* group of symmetries of a rectangle." We use the notation S(☐) for this group.

ii. The Cayley tables for the group S(☐), and the group, Z_4, of the integers 0, 1, 2, 3 with addition modulo 4, are shown below.

∘	e	a	h	v
e	e	a	h	v
a	a	e	v	h
h	h	v	e	a
v	v	h	a	e

S(☐)

$+_4$	0	1	2	3
0	0	1	2	3
1	1	2	3	0
2	2	3	0	1
3	3	0	1	2

Z_4

These tables have different patterns. But we must not make the mistake of immediately jumping to the conclusion that the groups are not isomorphic. We were not forced to list the symmetries of the rectangle in the order e, a, h, v. Perhaps if we had put them in the table in a different order the two tables *would* have the same pattern. To avoid checking all possible arrangements of the symmetries, we need an argument to rule out the possibility of ever getting the patterns to match. The key observation is that all the symmetries of the rectangle satisfy the equation $x \circ x = e$ whereas only 0 and 2 in Z_4 satisfy this equation. This means that, however its elements are arranged, in the Cayley table for S(☐) the identity element occurs four times on the main diagonal, whereas in the Cayley table for Z_4, the identity element will occur only twice on the main diagonal. Therefore the tables cannot have the same pattern, and so the groups cannot be isomorphic.

11.3.1A i. The set $\{e,a,b,c\}$ is a subgroup. Its Cayley table forms the top left-hand quadrant of the Cayley table given in Table 11.2. From this it can be seen that the set $\{e,a,b,c\}$ satisfies the subgroup properties.

ii. The set $\{e, a, b, c, h, v\}$ is not a subgroup. The symmetries a and h are in this set, but $a \circ h = r$, which is not in this set. So the closure property does not hold.

iii. The set $\{h,v,r,s\}$ is not a subgroup as it does not include the identity element (it also fails to satisfy the closure property).

iv. The set $\{e,r\}$ is a subgroup, as is shown by its Cayley table below, from which we can readily see that it satisfies the subgroup properties.

	e	r
e	e	r
r	r	e

11.3.2A Suppose $m, n \in H$. Then there are integers k, l such that $m = 5k$ and $n = 5l$. Hence $m + n = 5(k + l)$, and as $k + l$ is an integer, $m + n \in H$. So H satisfies the closure property. The identity element, 0, of Z is in H. Finally, $-m = 5(-k)$, and, as $-k$ is an integer, $-m \in H$. So H is a subgroup of Z. H has five cosets, consisting of H itself and, for $1 \le r \le 4$, the set of integers that have remainder r when divided by 5.

11.4.1A To answer this question it is best to think geometrically. The order of a symmetry is the number of times you need to carry out the symmetry in order to return each point of the figure to its original position. For example, the identity symmetry has order 1 because it leaves every point in its original position. Also, each reflection has order 2 because if a figure is reflected twice in the same line, or plane, of symmetry, each point is returned to its original position. Similarly a rotation through $1/k$ th of a turn has order k.

i. Using these facts, we can list the orders of the symmetries of a square as follows:

Symmetry	e	a	b	c	h	v	r	s
Order	1	4	2	4	2	2	2	2

ii. The rotational symmetries of a cube are given in the solution to Exercise 11.2.3A. We set out the order of these symmetries in the following table:

Axis of Rotation	Angle of Rotation	Number of these Symmetries	Order of these Symmetries
—	Identity	1	1
Line joining midpoints of opposite faces	$\pm \frac{1}{2}\pi$	6	4
Line joining midpoints of opposite faces	π	3	2
Line joining midpoints of opposite edges	π	6	2
Line joining diagonally opposite vertices	$\pm \frac{2}{3}\pi$	8	3

Thus, in addition to the identity, there are nine rotations of order 2, eight of order 3, and six of order 4.

11.4.2A The order of a number n in the group Z_{12} is the smallest number of n's we need to add up modulo 12 to obtain 0. That is, it is the least k such that $kn \equiv 0 \pmod{12}$. It is not difficult to see that, for $n \neq 0$, k is equal to 12 divided by the greatest common divisor of n and 12. So the orders are as given in the following table.

Element	0	1	2	3	4	5	6	7	8	9	10	11
Order	1	12	6	4	3	12	2	12	3	4	6	12

Note that, in particular, the elements of order 12 correspond to the elements, n, such that $1 \leq n \leq 11$ and the numbers n and 12 are coprime. The number of such elements is 4. Note that this is $\phi(12)$, where ϕ is Euler's ϕ-function, as described in Exercise 4.1.4.

11.4.3A We can represent e as g^0, and g as g^1. Thus $H = \{g^r : 0 \leq r \leq k-1\}$. For $0 \leq r, s \leq k-1$, if $r + s \leq k - 1$, we have $g^r \circ g^s = g^{r+s}$. Otherwise, $k \leq r + s \leq 2k - 2$, $g^r \circ g^s = g^{r+s} = g^k \circ g^{r+s-k} = e \circ g^{r+s-k} = g^{r+s-k}$, where $0 \leq r + s - k < k - 1$. So in either case $g^r \circ g^s \in H$ and so the closure condition is satisfied. Indeed, this calculation shows that $g^{r+s} = g^t$, where $r + s \equiv t \pmod{k}$ and $0 \leq t \leq k-1$. Hence the mapping $\theta: H \to Z_k$ given by $\theta(g^k) = k$ is an isomorphism. It follows that H forms a subgroup that is isomorphic to the group Z_k.

11.4.4A Our search for all the subgroups of the group $S(\square)$ is guided by Lagrange's theorem and its corollary. Lagrange's theorem tells us that the only possible orders of subgroups are the divisors of 8, that is, 1, 2, 4, and 8. The only subgroup of order 1 is the trivial subgroup $\{e\}$, and the only subgroup of order 8 is the entire group itself. This leaves us looking for subgroups of orders 2 and 4.

Subgroups of order 2. Such a subgroup contains two elements, one of which must be the identity element and the other, by Lagrange's corollary, must be of order 2. Conversely, if x has order 2, it can be seen from its Cayley table that $\{e, x\}$ is a subgroup.

	e	x
e	e	x
x	x	e

Thus the subgroups of order 2 correspond to the elements of order 2. So, from the solution to Exercise 11.4.1A, we see that the subgroups of order 2 are $\{e,b\}$, $\{e,h\}$, $\{e,v\}$, $\{e,r\}$, and $\{e,s\}$.

Subgroups of order 4. From the solution to Exercise 11.4.3A, we see that if g is an element of order 4, then $\{e, g, g^2, g^3\}$ is a subgroup of order 4. As there are two elements of order 4, it looks as though there are two subgroups of order 4 of this form. However, $\{e, a, a^2, a^3\} = \{e, a, b, c\}$ and $\{e, c, c^2, c^3\} = \{e, c, b, a\}$, so these two subgroups are the same. From the solution to Exercise 11.2.5A, we see that it is also possible to have a group of order 4 made up of the identity and three elements each of order 2, say x, y, and z, such that when any two of these elements are combined, we obtain the third element. From the Cayley table for $S(\square)$ we see that the only cases where two elements of order 2 combine to give another element of order 2 are $b \circ h = v = h \circ b$, $b \circ v = h = v \circ b$, $b \circ r = s = r \circ b$, $b \circ s = r = s \circ b$, $h \circ v = b = v \circ h$, and $r \circ s = b = s \circ r$. It follows that there are just two subgroups of order 4 of this form, $\{e,b,h,v\}$ and $\{e,b,r,s\}$. So there are altogether three different subgroups of order 4.

11.5.1A a. The disjoint cycles making up the permutation have lengths 2, 3, 4, and 5. So its order is $lcm(2,3,4,5) = 60$.

b. Similarly, here the order is $lcm(3,4,6) = 12$.

11.5.2A Since the $lcm(3,5,7) = 105$, any permutation in S_{15} made up of cycles of lengths 3, 5, and 7 will have order 105. Since $3 + 5 + 7 = 15$, it is possible to find permutations in S_{15} of this order. For example, (1 2 3)(4 5 6 7 8)(9 10 11 12 13 14 15) is a permutation in S_{15} of order 105.

11.5.3A The cycle types of the permutations in S_n correspond to the partitions of n, and the orders of the permutations are determined by their cycle types. So in the tables below, we list the partitions of n, for $n = 4,5,6$ and the orders of the corresponding permutations.

S_4

Partition	Order
4	4
3+1	3
2+2	2
2+1+1	2
1+1+1+1	1

S_5

Partition	Order
5	5
4+1	4
3+2	6
3+1+1	3
2+2+1	2
2+1+1+1	2
1+1+1+1+1	1

	Partition	Order
	6	6
	5+1	5
	4+2	4
	4+1+1	4
S_6	3+3	3
	3+2+1	6
	3+1+1+1	3
	2+2+2	2
	2+2+1+1	2
	2+1+1+1+1	2
	1+1+1+1+1+1	1

We thus see that the orders of the elements of these groups are as follows.

S_4 1, 2, 3, and 4

S_5 1, 2, 3, 4, 5, and 6

S_6 1, 2, 3, 4, 5, and 6

11.5.4A In bracket notation the bottom riffle shuffle is

$$\begin{pmatrix} 1 & 2 & 3 & . & . & 24 & 25 & 26 & 27 & 28 & 29 & . & . & 50 & 51 & 52 \\ 2 & 4 & 6 & . & . & 48 & 50 & 52 & 1 & 3 & 5 & . & . & 47 & 49 & 51 \end{pmatrix}.$$

In cycle notation this is

(1 2 4 8 16 32 11 22 44 35 17 34 15 30 7 14 28 3 6 12 24 48 43 33 13 26 52 51 49 45 37 21 42 31 9 18 36 19 38 23 46 39 25 50 47 29 5 10 20 40 27);

that is, it is made up of a single cycle of length 52, and hence has order 52.

11.5.5A The largest order of a permutation in S_{52} is 180,180. This corresponds to the partition 13+11+9+7+5+4+1+1+1 of 52.

CHAPTER 12

12.1.1A i. An 8×8 chessboard has 64 squares, and hence 2^{64} colorings using two colors. [As $2^{10} = 1024$, which is approximately equal to 10^3, $2^{64} = (2^{10})^6 \times 2^4$ is approximately equal to $(10^3)^6 \times 2^4 = 1.6 \times 10^{19}$.]

ii. A cube has six faces and hence $3^6 = 729$ colorings using three colors.

12.1.2A We indicate the three colors by using shading.

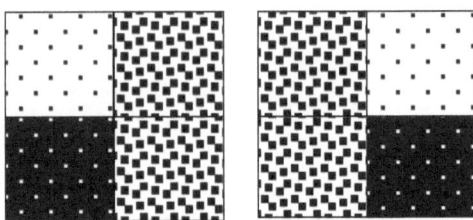

The coloring on the right may be obtained from the coloring on the left by reflecting the square in the vertical axis of symmetry but cannot be obtained from it by any rotation.

12.2.1A It is clear that the identity element, 0, of the group $(R, +)$ satisfies GA1 as rotation through an angle 0 leaves each point in its original position. That is, for each point $(x,y) \in R^2$, $0 \triangleright (x, y) = (x, y)$. A rotation through an angle θ followed by a rotation through an angle ϕ has the same effect as a rotation through an angle $\phi + \theta$. That is, $\phi \triangleright (\theta \triangleright (x, y)) = (\phi + \theta) \triangleright (x, y)$. So GA2 is also satisfied.

12.2.2A Let id be the identity element of S_n. Then for all $v_i, v_j \in V_n$, $id \triangleright (v_i, v_j) = (v_{id(i)}, v_{id(j)}) = (v_i, v_j)$. Thus GA1 holds. For all $\sigma, \tau, \in S_n$ and all $v_i, v_j \in V_n$ we have

$$\sigma \triangleright (\tau \triangleright (v_i, v_j)) = \sigma \triangleright (v_{\tau(i)}, v_{\tau(j)}) = (v_{\sigma(\tau(i))}, v_{\sigma(\tau(j))}) = (v_{\sigma\circ\tau(i)}, v_{\sigma\circ\tau(j)}) = (\sigma \circ \tau) \triangleright (v_i, v_j).$$

So GA2 also holds.

12.2.3A Suppose $g \triangleright x = h \triangleright x$. Then $g^{-1} \triangleright (g \triangleright x) = g^{-1} \triangleright (h \triangleright x)$. Hence, by GA2 and GA1, $(g^{-1}h) \triangleright x = g^{-1} \triangleright (h \triangleright x) = g^{-1} \triangleright (g \triangleright x) = (g^{-1} g) \triangleright x = e \triangleright x = x$. The converse is proved similarly.

12.3.1A The points obtained by rotating the point $(1,0)$ about the origin consist of all the points on the circle with center $(0,0)$ and radius 1. So the orbit of $(1,0)$ is this circle. A rotation about the origin leaves the point $(0,0)$ in its original position. So the orbit of this point is just the set $\{(0,0)\}$ consisting of the point $(0,0)$ by itself.

12.3.2A For all points $(x, y) \in R^2$, $0 \triangleright (x, y) = (x + 0, y + 2(0)) = (x, y)$, so GA1 is satisfied. Also, for all $t, u \in R$, $t \triangleright (u \triangleright (x, y)) = t \triangleright (x + u, y + 2u) = ((x + u) + t, (y + 2u) + 2t) = (x + (u + t), y + 2(u + t)) = (t + u) \triangleright (x, y)$, and hence GA2 is also satisfied.

The orbit of the point (0,0) is the set of points

$$\{t \triangleright (0,0) : t \in R\} = \{(0+t, 0+2t) : t \in R\} = \{(t, 2t) : t \in R\}.$$

This set makes up the line with equation $y = 2x$. Similarly, $Orb((0,1)) = \{(0+t, 1+2t) : t \in R\} = \{t, 1+2t) : t \in R\}$. This set makes up the line with equation $y = 2x + 1$. Also, $Orb((1,2)) = \{(1+t, 2+2t) : t \in R\}$, and this set makes up the line $y = 2x$.

12.4.1A Note that in this solution we omit the symbol, •, for the group operation.
  i. Suppose that $g, h \in Z(G)$. Then for each $x \in G$, we have that $gx = xg$ and $hx = xh$, and hence $(gh)x = g(hx) = g(xh) = (gx)h = (xg)h = x(gh)$. Therefore $gh \in Z(G)$. So $Z(G)$ satisfies the closure property. We have already noted that $e \in Z(G)$. Finally, suppose $g \in Z(G)$, and $x \in G$. Then, $gx = xg$. Hence $g^{-1}gxg^{-1} = g^{-1}xgg^{-1}$, and hence $xg^{-1} = g^{-1}x$. Therefore $g^{-1} \in G$. It follows that $Z(G)$ is a subgroup of G.

  ii. Suppose $g \in Z(G)$, then for each $x \in G$, $xgx^{-1} = (xg)x^{-1} = (gx)x^{-1} = g(xx^{-1}) = ge = g$. So the conjugacy class of g, that is, $\{xgx^{-1} : x \in G\}$, is just $\{g\}$. Conversely, suppose that the conjugacy class of g is the set $\{g\}$. It follows that for each $x \in G$, $xgx^{-1} = g$ and hence $xgx^{-1}x = gx$, that is, $xg = gx$, and so $g \in Z(G)$.

  iii. We know that the conjugacy classes of G partition G, and that number of elements in each conjugacy class divides the order of G. Thus if $\#(G) = p^n$, where p is a prime number and n is some positive integer, the possibilities for the number of elements in a conjugacy class are $1, p, p^2, \ldots, p^{n-1}$. Consequently, if k is the number of conjugacy classes of size 1, and there are t other conjugacy classes with sizes $k_1, k_2, \ldots, k_t$ we have that for $1 \leq s \leq t$, p is a divisor of k_s, and $p^n = k + k_1 + k_2 + \ldots + k_t$. Hence, $k = p^n - k_1 - k_2 - \ldots - k_t$, and so p is also a divisor of k. By part (b), $\#(Z(G)) = k$. We deduce that $\#(Z(G))$ is a multiple of p and hence $\#(Z(G)) \geq p$.

CHAPTER 13

13.1.1A We use the notation introduced in the solution to Exercise 11.2.2A for the six symmetries of an equilateral triangle. That is, e is the identity, a and b are the rotations through one-third of a turn, counterclockwise and clockwise, and r, s, and t are the reflections in axes joining one vertex to the midpoint of the opposite side.

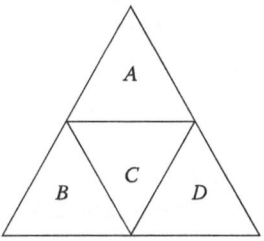

The large triangle is divided into four smaller triangles, labeled A, B, C, and D in the diagram. Hence there are $2^4 = 16$ colorings in which each of these triangles is colored black or white. The identity fixes every coloring, so #(Fix (e)) = 16. The rotations fix a coloring if and only if the triangles A, B, and D have the same color. So for a coloring to be fixed by either a or b, there are two choices for the common color of these triangles and two choices for the color of the triangle C. Hence #(Fix (a)) = #(Fix (b)) = 2^2 = 4. The reflection, r, in the vertical axis fixes a coloring provided B and D are the same color. So to get a coloring that is fixed by r we have two choices for the color of B and D, two choices for the color of A, and two choices for the color of C. Therefore #(Fix (r)) = 2^3 = 8. Similarly #(Fix (s)) = #(Fix (t)) = 8. Therefore, by Frobenius's counting theorem, there are $\frac{1}{6}(16+4+4+8+8+8) = 8$ different orbits.

13.2.1A We use the same notation for the symmetries of an equilateral triangle as in the solution to Exercise 13.1.1A. If we label the three edges of an equilateral triangle 1, 2, and 3, as shown, we can represent the symmetries as permutations of these sides and hence calculate the number of colorings fixed by each of them. The identity, e, fixes every coloring, so #(Fix (e)) = 4^3. The rotation, a, corresponds to the permutation (1 2 3). As this consists of just one cycle #(Fix (a)) = 4^1. Similarly, #(Fix (b)) = 4^1. The reflection, r, 1 in the vertical axis, corresponds to the permutation (1)(2 3). This has two cycles. So #(Fix (r)) = 4^2.

Similarly, #(Fix (s)) = #(Fix (t)) = 4^2. Therefore, by Frobenius's counting theorem, the number of different patterns, that is, the number of different orbits, is $\frac{1}{6}(4^3 + 4^1 + 4^1 + 4^2 + 4^2 + 4^2) = 20$.

13.2.2A We label the squares of a 5×5 chessboard with the numbers 1 to 25, as shown below,

1	2	3	4	5
6	7	8	9	10
11	12	13	14	15
16	17	18	19	20
21	22	23	24	25

and we use the standard symbols, e, a, b, c, h, v, r and s, for the eight symmetries of the square. There are 3^{25} colorings of the board using 3 colors, and the identity fixes them all, so #(Fix(e)) = 3^{25}. The quarter-turn clockwise rotation corresponds to the permutation

(1 5 25 21)(2 10 24 16)(3 15 23 11)(4 20 22 6)(7 9 19 17)(8 14 18 12)(13)

of the squares. This permutation is made up of 7 cycles. Hence #(Fix(a)) = 3^7. Similarly, #(Fix)(c) = 3^7. The half-turn rotation, b, interchanges 12 pairs of squares and leaves square 13 fixed. So the corresponding permutation has cycle type $x_1 x_2^{12}$. Since there are 13 cycles in this permutation, #(Fix(b)) = 3^{13}. The reflection h corresponds to a permutation of cycle type $x_1^5 x_2^{10}$, since it keeps the squares 11, 12, 13, 14, and 15 fixed and interchanges the other 20 squares in pairs. The reflections v, r, and s also correspond to permutations of the squares of cycle type $x_1^5 x_2^{10}$. So each of the reflections fixes 3^{15} colorings. Hence, by Frobenius's counting theorem, the number of different patterns, using three colors, is $\frac{1}{8}(3^{25}+2(3^7)+3^{13}+4(3^{15}))= 105{,}918{,}450{,}471$.

13.2.3A The rotations of a cube are given in the solution to Exercise 11.2.3A. There are 24 rotations in all. We list them in the following table, together with the cycle types of the corresponding permutations of the faces of the cube, and the number of colorings fixed by each of them.

Type of Rotation	Number of this Type	Cycle Type	#(Fix(g))
Identity	1	x_1^6	3^6
Rotations through $\pm \frac{1}{2}\pi$ about axes joining midpoints of opposite faces	6	$x_1^2 x_4$	3^3
Rotations through π about axes joining midpoints of opposite faces	3	$x_1^2 x_2^2$	3^4
Rotations through π about axes joining midpoints of opposite edges	6	x_2^3	3^3
Rotations through $\frac{1}{3}\pi$ about axes joining opposite vertices	8	x_3^2	3^2

It therefore follows from Frobenius's counting theorem that the number of different ways of coloring the cube using the three colors red, white, and blue is

$$\frac{1}{24}(3^6+6(3^3)+3(3^4)+6(3^3)+8(3^2))=57.$$

13.2.4A For each of the Platonic regular solids we list the rotational symmetries, the cycle types of the corresponding permutations of the faces, and hence the number of colorings that they fix.

First, the *regular octahedron*:

Type of Rotation	Number of this Type	Cycle Type	#(Fix(g))
Identity	1	x_1^8	c^8
Rotations through $\pm \frac{2}{3}\pi$ about axes joining the midpoints of opposite faces	8	$x_1^2 x_3^2$	c^4
Rotations through π about axes joining the midpoints of opposite edges	6	x_2^4	c^4
Rotations through $\pm \frac{1}{2}\pi$ about axes joining opposite vertices	6	x_4^2	c^2
Rotations through π about axes joining opposite vertices	3	x_2^4	c^4

It follows from Frobenius's counting theorem that the number of different colorings is

$$\frac{1}{24}(c^8 + 8c^4 + 6c^4 + 6c^2 + 3c^4) = \frac{1}{24}(c^8 + 17c^4 + 6c^2).$$

Next, the *regular dodecahedron*:

Type of Rotation	Number of this Type	Cycle Type	#(Fix(g))
Identity	1	x_1^{12}	c^{12}
Rotations through $\pm \frac{2}{5}\pi$ and $\pm \frac{4}{5}\pi$ about axes joining the midpoints of opposite faces	24	$x_1^2 x_5^2$	c^4
Rotations through π about axes joining the midpoints of opposite edges	15	x_2^6	c^6
Rotations through $\pm \frac{2}{3}\pi$ about axes joining opposite vertices	20	x_3^4	c^4

It follows from Frobenius's counting theorem that the number of different colorings is

$$\frac{1}{60}(c^{12} + 24c^4 + 15c^6 + 20c^4) = \frac{1}{60}(c^{12} + 15c^6 + 44c^4).$$

Finally, *the regular icosahedron*:

Type of Rotation	Number of this Rype	Cycle Type	#(Fix(g))
Identity	1	x_1^{20}	c^{20}
Rotations through $\pm \frac{2}{3}\pi$ about axes joining the midpoints of opposite faces	20	$x_1^2 x_3^6$	c^8
Rotations through π about axes joining the midpoints of opposite edges	15	x_2^{10}	c^{10}
Rotations through $\pm \frac{2}{5}\pi$ and $\pm \frac{4}{5}\pi$ about axes joining opposite vertices	24	x_5^4	c^4

It follows from Frobenius's counting theorem that the number of different colorings is

$$\frac{1}{60}(c^{20}+15c^{10}+20c^8+24c^4).$$

CHAPTER 14

14.1.1A $f \circ \pi^{-1}(1) = f(\pi^{-1}(1)) = f(4) = 1;\ f \circ \pi^{-1}(2) = f(\pi^{-1}(2)) = f(3) = 0;$

$f \circ \pi^{-1}(3) = f(\pi^{-1}(3)) = f(5) = 2;\ f \circ \pi^{-1}(4) = f(\pi^{-1}(4)) = f(1) = 1;$ and

$f \circ \pi^{-1}(5) = f(\pi^{-1}(5)) = f(2) = 2$.

14.1.2A We have that

$\pi_1 \triangleright (\pi_2 \triangleright f) = \pi_1 \triangleright (f \circ \pi_2^{-1})$, by the definition of the group action,

$= (f \circ \pi_2^{-1}) \circ \pi_1^{-1}$, by the definition of the group action,

$= f \circ (\pi_2^{-1} \circ \pi_1^{-1})$, because the operation $\circ$ is associative,

$= f \circ ((\pi_1 \circ \pi_2)^{-1})$, by Lemma 14.1, and

$= (\pi_1 \circ \pi_2) \triangleright f$, by the definition of the group action.

14.2.1A The store enumerator is $\sum_{c \in C} w(c) = b+b+b+r+r = 3b+2r$. The inventory of all the mappings from D to C is $(3b+2r)^5$.

14.3.1A The symmetries of a square, the corresponding permutations of the vertices, and their cycle types are listed in the table below.

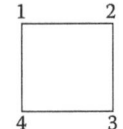

e	$(1)(2)(3)(4)$	x_1^4
a	$(1\ 2\ 3\ 4)$	x_4
b	$(1\ 3)(2\ 4)$	x_2^2
c	$(1\ 4\ 3\ 2)$	x_4
h	$(1\ 4)(2\ 3)$	x_2^2
v	$(1\ 2)(3\ 4)$	x_2^2
r	$(1)(2\ 4)(3)$	$x_1^2 x_2$
s	$(1\ 3)(2)(4)$	$x_1^2 x_2$

We therefore see that the cycle index of this group of permutations is

$$\frac{1}{8}(x_1^4 + 2x_1^2 x_2 + 3x_2^2 + 2x_4).$$

14.3.2A The rotations of a cube and the cycle types of the corresponding permutations of the faces of a cube are given in the solution to Exercise 13.2.3A. From this we see that the cycle index of this group of permutations is $(1/24)(x_1^6 + 6x_1^2 x_4 + 3x_1^2 x_2^2 + 6x_2^3 + 8x_3^2)$.

14.3.3A In the tables below we have listed the partitions of 5 and 6, and the cycle types of the permutations they correspond to. In the third column we have given the number of permutations of each cycle type, as calculated from the formula the expression 14.7.

$$S_5$$

Partition	Cycle Type	Number of this Type
5	x_5	24
4+1	$x_1 x_4$	30
3+2	$x_2 x_3$	20
3+1+1	$x_1^2 x_3$	20
2+2+1	$x_1 x_2^2$	15
2+1+1+1	$x_1^3 x_2$	10
1+1+1+1+1	x_1^5	1

Therefore the cycle index of S_6 is

$$\frac{1}{120}(x_1^5 + 10x_1^3 x_2 + 15x_1 x_2^2 + 20x_1^2 x_3 + 20x_2 x_3 + 30x_1 x_4 + 24x_5).$$

	S_6	
Partition	Cycle Type	Number of this Type
6	x_6	120
5+1	$x_1 x_5$	144
4+2	$x_2 x_4$	90
4+1+1	$x_1^2 x_4$	90
3+3	x_3^2	40
3+2+1	$x_1 x_2 x_3$	120
3+1+1+1	$x_1^3 x_3$	40
2+2+2	x_2^3	15
2+2+1+1	$x_1^2 x_2^2$	45
2+1+1+1+1	$x_1^4 x_2$	15
1+1+1+1+1+1	x_1^6	1

Thus the cycle index of S_6 is

$$\frac{1}{720}(x_1^6 + 15x_1^4 x_2 + 45x_1^2 x_2^2 + 15x_2^3 + 40x_1^3 x_3 + 120 x_1 x_2 x_3 + 40 x_3^2$$

$$+ 90 x_1^2 x_4 + 90 x_2 x_4 + 144 x_1 x_5 + 120 x_6).$$

14.3.4A In each case we use the formula in expression 14.7 for the number of permutations in S_n that have a specified cycle type.

i. $x_2^3 x_6$. Here, in the notation of expression 14.7, $t = 2$, $k_1 = 2$, $r_1 = 3$, $k_2 = 6$, $r_2 = 1$. So the number of permutation of this type is $12!/(3!2^3 1!6^1) = 1{,}663{,}200$.

ii. $x_1^4 x_3 x_5$. There are $12!/(4!1^4 1!3^1 1!5^1) = 1{,}330{,}560$ permutations of this type.

iii. $x_1 x_2^2 x_3 x_4$. There are $12!/(1!1^1 2!2^2 1!3^1 1!4^1) = 4{,}989{,}600$ permutations of this type.

14.4.1A From the solution to Exercise 13.2.2A we see that the cycle index of the group of permutations of the 25 squares of a 5×5 chessboard corresponding to the symmetries of a square is $\frac{1}{8}(x_1^{25} + 2x_1 x_4^6 + x_1 x_2^{12} + 4x_1^5 x_2^{10})$. Hence, by Pólya's theorem, the pattern inventory when the squares are colored using *black* and *white* is

$$\frac{1}{8}((b+w)^{25} + 2(b+w)(b^4 + w^4)^6 + (b+w)(b^2 + w^2)^{12} + 4(b+w)^5(b^2 + w^2)^{10}).$$

We seek the coefficient of $b^{15} w^{10}$. We take the terms one by one. The coefficient of $b^{15} w^{10}$ in $(b+w)^{25}$ is $C(25,15) = 3{,}268{,}760$. There is no term involving $b^{15} w^{10}$ in $(b+w)(b^4 + w^4)^6$. The term involving $b^{15} w^{10}$ in $(b+w)(b^2 + w^2)^{12}$ comes from multiplying b from the first bracket by $b^{14}w^{10}$ from the second, and, as $b^{14}w^{10} = (b^2)^7(w^2)^5$, its coefficient is $C(12,7) = 792$. Terms involving $b^{15}w^{10}$ in $(b+w)^5(b^2+w^2)^{10}$ can arise in three ways by multiplying a term from $(b+w)^5$ by a term from $(b^2+w^2)^{10}$ as set out in the following table,

Term From $(b+w)^5$	Term From $(b^2+w^2)^{10}$	Coefficient
b^5	$b^{10}w^{10}$	$C(5,5) \times C(10,5) = 1 \times 252 = 252$
b^3w^2	$b^{12}w^8$	$C(5,3) \times C(10,6) = 10 \times 210 = 2100$
bw^4	$b^{14}w^6$	$C(5,1) \times C(10,7) = 5 \times 120 = 600$

from which it follows that the required coefficient is $252 + 2100 + 600 = 2952$. It follows that the coefficient of $b^{15}w^{10}$ in the pattern inventory is $\frac{1}{8}(3,268,760 + 0 + 792 + 4 \times 2952) = 410,170$. This is the number of different patterns with 15 black squares and 10 white squares.

14.4.2A The cycle index of the group of permutations of the eight small triangles in the figure corresponding to the symmetries of the square is $\frac{1}{8}(x_1^8 + 2x_4^2 + 5x_2^4)$. Hence the pattern inventory for colorings using *red, white,* and *blue*, to which we assign the weights r, w, and b, respectively, is $\frac{1}{8}((r+w+b)^8 + 2(r^4+w^4+b^4)^2 + 5(r^2+w^2+b^2)^4)$. We seek the coefficient of $r^2w^2b^4$ in this pattern inventory. This coefficient is

$$\frac{1}{8}\left(\frac{8!}{2!2!4!} + 2 \times 0 + 5 \times \frac{4!}{1!1!2!}\right) = \frac{1}{8}(420 + 0 + 60) = 60.$$

Hence there are 60 patterns with two red, two white, and four blue triangles.

14.4.3A From the solution to Exercise 13.2.4A we see that the cycle index for the group of permutations of the faces of a rectangular octahedron corresponding to the rotational symmetries of the octahedron is $(1/24)(x_1^8 + 8x_1^2x_3^2 + 9x_2^4 + 6x_4^2)$. Hence the pattern inventory of the colorings using *red, white,* and *blue* is

$$\frac{1}{24}((r+w+b)^8 + 8(r+w+b)^2(r^3+w^3+b^3)^2 + 9(r^2+w^2+b^2)^4 + 6(r^4+w^4+b^4)^2).$$

It follows that the coefficient of $r^4w^2b^2$ is

$$\frac{1}{24}\left(\frac{8!}{4!2!2!} + 8 \times 0 + 9 \times \frac{4!}{2!1!1!} + 6 \times 0\right) = \frac{1}{24}(420 + 0 + 108 + 0) = 22.$$

Hence there are 22 patterns with four red, two white, and two blue faces.

14.6.1A In each case we make use of the formula given by Table 14.6, setting out the answers in the way used in the solution to Problem 14.6(b) and (c).

i. π has cycle type $x_1 x_2 x_5$.

π	x_1	x_2	x_5	x_1, x_2	x_1, x_5	x_2, x_5
π^*	–	x_1	x_5^2	x_2	x_5	x_{10}

Hence π^* has cycle type $x_1 x_2 x_5^3 x_{10}$.

ii. π has cycle type x_2^4.

π	x_2	x_2, x_2
no	4	6
π^*	x_1	x_2^2

Hence π^* has cycle type $(x_1)^4 \times (x_2^2)^6$, that is, $x_1^4 x_2^{12}$.

iii. π has cycle type $x_2 x_6$.

π	x_2	x_6	x_2, x_6
π^*	x_1	$x_3 x_6^2$	x_6^2

Hence π^* has cycle type $x_1 x_3 x_6^4$.

iv. π has cycle type $x_2^2 x_4$.

π	x_2	x_4	x_2, x_2	x_2, x_4
no	2	1	1	2
π^*	x_1	$x_2 x_4$	x_2^2	x_4^2

Hence π has cycle type $(x_1)^2 \times (x_2 x_4)^1 \times (x_2^2)^1 \times (x_4^2)^2$, that is, $x_1^2 x_2^3 x_4^5$.

14.6.2A The cycle types of the permutations in S_4 are given in Table 14.2. We now extend this table to include the cycles of the corresponding permutations in S_4^*.

Cycle type in S_4	x_1^4	$x_1^2 x_2$	x_2^2	$x_1 x_3$	x_4
Number	1	6	3	8	6
Cycle type in S_4^*	x_1^6	$x_1^2 x_2^2$	$x_1^2 x_2^2$	x_3^2	$x_2 x_4$

It follows that the cycle index of S_4^* is

$$\frac{1}{24}(x_1^6 + 9 x_1^2 x_2^2 + 8 x_3^2 + 6 x_2 x_4).$$

Hence the pattern inventory for simple graphs with four vertices is

$$\frac{1}{24}((1+c)^6 + 9(1+c)^2(1+c^2)^2 + 8(1+c^3)^2 + 6(1+c^2)(1+c^4)),$$

that is, $1 + c + 2c^2 + 3c^3 + 2c^4 + c^5 + c^6$.

We thus see that there is one graph with zero edges, one with one edge, two with two edges, three with three edges, two with four edges, one with five edges and one with six edges. Of course, we already know this from Problem 9.5 where all the different graphs with four vertices are listed.

14.6.3A We list all the different simple graphs with five vertices according to their number of edges.

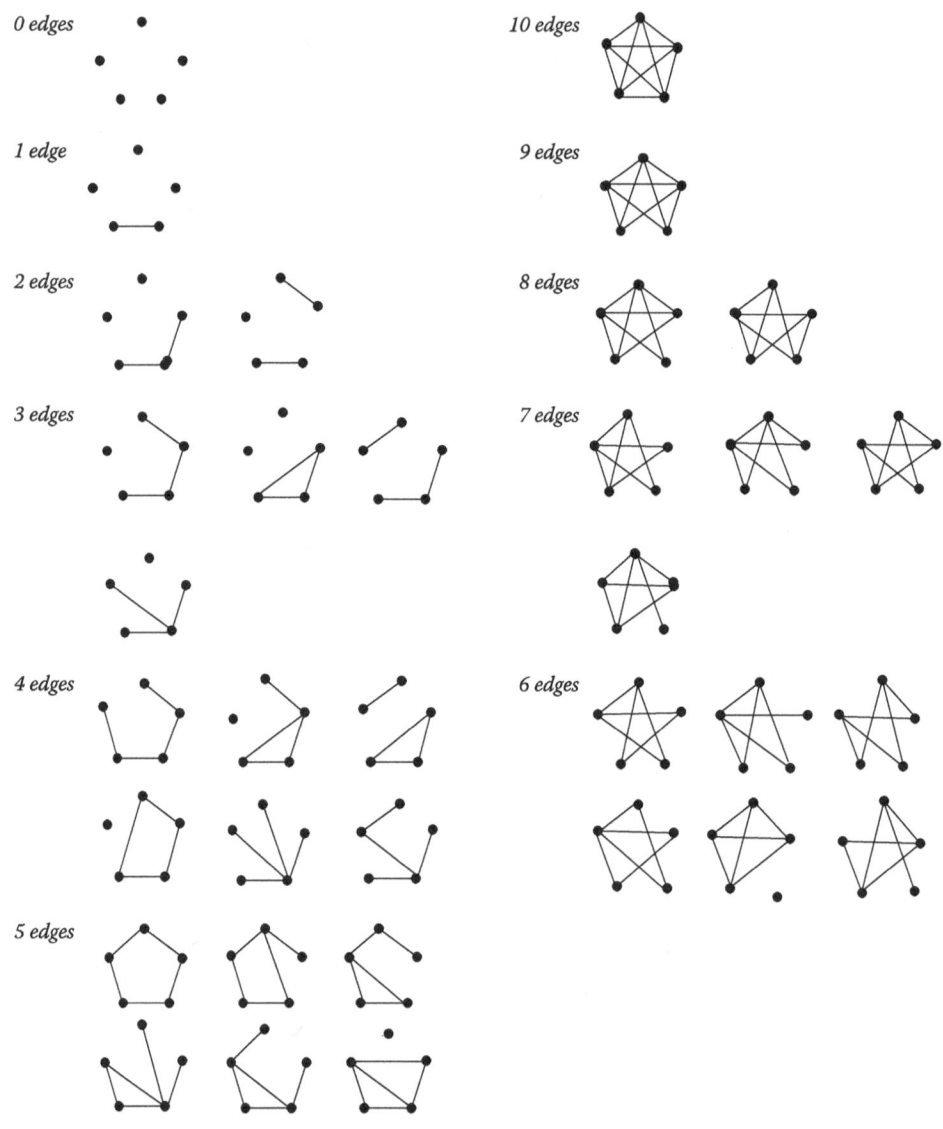

Notice that the graphs in the right-hand column are the *duals* of the graphs in the left-hand column, in the sense explained in Problem 9.3.2B.

14.6.4A We are given that π_1 has cycle type $x_1^{n-2r}x_2^r$. We can calculate the cycle type of π_1^* as follows:

π_1	x_1	x_2	x_1,x_1	x_2,x_2	x_1,x_2
Number	$n-2r$	r	$\frac{1}{2}(n-2r)(n-2r-1)$	$\frac{1}{2}r(r-1)$	$(n-2r)r$
π_1^*	—	x_1	x_1	x_2^2	x_2

It follows that π_1^* has cycle type $x_1^{r+(1/2)(n-2r)(n-2r-1)}x_2^{r(r-1)+(n-2r)r}$. Similarly π_2^* has cycle type $x_1^{r+(1/2)(n-2s)(n-2s-1)}x_2^{s(s-1)+(n-2s)s}$. Therefore they have the same cycle type if and only if

$$r+\frac{1}{2}(n-2r)(n-2r-1)=s+\frac{1}{2}(n-2s)(n-2s-1) \tag{1}$$

and

$$r(r-1)+(n-2r)r=s(s-1)+(n-2s)s. \tag{2}$$

If you do the algebra, you can check that both Equations 1 and 2 are equivalent to

$$(r-s)(r+s+1-n)=0. \tag{3}$$

From Equation 3, either $r = s$ or $r + s = n - 1$. Thus π_1 and π_2 have different cycle types, but π_1^* and π_2^* have the same cycle type if and only if $r \neq s$ and $r + s = n - 1$.

14.6.5A i. Suppose that the graphs $G_1 = (V_1, E_1)$ and $G_2 = (V_2, E_2)$ are isomorphic. It follows that there is an isomorphism, say ϕ, between them. Then $\phi: V_1 \to V_2$ is a bijection, and for each two-element subset $\{u,v\}$ of V_1, $\{u,v\} \in V_1 \Leftrightarrow \{\phi(u),\phi(v)\} \in V_2$. Hence, $\{u,v\} \notin E_1 \Leftrightarrow \{\phi(u),\phi(v)\} \notin E_2$, that is, $\{u,v\} \in E_1^* \Leftrightarrow \{\phi(u),\phi(v)\} \in E_2^*$. Hence ϕ is an isomorphism between the dual graphs $G_1^* = (V_1, E_1^*)$ and $G_2^* = (V_2, E_2^*)$, which are therefore isomorphic. Conversely, if G_1^* and G_2^* are isomorphic, then so also are G_1 and G_2.

ii. If a graph has n vertices, there are $\frac{1}{2}n(n-1)$ pairs of vertices that may or may not be joined by an edge. So a graph G has e edges if and only if its dual, G^*, has $\frac{1}{2}n(n-1)-e$ edges. So, by (i), the number of different graphs with e is the same as the number of different graphs with $\frac{1}{2}n(n-1)-e$ edges.

CHAPTER 15

15.2.1A For example, 11, 31, 54, 83, 84 is an increasing subsequence of length 5. There is no decreasing subsequence of length 5.

15.2.2A The people with surname Xerophyte have either one or two forenames. There are 26 possible different initials for those with just one forename, and $26^2 = 676$ possible initials for those with two forenames. So there are altogether $26 + 676 = 702$ possible sets of initials for people with the surname Xerophyte. As there are 777 people with this surname, the pigeonhole principle tells us that there must certainly be a pair of Xerophytes with identical initials.

15.2.3A The *lattice points*, (x, y), may be divided into four sets, namely, the sets of lattice points where (i) x and y are both even; (ii) x and y are both odd; (iii) x is even and y is odd; and (iv) x is odd and y is even, respectively. So given five lattice points, there must be two, say (x, y) and (u, v), from the same one of these sets. Thus x, u are both even or both odd, and in either case $x + u$ is even. Similarly, $y + v$ is even. Hence $(x + u)/2$ and $(y + v)/2$ are both integers. The midpoint, $((x + u)/2, (y + v)/2)$, of the line segment from (x, y) to (u, v) is therefore a lattice point.

15.2.4A If there are k possible years in which the people can have been born, then we need to have $2k + 1$ people to ensure that at least three were born in the same year. (With at most two people born in each of k years, there can be at most $2k$ people.) We just have to be a little careful about the number of possible birth years for people whose ages are between 18 and 30 (inclusive). For example, someone whose age is 18 on June 1, 2009, must have been born between June 2, 1990, and June 1, 1991, and so has two possible birth years, 1990 and 1991, and anyone whose age is 30 could have been born in either 1978 or 1979. Thus people whose ages are from 18 to 30 could have been born in any of the 14 years from 1978 to 1991 inclusive. Thus $k = 14$, and thus we need to invite at least 29 people to ensure that there are at least three who were born in the same year.

15.2.5A i. This is a generalization of Theorem 15.4, and so we repeat the proof of this theorem, but changing (with care) one of the n's to an m. Let $a_1, a_2, \ldots, a_{mn+1}$ be a sequence of $mn + 1$ distinct numbers. We shall assume that there is no increasing subsequence of length greater than m, and show how to deduce that there must be a decreasing subsequence of length at least $n + 1$.

For each i, with $1 \leq i \leq mn + 1$, we let s_i be the length of the longest increasing subsequence that begins at a_i. By our assumption, for each i, $s_i \leq m$. So we can put the $mn + 1$ numbers, a_i, in m numbered boxes, where a_i is put in box k, if $s_i = k$. By the second version of the pigeonhole principle, there must be at least one box containing at least $n + 1$ terms. So, for some k, there are at least $n + 1$ numbers in the sequence, say $a_{i_1}, a_{i_2}, \ldots, a_{i_{n+1}}$, with $i_1 < i_2 < \ldots < i_{n+1}$ and $s_{i_1} = s_{i_2} = \ldots = s_{i_{n+1}} = k$. Then, just as in the proof of Theorem 15.4, we can show that the numbers $a_{i_1}, \ldots, a_{i_{n+1}}$ form a decreasing sequence. It follows that there is a decreasing subsequence of length $n + 1$.

ii. The sequence 5, 4, 3, 2, 1, 10, 9, 8, 7, 6, 15, 14, 13, 12, 11 has no decreasing subsequence of length 6 and no increasing subsequence of length 4.

15.3.1A Suppose that the numbers in A are $a_1, \ldots, a_n$, where $a_1 \leq a_2 \leq \ldots \leq a_n$. In particular, we have $a_{n-k+1} \leq a_{n-k+2} \leq \ldots \leq a_n$ and, as the sum of these k numbers from A is at most t, not all of these numbers are greater than t/k. Hence $a_{n-k+1} \leq t/k$, and hence for $i < n - k + 1$, we also have $a_i \leq t/k$. Thus we have

$$\sum_{i=1}^{n} a_i = \sum_{i=1}^{n-k} a_i + \sum_{i=n-k+1}^{n} a_i \leq (n-k)\left(\frac{t}{k}\right) + t = \frac{nt}{k}.$$

Suppose that the sum of the ages of each group of 13 residents is at most 1066. Then, by what we have just shown, the total of all the ages of the residents is at most $(85 \times 1066)/13 = 6970$, contradicting the fact that the sum of their ages is more than 7000. Therefore, there must be a group of 13 residents, the sum of whose ages is more than 1066.

15.3.2A When an integer is divided by n, there are n possible values for the remainder, namely, $0, 1, \ldots, n - 1$. So, given a set of $n + 1$ distinct integers, by the pigeonhole principle there are at least two integers in the set, say a and b, that have the same remainder when they are divided by n. It follows that $a - b$ is divisible by n.

15.3.3A Let T be the equilateral triangle. Divide T into nine equilateral triangles each with side length $\frac{1}{3}$ as shown.

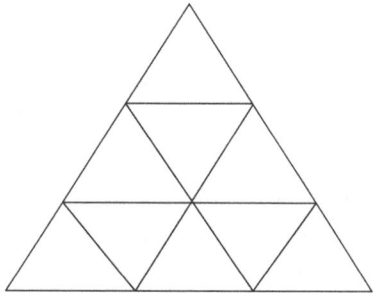

Given 10 points inside T, by the pigeonhole principle one of the smaller triangles, say T_0, contains two of these points. It cannot be that both of these points are vertices of T_0, as in that case at least one of them would be on the boundary of T. Hence the distance between these points is less than one-third.

15.3.4A

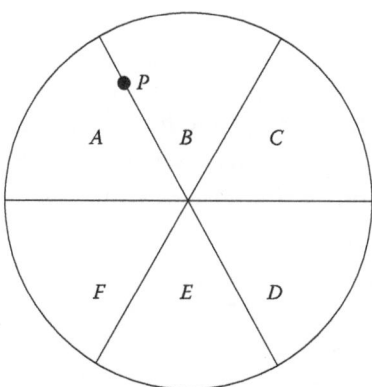

We note first that none of the six points lies on the circumference of the circle and that if there is a point at the center of the circle, its distance from all the other five points will be less than 1 unit. So from now on, we assume that none of the points is at the center. Divide the circle into six congruent sectors, A, B, C, D, E, and F, by diameters, one of which goes through one of the points, say P. Assuming A, B are the two sectors on either side of P, if there is any other point in either of the sectors A and B, it is distant less than one unit from P. Otherwise, five of the points are in the four regions C, D, E, and F. Hence, by the pigeonhole principle one of these sectors contains two of the points, and these two points will be less than one unit apart.

15.3.5A Since the student works for at most five hours in any three consecutive days, he works at most $(27/3) \times 5 = 45$ hours altogether. Let h_i, for $1 \leq i \leq 27$, be the total number of hours the student has worked after i days. We have that $1 \leq h_1 < h_2 < \ldots < h_{27} \leq 45$ and hence that $9 \leq h_1 + 8 < h_2 + 8 < \ldots < h_{27} + 8 \leq 53$. Thus $h_1, h_2, \ldots, h_{27}, h_1 + 8, h_2 + 8, \ldots, h_{27} + 8$ are 54 integers in the range from 1 to 53. Hence, by the pigeonhole principle, two of them must be equal. Consequently, for some s, r, we have $h_s + 8 = h_r$, so $h_r - h_s = 8$, and thus the student works for exactly eight days on the consecutive days $s + 1, s + 2, \ldots, r$.

15.3.6A We suppose, for simplicity, that $x > 0$. We know that there are integers p, q such that $|qa - p| < h$. As a is irrational, $a \neq p/q$ and hence $|qa - p| \neq 0$. We put $r = |qa - p|$. Let t be the least integer such that $(t - 1)r < x \leq tr$. Therefore, $0 \leq tr - x < r < h$ and, also, $0 \leq x - (t - 1)r < r < h$.

If $qa - p > 0$, then $r = qa - p$, and, as $0 \leq tr - x < h$, we have $0 \leq t(qa - p) - x < h$. Thus, if we put $u = tp$ and $v = tq$, we have $|(va - u) - x| < h$. If $qa - p < 0$, then $r = p - qa$ and, as $0 \leq x - (t - 1)r < h$, we have $0 \leq x - (t - 1)p + (t - 1)qa < h$. It follows that if we put $u = (t - 1)p$ and $v = (t - 1)q$, we have $|(va - u) - x| < h$.

15.3.7A The integers in X have sum at most $52 + 53 + 54 + 55 + 56 + 57 + 58 + 59 + 60 = 504$, but X has $2^9 - 1 = 511$ nonempty subsets. Hence there are two nonempty subsets of X, say Y' and Z', whose elements have the same sum. If Y', Z' are disjoint, we can take $Y = Y'$ and $Z = Z'$, and otherwise we remove the elements common to both sets; that is, we let $Y = Y' \setminus (Y' \cap Z')$ and $Z = Z' \setminus (Y' \cap Z')$.

CHAPTER 16

16.1.1A Let the vertices of K_6 be labeled a, b, c, d, e, and f. By the solution to Problem 16A, there is a monochromatic triangle. We may assume, without any loss in generality, that the edges ab, bc, and ca are all red. It may help to imagine the triangle abc lying below the triangle def and to call the nine edges joining d, e, and f to a, b, and c "struts."

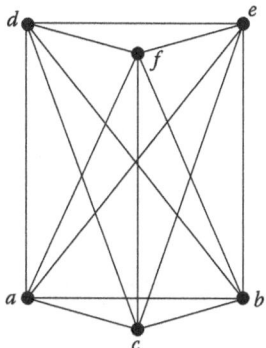

There are three struts from d to the vertices a, b, and c. If any two of these struts are red, then we immediately have our two monochromatic triangles—both red. Similar remarks apply to the vertices e and f. Hence, the only case we have left to consider is when at most one strut from each of the vertices d, e, and f to the vertices a, b, and c is red, giving a total of at most three red struts. Now suppose that the vertex a is the end vertex of two (or more) red struts. Then at least one of b and c would have to be an end vertex for three blue struts. Without loss of generality we can assume that the struts bd, be, and bf are all blue. Then either one of the edges de, df, and ef is blue, in which case there is a blue triangle, or they are all red, so that we have a second red triangle. Similar remarks apply if b or c is the end vertex of two red struts. This leaves the case where there are at most three red struts and no two of these meet. By the symmetry of the figure we can therefore suppose that the three red struts are some the edges ad, be, and cf, with all the other struts being blue.

Now consider triangle def. Either this forms a red triangle, in which case we again have two monochromatic triangles, or one of its edges is blue. In the latter case, without loss of generality, we can suppose that the edge de is blue. Then since the struts cd and ce are blue, cde is a blue triangle. So in any case we have a second monochromatic triangle. This completes the proof.

16.1.2A Suppose that among Mr. Friend's six friends there are not three mutual strangers. Then, by the result of Problem 16A, there are three mutual friends. These friends together with Mr. Friend form a foursome of mutual friends.

16.1.3A Consider the complete graph, K_{3n-1}, with vertices 1, 2, 3, ..., $3n-2$, $3n-1$. We color the edge ij red if $|i-j| \equiv 1(\mathrm{mod}3)$ and blue otherwise. Suppose that the edges ij and jk are both red. Then $|i-j| \equiv 1(\mathrm{mod}3)$ and $|j-k| \equiv 1(\mathrm{mod}3)$. It follows that either $|i-k| \equiv 0(\mathrm{mod}3)$ or $|i-k| \equiv 2(\mathrm{mod}3)$, and so the edge ik is blue. So there is no red triangle, that is, no red K_3.

Suppose we have a complete subgraph, say G, all of whose edges are blue. Because of the symmetry of the coloring, we can assume that this subgraph includes the vertex 1.

Hence it cannot include any of the vertices 2, 5, 8, ..., $3n - 1$, as the edge from 1 to any of these vertices is red. The edges joining the vertices in each of the $n - 1$ pairs, 3 and 4; 6 and 7; 9 and 10; ...; $3n-3$ and $3n-2$ are red. So G can include at most one vertex from each of these pairs. Consequently, the subgraph G has at most n vertices. So there is no blue K_{n+1}.

We have therefore seen that we can color the edges of K_{3n-1} red and blue so that there is no blue K_{n+1} and no red K_3. Hence $R(n + 1, 3) \geq 3n$.

16.2.1A i. By Theorem 16.6, $R(5,5) \geq 17$ and $R(6,6) \geq 26$. Theorem 16.8 gives

$$R(5,5) \geq \frac{5(\sqrt{2})^4}{e} = \frac{20}{e} = 7.35... \quad \text{and} \quad R(6,6) \geq \frac{6(\sqrt{2})^5}{e} = \frac{24\sqrt{2}}{e} = 12.48....$$

So in these cases Theorem 16.6 gives better lower bounds.

ii. Theorem 16.8 gives the lower bound $R(p,p) \geq (p-1)^2 + 1$. For all $p \geq 1$, $(p-1)^2 + 1 \leq p^2$, whereas $p(\sqrt{2})^{p-1}/e$ grows exponentially with p and so must eventually be larger than p^2. We leave it to the reader to make this argument more precise. It is not difficult to prove that for every integer $p \geq 10$,

$$(p-1)^2 + 1 < \frac{p(\sqrt{2})^{p-1}}{e}.$$

16.2.2A We prove by mathematical induction that, for each integer $n \geq 2$,

$$\frac{2}{3} 3^n \leq C(2n, n) \leq \frac{3}{8} 4^n. \tag{1}$$

For $n = 2$, the inequality 1 becomes $6 \leq C(4,2) \leq 6$, which is true since $C(4,2) = 6$. Now suppose that the inequality 1 holds for $n = k$, with $k \geq 2$, that is,

$$\frac{2}{3} 3^k \leq C(2k, k) \leq \frac{3}{8} 4^k. \tag{2}$$

Now

$$C(2k+2, k+1) = \frac{(2k+2)!}{(k+1)!(k+1)!} = \frac{(2k+2)(2k+1)}{(k+1)(k+1)} \cdot \frac{(2k)!}{k!k!} = \frac{4k+2}{k+1} C(2k, k), \tag{3}$$

and hence it helps to estimate the value of $(4k+2)/(k+1)$. For all $k \geq 1$, $3k + 3 \leq 4k + 2 \leq 4k + 4$, and hence

$$3 \le \frac{4k+2}{k+1} \le 4. \qquad (4)$$

By Equations 3 and 4,

$$3C(2k,k) \le C(2k+2,k+1) \le 4C(2k,k),$$

and hence by Equation 2,

$$\frac{2}{3}3^{k+1} \le C(2k+2,k+1) \le \frac{3}{8}4^{k+1},$$

and so Equation 1 holds also for $n = k + 1$. Therefore, by mathematical induction, Equation 1 holds for all positive integers $n \ge 2$.

16.3.1A i. Let $L = R(n_1,\ldots,n_k)$ $M = R(L,n_{k+1})$, and suppose that the edges of the complete graph G_M are colored using the $k + 1$ colors $C_1,C_2,\ldots,C_k,C_{k+1}$. Then either there is a complete subgraph with n_{k+1} vertices all of whose edges are colored C_{k+1} or a complete subgraph, say G', with L vertices each of whose edges is colored using one of the colors $C_1,\ldots,C_k$. Since $L = R(n_1,\ldots,n_k)$, it follows that for some i, $1 \le i \le k$, G', and hence G_M has a complete subgraph with n_i vertices all of whose edges are colored using C_i. It follows that $R(n_1,\ldots,n_k,n_{k+1}) \le R(L,n_{k+1})$, as required. In particular, $R(p, q, r) \le R(R(p, q), r)$ and, similarly, $R(p, q, r) \le R(R(p, r), q)$ and $R(p, q, r) \le R(R(q, r), p)$. Hence $R(p, q, r) \le \min\{R(R(p, q), r), R(R(p, r), q), R(R(q, r), p)\}$.

ii. Let $n = R(p - 1, q, r) + R(p, q - 1, r) + R(p, q, r - 1) - 1$, and suppose that the edges of the complete graph, G_n, with n vertices are colored using three colors, say *red*, *green*, and *blue*. Let v_0 be one vertex of G_n. This vertex is joined to $n - 1 = R(p - 1, q, r) + R(p, q - 1, r) + R(p, q, r - 1) - 2$ other vertices. Hence it must either be joined to at least $R(p - 1, q, r)$ vertices by red edges, or to at least $R(p, q - 1, r)$ vertices by green edges, or to at least $R(p, q, r - 1)$ vertices by blue edges. Suppose v_0 is joined to at least $R(p - 1, q, r)$ vertices by red edges. It follows that subgraph with these vertices had either a red G_{p-1} or a green G_q or a blue G_r. The red G_{p-1}, together with the vertex v_0, forms a red G_p. Similarly, in the other cases, it follows that G_n has either a red G_p or a green G_q or a blue G_r. It follows that $R(p, q, r) \le n$, as required.

16.4.1A By the result of Problem 16.1 we can find three points, say P, Q, and R, all with the same color and where Q is the midpoint of PR. Without loss of generality, we can assume that P, Q, and R are all red. Let S, T, U, and V be points on the line so that $PS = ST = TQ = QU = UV = VR$, as shown.

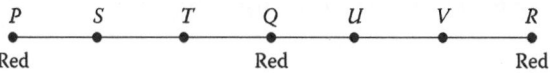

If S were red, then we would have $PS:SQ = 1:2$ with P, S, Q all red. If T were red, we would have $PT:TR = 1:2$ with P, T, R all red. If U were red, we would have $QU:UR = 1:2$ with Q, U, and R all red. Otherwise, we have $ST:TU = 1:2$, where S, T, U are all blue. So in every case we have three points, A, B, C, having the same color with $AB:BC = 1:2$.

16.4.2A Draw four parallel horizontal lines and 19 vertical lines crossing them. Each vertical line meets the horizontal lines in four points, each of which is red, green, or blue. Consider the quadruplets, such as (*red, green, blue, green*), giving the colors of these points from top to bottom. At least one color must occur twice in each quadruplet. As there are 19 quadruplets and 3 colors, at least one color, say *red*, must occur at least twice in at least 7 of these quadruplets. There are only C(4,2) = 6 different ways in which two '*red*'s can occur in a quadruplet. So at least two quadruplets must contain two '*red*'s in the same two positions. This gives a monochromatic rectangle.

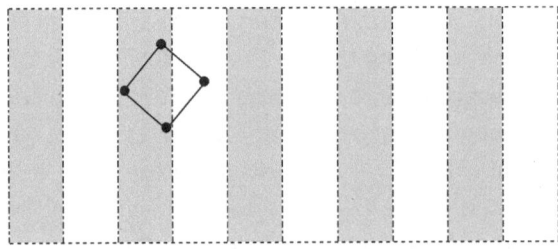

16.4.3A We color a point (x,y) red if the integer part of x is even, and green if the integer part of x is odd. We need to show that if a square of side length 1 is placed in the plane, not all its vertices can be the same color. Consider such a square that is placed

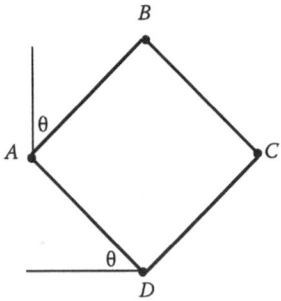

so that AB makes an angle θ with the vertical, then the horizontal distance from A to C is $|\sin\theta + \cos\theta|$ and the horizontal distance from B to D is $|\sin\theta - \cos\theta|$. Now $|\sin\theta + \cos\theta| = \sqrt{2}|\cos(\theta - \frac{1}{4}\pi)|$ and $|\sin\theta - \cos\theta| = \sqrt{2}|\cos(\theta + \frac{1}{4}\pi)|$. For $0 \le \theta \le \frac{1}{2}\pi$, $-\frac{1}{4}\pi \le \theta - \frac{1}{4}\pi \le \frac{1}{4}\pi$, and hence $\sqrt{2}|\cos(\theta - \frac{1}{4}\pi)| \ge 1$, and similarly for $\frac{1}{2}\pi < \theta \le \pi$, $\sqrt{2}|\cos(\theta + \frac{1}{4}\pi)| \ge 1$. It follows that either A and C are different colors, or B and D have different colors. So there is not a monochromatic square.

CHAPTER 17

17.1.1A i. One rook can be placed on any of the white squares, so there are $C(3,1) = 3$ ways in which they can be placed. Two rooks can be placed on any two of the squares, so there are $C(3,2) = 3$ ways to place them. Three rooks can be placed in only one way, and, very obviously, we cannot place more than three rooks on the white squares. Hence the rook polynomial is $1 + 3x + 3x^2 + x^3$, that is, $(1 + x)^3$.

ii. A similar argument shows that this board also has rook polynomial $(1 + x)^3$.

iii. One rook may be placed on any of the five white squares. Two rooks may be placed on any two diagonally adjacent white squares, and so may be placed in four different ways. It is not possible to place three or more nonattacking rooks on the white squares. So the rook polynomial is $1 + 5x + 4x^2$.

iv. One rook may be placed on any of the nine squares. To place two nonattacking rooks we need to choose two of the three rows and two of the three columns, which may be done in nine ways, and then the rooks can be placed in these two rows and columns in two ways. So there are 18 ways to place two nonattacking rooks. To place three nonattacking rooks on the board we must put one in each row and each column. There are three choices for the column for the rook in the first row, leaving two choices for the rook in the second row, after which there position of the third rook is automatically determined. So three rooks may be placed in $3 \times 2 \times 1 = 6$ ways. It is not possible to place more than three nonattacking rooks on the board. Hence the rook polynomial is $1 + 9x + 18x^2 + 6x^3$.

17.1.2A Here, instead of a direct calculation, we use the complementary board theorem. In the notation of this theorem, $m = n = 4$. Let B be the all-white 4×4 board. Then B^C is the all-black 4×4 board. Hence $r_k(B^C) = 1$ if $k = 0$, and $r_k(B^C) = 0$, for $k > 0$. Hence, by the complementary board theorem, we have $r_k(B) = C(4, k)C(4, k)k!$ It follows that

$$r_0(B) = 1, r_1(B) = 16, r_2(B) = 72, r_3(B) = 96, r_4(B) = 24, \text{ and, for } k > 4, r_k(B) = 0.$$

So the rook polynomial for the all-white 4×4 board is $1 + 16x + 72x^2 + 96x^3 + 24x^4$.

17.1.3A We are asked to calculate the values of $r_k(B)$, for $k = 0,1,2,3$, and 5, where, from the complementary board theorem, we have that

$$r_k(B) = \sum_{j=0}^{k} (-1)^j C(5-j, k-j)^2 (k-j)! r_j(B^C),$$

and we have calculated the values of $r_j(B^C)$ to be as shown in the following table.

k	0	1	2	3	4	5
$r_k(B^C)$	1	10	34	46	21	2

This gives

$r_0(B) = C(5,0)^2 = 1$,

$r_1(B) = C(5,1)^2 - 10C(4,0)^2 = 25 - 10 = 15$,

$r_2(B) = C(5,2)^2 2! - 10C(4,1)^2 + 34C(3,0)^2 = 200 - 160 + 34 = 74$,

$r_3(B) = C(5,3)^2 3! - 10C(4,2)^2 2! + 34C(3,1)^2 - 46C(2,0)^2 = 600 - 360 + 306 - 46 = 500$, and ,

$r_5(B) = C(5,5)^2 5! - 10C(4,4)^2 4! + 34C(3,3)^2 - 46C(2,2)2! + 2!C(1,1)^2 - 2!C(0,0)^2$

$= 120 - 240 + 204 - 92 + 21 - 2 = 11$.

17.1.4A i. By the disjoint subboards theorem, the rook polynomial of this board is

$$R(x, B_1)R(x, B_2),$$

where B_1 and B_2 are the boards shown below.

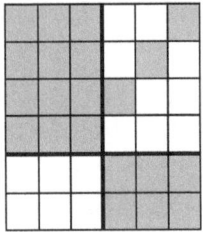

By direct calculation, we have $R(x, B_1) = 1 + 6x + 6x^2$. From the answer to Exercise 17.1.1A(a), we see that $R(x, B_2^C) = 1 + 3x + 3x^2 + x^3$, and hence, using the complementary board theorem, we can deduce that $R(x, B_2) = 1 + 9x + 21x^2 + 11x^3$. It follows that the rook polynomial is $(1 + 6x + 6x^2)(1 + 9x + 21x^2 + 11x^3)$, that is, $1 + 15x + 81x^2 + 191x^3 + 192x^4 + 66x^5$.

ii. If we first carry out the permutation (2 5) on the rows of the board and then the permutation (2 4 6)(3 5) on the columns, the board becomes the following. We then see, using Theorem 17.2 and the disjoint subboards theorem, that the given board has the same rook polynomial as in (i).

iii. This board is the complement of the board of (ii), so we may calculate its rook polynomial using the complementary board theorem in the case $m = n = 6$. This gives

$$r_k(B) = \sum_{j=0}^{k} (-1)^j C(6-j, k-j)^2 (k-j)! \, r_j(B^C),$$

where the values of the coefficient, $r_j(B^C)$, in the rook polynomial of the complementary board are as given in the common solutions for parts (i) and (ii). This gives

$r_0(B) = 1$,

$r_1(B) = C(6,1)^2 - 15C(5,0)^2 = 36 - 15 = 21$,

$r_2(B) = C(6,2)^2 2! - 15C(5,1)^2 + 81C(4,0)^2 = 450 - 375 + 81 = 156$,

$r_3(B) = C(6,3)^2 3! - 15C(5,2)^2 2! + 81C(4,1)^2 - 191C(3,0)^2$

$\qquad = 2400 - 3000 + 1296 - 191 = 505$,

$r_4(B) = C(6,4)^2 4! - 15C(5,3)^2 3! + 81C(4,2)^2 2! - 191C(3,1)^2 + 192C(2,0)^2$

$\qquad = 5400 - 9000 + 5832 - 1719 + 192 = 705$,

$r_5(B) = C(6,5)^2 5! - 15C(5,4)^2 4! + 81C(4,3)^2 3! - 191C(3,2)^2 2! + 192C(2,1)^2 - 66C(1,0)^2$

$\qquad = 4320 - 9000 + 7776 - 3438 + 768 - 66 = 360$,

and

$r_6(B) = C(6,6)^2 6! - 15C(5,5)^2 5! + 81C(4,4)^2 4! - 191C(3,3)^2 3!$

$\qquad + 192C(2,2)^2 - 66C(1,1)^2 + 0C(0,0)^2$

$\qquad = 720 - 1800 + 1944 - 1146 + 384 - 66 + 0 = 36$.

Hence the rook polynomial of the board is

$$1 + 21x + 156x^2 + 505x^3 + 705x^4 + 360x^5 + 36x^6.$$

17.1.5A Suppose there is a board, B, whose rook polynomial is $1 + 4x + 7x^2 + 2x^3 + x^4$. Since $r_1(B) = 4$, the board has four white squares. Hence there are $C(4,2) = 6$ ways in which two rooks can be put on the board, possibly including cases where they are in the same row or the same column. Hence, $r_2(B) \leq 6$. It follows that the term $7x^2$ cannot occur in the rook polynomial of B.

17.1.6A We show that if one 2×2 board cannot be obtained from a second 2×2 board by a rotation, then the boards have different rook polynomials. We already know that there are six different 2×2 boards with black and white squares such that none of them can be obtained from another by a rotation. These were listed in Chapter 12. All we need do is list these together with their rook polynomials.

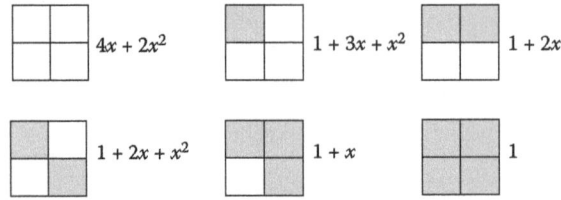

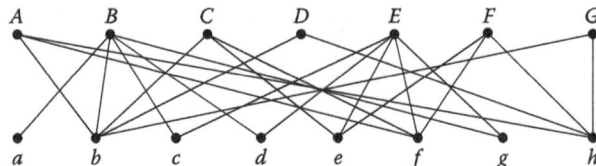

It is readily seen that these boards have different rook polynomials.

17.2.1A It can easily be seen that if C is also friendly with X, then A could marry U, B could marry V, C could marry X, D could marry Y, and E could marry Z, and then each man would be married to a woman he is friendly with. There are other solutions.

17.2.2A The bipartite graph representing the given friendships may be drawn as follows.

Between them the five women A, C, D, F, and G are friendly with just the four men b, e, f, and h. Hence it is not possible for each woman to marry a man she is friendly with.

Books for Further Reading

We list here books from which we have learned a great deal, or which will enable the reader to pursue further the topics introduced in this book, or both. It is not intended to be a comprehensive bibliography of combinatorics, and many excellent books are not included.

The first two books cover (more than) the assumed mathematical background of our book. We have placed a lot of emphasis on proofs, and both these books cover reading and understanding proofs as well as writing them.

MATHEMATICAL BACKGROUND

1. R. B. J. T. Allenby, *Numbers and Proofs*, Arnold, London, 1997.
2. Kevin Houston, *How to Think like a Mathematician*, Cambridge University Press, Cambridge, 2009.

COMBINATORICS IN GENERAL

3. Victor Bryant, *Aspects of Combinatorics*, Cambridge U. P., Cambridge, 1993.
4. R. L. Graham, M. Grötschel, and L. Lovász (editors), *Handbook of Combinatorics*, North-Holland, Amsterdam, 1995 (2 volumes).
5. Ralph P. Grimaldi, *Discrete and Combinatorial Mathematics*, Addison-Wesley, Reading, Massachussetts, 1999.
6. L. Lovász, *Combinatorial Problems and Exercises*, AMS, Chelsea, Rhode Island, 2007.
7. Fred S. Roberts and Barry R. Tesman, *Applied Combinatorics*, 2nd edition, Pearson/Prentice Hall, Englewood Cliffs, New Jersey, 2005.
8. Ioan Tomescu, *Problems in Combinatorics and Graph Theory*, translated by Robert A. Melter, Wiley Interscience, New York, 1985.
9. Alan Tucker, *Applied Combinatorics*, 4th edition, John Wiley and Sons, New York, 2002.

GRAPH THEORY

10. Norman L. Biggs, E. Keith Lloyd, and Robin J. Wilson, *Graph Theory 1736–1936*, Clarendon Press, Oxford, 1976.
11. Rudolf Fritsch and Gerda Fritsch, *The Four-Color Theorem*, Springer, New York, 1998.
12. Ronald Gould, *Graph Theory*, Benjamin/Cummings, Menlo Park, California, 1988.
13. Nora Hartsfield and Gerhard Ringel, *Pearls in Graph Theory*, Academic Press, San Diego, 1990.
14. James A. McHugh, *Algorithmic Graph Theory*, Prentice-Hall, Englewood Cliffs, New Jersey, 1990.
15. Robin J. Wilson, *Introduction to Graph Theory*, 4th edition, Longman, Harlow, Essex, 1996.

GROUP THEORY

16. R. B. J. T. Allenby, *Rings, Fields and Groups*, Arnold, London, 1983.

PÓLYA COUNTING

17. G. Pólya and R. C. Read, *Combinatorial Enumeration of Groups, Graphs, and Chemical Compounds*, Springer, New York, 1987.

RAMSEY THEORY

18. Ronald L. Graham, Bruce L. Rothschild, and Joel H. Spencer, *Ramsey Theory*, 2nd edition, Wiley, New York, 1990.

Index of Notation

This list indicates the page(s) where the notation is explained. Generally, it does not include notation which is used only in the section where it is introduced. Since an alphabetical list is not possible, the list is in page order.

$p(n)$, 6, 81
$A \cup B$, $A \cap B$, $A \setminus B$, 14
$\#(X)$, 14
$\mathbb{N}, \mathbb{N}^+, \mathbb{Z}, \mathbb{Q}, \mathbb{R}, \mathbb{C}$, 14
$\mathbb{R}^2$, 14
$x \mapsto x^2$, 14
$f: D \to C$, 14
$\sum_{i=1}^{n} a_i$, 15
$\prod_{i=1}^{n} a_i$, 15
$P(n,r)$, 19
$C(n,r)$, 22
$\begin{pmatrix} 1 & 2 & 3 & 4 & 5 & 6 \\ 6 & 1 & 3 & 5 & 4 & 2 \end{pmatrix}$, 37
$(1\ 6\ 2)(3)(4\ 5)$, 37
$S(n,k)$, 45
$p_k(n)$, 45, 81-2
$\theta(n,s)$, 56
$\phi(n)$, 57
$[x]_k$, 64
$s(n,k)$, 66
$\text{Perm}(n,k)$, 68
$p(n,k)$, 68
C_n, 71
$q_k(n)$, 82
$\lfloor x \rfloor$, 85
$\{a_n\}$, 98
$u_{e,d}(n), u_{o,d}(n)$, 133
$G = (V,E)$, 153
$xy\ [\ =\{x,y\}]$, 153
$\delta(v)$, 157
$d_1, d_2, ..., d_k \Rightarrow d_1', ..., d_l'$, 160
$d_1, d_2, ..., d_k \sim d_1', ..., d_l'$, 160

$v_0 \to v_1 \to v_2 \to ... \to v_k$, 165
K_n, 169
$K_{m,n}$, 173
$d(x,y)$, 203, 217
$\mu(e), \mu(G)$, 212
(G, μ), 212
$\mu(P)$, 217
S_n, 223
$\sigma \circ \tau$, 224
$S(X)$, 225
ι_X, 225
Z_n, 227
$d(p,q)$, 230
$S(\square)$, 231
gH, 237
$o(g)$, 240
$\text{lcm}(k_1,...,k_s)$, 243
$x_{k_1}^{r_1} .... x_{k_s}^{r_s}$, 243
$g \triangleright x$, 247
$\sim_G$, 249
$\text{Orb}(x)$, 250
$\text{Stab}(x)$, 251
$\text{Fix}(g)$, 258
$w(c)$, 271
$W(f)$, 271
$ct(\pi)$, 274
$CI(G)$, 274
D_n, 286
$R(p,q)$, 306
$r_k(B)$, 320
$r(x,B)$, 322
$S(\square)$, 391

Index

A

Abelian group, 229
Abstract groups, 227
Acts on, 247
Addition of sequences, 96
Adjacency matrix, 215
Adjacent, edge to a vertex, 157
Aiyar, Seshu, 148, 149
Algebraic proof, 23
Algorithm,
    concept of, 215
    to decide if a graph is connected, 375
    Dijkstra's for shortest paths, 218–222, 387–388
    Kruskal's for minimal connectors, 213–215, 385–387
    Prüfer's for labeled trees, 207–210
Allenby, R. B. J. T., 15, 227, 236, 239, 419
André, Désiré, 73
Appel, Kenneth, 9, 195
Approximating irrational numbers, 293
Associated homogeneous recurrence relation, 122
Associativity, 226
    problem of Catalan, 74
Auxiliary equation of a recurrence relation, 112
    distinct roots, 113, 121
    multiple (repeated) roots, 123–124

B

Ball, W. W. Rouse, 119, 184
Bijective function, 14–15
Binomial coefficient, 23
Binomial theorem, 24
Bipartite graph, 173
Birthdays problem, 3, 32, 341
Blackpool football club, 314
Board, 319
    complementary, 327
    pairwise disjoint subboards, 322
    subboard, 322
Boruka, Osakar, 216
Bottom riffle shuffle, 242
Boundary of a region in a plane graph, 172
Box principle, 293
Bracket notation for permutations, 36–37, 223
Bridges of Königsberg, 8, 154, 179
British national lottery, 3, 29, 41
Brown, Alexander Crum, 204
Burnside, William, 258

C

Calculus, 66
Card(s),
    rank, 29
    shuffling problem, 10
    suit, 29
Carroll, Lewis, 47, 164
Catalan, Eugene Charles, 72, 74
    number(s), 5, 71, 125
        associativity problem, 74
        generating function for, 125
        recurrence relation for, 77
    sequence, 73
    sequences of 1s and −1s, 72
Cauchy, Augustine-Louis, 255
Cayley, Arthur, 10, 201, 205, 207, 228, 234
    table, 228
    theorem for groups, 234
    theorem on labeled trees, 207
Center of a group, 254
Chessboard, 319
    coloring, 245
Chocolates problem, 3, 40
Chromatic number, 190
Closed path, trail, walk, 165
Closure property, 226
Codomain, 14
Coefficient, binomial, 23
Coloring(s), 269
    of a chessboard, 10–11, 245
    of a cube, 11, 245, 267

423

of a graph, 189–190
and group actions, 267
inventory of, 271
as mappings, 268
of points in the plane, 13
Combination(s), 17, 21–22
Combinatorics, uses of, 1–2
Commutative group, 229, 254
Complementary board, 327
    theorem, 328
Complete graph, 169
Complex numbers, 14
Composition of permutations, 224
Concrete groups, 227
Conjugacy classes, 253
Connected,
    component, 167
    graph, 167
Convergence, 97
Corollary, Lagrange's, 241
Coset, 237
Counting simple graphs, 285–292
Covering, faultless, 12, 300
Cube,
    coloring of, 245, 267
    graph, 171
    symmetries, 235, 399
Cycle, 37
    index, 274
    notation, 37, 223
    type, 243, 274

D

Degree,
    sequence,
        multigraph, 163
        simple graph, 159
    of a vertex, 157
Delete-a-square theorem, 326
De Moivre, Abraham, 25
De Montmort, Pierre Remond, 25
De Morgan, Augustus, 188
Derangements, 58–60
Derbyshire, John, 238
Determinants, 114, 363
Diagonal Ramsey number, 311
Diagram,
    dot, 83
    dual dot, 83, 84
    Ferrers, 83
Difference of sets, 14
Digraph, 151

Dijkstra, Edsger Wybe, 218, 247
    algorithm for a shortest path, 10, 218–222, 387–388
Directed,
    graph, 151
    tree, 199
Dirichlet, Peter Gustav Lejeune, 293, 299
    pigeonhole principle, 12, 293–295
Disjoint cycle form, 37
Disjoint subboards theorem, 322
Distance in a graph, 217
Distinct roots of auxiliary equation, 113
Dodecahedron, 264
    game, 183
    regular, symmetries of, 400
Domain, 14
Dominoes, faultless placing, 12, 300
Dot diagram, 83
    proofs using, 92–94
Double counting, 51
Drawer principle, 293
Dropping vertices, 174
Dual
    dot diagram, 83, 84
    of a graph, 162–163
    partition, 83
Dudeney, Henry, 175

E

Edge, 153
    adjacent to a vertex, 157
    coloring, 286
    labeled graph, 210
    weight of, 212
Election votes, 80
Empty graph, 154
Ends of a walk, 165
Equality,
    of functions, 230–231
    of power series, 97
Equation,
    auxiliary, 112
    integer solutions of, 3, 39–43
Equilateral triangle, symmetries of, 235
Erdös, Paul, 63, 295, 297, 307, 311, 312
    number, 63
Euclidean Ramsey theory, 315–318
Eulerian graph, 180
Euler, Leonhard, 5, 8, 72, 136, 140, 154, 180, 181, 184
    formula for planar graphs, 170
    identity, 136
    phi function, 57
    polygonal decomposition, 5

Existence problem, 333
Expression, 74

F

Face, 170
Falling factorial polynomial, 63
Faultless covering problem, 12, 300
Ferrers, Norman Macleod, 83
    diagram, 83
Fibonacci (Leonardo of Pisa), 103
    numbers, 7, 95, 103–109
        formula for, 106
        generating function for, 104
    sequence, 6
Field, 227
Figure, symmetry of, 230
Finish of a walk, 165
Five color theorem, 191, 193
Fixed set, 258
Formal algebraic expression, 96
Formula,
    Euler's for planar graphs, 170
    Stirling's for $n!$, 314
Four Color Theorem, 9, 188
Franklin, Fabian 132
Frobenius, Georg Ferdinand, 258
    counting theorem, 257–259
        applications, 259–265
Function(s),
    bijective, 14–15
    Euler's phi, 57
    equality of, 230
    generating, 98
    injective, 14
    linear, 109
    one-one, 15
    onto, 15
    surjective, 14

G

Galois, Everiste, 72
Generating function(s), 7, 98
    for Catalan numbers, 125
    for Fibonacci numbers, 104
    for partition numbers, 128
    for partitions into distinct parts, 129, 130
    for partitions into odd parts, 129, 130
    for rook polynomials, 322
Geometrical figure, symmetry of, 230
Graph(s), 8, 153
    bipartite, 173
    chromatic number, 190
    coloring of edges, 286–287
    colorings, 189–190
    complete, 169
    connected, 167
    contains, 174
    counting simple, 285–292
    cube, 171
    degree sequence, 159, 163
    directed, 151
    dodecahedron, 183
    dropping vertices, 174
    dual, 162–163
    edge, 153
    edge-labeled, 210
    empty, 154
    Eulerian, 180
    Euler's formula, 170
    Hamiltonian, 183
    isomorphic, 205
    isomorphism, 155
    k-colorable, 190
    Kuratowski's Theorem, 174
    labeled, 204–205
    loop, 154
    minimal connector, 212
    octahedron, 177
    path, 165
    Petersen, 151, 174, 376, 380
    planar, 152, 168
    plane, 168
    representation in the plane, 154–155
    self-dual, 162
    semi-Eulerian, 180
    semi-Hamiltonian, 183
    simple, 154
        number of, 285–292
    subgraph, 174
    trail, 165
    triangulated, 191
    utilities, 9, 172–173
    vertex, 153
    walk, 165
    weighted, 184, 212
Group(s), 10, 223, 226
    Abelian, 229
    abstract, 227
    acting on a set, 247
    action, 247
        conditions, 247
        orbit of, 249–250
    Cayley's theorem, 234
    Cayley table, 228

center, 254
commutative, 229, 254
concrete, 227
conjugate elements, 248
cycle index, 274
examples, 227
identity element, 226
inverse element, 226
isomorphism, 233
latin square, 228
multiplication table, 227
order, 237
properties, 226
subgroup, 235
of symmetries of a figure, 231
table 228
Guthrie, Francis, 188, 190

H

Haken, Wolfgang, 9, 195
Hakimi, S. L., 159
Hall, Philip, 336
    marriage theorem, 14, 334
    transversal theorem, 336
Halmos, Paul Richard, 336
Hamiltonian
    graph, 183
    path, 183
Hamilton, William Rowan, 183
Handshaking lemma, 164
Hardy, Godfrey Harold, 6, 88, 146–150
Hardy-Ramanujan,
    formula, 145–147
    story of 147–150
Havel-Hakimi Theorem, 159
Havel, Vaclav, 159
Heawood, Percy John, 195
Hierholzer, Carl, 180
Homogeneous, 110
    linear recurrence relations, 109, 110, 112–113
    recurrence relation, associated, 122
Houston, Kevin, 15, 419

I

Icosahedron 264
    regular, symmetries of, 401
Identity,
    element of a group, 226
    Euler's, 136
    map, 225
    permutation, 69, 225

Inclusion-Exclusion,
    principle, 4
    theorem, 53
Infinite order, 240
Initial conditions for a recurrence relation, 103
Injective function, 14
In shuffle, 242
Integer part, 85, 295
Integers, 14
Integer solutions of equations, 3, 39–43
Intersection of sets, 14
Inventory, 271
Inverse of a group element, 226
Irrational numbers, rational approximation to, 12, 293, 299, 302
Isomers, 204, 205
Isometry, 230
Isomorphic,
    graphs, 205
    groups, 233
Isomorphism,
    graphs, 155, 205
    groups, 233
Iyer, Ramasawamy, 148

K

k-colorable graph, 190
Kempe, Alfred Bray, 151, 191, 194–195
    chain argument, 193
Kirchoff, Gustav Robert, 200
Knight's tour, 183–184
Königsberg, bridges of, 8, 154, 179
Kruskal, Joseph B., 213
    algorithm for minimal connectors, 213–215, 385–387
K_7, represented on a torus, 378
Kuratowski, Kazimierz, 174
    theorem on planar graphs, 174

L

Labeled graph, 204–205
Labeled tree(s), 10, 201, 204–205
    Cayley's theorem, 207
    problem, 10
Ladd, Christine, 132
Lagrange, Joseph Louis, 238
    corollary, 241
    theorem, 238
Landau, Edmund Georg Hermann, 148
Latin square, 228
    not a group, 229

Lattice point, 296
Lehmer, Derrick (Dick) Henry, 138
Lehmer, Derrick Norman, 138
Leibnitz, Gottfried Wilhelm, 147, 148
Lemma,
    handshaking 164
    meaning of, 164
Length,
    of a path, 217
    of a walk, 165
Leonardo of Pisa (Fibonacci), 103
Linear function, 109
Linear recurrence relations, theory of, 120–124
Listen with Mother, 15
Littlewood, John Edensor, 148
Loop, 154
Lottery problem, 3, 29, 32, 41
Love, Augustus Edward Hough, 148
Lower bound for $p(n)$, 89, 91
Lucas, Francois-Edouard-Anatole, 6

M

MacMahon, Percy Alexander, 87, 128
MacTutor History of Mathematics archive, ix
Map, identity, 225
Marriage,
    problem, 333
    theorem, 334
Matchings, 332
Mathematical induction, 15
Matrix,
    adjacency, 215
    groups, 227
    row reduction, 118, 364–365
Method of equating coefficients, 98
Ming Antu, 72
Minimal connector, 212
    Kruskal's algorithm, 213–215, 385–387
Miriamoff, D., 73
Moser, Leo, 175
Multigraph, 154
    degree sequence, 163
Multinomial Theorem, 34, 36
Multiple (repeated) roots of auxiliary equation, 124
Multiplication of choices principle, 19
Multiplication of sequences, 97
Multiplication table, 227
Myangat, 72

N

Natural numbers, 14
Netto, Eugen Otto Erwin, 72

Newton, Isaac, 1, 25, 147, 148,
Nonattacking rooks, 13, 20, 319
Nonhomogeneous linear recurrence relation, 114–120
Nonlinear recurrence relations, 124
Notation,
    bracket, 37
    cycle, 37
    pi, 15
    sigma, 15
Noughts and crosses, 267
Number(s),
    Catalan, 5, 71
    complex, 14
    Erdös, 63
    Fibonacci, 7, 95, 103–109
    natural, 14
    Ramsey, 306, 310, 315
        diagonal, 311
    rational, 14
    real, 14
    Stirling, 4, 45, 60, 63, 66
        of the first kind, 66
        of the second kind, 66

O

Octahedron, 264
    graph, 177
    regular, symmetries of, 400
One-one correspondence, 15
One-one function, 15
Onto function, 15
Opinion polls, 341
Optimization problem, 333
Orbit,
    as an equivalence class, 250
    of a group action, 249–250
Orbit-Counting Theorem, 253
Orbit-Stabilizer Theorem, 10, 252
Order,
    of a group, 237
    of a group element, 240
    infinite, 240
    of a permutation, 242–244
Ore, Oystein, 186
    property, for Hamiltonian graphs, 186
Out shuffle, 242

P

Pack of cards, 29
    suit distribution, 30

Pairwise disjoint sets, 32
Pairwise disjoint subboards, 322
Paraffins, 205
Partial fractions, 105
Particular solution of a recurrence relation, 116
Partition(s), 7–8, 81–82
    and cycle types, 243
    into distinct odd parts, 138
    into distinct parts, 129, 133
    and dot diagrams, 83
    dual, 83
    with an even/odd number of distinct parts, 133, 142, 371
    with an even/odd number of even parts, 138
    into odd parts, 129
    parts of, 81–82
Partition numbers,
    generating function, 127–129
    lower bound, 89, 91
    recurrence relation, 136
    some values, 86, 88, 137, 138, 147
    upper bound, 142–145
Pascal, Blaise, 25
    triangle, 25–26
Path, 165
    closed, 165
    Hamiltonian, 183
    length, 217
    shortest, 217
Pattern counting problems, 11
Pattern inventory, 271
Pentagon, regular, symmetries of, 235
Permutation(s), 17, 19, 36, 68,
    bracket notation, 36–37, 223
    composition, 224
    cycle notation, 37, 223
    cycle type, 243, 274
    disjoint cycle form, 37
    form a group, 223–226
    formula for number of, 276
    identity, 69, 225
    inverse, 225
    order, 242–244
    as products of cycles, 223–226
Petersen, Julius Peter Christian, 151
    graph, 151
        not Hamiltonian, 187, 380
        not planar, 174, 177, 376
Phi function, Euler's, 57
Pigeonhole principle,
    Dirichlet's, 12, 293
        version 1, 294
        version 2, 295

Pi notation, 15
Planar graph, 152, 168
    Euler's formula, 170
Plane graph, 168
Platonic solids, 264
Point, lattice, 296
Poker hands, 31, 33, 342–343
Pólya, George, 11, 148, 278, 420
    counting theorem, 11, 278, 281, 284
Polygon triangulation problem, 5–6, 77, 78
Polynomial(s), rook, 14, 319, 322
Possible edges in a graph, 286
Power series, equality of, 97
Principle,
    Inclusion-Exclusion, 4
    Multiplication of Choices, 19
    pigeonhole, version 1, 294
    pigeonhole, version 2, 295
Probability, 2, 28
Problem,
    birthdays, 3, 32, 341
    bridges of Königsberg, 8, 154, 179
    card shuffling, 10, 244, 395
    chocolates, 3, 40
    existence, 333
    faultless covering, 12, 300
    four color, 9
    marriage, 333
    nonattacking rooks, 13, 20, 319
    optimization, 333
    pattern counting, 11
    polygon triangulation, 5, 77, 78
    selecting cards, 14, 336
    shuffling cards, 10, 244, 395
    Snap, 4, 58
    traveling salesman, 185, 220, 222
    utilities, 9, 168, 172
Proof by Mathematical Induction, 15
Properties,
    group, 226
    subgroup, 236
Prüfer, Heinz, 207
    algorithm for labeled trees, 207–210
    code, 207

R

Radian, 229
Radziskowski, Stanislaw, 309
Ramanujan, Srinivasa, 6, 88, 146–150
Ramsey, Arthur, 303
Ramsey, Frank Plumpton, 303

number(s), 306, 310, 315
    diagonal, 311
    theorem, 303, 314
    theory, 12
        Euclidean, 315–318
Rank of a card, 29
Rational number, 14
    approximation to an irrational number, 12, 293, 299, 302
Real numbers, 14
Rectangle, symmetries of, 235, 391
Recurrence relation(s), 7, 101, 103
    associated homogeneous, 122
    homogeneous linear, 109–113
    initial conditions, 103
    linear, method of solution, 109, 112
    nonhomogeneous linear, 114–120
    nonlinear, 124–125
    particular solution, 116
    for partition numbers, 136
    solution using generating functions, 104–109
    theory of, 120–124
    and vector spaces, 123
Redfield, J. H., 278
Rees, Nigel, 15, 83
Reflection method, 73
Region, 169
    boundary of, 172
Representation of a graph in a plane, 154, 155
Richardson, Richard Benjamin, 114
Riffle shuffle, 242
Rook polynomial(s), 14, 319, 322
    Complementary Board Theorem, 328
    Delete-a-Square Theorem, 326
    Disjoint Boards Theorem, 322
    generating functions, 322
    permuting rows and columns, 324–325
Rooks, nonattacking, 20, 319
    problem of, 13, 20, 319
Row reduction, 118, 364–365

S

Sampling with replacement, 55
Schubfachprinzip, 293
Selecting cards problem, 14, 336
Self-dual, graph, 162
Semi-Eulerian graph, 180
    necessary and sufficient conditions, 180
Semi-Hamiltonian graph, 183
Sequence(s),
    addition, 96
    Catalan, 73
    Fibonacci, 6
    multiplication, 97
Series, convergence, 97
Set(s),
    difference, 14
    fixed, 258
    intersection, 14
    pairwise disjoint, 32
    union, 14
Shortest path, 217
Shortest path problem, Dijkstra's algorithm for, 218–222, 387–388
Shuffle,
    bottom riffle, 242
    in, 242
    out, 242
    top riffle, 242
Shuffling cards problem, 10, 244, 395
Sigma notation, 15
Simple graph(s), 154
    degree sequence, 159, 163
    number of, 285–292
Snap problem, 4, 58
Span, 210–211
Spanning tree, 200, 210–211
Square, symmetries of, 231
Stabilizer of an element, 251
Start, of a walk, 165
Stirling, James, 45, 63
    formula for $n!$, 314
    numbers, 4, 45, 60, 63, 66
        of the first kind, 66
        formulas for, 50, 61
        of the second kind, 66
        tables of, 49, 64, 65, 67, 345
Store enumerator, 271
Subboard(s), 322
    disjoint, 322
    pairwise disjoint, 322
    splitting into, 322
Subgraph, 174
Subgroup, 232, 235
    coset, 237
    examples, 236
    order, 237
    properties, 236
    trivial, 236
Suit,
    of a card, 29
    distribution, 30
Surjective function, 14
Sylow, Ludwig, 239

Symmetries,
    of a cube, 235, 399
    of an equilateral triangle, 235
    of a figure, 230
    form a group, 231
    of a rectangle, 235
    of a (regular) dodecahedron, 400
    of a (regular) icosahedron, 401
    of a (regular) octahedron, 400
    of a (regular) pentagon, 235
    of a (regular) tetrahedron, 235, 262–263
    of a square, 231
Symmetry, 230
Symmetry groups, 229
System of distinct representatives, 335
Szekeres, George, 295, 297, 307

T

Tangerine argument, 314
Tartaglia, Nicolo, 25
Taxi Cab No. 1729, 149
Tetrahedron, 261
    regular, symmetries of, 235, 262–263
Theorem,
    binomial, 24
    Cayley's for groups, 234
    Cayley's on labeled trees, 207
    complementary board, 328
    delete-a-square, 326
    disjoint subboards, 322
    five color, 191, 193
    four color, 9, 188
    Frobenius's counting, 257–259
    Hall's marriage, 334
    Hall's transversal, 336
    Havel-Hakimi, 159
    Inclusion-exclusion, 53
    Kuratowski's, 174
    Lagrange's, 238
    multinomial, 34, 36
    orbit-counting, 253
    orbit-stabilizer, 252
    Polya's counting, 11, 278, 281, 284
    Ramsey's, 303, 314
    Wagner's, 191, 196
Tic-tac-toe, 267
Top riffle shuffle, 242
Torus, representing K_7 on, 118, 378
Tour, knight's, 183–185

Tower of Hanoi, 6, 118, 119, 365
Trail, 165
    closed, 165
Transversal, 336
Traveling salesman problem, 185, 220, 222
Tree(s), 10, 199
    directed, 199
    equivalent conditions, 201
    labeled, 10, 201, 204–205
    spanning, 200, 210–211
Triangulated graph, 191
Triangulation,
    of a polygon, 5, 77
    of a polygon problem, 6, 78

U

Union of sets, 14
    Inclusion-Exclusion theorem, 53
United Kingdom Mathematics Trust, xi, 25
Upper bound for $p(n)$, 142–145
Utilities problem, 9, 168, 172–173

V

Valence, 157
Vector space(s), 121, 123
Venn, John, 83
Vertex, 153
    adjacent to edge, 157
    degree, 157
Vertices, dropping, 174
Von Segner, J. A., 72
Voting, 80

W

Wagner, Klaus, 196
    theorem, 191, 196
Walk, 165
Weight,
    of a coloring, 271
    of an edge, 184, 212
    function, 271
Weighted graph, 184, 212
Wilson, Robin J., 188, 204, 419

Y

Yang Hui's triangle, 25